ADVANCES IN
MOLECULAR AND
CELL BIOLOGY

Volume 14 • 1996

PHYSIOLOGICAL FUNCTIONS OF CYTOCHROME P450
IN RELATION TO STRUCTURE AND REGULATION

ADVANCES IN MOLECULAR AND CELL BIOLOGY

PHYSIOLOGICAL FUNCTIONS OF CYTOCHROME P450

IN RELATION TO STRUCTURE AND REGULATION

Series Editor: E. EDWARD BITTAR
> *Department of Physiology*
> *University of Wisconsin*
> *Madison, Wisconsin*

Guest Editor: COLIN R. JEFCOATE
> *Department of Pharmacology and*
> *Environmental Toxicology Center*
> *University of Wisconsin*
> *Madison, Wisconsin*

VOLUME 14 • 1996

 JAI PRESS INC.

Greenwich, Connecticut   *London, England*

Copyright © 1996 by JAI PRESS INC.
55 Old Post Road, No. 2
Greenwich, Connecticut 06836

JAI PRESS LTD.
38 Tavistock Street
Covent Garden
London WC2E 7PB
England

ISBN: 0-7623-0113-9

Transferred to digital printing 2006

Printed and bound by CPI Antony Rowe, Eastbourne

CONTENTS

LIST OF CONTRIBUTORS

H. James Armbrecht

Geriatric Research, Education, and
Clinical Center
VA Medical Center
St. Louis, Missouri

Hugues Bernard

Department of Biochemistry
Faculty of Medicine
Sherbrooke University
Sherbrooke, Quebec, Canada

Serdar E. Bulun

Department of Biochemistry
Faculty of Medicine
Sherbrooke University
Sherbrooke, Quebec, Canada

Jorge H. Capdevila

Departments of Medicine and
Biochemistry
Vanderbilt University School of Medicine
Nashville, Tennessee

John Y. L. Chiang

Department of Biochemistry and
Molecular Pathology
Northeastern Ohio Universities
College of Medicine
Rootstown, Ohio

Bon-chu Chung

Institute of Molecular Biology
Academia Sinica
Nankang, Taipei, Taiwan, China

Lyne Ducharme

Department of Biochemistry
Faculty of Medicine
Sherbrooke University
Sherbrooke, Quebec, Canada

John R. Falck   Department of Molecular Genetics
                                 Southwestern Medical Center
                                 Dallas, Texas

Sandra E. Graham-Lorence   Department of Biochemistry
                                 The University of Texas Southwestern
                                 Medical Center at Dallas
                                 Dallas, Texas

Israel Hanukoglu   Department of Hormone Research
                                 Weizmann Institute of Science
                                 Rehovot, Israel

Margaret M. Hinshelwood   Cecil H. and Ida Green Center for
                                 Reproductive Sciences
                                 University of Texas Southwestern
                                 Medical Center
                                 Dallas, Texas

Colin R. Jefcoate   Department of Pharmacology and
                                 Environmental Toxicology Center
                                 University of Wisconsin
                                 Madison, Wisconsin

Armando Karara   Department of Medicine
                                 Vanderbilt University School of Medicine
                                 Nashville, Tennessee

Diane S. Keeney   Department of Medicine
                                 Vanderbilt University School of Medicine
                                 Nashville, Tennessee

Michael W. Kilgore   Cecil H. and Ida Green Center for
                                 Reproductive Sciences
                                 University of Texas Southwestern
                                 Medical Center
                                 Dallas, Texas

Andrée Lefebvre   Department of Biochemistry
                                 Faculty of Medicine
                                 Sherbrooke University
                                 Sherbrooke, Quebec, Canada

Jean-Guy LeHoux Department of Biochemistry
Faculty of Medicine
Sherbrooke University
Sherbrooke, Quebec, Canada

Mala S. Mahendroo Department of Molecular Genetics
University of Texas
Southwestern Medical Center
Dallas, Texas

Gary D. Means Fred Hutchinson Cancer Research Center
Basic Sciences Division
Seattle, Washington

Carole R. Mendelson Department of Biochemistry
University of Texas
Southwestern Medical Center
Dallas, Texas

Rama K. Nemani Geriatric Research, Education, and
Clinical Center
VA Medical Center
St. Louis, Missouri

Julian A. Peterson Department of Biochemistry
The University of Texas Southwestern
Medical Center at Dallas
Dallas, Texas

Dennis Shapcott Department of Biochemistry
Faculty of Medicine
Sherbrooke University
Sherbrooke, Quebec, Canada

Evan R. Simpson Cecil H. and Ida Green Center for
Reproductive Sciences
University of Texas Southwestern
Medical Center
Dallas, Texas

André Tremblay Department of Biochemistry
Faculty of Medicine
Sherbrooke University
Sherbrooke, Quebec, Canada

Steeve Véronneau   Department of Biochemistry
                             Faculty of Medicine
                             Sherbrooke University
                             Sherbrooke, Quebec, Canada

Z. Reno Vlahcevic   Division of Gastroenterology
                             Department of Internal Medicine
                             Medical College of Virginia
                             Richmond, Virginia

Michael R. Waterman   Department of Biochemistry
                             Vanderbilt University School of Medicine
                             Nashville, Tennessee

David J. Waxman   Division of Cell and Molecular Biology
                             Department of Biology
                             Boston University
                             Boston, Massachusetts

N. Wongsurawat   Departments of Medicine, Biochemistry,
                             and Pharmacology
                             St. Louis University School of Medicine
                             St. Louis, Missouri

Darryl Zeldin   Department of Medicine
                             Vanderbilt University School of Medicine
                             Nashville, Tennessee

PREFACE

Research on the cytochrome P450 family of genes has traditionally been dominated by forms participating in drug metabolism. This has occurred in spite of early discovery of steroid hydroxylase P450 cytochromes in the adrenal gland. More recently, contributions on the characterization and regulation of P450 cytochromes involved in biosynthetic reactions have been found at the international meetings on cytochrome P450 and in the several books on the field. Key recognition that P450 cytochromes should be recognized in a physiological context was provided by an international meeting in Jerusalem in 1991 and the subsequent publication of the proceedings in the *Journal of Steroid Biochemistry and Molecular Biology* (43, number 8, 92). Like this meeting, this book seeks to place equal weight on the physiological processes that are controlled by the products of reactions at usually very selective cytochrome P450 forms. Each of the authors was asked to discuss the molecular regulation of these P450 forms in the light of these physiological processes. In some cases the physiological role of the cytochrome P450 and even the natural substrate are unresolved, but a pattern of strong endocrine regulation is indicative of a hidden function. As more and more low abundance P450 genes are uncovered, the need to address potential physiological activities becomes more pressing. It is almost infinitely more difficult to identify a physiological substrate than to clone a new form.

I am grateful for the considerable time and energy put into this book by the distinguished contributors who each have many competing uses for their time—

xi

most particularly, to maintain funding for the research they have written about. I also need to acknowledge my editorial assistant, Jane Lambert, who has patiently kept the flow of faxes, FEDEX's, and telephone calls under control during the lengthy process of preparing this book.

Colin R. Jefcoate
Guest Editor

P450 CYTOCHROMES AND HORMONAL REGULATION:

AN OVERVIEW

Colin R. Jefcoate

Advances in Molecular and Cell Biology
Volume 14, pages 1–28.
Copyright © 1996 by JAI Press Inc.
All rights of reproduction in any form reserved.
ISBN: 0-7623-0113-9

I. DIVERSITY OF HEME THIOLATE AND OXYGENASE CHEMISTRY

The cytochrome P450 superfamily has now expanded to over 200 members, and with the advent of the polymerase chain reaction and the exploration of many more species, the number will continue to steadily increase. P450 cytochromes are members of a larger class of heme thiolate proteins which exploit the special chemistry of heme iron coordinated to a cysteine sulfhydryl in several ways in addition to the typical cytochrome P450 mono-oxygenase reaction: notably in peroxide isomerases [thromboxane, prostacyclin and allene oxide synthases (Hecker and Ullrich, 1989) (Song et al., 1993)], chloro peroxidase, and the several forms of nitric oxide synthase (Marletta, 1993). The first group probably represent divergent evolution with a loss of electron transfer function, the last pair convergent evolution in which heme thiolate/oxygen chemistry has been acquired by very different proteins.

P450 cytochromes provide the most flexible and effective means in living cells of activating dioxygen for the purpose of transferring one oxygen atom to an organic substrate while utilizing the energy gained from reducing the second atom to water. This so-called mixed function oxidase reaction can also be catalyzed through organic flavin co-factors (Massey, 1994). These proteins form hydroperoxide intermediates which donate an electrophilic oxygen to electron-rich nucleophilic orbitals (benzene, N, S). In several key aryl hydroxylases, a structurally related heterocyclic biopterin co-factor acts in conjunction with stoichiometric Fe localized in close proximity in the enzyme. The active species may be a biopterin hydroper-oxide, an "FeO" complex, or a tertiary complex involving both co-factors (Gibbs et al., 1993). While this chemistry is used by a diversity of mixed function oxidases reacting at benzene rings, no enzymes of this type seem to insert oxygen atoms into the much less reactive aliphatic "C-H" bonds. Dioxygen is also activated by a bis copper complex in dopamine β-hydroxylase but here, as with flavin and biopterin reactions, the CH bond is significantly activated (Klinman et al., 1984).

The hydroxylation of unreactive "C-H" bonds is the hallmark of cytochrome P450 biology, while chemically easier mono-oxygenation reactions are also well within the scope of these enzymes. This chemistry is not, however, unique to P450 cytochromes. Non-heme iron enzymes are fully capable of achieving the necessary activation of dioxygen for metabolism of unreactive "C-H" bonds as evidenced by processes such as fatty acid ω-hydroxylation and hydrocarbon hydroxylases such as methane oxidase. Here, two Fe atoms may be bridged by oxygen in a mixed-valence complex (Hendrich et al., 1992). In bacteria, hydrocarbon hydroxylation reactions are as likely to be carried out by non-heme Fe proteins as by cytochrome P450. Reactions catalyzed by bacterial cytochrome P450 may be highly specific and initiate use of a carbon source for energy production (Gunsalus et al., 1974) or provide a key step in synthesis of a toxin that is necessary for survival of the organism (Andersen and Hutchison, 1992). They may also mediate detoxification of multiple chemicals that otherwise impair bacterial growth (e.g., in streptomyces) (Omer et al., 1990). While several mechanisms of dioxygen activation are available, mono-oxygenation in more complex life forms (plants, insects, and all animals) is dominated by P450 cytochromes (Durst, 1994).

From the first discovery, P450 cytochromes have divided into 2 types: (1) multi-substrate forms involved in the excretory metabolism of diverse foreign and endogenous organic chemicals, and (2) forms that catalyze a specific step in a biosynthetic pathway. The first group of P450s has its experimental origins in the original discovery of cytochrome P450 in rat liver microsomes (Omura and Sato, 1962). This discovery explained previous observations that many drugs were metabolized in the liver through reactions dependent on NADPH and oxygen. The 1965 discovery by Cooper, Narasimulu, and Estabrook of adrenal microsomal cytochrome P450 that catalyzed specific steroid 21-hydroxylation established the alternative biosynthetic type of reaction. In this case, cytochrome P450 provides a specific step in hormonally controlled glucocorticoid synthesis (Cooper et al., 1965). In 1967, Simpson and Boyd established that adrenal mitochondria contained a cytochrome P450 that catalyzed the complete side-chain cleavage of cholesterol. This provided the first example of a mitochondrial cytochrome P450, and also of

Table 1.

Product	Tissue	Genes	Stimulation	Feedback
Cortisol (C)	Adrenal [Fasciculata]	11A1 11B1 17 21	ACTH via cAMP	C lowers ACTH. Suppresses CYP11A1
Aldosterone (A)	Adrenal [Glomerulosa]	11A1 11B1 11HB2 21	Angiotensin /K^+ via Ca^{++} [CYP11A1, CYP11B2]	A lowers Angiotensin /K^+
Testosterone (T)	Testis [Leydig]	11A1 17	LH via cAMP	T lowers LH
Estradiol (E)	Ovary [Interstitial, Follices]	11A1 17 19	LH/FSH cAMP, Ca^{++}	E lowers LH, FSH
Progesterone	Ovary [Corpus luteum]	11A1	LH, $PGF_{2\alpha}$	
1,25 di-HO-Vit. D	Liver Kidney	P45025(2C?) P450$_{1\alpha}$	PTH low Ca^{++}	1,25 di-HO- Vit. D lowers Ca++ and PTH
Bile acids	Liver	7 P45012 27	high cholesterol low bile acids	bile acids suppress CYP7
Arachidonic acid- epoxides (EET) 20-HO-Arachi- donic acid	Kidney [tubules]	2C form 4A	high salt ?	?

one capable of multiple sequential mono-oxygenase steps (Simpson and Boyd, 1967). The separation of adrenal mitochondrial $P450_{scc}$ and $P450_{11\beta}$, first by steroid binding studies and then by purification and characterization (Hume et al., 1970), set the stage for the characterization of multiple hepatic xenobiotic metabolizing forms which has dominated the cytochrome P450 field for the last 25 years.

The multiple forms of specific endocrine-related P450 cytochromes that have been identified during this period are shown in Table 1.

The possibility of multiple consecutive reactions is realized in several biosynthetic P450-dependent reactions. In such processes the rate of further oxygenation of an intermediate product exceeds the rate of dissociation from the catalytic site. This clearly requires much more of the catalytic site, since typically a more polar hydroxylated product dissociates readily. This retention of intermediates is realized in cholesterol side chain cleavage with high affinity binding of 20 R-hydroxycholesterol and 20 R, 20 S-dihydroxycholesterol (Chapter 5). Coordinated multistep reactions without release of intermediates also occur in the aromatase reaction (Chapter 8), in 17, 20 lyase (Chapter 7), and in aldosterone synthase (Chapter 6). Individual P450 enzymes have evolved for each of these specialized biosynthetic functions.

II. THE P450 MONO-OXYGENASE REACTION

For all living systems, mono-oxygenation requires reducing energy from NADH or, more typically, from NADPH, irrespective of the mechanism of dioxygen activation. As will be described in Chapter 2, transfer of electrons (e) from these hydride (2e) donors to cytochrome P450 requires an intervening process that accepts 2e but then transfers this reducing energy 1e at a time. For P450 cytochromes three options are used each involving FAD as the initial acceptor. This co-enzyme, through availability of the semiquinone state, can donate single electrons. For bacterial and mitochondrial P450s a simple flavoprotein transfers single electrons to P450 cytochromes through a low redox potential ferredoxin. P450 cytochromes located in the endoplasmic reticulum and nuclear envelope are reduced by P450 oxidoreductase in which FAD transfers electrons within the same protein to FMN and then to the P450 heme. The oxidoreductase and P450 may be fused into the same protein as discussed in Chapter 3 for P450 BM3 (Ruttinger et al., 1983). This bifunctionality parallels the structure of heme thiolate NO-synthases which also retain a complete oxidoreductase unit (Marletta, 1993). Recent work has shown that artificial oxidoreductase-P450 fusion proteins are effective for a diversity of P450 proteins and linker sequences (Fisher et al., 1992). A fourth class of P450 peroxide isomerases cannot be reduced by either process of reduction and mediate internal oxygen transfer reactions in which the peroxide reacts directly with heme.

For each of these reaction categories, a first electron reduction, generally facilitated by the presence of substrate, leads to formation of a dioxygen heme complex analogous to that in hemoglobin. Donation of a second electron in conjunction with proton transfer initiates a splitting of the dioxygen bond and liberation of water together with formation of a complex between a single oxygen and a high oxidation state iron (see Figure 1). This second electron transfer requires simultaneous passage of protons into the heme pocket to obtain dioxygen bond fission rather than formation of hydrogen peroxide (Sligar et al., 1994; Shimada et al., 1994). This step is mediated by a structural feature which further defines P450 cytochromes: adjacent hydroxy (T/S) and acidic amino acids (D/E) in the center of the helix which borders the heme-substrate site. This hydroxyl (usually threonine) donates a proton to the dioxygen while receiving one from the neighboring carboxyl (usually aspartate). This charge relay is linked by bridging water molecules to the outside of the protein. When this process is compromised by mutation of these amino acids or by a substrate which does not fully occupy the site, hydrogen peroxide forms rather than active "FeO." This complex chemistry is critically facilitated by the structural organization of the protein (Chapters 2 and 3) and the fit of the substrate in the active site.

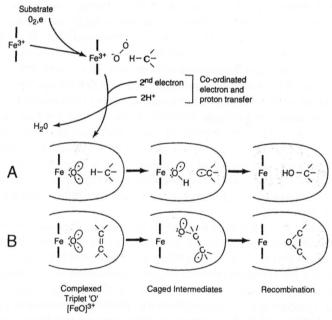

Figure 1. Mechanism for active oxygen transfer from heme to "C-H" and olefinic bonds. Complexed perferryl iron FeO^{3+} is envisioned as reacting like a triplet oxygen (2 unpaired electrons). The first "caged intermediate" undergoes recombination to form final products at rates dependent on the stability of the intermediate caged radicals.

Many questions remain open about the heme iron chemistry and particularly the contribution of the heme thiolate. Presumably this provides added stability to the protoporphyrin radical cation-Fe(IV) complex which probably best characterizes the reactive oxygen complex. The chemistry of oxygen insertion which is outlined in Figure 1 is based on free radical chemistry. The FeO complex functions in many respects as a triplet oxygen atom (2 unpaired electrons) which adds to π-bonds in olefinic or benzenoid compounds to form a biradical (thus conserving unpaired electrons). This biradical is discharged through cyclization to an epoxide, rearrangement to a ketone, or internal addition to a porphyrin N-atom (Ortiz de Montellano, 1994) (Figure 1). The FeO complex also abstracts a hydrogen from a C-H bond to form 2 radicals caged within the active site (again, possibly retaining unpaired spins) (Groves et al., 1978; White, 1991). Critical questions of stereochemistry and product selectivity are determined by the rates of changes in the radical (bond rotation, rotation of the molecule or further reaction) relative to recombination of the radical pairs. This is also determined by the constraints of the active site. The potential for movement of the substrate in the active site also determines the accessibility of different substrate atoms to this very reactive oxygen. Thus accessibility to FeO is combined with the relative reactivity of these centers in determining the relative reaction rates at alternative molecular sites.

A necessary characteristic of physiological mono- oxygenase reactions is that they produce a single product. This requires complete specificity in the point of attack and stereospecificity of the reaction. For cytochrome $P450_{CAM}$ this is reflected in the binding of the substrate to the protein. This is sufficiently tight that water is excluded from the heme and is oriented such that the reactive oxygen is adjacent to the point of attack. The reactions are also tightly coupled with little reduction of oxygen to hydrogen peroxide. Many of these mechanistic details including reaction selectivity (positional and stereo) have been evaluated from physiochemical studies using bacterial cytochrome $P450_{CAM}$ (Gunsalus et al., 1974). In particular, the crystal of this protein have allowed detailed structural interpretation at the atomic level (Raag and Poulos, 1991). Chapter 3 extends these arguments to 2 further P450 cytochromes where structural information is available from X-ray structures.

The substrate is not a neutral party to these changes. The substrate optimally changes the structure of the protein in such a way as to increase the redox potential and to facilitate first electron transfer (Raag and Poulos, 1991; Heyl et al., 1986). It is likely that this change in structure also facilitates second electron transfer and oxygen transfer processes. Even the electron transfer proteins (oxidoreductase, ferredoxin, cytochrome b_5) form complexes with P450 cytochromes which not only optimize electron transfer but produce structural changes that favor these other steps. Chapter 2 provides a detailed account of the relationship between protein structure and electron transfer. Charge distribution in the protein provides a more subtle structural component in these reactions. The formation of protein complexes, the transfer of electrons to dioxygen, and interprotein electron transfer are determined not just by localized ion-pair interactions but by regional charge and hydrophobicity effects, and are greatly affected by changes in intervening dielectric properties within the proteins (Vosnesensky and Schenkman, 1992, 1994).

Each of the specialized biosynthetic P450 cytochromes resembles $P450_{CAM}$ and is distinguished from the vast majority of low specificity xenobiotic forms by the extensive effect of substrate binding on the properties of the protein. This presumably reflects the need for strict reaction selectivity as much as for effective oxygen activation. In the presence of substrate, biosynthetic forms ($P450_{scc}$, $P450_{11\beta}$, $P450_{C17}$, $P450_{C21}$, $P450_{C19}$, etc.) change optical and magnetic properties from pure low spin (heme Fe^{3+} unpaired 3d-electrons) to high spin (5 unpaired 3d-electrons). This was originally recognized by optical spectra ($Soret_{\lambda_{max}}$ 418–390 nm) and, more directly, by electron spin resonance spectra (Jefcoate et al., 1976). The explanation for this change was also provided by spectroscopic analysis of the P. Putida cytochrome $P450_{CAM}$ in comparison with X-ray structures (Raag and Poulos, 1991). The low spin state corresponds to hexacoordinate Fe^{3+} bound by heme (xy positions), cysteine-SH and water (z-axis positions). The high spin substrate complex involves a tight fit of the substrate to the active site, which excludes the water ligand and pushes Fe^{3+} to a pentacoordinate configuration involving out of plane binding to cysteine SH.

P450 cytochromes which metabolize diverse xenobiotic compounds need to bind a broad range of substrates. As a necessary compromise to meet this diversity of substrates, a poor fit to the active site can be accommodated but with only low conversion to the high spin state. Analysis of $P450_{CAM}$ structures involving substrates less good than camphor shows only partial displacement of water and formation of mixed spin complexes (Raag and Poulos, 1991). This poorer fit is associated with uncoupled conversion of dioxygen to hydrogen peroxide and typically lower reaction rates resulting from suboptimal electron and proton transfer mechanisms. Interestingly, $P450_{CAM}$ and $P450_{BM3}$ metabolize many small xenobiotics at rates that are better than those typically seen for xenobiotic metabolizing P450s (Fruetel et al., 1992; Ortiz de Montellano et al., 1994). This raises the possibility that even these P450s have unidentified specific substrates.

III. P450 CYTOCHROMES AS MEMBRANE PROTEINS

With the exception of a majority of bacterial forms, all P450 cytochromes are targeted to cell membranes: notably endoplasmic reticulum (ER), inner mitochondria, and nuclei. Mitochondrial forms are synthesized with inner membrane targeting sequences, and associate with their host membrane after specific proteolytic processing. For biosynthetic P450 cytochromes such as $P450_{scc}$ this processing seems to be tissue specific (Matocha and Waterman, 1984). Electron transfer to purified ER P450 forms is facilitated by association of both oxidoreductase and cytochrome with phospholipid vesicles. Each protein attaches to the membrane via a short membrane spanning amino-terminal sequence which probably facilitates optimal orientation of reductase and P450 for electron transfer. Although complex formation may involve negative charges on the reductase and a positively charged domain on P450 (Shen and Strobel, 1992), the interaction appears to be dominated by hydrophobic contacts (Vosnesensky and Schenkman, 1992; 1994). Much slower but nevertheless successful complex formation can occur in the absence of phospholipids. Phospholipids also play a role in the function of cytochrome b_5, an alternative second electron donor. Cytochrome b_5 probably forms a 1:1 complex with many forms of cytochrome P450, possibly allowing the hetero dimer to function more efficiently by accepting 2 electrons as a unit (Tambarini and Schenkman, 1987; Schenkman, private communication). The structural impact on the core protein of these interactions seems to be minimal but the stringency of the mixture of phospholipids for effective reconstitution and the role of cytochrome b_5 is highly variable between forms and may even depend on the substrate. Some combinations of substrate and cytochrome P450 show little effect of either phospholipids, or added cytochrome b_5; some require a mixture reflecting the composition of the ER or are almost totally dependent on cytochrome b_5 (Gorsky and Coon, 1986). These interactions are particularly important for the activity of CYP17 where cytochrome b_5 enhances lyase activity relative to 17- hydroxylation (Yanagibashi and Hall, 1986).

The mitochondrial forms present a further problem that probably also sheds light on the interaction of P450 forms with ER: there is no membrane spanning N-terminus after processing to the mature form (see Chapter 5). The membrane interaction is particularly remarkable for $P450_{11\beta}$ (CYP11B1) which is only solubilized by very high concentrations of detergent, indicating a very strong protein-lipid interaction. Comparison of this sequence with that of $P450_{scc}$ (CYP11A1), which is much more weakly bound, provides no obvious explanation for this strong interaction. Interestingly, recent expression in *E. coli* of recombinant microsomal forms (1A, 2B, 2E, 3A), which are constructed without an N-terminal transmembrane sequence (deletion of 20–25 amino acids), shows that such forms can be fully active (Porter and Larson, 1991; Sakaki et al., 1985; Gillam et al., 1993). Interaction with cell membranes is variable and recently has been correlated with the hydropathy of the N-terminal region (Pernecky et al., 1993). However, expression of a truncated human CYP17 leads to a fully active membrane bound protein indicating the importance of other regions of integration (Sagara et al., 1993). Polar and hydrophobic phospholipid groups may therefore bind P450 cytochromes to the membrane, possibly by intercalating between the polar surfaces provided by core protein helices.

The targeting of P450 cytochromes to mitochondria and nuclear envelope still leaves many problems. Notably, counterparts to several ER forms (1A and 2B families) are expressed at lower levels in the mitochondria and are then reduced by ferredoxins but not oxidoreductase (Niranjan et al., 1988; Niranjan et al., 1984). Interestingly, the mitochondrial forms, but not ER forms, can be reduced by *both* oxidoreductase and mitochondrial ferredoxins. Immunological studies indicate substantial sequence identity between the mitochondrial form and a similarly induced ER form. However, the genetic origin for these anomalies remains obscure (different genes, alternative splicing variants, etc.?).

In all microsomal P450 cytochromes the hydrophobic transmembrane segment is followed by a positively charged domain and then a region rich in proline and glycine. The N-terminus is often, but not always, adjacent to a negatively charged residue. The location of positive and negative charges affects the configuration of this N-terminal region. The N-terminal negative charge may facilitate location of these residues in the lumen of the endoplasmic reticulum while the positively charged domain interacts with negatively charged lipid head groups on the cytosolic face (Sakaguchi et al., 1992; Szczesna-Skorupa et al., 1993). The proline-rich region (PPGPXPXXG) provides a more rigid linear segment which appears to be critical for correct folding of the protein (Sakaguchi et al., 1994). The luminal N-terminal segment may undergo further glycosylation as in CYP19 (Shimozawa et al., 1993).

IV. ROLE OF SPECIFIC P450s IN HORMONAL REGULATION

P450 cytochromes involved in these biosynthetic reactions exhibit expression which is typically under cell-specific hormonal regulation. For the most part a peptide hormone binds to a cell-type specific plasma membrane receptor which then provides a signal to P450 genes via regulatory elements associated with increases in cAMP, cytosolic Ca^{++}, or the more complex cascades initiated by receptor tyrosine kinase activity. The link between these signals and nuclear transcription is described in Chapter 4 and in detail in several other chapters. These signaling pathways are far from independent and involve both feedback and crossover processes. Systemic hormonal signals are frequently modulated by a variety of local regulators, particularly eicosanoids, cytokines, and other growth factors which can augment or antagonize the primary hormonal signals. This provides a localized fine-tuning in response to physiological changes, infection, or injury and coordination between the multiple cells of an organ. This complex coordination between individual cell types is well exemplified by the cyclical changes in the ovary which involve a highly cell-specific and temporally pro-grammed expression of biosynthetic P450 cytochromes (see Chapter 5).

The predominant source of hormonal regulation of steroidogenic P450 cyto-chromes is the pituitary gland under the control of the hypothalamus which provides a connection with the brain. These biosynthetic processes are always subject to feedback regulation which can occur via direct suppression of the receptor activity which stimulates the pathway (short loop), or through feedback suppression by the endocrine product of the release of the stimulatory hormone (long loop). Other feedback loops pair the liver and intestines (bile acids) and parathyroid and kidney (Vitamin D). These features of P450 regulation are summarized in Figure 2. ACTH is suppressed by glucocorticoids, LH and FSH by androgens and estrogens, angiotensin by aldosterone and PTH by 1,25 Vitamin D_3. Examples of short loop suppression are provided by glucocorticoid suppression of adrenal $P450_{scc}$ (Chapter 5) and suppression by 1,25-Vitamin D_3 of renal $P450_{1H}$ (Chapter 9). Thyroid hormone and growth hormone play key roles in either stimulating or suppressing P450 cytochromes in the liver (select members of the CYP2, CYP3, and CYP4 families) that metabolize steroids (Chapter 12) and arachidonic acid (Chapter 11). The activity of growth hormone is then also dependent on sex steroids, which affect the diurnal fluctuations of the hormone (Gustafsson, Waxman et al., 1991), and arachidonic acid, which may in part mediate the intracellular effect of GH, probably after metabolism by cytochrome P450 (Tollet et al., 1995).

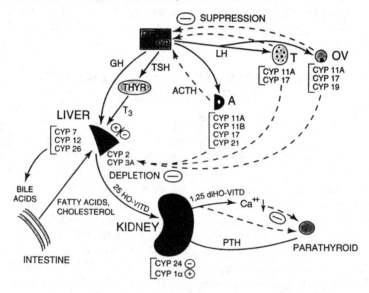

Figure 2. Feedback regulation of P450 cytochromes in the endocrine system. The pituitary secretes luteinizing hormone (LH), adrenocorticotropin (ACTH), thyroid stimulating hormone (TSH) and growth hormone (GH). These hormones exhibit effects on, respectively, testis or ovary [T, OV/LH], adrenal [A/ACTH], thyroid [TSH] and liver [GH]. The secretion of LH and ACTH is suppressed by steroids secreted by T, OV, and A. The P450s (CYP2,3A) of the liver are controlled both positively and negatively by GH and T_3 (from thyroid)—these P450s metabolize steroids produced by T, OV, and A.

Liver is also involved in bile acid synthesis (CYP7, CYP12, and CYP26) which is regulated positively by cholesterol intake and negatively by bile acids. Kidney activates 25 HO-vitamin D_3 from liver through 1α-hydroxylation and partitions metabolism to inactive products through 24-hydroxylation. The balance between these activities is determined by Ca^{++} and parathyroid hormone (PTH) from parathyroid gland. The latter is negatively repressed by 1, 25 dihydroxy vitamin D3 and Ca^{++}. Suppression pathways are shown as dashed lines. Stimulatory pathways are shown as solid lines.

The characterization of many genes has led to the recognition that hormonal regulation involves a complex array of nuclear binding proteins which interact with defined DNA sequences at positions either 5′ or 3′ to the transcriptional start site. Regulation in the 5′ region is easier to study and has received more attention, but even here regulation far removed from the start of transcription is likely to be underestimated. The primary site of regulation may be far removed from the transcriptional start site (> 2kb) and may be necessary for opening up the proximal promoter region where most regulatory proteins bind (Ramsden et al., 1993; Trottier et al., 1995; Wu and Whitlock, 1992). Nuclear regulatory proteins provide the means to transfer the hormonal signal to the RNA polymerase complex and also to provide additional tissue and developmentally specific regulatory controls. The conceptual basis for these mechanisms is described in Chapter 4 and is also addressed with respect to individual P450s in several chapters. Many different nuclear signaling mechanisms are used even for the same hormonal process (e.g., cAMP stimulation of transcription of the multiple genes participating in adrenal steroid synthesis). Several genes that show diverse cell regulation utilize multiple first exons in a cell-specific manner, each with distinct regulatory sequences. Aromatase provides the first example of a P450 gene regulated in this way (Chapter 8).

V. CO-EVOLUTION OF P450 CYTOCHROMES AND ZN-FINGER RECEPTORS

The role of P450 enzymes in hormone synthesis has most probably evolved along with the superfamily of ligand-regulated DNA-binding proteins which mediate the physiological activities of these hormones. Recent work indicates an expanding group of lipophilic ligands and also of orphan receptors that lack identified ligands (Mangelsdorf and Evans, 1995). These receptors also play a key role in regulation expression of P450 genes (see Chapter 4). For each of these proteins typified by steroid receptors, ligand binding in the C-terminal domain causes a conformational change which effects dissociation of chaperone proteins. These proteins (HSP90) are probably critical in maintaining the ligand-binding conformation of the protein, but also mask the DNA-binding Zn-finger region (Pratt, 1993). This step provides the necessary activation for intercalation of the receptor into recognition sites in the responsive genes (Zillacus et al., 1994) (see Figure 3). These recognition sequences typically comprise pairs of 6-base sequences that are often remarkably similar for distinct receptors. These pairs may be palindromic inverted repeats (for all steroid receptors), direct repeats or inverted repeats (Mangelsdorf and Evans, 1995). Several receptors use the same palindromic elements with differing spacings between the halves (2–5 bases) which presumably place different steric constraints on dimerization (Zillacus et al., 1994). Interaction with this palindrome involves cooperative binding of pairs of these proteins linked not only through the DNA interaction but also through protein- dimerization sequences. Several receptors

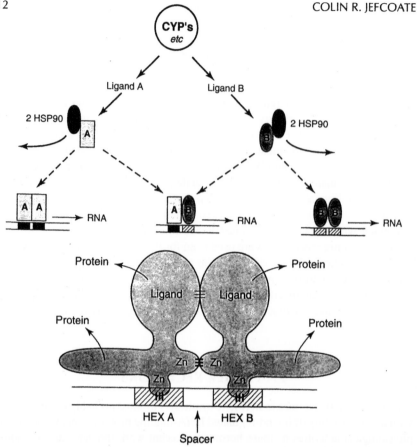

Figure 3. **A.** P450 cytochromes and other metabolic enzymes synthesize (or remove) ligands that bind to ligand-activated receptors (**A** and **B**). This binding to a C-terminal domain dissociates a 90kDa heat shock protein and stimulates binding to nuclear recognition elements. These are activated by formation of either homodimers (**AA**, **BB**) or heterodimers (**A,B**), each combination exhibiting a unique regulatory activity. **B.** Ligand-activated Zn-finger receptors form dimers that recognize 6-base recognition elements (HEXA and HEXB) that are separated by a spacer (up to 5 bases). This spacer depends on the receptor complex and modulates the conformation of the heterodimer by controlling dimerization compression. Transactivating signals are transmitted to other nuclear proteins acting on the gene (arrows).

(thyroid, vitamin D, peroxisome proliferator) also function at direct repeats as heterodimers, particularly with a set of RXR receptors which seem to be activated by 9-*cis* retinoic acid or act directly without the ligand. In this complex RXR is always the 5′ partner and may directly bind the transcription complex. CYP4A1 is induced by the peroxisome proliferator receptor in this way (Gearing et al., 1993).

Coup receptors are active as dimer partners but repress transcription (Qiu et al., 1994).

A second level of signaling is provided by domains in the N- and C-terminal regions that interact with other nuclear regulatory proteins (Danielian et al., 1992; Tsai and O'Malley, 1994). These domains may be additionally modified by phosphorylation of the receptor through one or more of the kinase cascades (Arnold et al., 1995). Glucocorticoid receptors (and presumably others) may either activate or inhibit, again depending on the particular regulatory site. Thus, these receptors can activate or inhibit transcriptional processes controlled through AP-1 DNA elements depending on the other proteins binding at this site (fos, jun, etc.) (Diamond et al., 1990). *P450* genes are also regulated by these mechanisms; for example, glucocorticoid receptors stimulate aromatase (Chapter 8) and suppress $P450_{scc}$ (Chapter 5). Recent work indicates that glucocorticoids may even enhance transcription of CYPBAI through a distinct regulatory protein acting at a DNA element very different from the typical glucocorticoid regulatory element (Huss et al., 1996). Other P450 cytochromes may be directly controlled by a standard glucocorticoid receptor interaction in conjunction with other inducers (Mathis et al., 1989; Waxman and Azaroff, 1992; Zhao et al., 1995).

Very small changes in protein structure in these receptors provide an enormous variety of gene regulation. The Zn-finger sequences are highly conserved between receptors responding to different ligands, but nevertheless differ at critical sites. The amino-terminal finger provides a helical region which fits in the major grove of the DNA helix and primarily determines recognition of the sequence of six bases. The other finger provides a surface for dimerization (Zillacus et al., 1994). This has been elegantly shown by expression of chimeric proteins where the ligand-binding domain from one receptor has been combined with the DNA-binding domain of a second receptor. When expressed in cells that typically express the receptor, the hybrid protein introduces the stimulation of genes that normally respond to the natural agonist (by set activations by the second hormone). Recent work shows that two amino acids in a so-called P-box at the foot of the upstream Zn-finger are even sufficient to distinguish glucocorticoid and estrogen receptor recognition (Zillacus et al., 1994).

This highly tuned signaling system must have been dependent on the evolution of specific P450 cytochromes that can synthesize the appropriate ligands. Since very small amounts of ligand (10^{-11}–10^{-9} M) are needed to trigger these responses, as-yet-unknown ligands and associated biosynthetic P450 cytochromes may regulate some of the many orphan receptors that have been currently identified (proteins homologous to ligand regulated proteins without known regulatory ligands). Recently identified examples include HNF4, COUP, EAR, and TR proteins (Chang et al., 1994). Since the structures of these dimer-forming Zn-finger proteins provide many potential regulatory possibilities, it is certainly possible that some may have lost the capacity to bind ligands while retaining DNA-binding, dimer formation, and trans-activation. Activation may be limited to receptor phosphorylation which probably plays a role in regulating even steroid receptors. The presence of multiple P450 cytochromes and Zn-finger receptors in insects and plants points to co-evolution of this regulatory system for at least 500 million years (Laudet et al., 1992; Mangelsdorf et al., 1995). A diversity of terpenoid products may participate in ligand-activation of these receptors, probably requiring many more specific P450-mediate oxygenation reactions to sculpt the carbon skeletons. The identification of a new farnesol-activated receptor in mammalian tissues emphasizes the possibility of new ligands of this type (Forman et al., 1995). The activation of PPAR receptors by PGD_2 and PGJ_2 also points to further developments in ligand identification (Kliewer et al., 1995).

A scheme representing a co-evolution of P450 cytochromes and Zn-finger receptors is presented in Figure 4. The availability of hundreds of amino acid and gene sequences in the P450 superfamily has led to the assemblage of these sequences in families and subfamilies. P450 protein sequences within a single gene family are defined as having $\geq 40\%$ amino acid identity to each other. P450 proteins with sequences that are far more closely related ($\geq 65\%$ identity) are placed within the same sub-family (Nelson et al., 1993). Multiplicity within a gene family is very common for polysubstrate microsomal P450 cytochromes (families I–IV) but has only been demonstrated at this time for the CYP11 family among biosynthetic P450s (11A1, 11B1, 11B2) (see Chapters 5 and 6). It has been recognized that this evolution has taken place over the past 400 million years with separation of new families as new species evolve more complex physiological functions (biosynthetic enzymes) (Nebert and Feyereisen, 1994) or dietary needs (xenobiotic enzymes). This phylogenetic evolution of the P450 superfamily has recently been reviewed (Nelson et al., 1993).

Steroids are relative rigid hydrophobic cylindrical molecules whose interactions with protein binding sites are highly sensitive to the position of polar substituents such as hydroxyl groups. Reactions at cytochrome-P450 can diminish the spatial dimensions of the steroid through C-C cleavage reactions (CYP11A1, CYP17, and CYP19), introduction of hydroxyls at the beginning (Vitamin D 1–H), middle ($P450_{11b}$), or end ($P450_{21}$) of the molecule, and also can increase the rigidity of the structure (aromatization). These relatively gross structural changes can select very different receptor binding sites. For example, a receptor site might evolve an optimally placed polar pocket that can accommodate a new or differently placed hydroxyl group on the steroid. Recent work shows that even mutation of a single critical amino acid can greatly affect substrate selectivity (Iwasaki et al., 1993). It seems likely that there has been a co-evolution of steroid receptors and P450 cytochromes which generate their activating ligands. Steroid receptors are closely related within a single subfamily (I) while receptors for 1,25-dihydroxy vitamin D are more closely related to steroid receptors or to retinoid and thyroid receptors, depending on which domains are compared (Laudet et al., 1992). The receptors for family I may have evolved from a common precursor protein active in a primitive organism that first derived advantage from this more complex nuclear signaling.

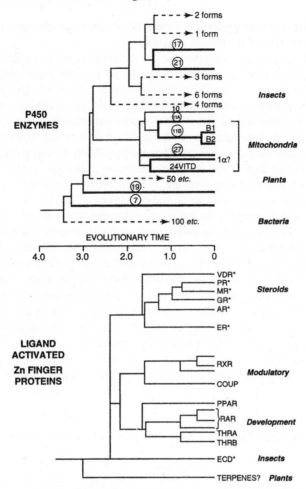

Figure 4. Co-evolution of cytochrome *P450* genes and ligand-activated nuclear receptors that determine the function of P450 products. Heavy lines for *P450* genes are used to designate biosynthetic enzymes.

The P450 cytochromes could have similarly diversified from an initial isoprenoid-modifying cytochrome which generated the ligand for this prototype receptor. Plants, insects, and higher animals have evolved distinct ligand receptor strategies, particularly with respect to nuclear signaling through hydroxylated isoprenoids (Feyereisen et al., 1994). This C27 sterol side chain is retained but oxidatively modified in ecdysone, the insect molting hormone, but is cleaved for all steroids (Keighlen et al., 1990). Plant hormones are formed by hydroxylation and other reactions based on smaller isoprenoids (Durst et al., 1994). Such alternative pathways may precede evolution of cholesterol side chain cleavage (CYP11A1) which initiates steroid production in vertebrates.

Presumably enzymes dependent on the production of progesterone or later steroids evolved after the enzyme necessary to produce their substrate. Thus CYP17, CYP21, and CYP11B1 probably evolved after CYP11A1 as a greater diversity of steroid receptors became advantageous by providing more refined gene regulation. It seems logical that CYP19 (aromatase) should evolve after CYP17, which provides its substrate. However, the distant relationship of these sequences indicates parallel evolution and provokes the possibility that an aromatase evolutionary precursor functioned on a very different substrate—perhaps a smaller terpene with a similar A-ring structure. The origins of the aromatase gene are also discussed in this volume by Simpson and co-workers. CYP11B1 has evolved from CYP11A1, which provides the necessary substrate, while CYP21 probably appeared after CYP17, assuming an earlier need for androgens and estrogens. The CYP11B subfamily is in a current state of functional evolution since some species have separate $P450_{11\beta1}$ and $P450_{11\beta2}$ proteins which provide specialization in the formation of, respectively, glucocorticoids and aldosterone, while in other species only one gene is apparent with regulation of product selectivity determined at the protein level (Chapter 6). Changes of this type appear to arise via gene duplications at adjacent positions in the same chromosome. Sometimes this duplication results in an inactive pseudogene that is nevertheless retained as in the case of $P450_{C21}$ (A & B) (Gitelman et al., 1992). When the separation of steroid functions is more ancient, relocation of genes to distinct chromosomes appears to be the norm.

Table 2. Characteristics of Biosynthetic *P450* Genes and Gene Products

Gene	Chromosome [Human]	Molecular Size		
		Gene (kb)	mRNA (kb)	protein (AA)
Mitochondrial				
CYP11A1 ($P450_{SCC}$)	15	>20	2.0	482 (521)[b]
CYP11B1 ($P450_{11\beta}$)	8	7[a]	4.1	479 (503)[b]
CYP11B2 ($P450_{C18}$)	8	7[a]	2.6	479 (503)[b]
CYP27 (cholesterol, bile acid)	2			
CYP24 (25-HO vit. D)				
Microsomal				
CYP21	6	3.3	2.0	495
CYP17 (17α, lyase)	10	6.6	1.9	508
CYP19 (aromatase)	15	>70kb	3.4	503
CYP7 (cholesterol)	8		3.6	

Notes: [a] from other species
         [b] length of presequences

The chromosomal locations of the various biosynthetic *P450* genes for humans are shown in Table 2 together with other characteristics (from Nebert et al., 1994). The genes vary dramatically in size (3.3 → 70 kb) and even mRNA lengths differ by over 2-fold (1.8–4.1 kb) because of variable non-coding and presumably regulatory sequences.

VI. P450-ACTIVITY AND RECEPTOR BINDING

Adjustments in cytochrome P450 activity can play a key part in regulating receptor signaling mediated by lipophilic hormones. Optimal receptor signaling is achieved at low levels of binding (< 50%) where the response is linearly dependent on the ligand concentration. Consequently, there is typically a feedback down-regulation response to excessive levels of regulators that may utilize several different mechanisms. Possible adjustments by the animal include decreased synthesis of ligand, increased metabolic degradation of the ligand, alternative ligand binding, or a decrease in the affinity of the receptor. The target cell may also adjust by lowering the levels of receptor and receptor response. A decrease in ligand synthesis may be produced by three types of inhibition of the biosynthetic P450 cytochromes: (i) decreased availability of substrate (ii) lower intrinsic mono-oxygenase activity, or (iii) down-regulation of the enzyme (protein synthesis or degradation). Where rapid fluctuations in activity occur (within 1–2 hours) control of substrate availability seems to be preferred. Although few examples of turnover modification have been documented for biosynthetic P450 cytochromes, several hepatic forms (CYP2B and CYP3A) exhibit substrate stabilization (Watkins et al., 1986) and also destabilization following phosphorylation (Eliasson et al., 1990; 1992). Adjustment in the turnover of cytochrome provides one mechanism for slower changes in hormone levels. Transcriptional regulation, however, is unlikely to be effective for fluctuations of much less than 12 h. Auxiliary regulation of receptor binding can also be provided by varying the level of a lower affinity ligand-binding protein. For example, retinoid binding proteins may be present in serum (Matarese and Lodish, 1993) or within the target cells (Rong et al., 1993). The levels and affinities of such proteins in conjunction with the activities of various ligand-modifying enzymes determine the free ligand concentration that is available to the receptor.

Chapter 5 points out that cholesterol availability to adrenal cytochrome P450$_{scc}$ is the primary means of *acute* regulation of glucocorticoid and mineralocorticoid hormones. The slower changes needed for androgen or estrogen synthesis are also achieved by hormonal regulation of transcription, often in a cell-type specific manner (see Chapters 7 and 8). Additional regulation may determine the division of steroid biosynthesis between the different final products. For example, the activity of CYP17 is crucial in determining whether pregnenolone is converted to cortisol (needs CYP17) or aldosterone (CYP11B2 does not metabolize 17 hydroxy steroids). Significantly, CYP17 is expressed in adrenal fasciculata but not glomerulosa cells. In the liver, an increase in cholesterol substrate stimulates the transcriptional regulation of P450$_{7H}$ (Chapter 10) while bile acids, the ultimate products, cause repression. Here the regulatory demands can be met relatively rapidly by biosynthesis because the turnover of P450$_{7H}$ is very fast (t$_{1/2}$ 2–3 h). Many of these enzymes show diurnal fluctuations in activity over the course of 6–12 h which parallel changes in the blood levels of regulatory hormones. Recent studies of CYP7 in liver suggest that much of this change in transcriptional activity can be directly linked to specific nuclear transcription factors (Lee et al., 1994).

In each of these pathways which utilize cholesterol, regulation is met by the initial enzyme in the pathway. This initial enzyme in each case is rate limiting. This minimizes wasted biosynthetic effort and the accumulation of unwanted intermediates that would occur with a slow step in the middle of the pathway. Nevertheless limiting steps may occur at the end of pathway. Thus aldosterone production is regulated in adrenal glomerulosa cells both by stimulation of cholesterol metabolism in adrenal glomerulosa cells and by stimulation of the last step, C$_{18}$ corticosterone oxidation by CYP11B1 or CYP11B2 (Chapter 6). Similarly 1,25 dihydroxy Vitamin D$_3$ formation is regulated by activation of 1-hydroxylation of 25-hydroxy Vitamin D$_3$ in the kidney rather than through hepatic 25-hydroxylation of Vitamin D$_3$. This pathway offers several novel regulatory features: the initial and final steps occur in different organs (liver and kidney), the first step is apparently not regulated, and formation of the active product is determined by a switch in kidney proximal tubule mitochondria from activation by 1-hydroxylation (P450$_{1H}$) to inactivation via 24-hydroxylation (P450$_{24H}$) (Chapter 9).

VII. XENOBIOTIC INDUCTION AND P450-MEDIATED INACTIVATION PROCESSES

P450-mediated reactions provide a major way of inactivating signaling ligands just as xenobiotic molecules are detoxified. An additional hydroxyl group, unless specifically placed, typically weakens the receptor-ligand interaction and provides either a molecule that can be directly excreted or a substrate for conjugation. These systemic inactivation processes are most actively carried out by hepatic P450 cytochromes, but may also occur in target organs as a means to control receptor-mediated processes. Many of these reactions and their regulation are reviewed in Chapter 12. Interestingly, these processes are hormonally regulated by steroids, growth hormone, and thyroxine (Levgraverand et al., 1992; Yamazoe et al., 1987; Ram and Waxman, 1991). The physiological significance of these interacting endocrine reactions remains to be fully evaluated—for example, the relative importance of P450-mediated metabolism versus direct conjugation. Effects here may be short-term (responses to 1–2 day hormone administration) or result from irreversible imprinting effects where hormonal levels during early development determine functions in the mature adult, including cytochrome P450 expression. Regulation including xenobiotic induction may be additionally dependent on diurnal hormonal secretion patterns, as seen for growth hormone (Shapiro et al., 1994).

A paradigm for regulation in target tissues by metabolism is provided by 11_β-hydroxy steroid dehydrogenase. This enzyme inactivates glucocorticoids which would otherwise activate mineralocorticoid receptors in target tissues for aldosterone. This steroid retains activity because of resistance to this inactivation (Funder, 1993). Selective aldosterone activity therefore requires co-expression of receptor and dehydrogenase. Thus it seems that an additional P450 activity (C-18 oxidase) has evolved because the product aldosterone selectively binds to a variant of the glucocorticoid receptor (mineralocorticoid receptor or Type II glucocorticoid). Localized expression of an inactivation enzyme has avoided the need to evolve a more selective mineralocorticoid receptor. The metabolism of 25-hydroxy Vitamin D_3 exemplifies a different strategy: alternative activation (1-hydroxylation) and inactivation (24-hydroxylation) reactions whose activities adjust to physiological needs for receptor activity.

Stimulation of these hepatic P450 cytochromes by xenobiotic molecules has been extensively described in many reviews (Gonzalez, 1989; Dennison and Whitlock, 1995). The best studied mechanism is provided by the induction of CYP1 family proteins: 1A1, 1A2, 1B1. While active in polycyclic aromatic hydrocarbon (PAH) metabolism, these forms are also active in metabolism of natural compounds such as bilirubin, estradiol and arachidonic acid. This induction is stimulated by the binding of PAHs to a recently cloned helix-loop-helix protein, the Ah-receptor (Burbach et al., 1992; Dolwich et al., 1993). Like glucocorticoid receptors, this protein binds to HSP90 in the cytosol and is released after binding to the PAH followed by transfer to the nucleus. Association with a related nuclear protein, the Ah-receptor nuclear translocation (ARNT) protein stimulates nuclear transfer and binding to specific DNA recognition elements which mediate enhanced gene transcription (Whitlock, 1994). Recent work suggests a constitutive function for Ah-receptor. Most notably, Ah receptor-deficient mice show abnormalities in the development of liver, skin, and the immune system (Fernandez-Salaguero et al., 1995). Ah-receptor activation may be highly dependent on the stage of the cell cycle (Scholler et al., 1991) and is seen in cultured hepatoma cells deficient in CYP1A1 activity (RayChaudhuri et al., 1990). Activation is associated with inhibition of HMG CoA reductase (Puga et al., 1992) and occurs when cells are forced to grow in suspension rather than attached to a surface (Sadek and Allen-Hoffmann, 1994). These observations suggest that there is a constitutive activator of the Ah-receptor that increases when CYP1A1 is very low at certain stages of the cell cycle or when cell adhesion decreases. A labile repressor has also been implicated in these processes through stimulation of many Ah-receptor-linked genes by protein synthesis inhibitors (Lusska et al., 1992). A recently cloned CYP1B1 (Savas et al., 1994; Sutter et al., 1994) shows different cell specificity of expression but similar Ah-receptor regulation. Coordinate control of endogenous regulatory molecules may occur through metabolism at, respectively, CYP1A1 in epithelial cells or

CYP1B1 in adjacent stromal fibroblasts. CYP1B1 is also expressed in steroidgenic cells such as the adrenal under cAMP control while this cytochrome metabolizes polycyclic aromatic compounds it also effectively metabolizes estradiol in humans (Spink et al.,). This raises a very general question concerning the likely physiological function of this form.

Phenobarbital and a host of dissimilar molecules stimulate transcription of other genes involved in both xenobiotic and hormone metabolism (Waxman and Azaroff, 1992). Unlike the Ah-receptor pathway this process is poorly characterized but apparently involves more complex regulation. Many genes regulated by phenobarbital in species as diverse as bacterial strains, to house flies and rats contain a 15 base sequence in the $5'$-proximal promoter region. In responsive bacteria this element interacts with a helix-loop-helix repressor protein that becomes inactive when bound by phenobarbital thus leading to stimulation of transcription (He and Fulco, 1991). However, other work points to more distal sites as primary sites of phenobarbital induction possible through opening up the more proximal promoter region to nuclear regulatory proteins (Ramsden et al., 1993; Trottier et al., 1995). Many phenobarbital inductions share a common endocrine sensitivity. They are suppressed by a coordinated process involving T_3, GH, and polyunsaturated fatty acids, which is reversed by testosterone and to a lesser extent by estradiol (Larsen et al., 1994) and also by cAMP (Sidhu and Omiecinski, 1995). Again this suggests an important physiological role for these enzymes as described in Chapters 11 and 12. Phenobarbital and organochlorine compounds induce P450 cytochromes in insects by an apparently similar mechanism. More limited sets of P450 inductions are effected through, respectively, elevated glucocorticoid activity (CYP3A and CYP2B sub-families) and binding to the peroxisome proliferator activator receptor (CYP4A). The latter clearly has a physiological role since CYP4A isoforms metabolize fatty acids which are in turn inducers of this receptor (Sundseth and Waxman, 1992; Gottlicher et al., 1992). Many of the effects of GH on liver, including CYP2C11 expression, may also be mediated by P450-dependent arachidonic acid metabolism (Tollet et al., 1995). This linkage of several pathways to fatty acids strongly suggests that even P450 cytochromes associated with xenobiotic metabolism may have important roles in more specific conversion of fatty acids to physiologically active metabolites.

VIII. ANOTHER FUNCTION FOR CHOLESTEROL METABOLITES

Yet another example of the biological versatility of hydroxylation reactions is provided by bile acid formation. In these molecules, the asymmetric distribution of polar carboxyl and hydroxyl groups around an otherwise hydrophobic organic compound generates a molecule capable of bridging lipophilic and aqueous environments that can form micellar structures in water. 7α-hydroxylation of the extremely hydrophobic cholesterol not only provides an additional way to regulate hepatic cholesterol content but is also the rate-limiting step in the production of bile acids. These metabolites function as detergents in controlling the excretion of large amphoteric molecules from the liver and in solubilizing fats in the intestines. Chapter 10 points out that the detergent characteristics are modified in subtle ways by the extent of hydroxylation. This can be critical to the free flow of bile and the effectiveness of this process in the excretion of metabolites generated by hydroxylation and conjugation of other organic molecules. Here again, feedback regulation plays a key role. When bile acids are decreased in the intestines by feeding an ion exchange resin which binds these anions, then the rate-limiting cholesterol 7α-hydroxylase is induced in the liver. Just as with steroidogenic cholesterol metabolism, this first step in the pathway is the key regulatory step. The P450 cytochrome which catalyzes this reaction is located only in liver endoplasmic reticulum, has evolved as a single member family, and undergoes a clear pattern of developmental and endocrine regulation mediated via appropriate hepatic regulatory factors (Lee et al., 1992). Other later hydroxylation steps play a role in determining the detergent characteristics of the bile acid.

IX. BIOSYNTHESIS BASED ON FATTY ACIDS

Arachidonic acid provides a second major core molecule for oxygenase-linked signaling which is described in Chapter 11. Here, cytochrome P450 has made a late entry to the metabolic scene. Nature has taken extensive advantage of the multiplicity of structural changes which can be brought about by interaction of oxygen with polyunsaturated fatty acids. Interestingly, many similar changes occur without enzyme intervention during lipid peroxidation. The first signaling molecules to be identified in this class were, of course, prostaglandins that are formed by a set of heme-catalyzed, oxygen-dependent free radical cyclization reactions. This concerted process is catalyzed PGH_2 synthase and forms endoperoxides. One of several enzymes then acts on PGH_2 to form individual prostaglandins in a generally cell-specific manner. Two of these reactions are catalyzed by P450 cytochromes which use the capacity of the heme thiolate to form a reactive oxy intermediate which effects peroxide rearrangement (Hecker and Ullrich, 1989). A similar mechanism is used in the rearrangement of plant hydroperoxides (Song et al., 1993). The second class of arachidonic acid regulators to be discovered were leukotrienes formed by 5-oxygenation and subsequent rearrangement of the hydroperoxide to an epoxide (leukotriene A). Further enzymatic conversions form an additional set of specialized signaling molecules (leukotrienes B to D). Hydroperoxide formation at 12- and 15- positions also leads to biologically active molecules.

Arachidonic acid epoxides and hydroxy arachidonic acids derived from cytochrome P450-dependent reactions provide the third signaling alternative (Chapter 11) (Capdevila et al., 1992) and a possible explanation of the sensitivity of hepatic P450 to dietary fatty acids. Arachidonic acid epoxides may have very different binding characteristics from hydroxy-derivatives; they are less polar and retain the rigid geometry of the original olefinic bonds. Chapter 11 presents evidence that epoxidation and hydroxylation at different positions in the C20 chain determine different biological activities. Like the prostaglandins and leukotrienes, these new signaling molecules seem to provide opposing activities. This is most notable in the regulation of blood pressure where 20-hydroxy arachidonic acid raises blood pressure while various epoxides lower blood pressure (McGiff, 1989, Chapter 11). Like prostaglandins these compounds appear to act near to the site of synthesis (unlike steroids). Thus the actual site of synthesis and localization of the contributing cytochrome P450 becomes very important. In the kidney, cell-specific expression of P450-dependent hydroxylases, epoxidases, and prostaglandin synthesizing enzymes is coordinately regulated in response to changes in blood pressure. These responses may be mediated by multiple contributors to blood pressure control, which are hard to separate: vascular stretch receptors or responses to aldosterone, peptide hormones (angiotensin and bradykinin), and ion concentrations (Na^+/K^+). Arachidonic acid epoxides may also react further to produce physiological activity through dihydrodiols (di HETEs), conjugates, or PG-synthase products.

The synthesis of arachidonic acid metabolites is primarily regulated, like steroid synthesis, by the availability of the starting substrate from inactive ester stores (availability of arachidonic acid is initiating instead of cholesterol). For arachidonic acid this usually involves hormonal regulation of phospholipase A_2 which is subject to stimulation by multiple growth factors and other cellular regulators (Lin et al., 1993). Additional regulation is provided by levels of expression of enzymes competing for arachidonic acid (PG synthases, leukotriene A synthase, P450 cytochromes) or later enzymes in the pathway.

The mechanism of action of these new arachidonic acid derivatives remains to be determined. Prostaglandins and leukotrienes transmit their signals to cells through a set of structure-specific cell surface receptors. These receptors are linked through G-proteins to adenylate cyclase (cAMP production) to phospholipases (IP_3/Ca^{++} regulation) and intracellular ion regulation through various channels and pumps. While other oxygenated arachidonic acid derivatives may act similarly they may also be targeted to specific enzymes or through incorporation into phospholipids causing localized membrane perturbations (see Chapter 11). This remains an area for future research.

X. PERSPECTIVE

The following chapters aim to present recent research which shows how new approaches in molecular biology have expanded understanding of the role of P450 cytochromes in physiological processes. Chapters 2 to 4 examine fundamental concepts that are general to all aspects of P450 function: electron transport, enzyme activity, and specificity in relation to structure and gene regulation in relation to hormonal regulation. Chapters 5 to 11 describe various specific P450 cytochromes and how they synthesize in response to defined regulatory molecules. Emphasis is placed on the regulation of their activity and new concepts arising from the most recent work. Chapter 12 looks at how P450 cytochromes play a regulatory role through controlled removal of these regulatory molecules. Biosynthesis and bio-degradation of lipophilic regulators are typically coordinately regulated by regulatory processes involving peptide hormones. Many excellent books and reviews have covered parts of this material. This book aims to emphasize concepts and allow presentation of provocative ideas that might not be seen in the refereed literature.

REFERENCES

Andersen, J.F., & Hutchinson, D.R. (1992) Characterization of *Saccharopolyspora erythraea* cyto-chrome *P-450* genes and enzymes, including 6-deoxyerythronolide B hydroxylase. J. Bacteriol. 174, 725–735.

Bradshaw, M.S., Tsai, S.Y., Leng, X., Dobson, A.D.W., Conneely, O.M., O'Malley, B.W., & Tsai, M-J. (1991). Studies on the mechanism of functional cooperativity between progesterone and estrogen receptors. J. Biol. Chem. 266, 16684–16690.

Burbach, K.M., Poland, A., & Bradfield, C.A. (1992). Cloning of the Ah-receptor cDNA reveals a distinctive ligand-activated transcription factor. Proc. Natl. Acad. Sci. USA 89, 8185–8189.

Capdevila, J.H., Falck, J.R., & Estabrook, R.W. (1992). Cytochrome P450 and the arachidonate cascade. FASEB J. 6, 731–736.

Chang, C., da Silva, S.L., Ideta, R., Lee, Y., & Yeh, S. (1994). Human and rat TR4 orphan receptors specify a subclass of the steroid receptor superfamily. Proc. Natl. Acad. Sci. USA 91, 6040–6044.

Cooper, D.Y., Levin, S.S., Rosenthal, O., & Estabrook, R.W. (1965). Photochemical action spectrum of the terminal oxidase of mixed function oxidase systems. Science 147, 400–402.

Danielian, P.S., White, R., Lees, J.A., & Parker, M.G. (1992). Identification of a conserved region required for hormone dependent transcriptional activation by steroid hormone receptors. EMBO J., 11, 1025–1033.

Denison, M.S., & Whitlock, J.P. (1995). Xenobiotic-inducible transcription of cytochrome P450 genes. J. Biol. Chem. 270, 18175–18178.

Diamond, M.I., Miner, J.N., Yoshinaga, S.K., & Yamamoto, K.R. (1990). Transcription factor interactions: Selectors of positive or negative regulation. Science 249, 1266–1271.

Dolwich, K.M., Swanson, H.I., & Bradfield, C.A. (1993). *In vitro* analysis of Ah receptor domains involved in ligand-activated DNA recognition. Proc. Natl. Acad. Sci. USA 90, 8566–8570.

Durst, F., Benveniste, I., Salaün, J.P, & Werck, D. (1994). Function and diversity of plant cytochrome P450. In: Cytochrome P450, 8th International Conference (Lechner, M.C., ed.), pp. 23–30. John Libbey Eurotext, Paris.

Eliasson, E., Mkrtchian, S., & Ingelman-Sundberg, M. (1992). Isozyme-specific and substrate-regulated intracellular degradation of cytochrome P450 (2E1) involving MgATP-activated, rapid proteolysis in the endoplasmic reticulum membranes. J. Biol. Chem., 257, 15765–15789.

Fernandez-Salguero, P., Pineau, T., Hilbert, D.M., & Gonzalez, F.J. (1995). Immune system impairment and hepatic fibrosis in mice lacking the dioxin-binding Ah-receptor. Science 268, 722–726.

Feyereisen, R., Andersen, J.F., Cariño, F.A., Cohen, M.B., Koener, J.G., Repecko, S., Scott, J.A., & Snyder, M.J. (1994). Insect cytochrome P450: Functions, regulation and diversity. In: Cytochrome P450, 8th International Conference (Lechner, M.C., ed.), pp. 31–36. John Libbey Eurotext, Paris.

Fisher, C.W., Shet, M.S., Caudle, D.L., Martin-Wixtrom, C.A., & Estabrook, R.W. (1992). High-level expression in *Escherichia coli* of enzymatically active fusion proteins containing the domains of mammalian cytochromes P450 and NADPH-P450 reductase flavoprotein. Proc. Natl. Acad. Sci. USA 89, 10817–10821.

Forman, B.M., Goode, E., Chen, J., Oro, A.E., Bradley, D.J., Perlman, T., Noonan, D.J., Burka, L.T., Evans, R.M., & Weinberger, C. (1995). Identification of a nuclear receptor that is activated by farnesol metabolites. Cell 81, 687–693.

Fulco, A.J. (1991). $P450_{BM-3}$ and Other Inducible Bacterial P450 Cytochromes: Biochemistry and Regulation. Annu. Rev. Pharm. Toxicol. 31, 177–203.

Fuller, P.J. (1991). The steroid receptor superfamily: Mechanisms of diversity. FASEB J. 5, 3092–3099.

Funder, J.W. (1993). Mineralocorticoids, glucocorticoids, receptors and response elements. Science. 259, 1132–1133.

Gearing, K.L., Göttlicher, M., Teboul, M., Widmark, E., & Gustafsson, J.-Å. (1993). Interaction of the peroxisome-proliferator-activated receptor and retinoid X receptor. Proc. Natl. Acad. Sci. USA 90, 1440–1444.

Gibbs, B.S., Wojochowski, D., & Benkovic, S.J. (1993). Expression of rat liver phenylalanine hydroxylase in insect cells and site-directed mutagenesis of putative non-heme iron-binding sites. J. Biol. Chem. 268, 8046–8052.

Gillam, E.M.J., Baba, T., Kim, B.R., Ohmori, S., & Guengerich, F.P. (1993). Expression of modified human cytochrome P450 3A4 in *Escherichia coli* and purification and reconstitution of the enzyme. Arch. Biochem. Biophys. 305, 123–131.

Gitelman, S.E., Bristow, J., & Miller, W.L. (1992). Mechanism and consequences of the duplication of the human C4/P450c21/Gene X locus. Mol. Cell. Biol. 12, 2124–2134.

Gonzalez, F.J. (1989). The molecular biology of cytochrome P450s. Pharmacol. Rev. 40, 244–280.

Gonzales, F.J., & Nebert, D.W. (1990). Evolution of the *P450* gene superfamily: Animal-plant "warfare," molecular drive, and human genetic differences in drug oxidation. Trends Genet. 6, 182–186.

Gorsky, L.D., & Coon, M.J. (1986). Effects of conditions for reconstitution with cytochrome b_5 on the formation of products in cytochrome P-450-catalyzed reactions. Drug Met. & Disp. 14, 89–95.

Göttlicher, M., Widmark, E., Li, Q., & Gustafsson, J.Å. (1992). Fatty acids activate a chimera of the clofibric acid-activated receptor and the glucocorticoid receptor. Proc. Natal. Acad. Sci. USA 89, 4653–4657.

Groves, J.T., McClusky, G.A., White, R.E., & Coon, M.J. (1978). Aliphatic hydroxylation by highly purified liver microsomal cytochrome P450. Evidence for a carbon radical intermediate. Biochem. Biophys. Res. Commun. 81, 154–160.

Gunsalus, I.C., Meeks, J.R., Lipscomb, J.D., Debrunner, P., & Munck, E. (1974). Bacterial monooxygenases—The P450 cytochrome system. In: Molecular Mechanisms of Oxygen Activation (Hayaishi, O., ed.), pp. 559–613.

Gustafsson, J.-Å., Tollet, P., & Mode, A. (1994). Growth hormone regulation of liver sexual differentiation. In: Cytochrome P450, 8th International Conference (M.C. Lechner, ed.), pp. 51–57. John Libbey Eurotext, Paris.

He, J.-S., & Fulco, A.J. (1991). A barbiturate-regulated protein binding to a common sequence in the cytochrome *P450* genes of rodents and bacteria. J. Biol. Chem. 266, 7864–7869.

Hecker, M., & Ullrich, V. (1989). On the mechanism of prostacyclin and thromboxane A_2 biosynthesis. J. Biol. Chem. 264, 141–150.

Hendrich, M.P., Fox, B.G., Andersson, K.K., Debrunner, P.G., & Lipscomb, J.D. (1992). Ligation of the diiron site of the hydroxylase component of methane monooxygenase: An electron nuclear double resonance study. J. Biol. Chem. 267, 261–269.

Heyl, B.L., Tyrrell, D.J., & Lambeth, J.D. (1986). Cytochrome $P-450_{scc}$-substrate interactions: Role of the 3b and side chain hydroxyls in binding to oxidized and reduced forms of the enzyme. J. Biol. Chem. 261, 2743–2749.

Huss, J.M., Wong, S.I., Astrom, A., McQuiddy, P., & Kaspar, C.B. (1996). Dexamethasome responsiveness of a major glucocorticoid-inducible CYP3A gene is mediated by elements unrelated to a glucocorticoid receptor binding motif. Proc. Natl. Acad. Sci. USA 93, 4666–4670.

Iwasaki, M., Darden, T.A., Pedersen, L.G., Davis, D.G., Juvonen, R.O., Sueyoshi, T., & Negishi, M. (1993). Engineering mouse P450coh to a novel corticosterone 15α-hydroxylase and modeling steroid-binding orientation in the substate pocket. J. Biol. Chem. 268, 759–762.

Jefcoate, C.R., Orme-Johnson, W.H., & Beinert, H. (1976). Cytochrome P-450 of bovine adrenal mitochondria: Ligand binding to two forms resolved by EPR spectroscopy. J. Biol. Chem. 251, 3706–3715.

Keightley, D.A., Lou, K.J., & Smith, W.A. (1990). Involvement of translation and transcription in insect steroidogenesis. Mol. Cell. Endocrinol. 74, 229–237.

Kliewer, J.A., Lenhard, J.M., Wilson, T.M., Patel, I., Morris, D.C., & Lehmann, J.M. (1995). A prostaglandin J2 metabolite binds peroxisome proliferator-activated receptor and promotes adipocyte differentiation. Cell 83, 813–819.

Klinman, J.P., Krueger, M., Brenner, M., & Edmondson, D.E. (1984). Evidence for two copper atoms/subunit in dopamine b-monooxygenase catalysis. J. Biol. Chem. 259, 3399–3402.

Larson, J.R., Coon, M.J., & Porter, T.D. (1991b). Purification and properties of a shortened form of cytochrome P-450 2E1: Deletion of the NH-2-terminal membrane-insertion signal peptide does not alter the catalytic activities. Proc. Natl. Acad. Sci. USA 88, 9141–9145.

Larsen, M., Ikegwuonu, R., Brake, P., & Jefcoate, C. (1994). Phenobarbital induction of P450 cytochromes in rat liver: Genetic and endocrine control of a common regulatory mechanism. In: Cytochrome P450, 8th International Conference (M.C. Lechner, ed.), pp. 121–124. John Libbey Eurotext, Paris.

Laudet, V., Hänni, C., Coll, J., Catzeflis, F., & Stéhelin, D. (1992). Evolution of the nuclear receptor gene superfamily. EMBO J. 11, 1003–1013.

Lee, Y-H, Alberta, J.A., Gonzalez, F.J., & Waxman, D.J. (1994). Multiple, functional DBP sites on the promoter of the cholesterol 7α-hydroxylase *P450* gene, *CYP7*. J. Biol. Chem. 269, 14681–14689.

Legraverend, C., Mode, A., Westin, S., Ström, A., Eguchi, H., Zaphiropoulos, P.G., & Gustafsson, J.A. (1992). Transcriptional regulation of rat P450 2C gene subfamily members by the sexually dimorphic pattern of growth hormone secretion. Mol. Endocrinology 259–266.

Lin, L.L., Wartmann, M., Lin, A.Y., Knopf, J.L., Seth, A., & Davis, R.J. (1993). cPLA$_2$ is phosphorylated and activated by MAP kinase. Cell. 72, 269, 278.

Lusska, A., Wu, L., & Whitlock, Jr., J.P. (1992). Superinduction of *CYP1A1* transcription by cycloheximide: Role of the DNA binding site for the liganded Ah receptor. J. Biol. Chem. 267, 15146–15151.

Mangelsdorf, D.J., & Evans, R.M. (1995). The RXR heterodimers and orphan receptors. Cell 83, 844–850.

Mangelsdorf, D.J., Thummel, C., Beato, M., Chambon, P., & Evans, R.M. (1995). The nuclear receptor superfamily: The second decade. Cell 83, 835–839.

Marletta, M.A. (1993). Nitric oxide synthase structure and mechanism. J. Biol. Chem. 268, 12231–12234.

Massey, V. (1994). Activation of molecular oxygen by flavins and flavoproteins. J. Biol. Chem. 269, 22459–22462.

Matarese, V., & Lodish, H.F. (1993). Specific uptake of retinol-binding protein by variant F9 cell lines. J. Biol. Chem. 268, 18849–18865.

Mathis, J.M., Howser, W.H., Bresnick, E., Cidlowski, J.A., Hines, R.N., Prough, R.A., & Simpson, E.R. (1989). Glucocorticoid regulation of the rat p450 1A1 gene: Receptor-binding with intron I. Arch. Biochem. Biophys. 269, 93–105.

Matocha, M., & Waterman, M.R. (1984). Discriminatory processing of the precursor forms of cytochrome P-450$_{scc}$ and adrenodoxin by adrenocortical and heart mitochondria. J. Biol. Chem. 259, 8672.

McGiff, J.C. (1991). Cytochrome P-450 metabolism of arachidonic acid. Ann. Rev. Pharmacol. Toxicol., 31, 339–69.

Miller, W.L. (1988). Molecular biology of steroid hormone synthesis. Endocrine Rev. 9, 295.

Nelson, D.R., Kamataki, T., Waxman, D.J., Guengerich, F.P., Estabrook, R.W., Feyereisen, R., Gonzalez, F.J., Coon, M.J., Gunsalus, I.C., Gotoh, O., Okuda, K., & Nebert, D.W. (1993). The P450 superfamily. Update on new sequences, gene mapping, accession numbers, early trivial names of enzymes, and nomenclature. DNA & Cell Biol. 12, 1–51.

Niranjan, B.G., Raza, H., Shayiq, R.M., Jefcoate, C.R., & Avadhani, N.G. (1988). Hepatic mitochondrial cytochrome P-450 system. J. Biol. Chem. 263, 575–580.

Oinonen, T., & Lindros, K.O. (1994). Growth hormone regiospecifically represses the expression of CYP2B1/2 and CYP3A in liver. In: Cytochrome P450, 8th International Conference (M.C. Lechner, ed.), pp. 833–835. John Libbey Eurotext, Paris.

Omiecinski, C.J., Hassett, C., & Costa, P. (1990). Developmental expression and in situ localization of the phenobarbital-inducible rat hepatic mRNAs for cytochromes CYP2B1, CYP2B2, CYP2C6, and CYP3A1. Mol. Pharmacol. 38, 462–470.

Omer, C.A., Lenstra, R., Litle, P.J., Dean, C., Tepperman, J.M., Leto, K.J., Romesser, J.A., & O'Keefe, D.P. (1990). Genes for two herbicide-inducible cytochromes P-450 from Streptomyces griseolus. J. Bacteriol. 172, 3335–3345.

Omura, T., & Sato, R. (1962). The carbon monoxide-binding pigment of liver microsomes. I. Evidence for its hemoprotein nature. J. Biol. Chem. 239, 2370–2385.

Ortiz de Montellano, P.R., Shirane, N., Sui, Z., Fruetel, J., Peterson, J.A., & De Voss, J.J. (1994). Cytochrome P450: Topology and catalysis. In: Cytochrome P450, 8th International Conference (Lechner, M.C., ed.), pp. 409–416. John Libbey Eurotext, Paris.

Pernecky, S.J., Larson, J.R., Philpot, R.M., & Coon, M.J. (1993). Expression of truncated forms of liver microsomal P450 cytochromes 2B4 and 2E1 in Escherichia coli: Influence of NH_2-terminal region on localization in cytosol and membranes. Proc. Natl. Acad. Sci. USA 90, 2651–2655.

Porter, T.D., & Larson, J.R. (1991). Expression of mammalian P450's in Escherichia coli. Methods Enzymol. 206, 108–116.

Pratt, W.B. (1993). The role of heat shock proteins in regulating the function, folding, and trafficking of the glucocorticoid receptor. J. Biol. Chem. 268, 21455–21458.

Probst, M.R., Reisz-Porszasz, S., Agbunag, R.V., Ong, M.S., & Hankinson, O. (1993). Role of the aryl hydrocarbon (Ah) receptor nuclear translocator protein (ARNT) in aryl hydrocarbon (dioxin) receptor action. Mol. Pharm. 44, 511–518.

Puga, A., RayChaudhuri, B., & Nebert, D.W. (1992). Transcriptional derepression of the murine Cyp1a-1 gene by mevinolin. FASEB J. 6, 777–785.

Raag, R., & Poulos, T.L. (1991). Crystal structures of cytochrome P-450$_{CAM}$ complexed with camphane, thiocamphor, and adamantane: Factors controlling P-450 substrate hydroxylation. Biochem. 30, 2674–2684.

Ram, P.A., & Waxman, D.J. (1991). Hepatic P450 expression in hypothyroid rats: differential responsiveness of male-specific P450 forms 2a (IIIA2), 2c (IIC11), and RLM2 (IIA2) to thyroid hormone. Molec. Endocrinology. 5, 13–20.

Ramsden, R., Sommer, K.M., & Omiecinski, C.J. (1993). Phenobarbital induction and tissue-specific expression of the rat CYP2B2 gene in transgenic mice. J. Biol. Chem. 268, 21722–21726.

RayChaudhuri, B., Nebert, D.W., & Puga, A. (1990). The murine Cyp1a-1 gene negatively regulates its own transcription and that of other members of the aromatic hydrocarbon-responsive [Ah] gene battery. Mol. Endocrinol. 4, 1773–1781.

Raza, H., & Avadhani, N.G. (1988). Hepatic mitochondrial cytochrome P-450 system. J. Biol Chem. 263, 9533–9541.

Rong, D., Lovey, A.J., Rosenberger, M., d'Avignon, A., Ponder, J., & Lit, Ellen. (1993). Differential binding of retinol analogs to two homologous cellular retinol-binding proteins. J. Biol. Chem. 268, 7929–7934.

Sadek, C.M., & Allen-Hoffmann, B.L. (1994). Cytochrome P450IA1 is rapidly induced in normal human keratinocytes in the absence of xenobiotics. 269, 16067–16074.

Sagara, Y., Barnes, H.J., & Waterman, J.R. (1993). Expression in *Escherichia coli* of functional cytochrome P450c17 lacking its hydrophobic amino-terminal signal anchor. Arch. Biochem. Biophys. 304, 272–278.

Sakaguchi, M., Mihara, K., & Omura, T. (1994). Biosynthesis and membrane topology of microsomal cytochrome P450. In: Cytochrome P450, 8th International Conference (M.C. Lechner, ed.), pp. 265–270. John Libbey Eurotext, Paris.

Sakaguchi, M., Tomiyoshi, R., Kuroiwa, T., Mihara, K., & Omura, T. (1992). Functions of signal and signal-anchor sequences are determined by the balance between the hydrophobic segment and the N-terminal charge. Proc. Natl. Acad. Sci. USA 89, 16–19.

Sakaki, T., Oeda, K., Miyoshi, M., & Ohkawa, H. (1985). Characterization of rat cytochrome $P450_{MC}$ synthesized in *Saccharamoyces cerevisiae*. J. Biochem. 98, 167–175.

Savas, Ü., Bhattacharyya, K.K., Christou, M., Alexander, D.L., & Jefcoate, C.R. (1994). Mouse cytochrome P-450EF, representative of a new *1B* subfamily of cytochrome P-450s: Cloning, sequence determination, and tissue expression. J. Biol. Chem. 269, 14905–14911.

Schöller, A., Hong, N.J., Bischer, P., & Reiners, Jr., J.J. (1994). Short and long term effects of cytoskeleton-disrupting drugs on cytochrome P450 *CYP1a-1* induction in murine hepatoma 1c1c7 cells: Suppression by the microtubule inhibitor nocodazole. Molec. Pharmac. 45, 944–954.

Shapiro, B.H., Pampori, N.A., Lapenson, D.P., & Waxman, D.J. (1994). Growth hormone dependent and independent sexually dimorphic regulation of phenobarbital-induced hepatic cytochromes p450 2B1 and 2B2. Arch. Biochem. Biophys. 312, 234–239.

Shimada, H., Makino, R., Unno, M., Horiuchi, T., & Ishimura, Y. (1994). Proton and electron transfer mechanism in dioxygen activation by cytochrome P450cam. In: Cytochrome P450, 8th International Conference (M.C. Lechner, ed.), pp. 299–306. John Libbey Eurotext, Paris.

Shimozawa, O., Sakaguchi, M., Ogawa, H., Harada, N., Mihara, K., & Omura, T. (1993). Core glycosylation of cytochrome P450(arom). Evidence for localization of N-terminus of microsomal cytochrome P450 in the lumen. J. Biol. Chem. 268 (in press).

Sidhu, J.S., & Omiecinski, C.J. (1995). cAmp-associated inhibition of phenobarbital-inducible cytochrome p450 gene expression in primary rat hepatocyte cultures. J. Biol. Chem. 270, 12762–12773.

Simpson, E.R., & Boyd, G.S. (1967). Cholesterol side-chain cleavage system of bovine adrenal cortex. Eur. J. Biochem. 2, 275–285.

Sligar, S.G., Aikens, J., Gerber, N., McLean, M., Suslick, K., & Benson, D. (1994). Electron transfer associated dioxygen activation in P450 systems. In: Cytochrome P450, 8th International Conference (M.C. Lechner, ed.), pp. 373–378. John Libbey Eurotext, Paris.

Song, W.-C., Funk, C.D., & Brash, A.R. (1993). Molecular cloning of an allene oxide syntase: A cytochrome P450 specialized for the metabolism of fatty acid hydroperoxides. Proc. Natl. Acad. Sci. USA 90, 8519–8523.

Sundseth, S.S., & Waxman, D.J. (1992). Sex-dependent expression and clofibrate inducibility of cytochrome P450 4A fatty acid ω-hydroxylases: Male specificity of liver and kidney CYP4A2 mRNA and tissue-specific regulation by growth hormone and testosterone. J. Biol. Chem. 267, 3915–3921.

Sutter, T.R., Tang, Y.M., Hayer, C.L., Wo, Y.Y.P., Jabs, E.W., Li, X., Yin, H., Cody, C.W., & Greenlee, W.F. (1994). Complete cDNA sequence of a human dioxin-inducible mRNA identifies a new gene subfamily of cytochrome P450 that maps to chromosome 2. J. Biol. Chem. 269, 13092–13099.

Szczesna-Skorupa, E., Straub, P., & Kemper, B. (1993). Deletion of a conserved tetrapeptide, PPGP, in P450 2C2 results in loss of enzymatic activity without a change in its cellular location. Arch. Biochem. Biophys. 304, 170–175.

Tamburini, P., & Schenkman, J.B.. (1987). Purification to homogeneity and enzymological characterization of a functional covalent complex composed of cytochromes P450 isozyme 2 and b5 from rabbit liver. Proc. Natl. Acad. Sci. 84, 11–15.

Tollet, P., Hamberg, M., Gustafsson, J.-A., & Mode, A. (1995). Growth hormone signalling leading to *CYP2C12* gene expression in rat hepatocytes involves phospholipase A2. J. Biol. Chem. 270, 12569–12577.

Trottier, E., Belzil, A., Stoltze, C., & Anderson, A. (1995). Localization of the phenobarbital-responsive element (PBRE) in the 5' flanking region of the CYP2B2 gene. Gene 158, 263–268.

Tsai, M.-J., & O'Malley, B.W. (1994). Molecular mechanisms of action of steroid/thyroid receptor superfamily members. Ann. Rev. Biochem. 63, 451–486.

Vergères, G., Winterhalter, K.H., & Richter, C. (1988). Identification of the membrane anchor of microsomal rat liver cytochrome P-450. Biochem. 28, 3650–3655.

Voznesensky, A.I., & Schenkman, J.B. (1992). Inhibition of cytochrome P450 reductase by polyols has an electrostatic nature. Eur. J. Biochem. 210, 741–746.

Voznesensky, A.I., & Schenkman, J.B. (1994). On the nature of the NADPH-cytochrome P450 reductase-cytochrome P450 electron transfer complex. In: Cytochrome P450, 8th International Conference (M.C. Lechner, ed.), pp. 349–356. John Libbey Eurotext, Paris.

Watkins, P.B., Wrighton, S.A., Scheutz, E.G., Maurel, P., & Guzelian, P.S. (1986). Macrolide antibiotics inhibit the degradation of the glucocorticoid-responsive cytochrome P450p in rat hepatocytes *in vivo* and in primary monolayer culture. J. Biol. Chem. 261, 6264–6271.

Waxman, D.J., & Azaroff, L. (1992). Phenobarbital induction of cytochrome *P450* gene expression. Biochem. J. 281, 577–592.

White, R.E. (1994). The importance of one-electron transfers in the mechanism of cytochrome P450. In: Cytochrome P450, 8th International Conference (M.C. Lechner, ed.), pp. 333–340. John Libbey Eurotext, Paris.

Whitlock, E. (1994). The aromatic hydrocarbon receptor, dioxin action and endocrine homeostasis. Trends Endocrinol. Metals 5, 183–188.

Wu, L., & Whitlock, Jr., J.P. (1992). Mechanism of dioxin action: Ah receptor-mediated increase in promoter accessibility *in vivo*. Proc. Natl. Acad. Sci. USA 89, 4811–4815.

Yamazoe, Y., Shimada, M., Murayama, N., & Kato, R. (1987). Suppression of levels of phenobarbital-inducible rat liver cytochrome P-450 by pituitary hormone. J. Biol. Chem. 262, 7423–7428.

Yanagibashi, K., & Hall, P.F. (1986). Role of electron transport in the regulation of the lyase activity of C_{21} side-chain cleavage P-450 from porcine adrenal and testicular microsomes. J. Biol. Chem. 261, 8429–8433.

Zhao, Y., Mendelson, C.R., & Simpson, E.R. (1995). Characterization of the sequences of the human CYP19 (aromatase) gene that mediates regulation by glucocortocoids in adipose stromal cells and fetal hepatocytes. Mol. Endocrinol. 9, 340–349.

Zilliacus, J., Carlstedt-Duke, J., Gustafsson, J.Å., & Wright, A.P.H. (1994). Evolution of distinct DNA-binding specificities within the nuclear receptor family of transcription factors. Proc. Natl. Acad. Sci. USA 91, 4175–4179.

ELECTRON TRANSFER PROTEINS OF CYTOCHROME P450 SYSTEMS

Israel Hanukoglu

I. INTRODUCTION

The reactions catalyzed by diverse P450 systems generally involve hydroxylations of substrates. These reactions are named monooxygenations because P450 cata-

---

Advances in Molecular and Cell Biology
Volume 14, pages 29–56.
Copyright © 1996 by JAI Press Inc.
All rights of reproduction in any form reserved.
ISBN: 0-7623-0113-9

lyzes incorporation of only one atom of molecular O_2 into substrate while reducing the second into H_2O with the following stoichiometry:

$$RH + O_2 + NAD(P)H + H_+ \rightarrow ROH + H_2O + NAD(P)^+$$

Thus, each cycle of monooxygenation requires two electrons that originate from pyridine nucleotides, NADH or NADPH. The function of the electron transport proteins of P450 systems is to accept the two electrons from NAD(P)H and to transfer them one at a time to the P450 during the monooxygenase reaction sequence (White and Coon, 1980; Black and Coon, 1987; Archakov and Bachmanova, 1990; Takemori et al., 1993; for a review of the history of the field see Omura, 1993a). The known P450 systems can be grouped into four categories based on their electron transport components:

1. Mitochondrial systems include two proteins, NADPH specific adrenodoxin reductase with FAD cofactor, and adrenodoxin[1], a [2Fe-2S] ferredoxin type iron-sulfur protein (previous reviews: Lambeth et al., 1982; Orme-Johnson, 1990; Lambeth, 1991; Hanukoglu, 1992). FAD can be reduced by two electrons from NADPH, which are transferred one at a time to adrenodoxin which is a one electron carrier. Both proteins are located on the matrix side of the inner mitochondrial membrane (Mitani, 1979; Hanukoglu, 1992). Whereas the P450s are anchored to the membrane, the electron transfer proteins are soluble and may have greater mobility in the matrix. They are encoded as larger precursors, and proteolytically processed to their mature sizes after transfer into mitochondria (Omura, 1993b).

2. Microsomal systems depend on a single NADPH specific P450 reductase that contains both FAD and FMN as cofactors (previous reviews: Black and Coon, 1987; Archakov and Bachmanova, 1990; Schenkman and Greim, 1993). FAD can accept two electrons from NADPH and FMN functions as the single electron carrier. FAD and FMN have midpoint potentials of -328 mV, and -190 mV. Thus, the route of electron transfer is NADPH $\rightarrow$ FAD $\rightarrow$ FMN $\rightarrow$ P450 (Vermillion et al., 1981; Oprian and Coon, 1982). Some microsomal P450s may receive the second electron from NADH through cytochrome b_5 reductase and cytochrome b_5 (previous reviews: Arinc, 1991; Borgese et al., 1993; Schenkman, 1993).

P450 reductase can reduce cytochrome c in addition to cytochromes P450 and b_5. Hence, in early work it was named "cytochrome c reductase," despite cytochrome c's being an artificial acceptor, normally located in the inter-membrane space of the mitochondria. The reductase and the P450s are located on the cytoplasmic side of the endoplasmic reticulum membrane, anchored to the membrane by their hydrophobic amino termini which are inserted into the membrane during translation. The amino termini include both the signal and membrane anchor sequences and are not cleaved after membrane insertion (Haniu et al., 1989; Pernecky et al., 1993; Sakaguchi and Omura, 1993; Tashiro et al., 1993).

3. Bacterial P450cam (from *Pseudomonas putida*, "cam" for camphor substrate) type systems include a ferredoxin reductase and a ferredoxin (named putidare-

doxin) that are functionally similar to the mitochondrial electron transfer proteins (Gunsalus and Sligar, 1978). P450cam and the electron transfer proteins are all soluble and do not appear to be membrane associated. The proteins are encoded by an operon in the CAM plasmid (Koga et al., 1989; Peterson et al., 1990).

4. Bacterial P450meg (from *Bacillus megaterium*) type system composed of one large soluble protein with a P450 reductase domain and a cytochrome P450 domain each of which are homologous to the microsomal P450 system components (Ruettinger et al., 1989; Oster et al., 1991; Boddupalli et al., 1992; Ravichandran et al., 1993).

In contrast to the multiplicity of cytochromes P450 (Nelson et al., 1993), there appears to be only one form of each of mitochondrial adrenodoxin reductase, adrenodoxin, and the microsomal P450 reductase, encoded by one or two similar nuclear genes in all animal species (for review see Hanukoglu, 1992). Thus, the electron transfer proteins are not specific to individual P450s and serve as electron donors for different cytochromes P450 in different tissues. While in some plant species there is a single gene for P450 reductase (Meijer et al., 1993), in others there are several related forms of P450 reductase (e.g., Lesot et al., 1992). The function and specificities of these remain to be determined.

This chapter presents a review of the structural and functional aspects of the flavoenzyme and ferredoxin type electron transfer proteins of the mitochondrial, microsomal and bacterial P450 systems. The studies summarized below established several general principles for P450 system electron transport chains: (1) The reductases are generally expressed at much lower levels than P450s, there being only one molecule of reductase per about 10 or more molecules of P450, (2) the protein components are independently mobile and do not form static multicomponent complexes, (3) proteins that are redox partners form transient high affinity 1:1 complexes during their random diffusions, in accordance with the principles of mass action. Dissociation constants of these protein-protein complexes are strongly influenced by the redox states of the proteins and other molecules in the environment, such as P450 substrate, ions, and phospholipids, and (4) the transfer of an electron between two redox partners depends on the formation of a specific high affinity 1:1 complex between the two proteins. In P450 systems electron transfer is not always coupled to substrate monooxygenation. P450s and their electron transfer proteins may transfer electrons to other acceptors, such as O_2. This type of "uncoupling" or "leaky electron transport" is observed in both mitochondrial and microsomal systems (Hornsby, 1989; Archakov and Bachmanova, 1990; Hanukoglu et al., 1993). The regulation of protein–protein complex formation generally enhances productive associations for monooxygenase activities and helps to minimize uncoupled reactions that produce harmful free radicals.

II. SEQUENCE AND STRUCTURAL RELATIONSHIPS

Although the reductases of the mitochondrial and microsomal P450 systems are all flavoenzymes that bind NADP, as summarized below, their sequences represent different superfamilies of enzymes with no apparent evolutionary relationship.

A. Adrenodoxin Reductase and Putidaredoxin Reductase

Adrenodoxin reductase (~50 kD) and putidaredoxin reductase (~45 kD) are both soluble proteins that contain FAD as a cofactor (Chu and Kimura, 1973; Hiwatashi et al., 1976; Gunsalus and Sligar, 1978). In immunoelectron microscopy of adrenal cells, adrenodoxin reductase appears as membrane associated (Mitani, 1979). However, its sequence does not have a hydrophobic membrane spanning segment (Fig. 1). Thus, it probably functions as a peripheral membrane protein associated with ionic interactions.

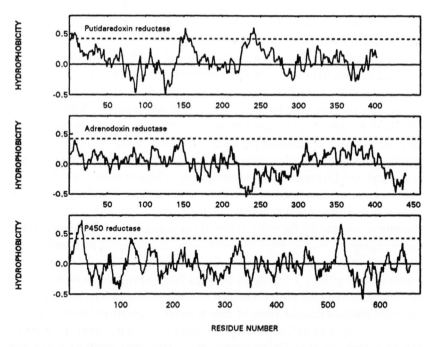

Figure 1. Hydrophobicity profiles of putidaredoxin reductase (Koga et al., 1989; Peterson et al., 1990), human adrenodoxin reductase (Solish et al., 1988), and human P450 reductase (Haniu et al., 1989). Each point is the average hydrophobicity of a 20 residue segment (Eisenberg et al., 1984). In the hydrophobicity scale, the cutoff for the prediction of a transmembrane segment is 0.42 (*dashed line*) (Eisenberg et al., 1984).

Despite functional similarities, adrenodoxin reductase shows no sequence homology with putidaredoxin reductase or other types of oxidoreductases (Hanukoglu and Gutfinger, 1989) (Fig. 2). The hydrophobicity profiles of these two enzymes are also highly different (Fig. 1). However, putidaredoxin reductase shows unequivocal sequence homology with rubredoxin reductase (Eggink et al., 1990), and the ferredoxin-NAD reductase component of benzene and toluene dioxygenases from P. putida (Irie et al., 1987; Zylstra et al., 1989), over the entire lengths of these enzymes (Fig. 2).

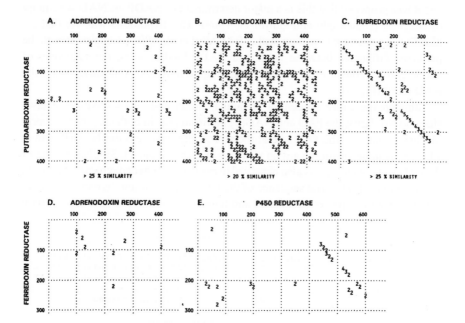

Figure 2. Matrix plot sequence comparisons: (*A, B, C*) Putidaredoxin reductase (Koga et al., 1989; Peterson et al., 1990) versus human adrenodoxin reductase (Solish et al., 1988), and (C) rubredoxin reductase (Eggink et al., 1990). (*D, E*) Spinach ferredoxin reductase (Karplus et al., 1984) versus (D) adrenodoxin reductase and (*E*) human P450 reductase (Haniu et al., 1989). The numbers on the axes correspond to the residue numbers of the mature enzyme. The parameters used in the analysis: Range = 30, compression factor = 10, cutoff point = 25% (20% in B). Thus, each digit on the plots indicates that within a 30 residue long segment to the right of the marked position at least 25% of the residues are identical in both sequences. The numbers indicate the percentile range of identical residues in the 30 residue segment, e.g., 3 = 30–39%, 4 = 40–49%. The diagonal lines indicate homologous sequences. If the cutoff for sequence similarity is lowered, instead of a diagonal line a random distribution of points appears.

The FAD and NAD(P) binding sites of both adrenodoxin reductase and putidare-doxin reductase were identified using an ADP dinucleotide binding site consensus motif (Hanukoglu and Gutfinger, 1989; Peterson et al., 1990). FAD and NAD(P) both have ADP as a common part of their structures. In most FAD or NAD(P) binding enzymes, the sites that bind this ADP portion also share a similar confor-mation of a βαβ-fold. The most highly conserved sequence in this fold is Gly-X-Gly-X-X-Gly/Ala which forms a tight turn between the first β-strand and the α-helix (Rossman et al., 1975; Wierenga et al., 1985; Hanukoglu and Gutfinger, 1989). Analyses of adrenodoxin reductase sequence led to the discovery that in NADP binding sites of this type, there is an Ala instead of the third Gly residue, and it was proposed that this is a major determinant of NADP vs. NAD specificity of enzymes (Hanukoglu and Gutfinger, 1989). This hypothesis was verified for glutathione reductase (Scrutton et al., 1990; Mittl et al., 1993). Consistent with the predicted difference between NAD and NADP binding sites, the sequences of putidaredoxin reductase and the other homologous ferredoxin reductases match the NAD motif (Fig. 3).

Although the entire sequences of putidaredoxin and adrenodoxin reductases share no similarity, the FAD and NAD(P) motifs appear in both enzymes at nearly identical positions: the FAD site at the amino terminus, and the NAD(P) site at 146–151 residues from the amino terminus (Figures 3 and 4). Similar spacing of the FAD and NADP sites is also observed in many other flavoenzymes (Hanukoglu and Gutfinger, 1989). The sequence of the FAD binding amino terminus is highly conserved across species (Yamazaki and Ichikawa, 1990). A bovine adrenodoxin

FAD binding site

Adrenodoxin R (Bovine)	1	STQEQTPQICVVGSGPAGFYTAQHLLKHHSRAHVDIYEKQLVPFGL
Adrenodoxin R (Human)	1	STQEKTPQICVVGSGPAGFYTAQHLLKHP QAHVDIYEKQPVPFGL
Putidaredoxin R	1	MNANDNVVIVGTGLAGVEVAFGLRASGWEGNIRLVGDATVIPHH
FAD consensus motif		±••••G G •G• A +••± G ±• ••
Secondary structure		TββββββTαααααααααααααααααTTββββββT

NAD(P) binding site

Adrenodoxin R (Bovine)	137	RELAPDLSCDTAVILGQGNVALDVARILLTPPDHLEKTDITEAALGALR
Adrenodoxin R (Human)	136	QELEPDLSCDTAVILGQGNVALDVARILLTPPEHLERTDITKAALGVLR
Putidaredoxin R	141	CIRRQLIADNRLVVIGGGYIGLEVAATAIKANMHVTLLDTAARVLERVT
NADP consensus motif		+••••G G •A••A •• •G +• •
NAD consensus motif		+• • G G G • • -
Secondary structure		TββββββTαααααααααααααααααTTββββββT

Figure 3. Predicted FAD and NAD(P) binding sites of adrenodoxin reductase (Solish et al., 1988; Hanukoglu & Gutfinger, 1989; Sagara et al., 1987) and putidaredoxin reductase (Koga et al., 1989; Peterson et al., 1990). The residues are numbered starting with the amino terminus of the mature protein. The FAD and NADP motifs are from Hanukoglu & Gutfinger (1989), and the NAD motif from Wierenga et al. (1986). (·) indicates a hydrophobic residue. (+), (–), and (±) indicate charged and hydrophilic residues respectively.

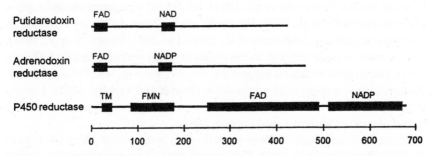

Figure 4. The localization of the FAD and NAD(P) binding sites of putidaredoxin and adrenodoxin reductases and the structural domains of P450 reductase. The FAD and NAD(P) binding sites of putidaredoxin and adrenodoxin reductases refer to the ADP portion of these coenzymes. The domains of P450 reductase is based on the prototype structure of ferredoxin reductase (Karplus et al., 1991), and on the model of Porter et al. (1990). *TM*: transmembrane segment.

reductase cDNA expressed in yeast with an extra segment encoding four additional residues at the N-terminus did not yield an active enzyme, suggesting that the four residues disrupted the incorporation of FAD into the apoprotein (Akiyoshi-Shibata, 1991).

The evolution of the enzymes with the FAD and NAD(P) binding domains noted above remains an enigma because of the lack of sequence homology outside of these very short domains. One explanation may be found in the observation that in the adrenodoxin reductase gene, the FAD and NADP binding sites are encoded by separate small exons (Lin et al., 1990). In this gene, exon 7 starts at the end of the α-helix of the βαβ fold. One alternative splicing product encodes six extra residues at this position in both bovine and human genes (Sagara et al., 1987; Solish et al., 1988). However, this form represents ~1% of the total reductase mRNA population (Brentano et al., 1992), and expression of its cDNA in *E. coli* did not yield active enzyme, suggesting that the extra residues disrupt the structure of the enzyme (Brandt and Vickery, 1992).

Whereas the bovine adrenodoxin reductase was reported to be glycosylated (Hiwatashi et al., 1976; Suhara et al., 1982), a recent study could not corroborate previous evidence for functional glycosylation of the bovine enzyme (Warburton and Seybert, 1994). The porcine enzyme is free of carbohydrate (Hiwatashi and Ichikawa, 1978). Adrenodoxin reductase expressed in *E. coli* functioned as well as the native enzyme in a reconstituted mitochondrial P450scc system. Thus, the apoprotein may be assembled to active holoenzyme without eukaryotic posttranslational modifications (Sagara et al., 1993).

Structure-function studies using chemical modification methods implicated the following roles for specific residues of adrenodoxin reductase: (a) FAD binding: a histidine residue (Yamazaki et al., 1992), (b) NADP binding: a histidine and a cysteine (Hiwatashi et al., 1976; Hiwatashi and Ichikawa, 1978), (c) adrenodoxin binding: a tryptophan (Sarkissova et al., 1990), Lys-243 (Hamamoto et al., 1988), and a cluster of basic residues in the amino terminal portion (Hara and Miyata, 1991). Site directed mutations suggest that Arg-239 and Arg-243 of human adreno-doxin reductase interact with adrenodoxin (Brandt and Vickery, 1993). Limited proteolysis cleaves the enzyme to yield two large fragments that may represent distinct domains (Warburton and Seybert, 1988).

Adrenodoxin reductase has been crystallized in several laboratories (Nonaka et al., 1985; Hanukoglu et al., 1992; Kuban et al., 1993). The elucidation of its crystal structure is necessary to increase our understanding beyond the sequence analyses, to identify in detail the cofactor pockets, and the electron acceptor ferredoxin binding sites, and to elucidate the routes of electron transfer.

B. Adrenodoxin and Putidaredoxin

Adrenodoxin (~14 kD) and putidaredoxin (~11.5 kD) are soluble [2Fe-2S] ferredoxin type iron-sulfur proteins that function as a single electron (for previous reviews see Estabrook et al., 1973; Gunsalus and Sligar, 1978). The sequences of adrenodoxin and putidaredoxin can be aligned over their entire lengths with only a few gaps (Cupp and Vickery, 1988). A recently cloned [2Fe-2S] ferredoxin from E. coli also shares 36% identity with adrenodoxin and putidaredoxin (Ta and Vickery, 1992). Waki and associates (1986) reported the purification of a [2Fe-2S] protein from bovine liver mitochondria that can support C-25 and C-27 hydroxy-lations of steroids, but with an amino terminus sequence completely different from adrenodoxin. The functional and structural relationship of this protein to adreno-doxin may be elucidated after cloning of its cDNA.

Bovine adrenodoxin is translated from multiple species of mRNA encoded by a single gene (Sagara et al., 1990). The protein sequences encoded by these mRNAs differ only in the C-terminus of the signal peptide and the first two residues of the mature sequence (Okamura et al., 1987). In the human genome there are two genes, but both encode the same protein product (Chang et al., 1990). In polyacrylamide gel electrophoresis of adrenodoxin purified from different tissues two major bands or a broad band may be observed around ~12–14 kD (Driscoll and Omdahl, 1986; Bhasker et al., 1987). This heterogeneity was considered to represent a multiplicity of tissue specific forms of ferredoxin. However, it apparently results from pro-teolytic cleavage of up to 14 residues from the carboxy terminus of the mature adrenodoxin during the purification process (Hiwatashi et al., 1986; Sakihama et al., 1988). Trypsin treatment of purified adrenodoxin produces a truncated form of adrenodoxin (des 116–128) which shows a lower K_m (higher affinity) in supporting P450 activity (Cupp and Vickery, 1989).

Putidaredoxin sequence is shorter than the adrenodoxin sequence at the carboxy terminus. The deletion or alteration of the terminal tryptophan residue adversely affects its interaction with P450cam (Davies and Sligar, 1992).

In the [2Fe-2S] center two iron atoms are coordinated to four cysteines and two labile sulfur atoms. Mature bovine adrenodoxin sequence includes five cysteines. Chemical modification and site-directed mutagenesis studies indicated that *Cys*-46, 52, 55, and 92 are involved in iron-sulfur coordination, whereas *Cys*-95 is free (Cupp and Vickery, 1988; Uhlmann et al., 1992). In putidaredoxin sequence, the corresponding cysteines 39, 45, 48, and 86 are conserved and function as the ligands to the iron atoms (Gerber et al., 1990). Putidaredoxin sequence includes two additional cysteines that are apparently free.

A cluster of negatively charged residues of bovine adrenodoxin have been implicated in complex formation with both adrenodoxin reductase and mitochondrial P450s in studies employing different approaches (Lambeth et al., 1984; Tuls et al., 1987; Hara and Miyata, 1991; Coghlan and Vickery, 1992). Site directed mutation of *Asp*-76 and *Asp*-79 showed that these residues are essential for adrenodoxin binding to both the reductase and P450scc (Coghlan and Vickery, 1992). In contrast, modification of all lysine or arginine residues did not affect adrenodoxin interactions with either of its redox partners, suggesting that these are not located at the binding site for either protein (Tuls et al., 1987). These findings support the conclusions based on kinetic studies (see below) that the binding sites of these two enzymes on adrenodoxin overlap. Yet, the differential effects of carboxy-terminal truncation at *Arg*-115 on interactions with adrenodoxin reductase and P-450 suggest that the sites are not identical (Cupp and Vickery, 1989). A mutated form of adrenodoxin missing six amino terminal residues supported only 60% of the activity of P450scc suggesting that the amino terminal residues may also play a role in P450scc binding or electron transfer (Sagara et al., 1992). Modification of *His*-56 reduced the binding of adrenodoxin to its redox partners (Miura et al., 1991), and *Tyr*-82 mutation affected binding of adrenodoxin to mitochondrial P450s without affecting cytochrome c reduction (Beckert et al., 1994). Proton NMR studies indicate that the reduction of adrenodoxin changes the conformation of the residues in the putative binding site for its redox partners (Miura and Ichikawa, 1991).

Bovine and chick adrenodoxins can be phosphorylated. This may affect their interaction with P450 and hence the activity of P450 (Monnier et al., 1987; Mandel et al., 1990). Adrenodoxin cDNAs expressed in *E. coli* encode proteins functionally as active as the native protein, indicating that the [2Fe-2S] centers of these proteins can be properly assembled in bacteria, and that eukaryote specific posttranslational modifications are not necessary for activity (Tang and Henry, 1993; Coghlan and Vickery, 1992; Uhlmann et al., 1992).

The structure of adrenodoxin crystals currently available cannot be solved because of their complexity (Marg et al., 1992). The elucidation of the complete structure of adrenodoxin awaits formation of different crystals suited for crystallographic analysis. Proton NMR studies indicate that the structure of adrenodoxin is similar to that of Spirulina platensis ferredoxin (Miura and Ichikawa, 1991).

C. P450 Reductase

P450 reductase (~77 kD) is the only known enzyme with both a FAD and a FMN cofactor. The enzyme is anchored to the endoplasmic reticulum membrane by a highly hydrophobic transmembrane segment at its amino terminus (Fig. 1). The expression of a P450 reductase cDNA with a deleted anchor sequence yields a cytosolic enzyme in the yeast (Urban et al., 1990). This hydrophobic domain can be cleaved with hydrolytic enzymes producing a large soluble peptide (~71 kD) that retains the FAD and FMN spectra of the native enzyme, can reduce cytochrome c as an artificial acceptor, but not P450 (Iyanagi et al., 1978; Strobel et al., 1980; Black and Coon, 1982; Lu, 1991). The expression of rat P450 reductase cDNA in E. coli yielded truncated forms in addition to the full size protein with FAD and FMN cofactors. These findings show that the holoenzyme can be assembled in bacteria and that the hydrophobic termini may be degraded inside the bacteria (Porter et al., 1987). Functional reductase was also obtained in large yields in a baculovirus expression system (Tamura et al., 1992).

The sequences of P450 reductases from mammalian species are highly conserved, yet the yeast reductase shares only 30–35% similarity with the mammalian enzymes (Yabusaki et al., 1988; Haniu et al., 1989; Ohgiya et al., 1992). The transmembrane segment, and FAD, FMN, NADPH, and P450 binding sites of the enzyme have been tentatively identified by sequence analyses and structure-function studies summarized below (Fig. 4). FMN binding domain of P450 reductase shares a very short segment of similarity with flavodoxins (Porter and Kasper, 1986). Although P450 reductase does not interact with a ferredoxin, its FAD and NADP domains are homologous to plant ferredoxin-NADP reductase (Porter and Kasper, 1986) (Fig. 2). These observations led to the suggestion that P450 reductase evolved by fusion of two genes encoding a flavodoxin and a ferredoxin reductase type enzyme (Porter and Kasper, 1986). The gene sequence of the rat reductase revealed a general correspondence between the exons and the predicted structural domains of the protein (Porter et al., 1990).

P450 reductase and plant ferredoxin reductase sequences are also homologous to NADPH-sulfite reductase, NADH cytochrome b_5 reductase, NADH nitrate reductase, and nitric oxide synthase (Karplus et al., 1991; Bredt et al., 1991). These enzymes constitute an evolutionarily related superfamily. P450 reductase was claimed to be homologous also to glutathione reductase (Haniu et al., 1986; Porter and Kasper, 1986; Porter et al., 1990). However, homology matrix plot analysis does not reveal any significant similarity between these two enzymes. The prototype tertiary structure of spinach ferredoxin-NADP reductase is completely different from that of glutathione reductase which includes the $\beta\alpha\beta$ NADP binding motif noted above (Karplus et al., 1991).

The Type I FAD and NAD(P) binding site motifs noted above for adrenodoxin reductase do not appear in P450 reductase and other related enzymes. The residues involved in binding FAD and NADP in the prototype structure of ferredoxin reductase are conserved in other members of this family and present a consensus sequence entirely different from the Type I FAD and NAD(P) motifs (Karplus et al., 1991). The relative orientations of the isoalloxazine ring of FAD and of nicotinamide ring of NADP are also suggested to be different in these families of flavoenzymes (Sem and Kasper, 1992).

By sequence alignment with a flavodoxin, *Tyr*-178 of rat P450 reductase was suspected to be involved in FMN binding, and its replacement, but not of *Tyr*-140, indeed abolished FMN binding (Shen et al., 1989). *Cys*-471, *Cys*-565, and *Lys*-601 of P450 reductase (aligned with *Lys*-244 of ferredoxin reductase) are probably located in the NADP binding site as these residues are protected against chemical modification in the presence of NADP$^+$ (Haniu et al., 1986; 1989; Vogel and Lumper, 1986; Slepneva and Weiner, 1988; Muller et al., 1990). However, the mutation of *Cys*-566 of rat reductase (corresponding to *Cys*-565 of human and porcine enzymes) to Ala or Ser did not abolish catalytic activity and NADPH binding. Moreover, the position that corresponds to *Cys*-565 has a Ser residue in yeast reductase (Yabusaki et al., 1988), indicating that this *Cys* is not essential for activity (Shen et al., 1991).

Kinetic analysis and site-directed mutagenesis have been used by Kasper and co-workers to further refine the mechanisms of electron transfer by P450 reductase. Kinetic analysis of reduction of cytochrome c indicates a mechanism that is very dependent on ionic strength (Sem and Kasper, 1995). NADPH and cytochrome c bind at independent sites. Isotope effects indicate that hydride transfer from NADPH is rate limiting only at high ionic strengths (0.2–0.75M). At lower, more physiological pH's, NADP$^+$ release becomes limiting. At very high ionic strength, a conformational change that intervenes between NADPH binding and hydride transfer may be rate determining. Site directed mutagenesis of two acidic clusters ^{207}Asp.Asp.Asp^{209} and ^{213}Glu.Glu.Asp^{215} demonstrates that the first cluster is critical for P450 reaction but not for cytochrome c reduction, while the reverse selectivity occurs for the second cluster (Shen and Kasper, 1995). Site directed mutagenesis has also been used to show that the positive charge of arginine 597 is a major site for binding of the 2'-phosphate of NADPH (Sem and Kasper, 1993). Serine 457 has also been implicated in the hydride transfer from NADPH to FAD. The S457A mutant shifts to hydride transfer as the slow step instead of NADP+ release (Shen and Kasper, 1996).

The binding of reductase to various P450s has been suggested to be mediated by both hydrophobic and complementary charge interactions (Schenkman, 1993; Shen and Strobel, 1993; Voznesensky and Schenkman, 1994). The amino terminal region of the reductase was implicated in binding to P450 (Black and Coon, 1982). Chemical modification studies showed that carboxyl residues in the range of 109–130, within the FMN binding domain, are involved in interaction with P450

(Nadler and Strobel, 1991). Quantitative analyses of ionic dependence of reactions indicated that different numbers of charged residues are involved in interactions between various P450s and the P450 reductase (Voznesensky and Schenkman, 1994). In the binding of P450 reductase to cytochrome b_5 only charge-pair interactions appear to be involved (Dailey and Strittmatter, 1980; Schenkman, 1993).

III. LEVELS OF EXPRESSION

A. Mitochondrial and Bacterial P450cam Systems

The mitochondrial electron transfer proteins adrenodoxin reductase and adrenodoxin are expressed in all human tissues examined (Brentano et al., 1992; Hanukoglu, 1992). Their highest levels of expression are observed in steroidogenic cells especially in adrenal cortex and ovarian corpus luteum (Hanukoglu, 1992). In these two tissues the molar ratios of adrenodoxin reductase, adrenodoxin and P450 were estimated as 1:(3–10):8, using specific antibodies against each component (Ohashi and Omura, 1978; Hanukoglu and Hanukoglu, 1986). The levels of these proteins show no significant sex, or interindividual variation in bovine adrenal cortex (Hanukoglu and Hanukoglu, 1986). Earlier spectroscopic measurements also showed that adrenodoxin and mitochondrial P450 are present at equal molar concentrations (Estabrook et al., 1973). These independent estimations contradict the findings of another report that mitochondrial P450 levels are similar to that of the adrenodoxin reductase and much lower than that of adrenodoxin (Hamamoto et al., 1986). Similar to mitochondrial systems, in *Pseudomonas putida* grown on camphor, putidaredoxin reductase, putidaredoxin and P450cam were found at a molar ratio of about 1:8:8 (Roome et al., 1983).

Consistent with the level of the protein, the level of adrenodoxin reductase mRNA is also about 10-fold lower than that of P450scc (Hanukoglu et al., 1987). This correlation between the relative levels of the enzymes and their mRNAs indicates that the low level of adrenodoxin reductase reflects its mRNA level, and does not result from lower translational efficiency of its mRNA. In contrast, the low level of putidaredoxin reductase apparently results from the low translatability of the mRNA that has an unusual GTG initiation codon; mutation of this to the normal ATG codon enhanced its expression to levels comparable to those of putidaredoxin and P450cam (Davies et al., 1990).

The levels of mitochondrial P450 system proteins in steroidogenic tissues are regulated by specific trophic hormones as part of the physiological mechanisms regulating the steroid output of these tissues (for review see Hanukoglu, 1992; and chapter by Waterman in this volume). The level of adrenodoxin reductase is correlated with the level of adrenodoxin and P450scc in both adrenal cortex and corpus luteum *in vivo* (Hanukoglu, 1986). In bovine adrenal cortex cells in primary culture grown to confluence without ACTH, the levels of the enzymes drastically decrease. ACTH increases the levels of all three enzymes and their mRNAs, but with some differences in their time courses (Hanukoglu et al., 1990a). In analogy to the regulation observed in steroidogenic tissues, the levels of some mitochondrial P450 system enzymes in other tissues may also be regulated by the relevant physiological factors (chapter by Armbrecht in this volume).

B. Microsomal Systems

The expression of microsomal electron transfer proteins, P450 reductase, and cytochrome b_5 has been detected in various tissues by Western blot analysis using specific antibodies and RNase protection assays using cDNA probes (Hamamoto et al., 1986; Katagiri et al., 1989; Arinc, 1991; Shephard et al., 1992). The highest level of P450 reductase is observed in the liver (Katagiri et al., 1989). The levels of P450 reductase mRNA show several fold differences among tissues and at most 3-fold difference among adult human liver samples (Shephard et al., 1992). The average ratio of P450 reductase to total cytochromes P450 in rat and human livers was estimated as 1:15 and 1:7 respectively (Shephard et al., 1983; McManus et al., 1987). In the bovine adrenal cortex the molar ratio of P450 reductase to cytochrome b_5 was estimated to vary between 1:0.6 and 1:4 in different zones (Hamamoto et al., 1986). The level of the cytochrome b_5 reductase is at least ten fold lower than that of cytochrome b_5 (Hamamoto et al., 1986; Archakov and Bachmanova, 1990; Borgese et al., 1993).

Some hormone and drug treatments that increase the expression of specific P450s, show little effect on reductase levels; thus the ratio of P450 reductase to various P450s in liver microsomes may vary from 1:15 to about 1:100 under different treatments (Shephard et al., 1983; Shiraki and Guengerich, 1984). The expression of P450 reductase in rat liver is strongly dependent on thyroxine stimulation (Waxman, 1992). In adrenocortical cells ACTH stimulation enhances P450 reductase expression similar to its effects on other steroidogenic enzymes. However, phenobarbital which induces P450 reductase in liver, shows no similar effect in adrenocortical cells (Dee et al., 1985). Thus, P450 reductase gene apparently possesses a repertoire of regulatory elements that can respond to a variety of stimuli in a cell specific manner and in concert with the responses of its redox partners.

IV. PROTEIN-PROTEIN INTERACTIONS

A. Mitochondrial and Bacterial P450cam Systems

The stoichiometry of about one reductase molecule per ten P450 precludes the structural organization of these enzymes in rigid complexes or arrays of electron transport chains. The findings summarized below indicate that during catalytic turnover adrenodoxin functions as a mobile electron carrier: its oxidized form binds to reductase, accepts one electron from it and dissociates, and then binds to P450 and unloads its electron.

The ferredoxin of both mitochondrial and P450cam systems can form a tight 1:1 complex ($K_d < 10^{-7}$) with either the reductase or P450 (Chu and Kimura, 1973; Gunsalus and Sligar, 1978; Hanukoglu and Jefcoate, 1980; Hanukoglu et al., 1981; Sakamoto et al., 1981; Lambeth et al., 1982; Lambeth, 1991). As noted above, the binding sites on adrenodoxin for its different electron transfer partners overlap; thus, the concurrent specific association of the reductase and the P450 with the ferredoxin is sterically impossible. In adrenal cortex there are two mitochondrial P450s, P450scc (CYP11A) and P450c11 (CYP11B), at nearly equal levels and distributed evenly across mitochondria (Hanukoglu, 1992). Studies in rat adrenal mitochondrial preparations suggest that adrenodoxin may interact preferentially with P450scc (Yamazaki et al., 1993).

Kinetic studies also show that a 1:1:1 ternary complex of adrenodoxin reductase, adrenodoxin, and P450 does not form under equilibrium or steady state turnover conditions (Hanukoglu and Jefcoate, 1980; Hanukoglu et al., 1981; Lambeth et al., 1982; Lambeth, 1991). Kido and Kimura (1979) reported that a ternary complex can form under equilibrium conditions; however, their results can be interpreted differently, refuting their conclusion of a ternary complex (see discussion by Lambeth et al., 1982). In contrast to interactions with P450, adrenodoxin can reduce the artificial acceptor cytochrome c while in a 1:1 complex with adrenodoxin reductase, showing that its binding sites for adrenodoxin reductase and cytochrome c are not overlapping (Lambeth et al., 1982). A covalently crosslinked complex of adrenodoxin reductase-adrenodoxin can reduce cytochrome c at maximal rates in absence of added free adrenodoxin; while the same complex requires excess free adrenodoxin for P450scc reduction (Hara and Kimura, 1989). These findings further indicate that the sites of intermolecular electron transfer between adrenodoxin and cytochromes c and P450 are different.

Turko et al. (1989) obtained a ternary complex by chemical crosslinking, yet it is difficult to ascertain that in such complexes the proteins are crosslinked at the high affinity binding sites identified in equilibrium binding studies. Harikrishna et al. (1993) designed and expressed in COS cells fused proteins including two or three components of the mitochondrial P450 system at different orientations. Cells expressing some of these constructs showed greater activity than cells expressing each component from separate cotransfected vectors. It should be noted however, that each construct may not necessarily function on its own transferring electrons within the fused multi-protein complex, as its individual components may interact with other fused proteins in the mitochondria.

The association of adrenodoxin with reductase and P450 is dependent on ionic interactions and can be affected strongly by changes in pH and ionic composition (Hanukoglu et al., 1981; Jefcoate, 1982; Lambeth and Kriengsiri, 1985; Hamamoto et al., 1993). These findings suggest that complementary charged residues are involved in the specific complex formations. Indeed, site directed mutations of human adrenodoxin reductase and adrenodoxin suggest that *Arg*-239 and *Arg*-243 of adrenodoxin reductase interact directly with *Asp*-76 and *Asp*-79 of adrenodoxin (Brandt and Vickery, 1993). Hydrophobic interactions do not appear to play a role in adrenodoxin-P450 complex formation as the K_d of the complex was unaffected by an uncharged detergent (Hanukoglu et al., 1981).

Adrenodoxin reductase-adrenodoxin couple can oxidize NADPH at a high steady rate even in absence of P450scc, and transfer electrons directly to O_2 to produce superoxide radical (Hanukoglu et al., 1993). This leakage of electrons to O_2 is enhanced by the addition of P450, showing that adrenodoxin can bind to substrate-free P450 and reduce it. However, electron leakage is reduced in the presence of P450 substrate. These findings suggest that to minimize the production of harmful superoxide in the cell, the supply of NADPH to the mitochondrial P450 systems may be regulated together with substrate availability (Hanukoglu et al., 1993).

The binding of adrenodoxin to P450 is strongly influenced by the binding of substrate and the redox states of the proteins. Cholesterol and adrenodoxin mutually enhance each other's affinity for P450scc (Lambeth et al., 1980; Hanukoglu et al., 1981; Jefcoate, 1982). Reduced adrenodoxin binds more strongly to oxidized rather than reduced P450scc (Lambeth, 1991). Similar effects have been also observed in the P450cam system (Davies and Sligar, 1992). These effects enhance the binding of adrenodoxin to P450-substrate complex and reduce the propensity of the system to function as a superoxide producing NADPH oxidase (Lambeth, 1991; Hanukoglu et al., 1993).

The monooxygenase activity of mitochondrial P450s shows Michaelis-Menten dependence on free reduced adrenodoxin (Hanukoglu and Jefcoate, 1980; Hanukoglu et al., 1981; Lambeth, 1991). Reductase competes with P450 for binding to the same site on adrenodoxin, and oxidized adrenodoxin can bind to P450 in competition with reduced adrenodoxin and consequently inhibit the catalytic activity of the P450 (Hanukoglu and Jefcoate, 1980; Hanukoglu et al., 1981; Lambeth, 1991). Thus, during monooxygenation the electron transport system would function most efficiently if most of the adrenodoxin molecules are maintained in reduced form and unbound to reductase. These biological design specifications are apparently met by the following characteristics of the system: (1) a relatively low concentration of reductase to minimize the competition of reductase with P450 for binding to adrenodoxin, (2) a much faster turnover rate of adrenodoxin reduction by reductase than the rate of adrenodoxin oxidation by P450 so that the low concentrations of reductase suffice for the system, and (3) redox equilibria that favor dissociation of adrenodoxin from reductase after reduction, and binding of adrenodoxin to P450-substrate complex.

Some of the characteristics of mitochondrial P450 systems noted above are similar to those of the mitochondrial oxidative phosphorylation system which includes four multienzyme complexes (I–IV) embedded in the inner mitochondrial membrane, and a small soluble protein (cytochrome c) located in the intermembrane space (Gupte et al., 1984; Hackenbrock et al., 1986). The stoichiometry of these five components (1:2:3:7:9) precludes rigid structural organization of enzymes on the membrane. Electron transfer between complexes I, II, and III is mediated by ubiquinone, and between complexes III and IV by cytochrome c. A functional 1:1:1 aggregate of complex III, cytochrome c, and cytochrome c oxidase apparently cannot form as the same domain on cytochrome c is involved in binding both to its reductase (complex III) and oxidase (complex IV) (Capaldi, 1982). Thus, like adrenodoxin, cytochrome c functions as an electron shuttle between its reductase and oxidase, and not as an "electron bridge" between the two enzymes. Studies with fluorescently labeled cytochrome c in intact mitochondria indicate that the protein diffuses in three dimensions and not only along the inner membrane surface (Cortese and Hackenbrock, 1993). Immuno-electron microscopy shows adrenodoxin molecules in association with the inner membrane, and in the matrix space (Hatano et al., 1989; Hanukoglu et al., 1990) suggesting that adrenodoxin may similarly diffuse in the mitochondrial matrix.

It is significant that both the mitochondrial P450 and oxidative phosphorylation systems include small soluble proteins as electron transporters for enzymes embedded in the inner mitochondrial membrane. A major reason for this may be that proteins may diffuse faster on the surface than in the plane of the inner mitochondrial membrane which has a particularly high protein content. Indeed the diffusion rate of cytochrome c is over ten-fold faster than those of the membrane embedded complexes (Gupte et al., 1984). Similarly, adrenodoxin may have greater mobility. Rotational mobility measurements in a mitochondrial preparation showed that only about 30% of P450 are mobile and that the proportion of mobile P450 increases after addition of adrenodoxin (Ohta et al., 1991). Yet, these findings should be interpreted with the caveat that the studies were carried out in viscous 60% sucrose or 80% glycerol. In artificial phospholipid vesicles, P450scc shows high mobility as assessed by different approaches (Dhariwal et al., 1991).

B. Microsomal Systems

The functions of microsomal P450 systems involve complex interactions among many different forms of P450, and the electron transfer proteins P450 reductase and cytochrome b_5 and its reductase, all of which are anchored to the endoplasmic reticulum membrane. The lopsided stoichiometry of these proteins, noted above, sterically would not permit their organization in rigid clusters in the membrane. Many different lines of evidence indicate that these proteins can move laterally in the membrane and transfer electrons after formation of high affinity 1:1 complexes. The biphasic kinetics of P450 reduction by reductase was interpreted as evidence for the organization of these proteins in clusters; however, alternative explanations suggest that these observations are consistent with non-rigid random distribution of the proteins (Peterson et al., 1976; Yang, 1977; Taniguchi et al., 1979; Archakov and Bachmanova, 1990; Kanaeva et al., 1992). In immunogold labeled sections of hepatocytes a uniform random distribution of microsomal P450s is observed. In microsomes ferritin labeled P450 antibodies occasionally appear in clusters, yet these may result from aggregation of molecules during microsome preparation (Tashiro et al., 1993). Studies with double immunogold labeling of both the reductase and the P450s are needed to further examine the relative organization of these proteins in the endoplasmic reticulum.

The reduction of P450 by the reductase is dependent on the formation of a specific high affinity 1:1 complex ($K_d < 10^{-7}$), between the two enzymes, though individual P450s differ in their affinity for the reductase (Miwa et al., 1979; Archakov and Bachmanova, 1990). Reconstitution studies using purified components indicated that the strength of this association is dependent on the phospholipid environment (Taniguchi et al., 1979; French et al., 1980; Ingelman-Sundberg et al., 1983). The binding of reductase to P450 is also modulated by P450 substrates which can decrease the K_d for the reductase-P450 complex, enhancing the affinity of the reductase to P450-substrate complex rather than substrate free P450 (French et al., 1980). Some substrates can increase P450 reduction potential, thus facilitating reduction by the reductase (Archakov and Bachmanova, 1990).

In addition to P450s, P450 reductase can also bind with high affinity to cytochrome b_5 and reduce it. This pathway of electron transport has been implicated in both P450 and non-P450 mediated reactions supported by cytochrome b_5 (Noshiro et al., 1980; Ilan et al., 1981; Fisher and Gaylor, 1982; Schenkman, 1993). As noted above, the binding of the reductase to different P450s and cytochrome b_5 may be mediated by electrostatic and hydrophobic interactions (Dailey and Strittmatter, 1980; Schenkman, 1993). Complementing these findings, chemical modification and site-directed mutation of P450s identified positively charged residues (Lys and Arg) on $P450_d$ (CYP1A2) that are involved in complex formation with the P450-reductase by charge pairing (Shimizu et al., 1991; Shen and Strobel, 1993). Some microsomal P450s may also form a tight 1:1 complex with cytochrome b_5 and receive the second electron from it during the usual catalytic cycle of monooxygenation (Ingelman-Sundberg and Johansson, 1980; Kuwahara and Omura, 1980; Bonfils et al., 1981; Hlavica, 1984; Schenkman, 1993).

The relative concentrations of P450 reductase and cytochrome b_5 can play major roles in determining reaction rates and specificities. One good example for this effect is P450c17 which is expressed in the adrenal cortex and some gonadal cells catalyzing steroid C17-hydroxylation and lyase (C21 side chain cleavage) reactions at rates characteristic of each cell type. *In vitro* studies showed that cytochrome b_5 can stimulate the lyase activity, and increasing P450 reductase concentration increased the lyase activity relative to hydroxylase. In testicular microsomes where the lyase activity predominates, the reductase activity is higher than in the adrenal. Hence, the *in vitro* findings are considered to reflect physiological regulatory mechanisms based on relative levels of reductase vs. cytochrome b_5 expression (Hall, 1991; Takemori and Kominami, 1991; Kominami et al., 1993).

Co-expression of P450 reductase, and various microsomal P450s in yeast cells and mammalian cell lines generally enhanced the activities of the P450s many fold, as the endogenous levels of the reductase may be limiting in these cells (Murakami et al., 1990; Peyronneau et al., 1992; Lin et al., 1993; Sawada et al., 1993; Truan et al., 1993). Fused enzymes expressed from constructs including P450 reductase and a microsomal P450 cDNAs also showed high P450 activity in yeast cells. Yet, the stability of the mRNAs and proteins encoded by the constructs may be lower than their individual components and depends on the length and sequence of the hinge region between the two enzymes (Sakaki et al., 1990; Shibata et al., 1990). The N-terminal region of the yeast reductase is highly divergent from that of mammalian reductases (Yabusaki et al., 1988). Hybrid constructs of yeast-rat P450 reductase with the N-terminal sequence of the yeast enzyme appeared to have higher stability in yeast cells (Bligh et al., 1992).

Chemical modification studies indicated that the interaction of cytochromes P450 and b_5 is dependent on carboxyl groups of the b_5 which are considered to be paired with complementary charges on the P450 (Tamburini et al., 1985; Schenkman, 1993). The binding sites on cytochrome b_5, for the b_5 reductase and P450 are overlapping as its reduction by NADH-cytochrome b_5 reductase is inhibited in the presence of P450. In contrast, the P4502B4 sites for binding P450 reductase and cytochrome b_5, appear to be different because addition of b_5 does not affect the K_m for the reductase (Tamburini and Schenkman, 1987; Schenkman, 1993). Both P450 reductase and cytochrome b_5 can reduce cytochromes c and P450, and it was suggested that their binding sites for these cytochromes share a short stretch of similar structure (Davydov et al., 1992). The mapping of the sites of interactions among these proteins awaits elucidation of the crystal structure of P450 reductase.

NOTE

1. Adrenodoxin was named reflecting its first isolation from the adrenal cortex (Kimura and Suzuki, 1967). In some reports it is referred to as "ferredoxin." In this review the term "adrenodoxin" is used because the term "ferredoxin" is also employed as a class name that does not distinguish among various ferredoxin type proteins (Beinert, 1990; Nomenclature Committee, 1979).

REFERENCES

Akiyoshi-Shibata, M., Sakaki, T., Yabusaki, Y., Murakami, H., & Ohkawa, H. (1991). Expression of bovine adrenodoxin and NADPH-adrenodoxin reductase cDNAs in Saccharomyces cerevisiae. DNA Cell Biol. 10, 613–621.

Archakov, A.I., & Bachmanova, G.I. (1990). Cytochrome P-450 and active oxygen. Taylor & Francis, Hants, U.K.

Arinc, E. (1991). Essential features of NADH dependent cytochrome b_5 reductase and cytochrome b_5 of liver and lung microsomes. In: Molecular aspects of monooxygenases and Bioactivation of toxic compounds (Arinc, E., Schenkman, J.B., & Hodgson, E., eds.), pp. 149–170, Plenum Press, New York.

Beckert, V., Dettmer, R., & Bernhardt, R. (1994). Mutations of tyrosine 82 in bovine adrenodoxin that affect binding to cytochromes P45011A1 and P45011B1 but not electron transfer. J. Biol. Chem. 269, in press.

Beinert, H. (1990). Recent developments in the field of iron-sulfur proteins. FASEB J. 4, 2483–2491.

Bhasker, C.R., Okamura, T., Simpson, E.R., & Waterman, M.R. (1987). Mature bovine adrenodoxin contains a 14-amino-acid COOH-terminal extension originally detected by cDNA sequencing. Eur. J. Biochem. 164, 21–25.

Black, S.D., & Coon, M.J. (1982). Structural features of liver microsomal NADPH-cytochrome P-450 reductase. Hydrophobic domain, hydrophilic domain, and connecting region. J. Biol. Chem. 257, 5929–5938.

Black, S.D., & Coon, M.J. (1987). P-450 cytochromes: Structure and function. Adv. Enzymol. Rel. Areas Mol. Biol. 60, 35–87.

Bligh, H.F., Wolf, C.R., Smith, G., & Beggs, J.D. (1992). Production of cytochrome P450 reductase yeast-rat hybrid proteins in *Saccharomyces cerevisiae*. Gene 110, 33–39.

Boddupalli, S.S., Oster, T., Estabrook, R.W., & Peterson, J.A. (1992). Reconstitution of the fatty-acid hydroxylation function of cytochrome-P-450Bm-3 utilizing its individual recombinant hemoprotein and flavoprotein domains. J. Biol. Chem. 267, 375–380.

Bonfils, C., Balny, C., & Maurel, P. (1981). Direct evidence for electron transfer from ferrous cytochrome b_5 to the oxyferrous intermediate of liver microsomal cytochrome P-450 LM2. J. Biol. Chem. 256, 9457–9465.

Borgese, N., D'Arrigo, A., DeSilvestris, M., & Pietrini, G. (1993). NADH-cytochrome b_5 reductase and cytochrome b_5—The problem of posttranslational targeting to the endoplasmic reticulum. Subcell. Biochem. 21, 313–341.

Brandt, M.E., & Vickery, L.E. (1992). Expression and characterization of human mitochondrial ferredoxin reductase in *Escherichia coli*. Arch. Biochem. Biophys. 294, 735–740.

Brandt, M.E., & Vickery, L.E. (1993). Charge pair interactions stabilizing ferredoxin-ferredoxin reductase complexes. Identification by complementary site-specific mutations. J. Biol. Chem. 268, 17126–17130.

Bredt, D.S., Hwang, P.M., Glatt, C.E., Lowenstein, C., Reed, R.R., & Snyder, S.H. (1991). Cloned and expressed nitric oxide synthase structurally resembles cytochrome P450 reductase. Nature 351, 714–719.

Brentano, S.T., Black, S.M., Lin, D., & Miller, W.L. (1992). cAMP post-transcriptionally diminishes the abundance of adrenodoxin reductase mRNA. Proc. Natl. Acad. Sci. USA 89, 4099–4103.

Capaldi, R.A. (1982). Arrangement of proteins in the mitochondrial inner membrane. Biochim. Biophys. Acta 694, 291–306.

Chang, C.Y., Wu, D.A., Mohandas, T.K., & Chung, B.C. (1990). Structure, sequence, chromosomal location, and evolution of the human ferredoxin gene family. DNA Cell Biol. 9, 205–212.

Chu, J.W., & Kimura, T. (1973). Studies on adrenal steroid hydroxylases. Complex formation of the hydroxylase components. J. Biol. Chem. 248, 5183–5187.

Coghlan, V.M., & Vickery, L.E. (1992). Electrostatic interactions stabilizing ferredoxin electron transfer complexes. Disruption by "conservative" mutations. J. Biol. Chem. 267, 8932–8935.

Cortese, J.D., & Hackenbrock, C.R. (1993). Motional dynamics of functional cytochrome c delivered by low pH fusion into the intermembrane space of intact mitochondria. Biochim. Biophys. Acta 1142, 194–202.

Cupp, J.R., & Vickery, L.E. (1988). Identification of free and [Fe2S2]-bound cysteine residues of adrenodoxin. J. Biol. Chem. 263, 17418–17421. (erratum in J. Biol. Chem. 264, 7760, 1989.)

Cupp, J.R., & Vickery, L.E. (1989). Adrenodoxin with a COOH-terminal deletion (des 116-128) exhibits enhanced activity. J. Biol. Chem. 264, 1602–1607.

Dailey, H.A., & Strittmatter, P. (1980). Characterization of the interaction of amphipathic cytochrome b_5 with stearyl coenzyme A desaturase and NADPH: Cytochrome P-450 reductase. J. Biol. Chem. 255, 5184–5189.

Davies, M.D., Koga, H., Horiuchi, T., & Sligar, S.G. (1990). Site-directed mutagenesis of the Pseudomonas cam operon. In: Pseudomonas: Biotransformations, Pathogenesis, and Evolving Biotechnology (Silver, S., et al., eds.), pp. 101–110, American Soc. for Microbiology, Washington, D.C.

Davies, M.D., & Sligar, S.G. (1992). Genetic variants in the putidaredoxin-cytochrome P450cam electron transfer complex: Identification of the residue responsible for redox-state-dependent conformers. Biochemistry 31, 11383–11389.

Davydov, D.R., Darovsky, B.V., Dedinsky, I.R., Kanaeva, I.P., Bachmanova, G.I., Blinov, V.M., & Archakov, A.I. (1992). Cytochrome-c (Fe2+) as a competitive inhibitor of NADPH-dependent reduction of cytochrome P450LM2: Locating protein-protein interaction sites in microsomal electron carriers. Arch. Biochem. Biophys. 297, 304–313.

Dee, A., Carlson, G., Smith, C., Masters, B.S., & Waterman, M.R. (1985). Regulation of synthesis and activity of bovine adrenocortical NADPH-cytochrome P-450 reductase by ACTH. Biochem. Biophys. Res. Commun. 128, 650–656.

Dhariwal, M.S., Kowluru, R.A., & Jefcoate, C.R. (1991). Cytochrome P-450-scc induces vesicle aggregation through a secondary interaction at the adrenodoxin binding sites (in competition with protein exchange). Biochemistry 30, 4940–4949.

Driscoll, W.J., & Omdahl, J.L. (1986). Kidney and adrenal mitochondria contain two forms of NADPH-adrenodoxin reductase-dependent iron-sulfur proteins. J. Biol. Chem. 261, 4122–4125.

Eggink, G., Engel, H., Vriend, G., Terpstra, P., & Witholt, B. (1990). Rubredoxin reductase of Pseudomonas oleovorans: Structural relationship to other flavoprotein oxidoreductases based on one NAD and two FAD fingerprints. J. Mol. Biol. 212, 135–142.

Eisenberg, D., Schwarz, E., Komaromy, M., & Wall, R. (1984). Analysis of membrane and surface protein sequences with the hydrophobic moment plot. J. Mol. Biol. 179, 125–142.

Estabrook, R.W., Suzuki, K., Mason, J.I., Baron, J., Taylor, W.E., Simpson, E.R., Purvis, J., & McCarthy, J. (1973). Adrenodoxin: An iron-sulfur protein of adrenal cortex mitochondria. In: Iron-sulfur Proteins (Lovenberg, W., ed.) 1, pp. 193–223, Academic Press, New York.

Fisher, G.J., & Gaylor, J.L. (1982). Kinetic investigation of rat liver microsomal electron transport from NADH to cytochrome P-450. J. Biol. Chem. 257, 7449–7455.

French, J.S., Guengerich, F.P., & Coon, M.J. (1980). Interactions of cytochrome P-450, NADPH-cytochrome P-450 reductase, phospholipid, and substrate in the reconstituted liver microsomal enzyme system. J. Biol. Chem. 255, 4112–4119.

Gerber, N.C., Hortuchi, T., Koga, H., & Sligar, S.G. (1990). Identification of 2FE-2S cysteine ligands in putidaredoxin. Biochem. Biophys. Res. Commun. 169, 1016–1020.

Gunsalus, I.C., & Sligar, S.G. (1978). Oxygen reduction by the P-450 monooxygenase systems. Adv. Enzymol. Relat. Areas Mol. Biol. 47, 1–44.

Gupte, S., Wu, E.S., Hoechli, L., Hoechli, M., Jacobson, K., Sowers, A.E., & Hackenbrock, C.R. (1984). Relationship between lateral diffusion, collision frequency, and electron transfer of mitochondrial inner membrane oxidation-reduction components. Proc. Natl. Acad. Sci. U.S.A. 81, 2606–2610.

Hackenbrock, C.R., Chazotte, B., & Gupte, S.S. (1986). The random collision model and a critical assessment of diffusion and collision in mitochondrial electron transport. J. Bioenerg. Biomembr. 18, 331–368.

Hall, P.F. (1991). Cytochrome P-450C21scc: One enzyme with two actions: Hydroxylase and lyase. J. Steroid Biochem. Mol. Biol. 40, 527–532.

Hamamoto, I., Hiwatashi, A., & Ichikawa, Y. (1986). Zonal distribution of cytochromes P-450 and related enzymes of bovine adrenal cortex—Quantitative assay of concentrations and total contents. J. Biochem. 99, 1743–1748.

Hamamoto, I., Kurokohchi, K., Tanaka, S., & Ichikawa, Y. (1988). Adrenoferredoxin-binding peptide of NADPH-Adrenoferredoxin reductase. Biochim. Biophys. Acta 953, 207–213.

Hamamoto, I., Kurokohchi, K., Tanaka, S., & Ichikawa, Y. (1993). Effects of ionic strength and pH on the dissociation constant between NADPH-adrenoferredoxin reductase and adrenoferredoxin. J. Steroid Biochem. Mol. Biol. 46, 33–37.

Haniu, M., Iyanagi, T., Miller, P., Lee, T.D., & Shively, J.E. (1986). Complete amino acid sequence of NADPH-cytochrome P-450 reductase in porcine hepatic microsomes. Biochemistry 25, 7906–7911.

Haniu, M., McManus, M.E., Birkett, D.J., Lee, T.D., & Shively, J.E. (1989). Structural and functional analysis of NADPH-cytochrome P-450 reductase from human liver: Complete sequence of human enzyme and NADPH binding sites. Biochemistry 28, 8639–8645.

Hanukoglu, I. (1992). Steroidogenic enzymes: Structure, function, and regulation of expression. J. Steroid Biochem. Mol. Biol. 43, 779–804.

Hanukoglu, I., & Gutfinger, T. (1989). cDNA sequence of adrenodoxin reductase: Identification of NADP binding sites in oxidoreductases. Eur. J. Biochem. 180, 479–484.

Hanukoglu, I., & Hanukoglu, Z. (1986). Stoichiometry of mitochondrial cytochromes P-450, adreno-doxin and adrenodoxin reductase in adrenal cortex and corpus luteum: Implications for membrane organization and gene regulation. Eur. J. Biochem. 157, 27–31.

Hanukoglu, I., & Jefcoate, C.R. (1980). Mitochondrial cytochrome P-450scc: Mechanism of electron transport by adrenodoxin. J. Biol. Chem. 255, 3057–3061.

Hanukoglu, I., Spitsberg, V., Bumpus, J.A., Dus, K.M., & Jefcoate, C.R. (1981). Adrenal mitochondrial cytochrome P-450scc: Cholesterol and adrenodoxin interactions at equilibrium and during turn-over. J. Biol. Chem. 256, 4321–4328.

Hanukoglu, I., Privalle, C.T., & Jefcoate, C.R. (1981). Mechanism of ionic activation of mitochondrial cytochromes P-450scc and P-45011β. J. Biol. Chem. 256, 4329–4335.

Hanukoglu, I., Gutfinger, T., Haniu, M., & Shively, J.E. (1987). Isolation of a cDNA for adrenodoxin reductase (ferredoxin - NADP$^+$ reductase): Implications for mitochondrial cytochrome P-450 systems. Eur. J. Biochem. 169, 449–455.

Hanukoglu, I., Feuchtwanger, R., & Hanukoglu, A. (1990a). Mechanism of ACTH and cAMP induction of mitochondrial cytochrome P450 system enzymes in adrenal cortex cells. J. Biol. Chem. 265, 20602–20608.

Hanukoglu, I., Suh, B.S., Himmelhoch, S., & Amsterdam, A. (1990b). Induction and mitochondrial localization of cytochrome P450scc system enzymes in normal and transformed ovarian granulosa cells. J. Cell Biol. 111, 1373–1382.

Hanukoglu, I., Rapoport, R., Schweiger, S., Sklan, D., Weiner, L., and Schulz, G. (1992). Structure and function of the mitochondrial P450 system electron transfer proteins, adrenodoxin reductase and adrenodoxin. J. Basic Clin. Physiol. Pharmacol. 3 (Suppl.), 36–37.

Hanukoglu, I., Rapoport, R., Weiner, L., & Sklan, D. (1993). Electron leakage from the mitochondrial NADPH-adrenodoxin reductase-adrenodoxin-P450scc (cholesterol side chain cleavage) system. Arch. Biochem. Biophys. 305, 489–498.

Hara, T., & Kimura, T. (1989). Active complex between adrenodoxin reductase and adrenodoxin in the cytochrome P-450scc reduction reaction. J. Biochem. 105, 601–605.

Hara, T., & Miyata, T. (1991). Identification of a cross-linked peptide of a covalent complex between adrenodoxin reductase and adrenodoxin. J. Biochem. 110, 261–266.

Harikrishna, J.A., Black, S.M., Szklarz, G.D., & Miller, W.L. (1993). Construction and function of fusion enzymes of the human cytochrome P450scc system. DNA Cell Biol. 12, 371–379.

Hatano, O., Sagara, Y., Omura, T., & Takakusu, A. (1989). Immunocytochemical localization of adrenodoxin in bovine adrenal cortex by protein A-gold technique. Histochemistry 91, 89–97.

Hiwatashi, A., & Ichikawa, Y. (1978). Crystalline reduced nicotinamide adenine dinucleotide phosphate-adrenodoxin reductase from pig adrenocortical mitochondria. Essential histidyl and cysteinyl residues of the NADPH binding site and environment of the adrenodoxin-binding site. J. Biochem. 84, 1071–1086.

Hiwatashi, A., Ichikawa, Y., Maruya, N., Yamano, T., & Aki, K. (1976). Properties of crystalline reduced nicotinamide adenine dinucleotide phosphate-adrenodoxin reductase from bovine adrenocortical mitochondria. I. Physicochemical properties of holo- and apo- NADPH-adrenodoxin reductase and interaction between non-heme iron proteins and the reductase. Biochemistry 15, 3082–3090.

Hiwatashi, A., Ichikawa, Y., Yamano, T., & Maruya, N. (1976). Properties of crystalline reduced nicotinamide adenine dinucleotide phosphate-adrenodoxin reductase from bovine adrenocortical mitochondria. II. Essential histidyl and cysteinyl residues at the NADPH binding site of NADPH-adrenodoxin reductase. Biochemistry 15, 3091–3097.

Hiwatashi, A., Sakihama, N., Shin, M., & Ichikawa, Y. (1986). Heterogeneity of adrenocortical ferredoxin. FEBS Lett. 209, 311–315.

Hlavica, P. (1984). On the function of cytochrome b_5 in the cytochrome P-450 dependent oxygenase system. Arch. Biochem. Biophys. 228, 600–608.

Hornsby, P.J. (1989). Steroid and xenobiotic effects on the adrenal cortex: Mediation by oxidative and other mechanisms. Free Radicals Biol. Med. 103–115.

Ilan, Z., Ilan, R., & Cinti, D.L. (1981). Evidence for a new physiological role of hepatic NADPH: Ferricytochrome (P-450) oxidoreductase. Direct electron input to the microsomal fatty acid chain elongation system. J. Biol. Chem. 256, 10066–10072.

Ingelman-Sundberg, M., & Johansson, I. (1980). Cytochrome b_5 as electron donor to rabbit liver cytochrome P-450$_{LM2}$ in reconstituted phospholipid vesicles. Biochem. Biophys. Res. Commun. 97, 582–589.

Ingelman-Sundberg, M., Blanck, J., Smettan, G., & Ruckpaul, K. (1983). Reduction of cytochrome P-450 LM2 by NADPH in reconstituted phospholipid vesicles is dependent on membrane charge. Eur. J. Biochem. 134, 157–162.

Irie, S., Doi, S., Yorifuji, T., Takagi, M., & Yano, K. (1987). Nucleotide sequencing and characterization of the genes encoding benzene oxidation enzymes of Pseudomonas putida. J. Bacteriol. 169, 5174–5179.

Iyanagi, T., Anan, F.K., Imai, Y., & Mason, H.S. (1978). Studies on the microsomal mixed function oxidase system: Redox properties of detergents-solubilized NADPH-cytochrome P-450 reductase. Biochemistry 17, 2224–2230.

Jefcoate, C.R. (1982). pH modulation of ligand binding to adrenal mitochondrial cytochrome P-450scc. J. Biol. Chem. 257, 4731–4737.

Kanaeva, I.P., Nikityuk, O.V., Davydov, D.R., Dedinskii, I.R., Koen, Y.M., Kuznetsova, G.P., Skotselyas, E.D., Bachmanova, G.I., & Archakov, A.I. (1992). Comparative study of monomeric reconstituted and membrane microsomal monooxygenase systems of the rabbit liver .2. Kinetic parameters of reductase and monooxygenase reactions. Arch. Biochem. Biophys. 298, 403–412.

Katagiri, M., Sugiyama, T., Tsutsukawa, N., Ishiguro, H., Miyoshi, N., Ishibashi, F., & Taniguchi, N. (1989). Comparative studies on microsomal NADPH-cytochrome P-450 reductase using a monoclonal antibody: Tissue distribution, specific activity and peptide mapping. Int. J. Biochem. 21, 1396–1405.

Kimura, T., & Suzuki, K. (1967). Components of the electron transport system in adrenal steroid hydroxylase: Isolation and properties of non-heme iron protein (adrenodoxin). J. Biol. Chem. 242, 485–491.

Karplus, P.A., Walsh, K.E., & Herriott, J.R. (1984). Amino acid sequence of spinach ferredoxin: NADP$^+$ oxidoreductase. Biochemistry 23, 6576–6583.

Karplus, P.A., Daniels, M.J., & Herriott, J.R. (1991). Atomic structure of ferredoxin: NADP$^+$ reductase. Prototype for a structurally novel flavoenzyme family. Science 251, 60–66.

Karuzina, I.I., & Archakov, A.I. (1994). The oxidative inactivation of cytochrome P450 in monooxygenase reactions. Free Radicals Bio. Med. 16: 78–97.

Kido, T., & Kimura, T. (1979). The formation of binary and ternary complexes of cytochrome P-450$_{scc}$ with adrenodoxin and adrenodoxin reductase-adrenodoxin complex. J. Biol. Chem. 254, 11806–11815.

Koga, H., Yamaguchi, E., Matsunaga, K., Aramaki, H., & Horiuchi, T. (1989). Cloning and Nucleotide sequences of NADH-putidaredoxin reductase gene (camA) and putidaredoxin gene (camB) involved in cytochrome P-450cam hydroxylase of Pseudomonas putida. J. Biochem. 106, 831–836.

Kominami S., Ogawa N., Morimune R., De-Ying H., & Takemori S. (1992). The role of cytochrome b_5 in adrenal microsomal steroidogenesis. J. Steroid Biochem. Mol. Biol. 42, 57–64.

Kuban, R.-J., Marg, A., Resch, M., & Ruckpaul, K. (1993). Crystallization of bovine adrenodoxin-reductase in a new unit cell and its crystallographic characterization. J. Mol. Biol. 234, 245–248.

Kuwahara, S., & Omura, T. (1980). Different requirement for cytochrome b5 in NADPH-supported O-dethylation of p-nitrophenetole catalyzed by two types of microsomal cytochrome P-450. Biochem. Biophys. Res. Commun. 96, 1562–1568.

Lambeth, J.D. (1991). Enzymology of mitochondrial side-chain cleavage by cytochrome P-450scc. Frontiers Biotransformation 3, 58–100.

Lambeth, J.D., Seybert, D.W., & Kamin, H. (1980). Phospholipid vesicle-reconstituted cytochrome P-450scc. Mutually facilitated binding of cholesterol and adrenodoxin. J. Biol. Chem. 255, 138–143.

Lambeth, J.D., Seybert, D.W., Lancaster, J.R., Salerno, J.C., & Kamin, H. (1982). Steroidogenic electron transport in adrenal cortex mitochondria. Mol. Cell. Biochem. 45, 13–31.

Lambeth, J.D., Geren, L.M., & Millett, F. (1984). Adrenodoxin interaction with adrenodoxin reductase and cytochrome P-450scc. Cross-linking of protein complexes and effects of adrenodoxin modification by 1-ethyl-3-(3-dimethylaminopropyl)carbodiimide. J. Biol. Chem. 259, 10025–10029.

Lambeth, J.D., & Kriengsiri, S. (1985). Cytochrome P-450scc-adrenodoxin interactions. Ionic effects on binding, and regulation of cytochrome reduction by bound steroid substrates. J. Biol. Chem. 260, 8810–8816.

Lesot, A., Benveniste, I., Hasenfratz, M.P., & Durst, F. (1992). Production and characterization of monoclonal antibodies against NADPH-cytochrome P-450 reductases from Helianthus tuberosus. Plant Physiol. 100, 1406–1410.

Lin, D., Shi, Y., & Miller, W.L. (1990). Cloning and sequence of the human adrenodoxin reductase gene. Proc. Natl. Acad. Sci. USA 87, 8516–8520.

Lin, D., Black, S.M., Nagahama, Y., & Miller, W.L. (1993). Steroid 17α-hydroxylase and 17,20-lyase activities of P450c17: Contributions of serine 106 and P450 reductase. Endocrinology 132, 2498–2506.

Lu, A.Y.H. (1991). NADPH-dependent cytochrome P450 reductase. In: Molecular aspects of monooxygenases and Bioactivation of toxic compounds (Arinc, E., Schenkman, J.B., & Hodgson, E., eds.), pp. 135–147, Plenum Press, New York.

Mandel, M.L., Moorthy, B., & Ghazarian, J.G. (1990). Reciprocal post-translational regulation of renal 1α- and 24-hydroxylases of 25-hydroxyvitamin D_3 by phosphorylation of ferredoxin. Biochem. J. 266, 385–392.

Marg, A., Kuban, R-J., Behlke, J., Dettmer, R., & Ruckpaul, K. (1992). Crystallization and X-ray examination of bovine adrenodoxin. J. Mol. Biol. 227, 945–947.

McManus, M.E., Hall, P.D., Stupans, I., Brennan, J., Burgess, W., Robson, R., & Birkett, D.J. (1987). Immunohistochemical localization and quantitation of NADPH-cytochrome P-450 reductase in human liver. Mol. Pharmacol. 32, 189–194.

Meijer, A.H., Lopes Cardoso, M.I., Voskuilen, J.T., De Waal, A., Verpoorte, R., & Hoge, J.H.C. (1993). Isolation and characterization of a cDNA clone from Catharanthus roseus encoding NADPH:cytochrome P-450 reductase, an enzyme essential for reactions catalyzed by cytochrome P-450 mono-oxygenases in plants. Plant J. 4, 47–60.

Mitani, F. (1979). Cytochrome P450 in adrenocortical mitochondria. Mol. Cell. Biochem. 24, 21–43.

Mittl, P.R.E., Berry, A., Scrutton, N.S., Perham, R.N., & Schulz, G.E. (1993). Structural differences between wild-type NADP-dependent glutathione reductase from *Escherichia coli* and a redesigned NAD-dependent mutant. J. Mol. Biol. 231, 191–195.

Miura, S., & Ichikawa, Y. (1991). Proton nuclear magnetic resonance investigation of adrenodoxin. Assignment of aromatic resonances and evidence for a conformational similarity with ferredoxin from Spirulina platensis. Eur. J. Biochem. 197, 747–757.

Miura, S., & Ichikawa, Y. (1991). Conformational change of adrenodoxin induced by reduction of iron-sulfur cluster. Proton nuclear magnetic resonance study. J. Biol. Chem. 266, 6252–6258.

Miura, S., Tomita, S., & Ichikawa, Y. (1991). Modification of histidine 56 in adrenodoxin with diethyl pyrocarbonate inhibited the interaction with cytochrome P-450scc and adrenodoxin reductase. J. Biol. Chem. 266, 19212–19216.

Miwa, G.T., West, S.B., Huang, M.T., & Lu, A.Y.H. (1979). Studies on the association of cytochrome P-450 and NADPH-cytochrome c reductase during catalysis in a reconstituted hydroxylating system. J. Biol. Chem. 254, 5695–5700.

Monnier, N., Defaye, G., & Chambaz, E.M. (1987). Phosphorylation of bovine adrenodoxin. Structural study and enzymatic activity. Eur. J. Biochem. 169, 147–153.

Muller, K., Linder, D., & Lumper, L. (1990). The cosubstrate NADP(H) protects lysine 601 in the porcine NADPH-cytochrome P-450 reductase against pyridoxylation. FEBS Lett. 260, 289–290.

Murakami, H., Yabusaki, Y., Sakaki, T., Shibata, M., & Ohkawa, H. (1990). Expression of cloned yeast NADPH-cytochrome P450 reductase gene in Saccharomyces cerevisiae. J. Biochem. 108, 859–865.

Nadler, S.G., & Strobel, H.W. (1991). Identification and characterization of an NADPH-cytochrome P450 reductase derived peptide involved in binding to cytochrome P450. Arch. Biochem. Biophys. 290, 277–284.

Nelson, D.R., Kamataki, T., Waxman, D.J., Guengerich, F.P., Estabrook, R.W., Feyereisen, R., Gonzalez, F.J., Coon, M.J., Gunsalus, I.C., Gotoh, O., Okuda, K., & Nebert, D.W. (1993). The P450 superfamily: Update on new sequences, gene mapping, accession numbers, early trivial names of enzymes, and nomenclature. DNA Cell Biol. 12, 1–51.

Nomenclature Committee of the IUB. (1979). Nomenclature of iron-sulfur proteins. Eur. J. Biochem. 93, 427–430.

Nonaka, Y., Aibara, S., Sugiyama, T., Yamano, T., & Morita, Y. (1985). A crystallographic investigation on NADPH-adrenodoxin oxidoreductase. J. Biochem. 98, 257–260.

Noshiro, M., Harada, N., & Omura, T. (1980). Immunochemical study on the route of electron transfer from NADH and NADPH to cytochrome P-450 of liver microsomes. J. Biochem. 88, 1521–1535.

Ohashi, M., & Omura, T. (1978). Presence of the NADPH-cytochrome P-450 reductase system in liver and kidney mitochondria. J. Biochem. 83, 249–260.

Ohgiya, S., Goda, T., Ishizaki, K., Kamataki, T., & Shinriki, N. (1992). Molecular cloning and sequence analysis of guinea pig NADPH-cytochrome P-450 oxidoreductase. Biochim. Biophys. Acta 1171, 103–105.

Ohta, Y., Yanagibashi, K., Hara, T., Kawamura, M., & Kawato, S. (1991). Protein rotation study of cytochrome P-450 in submitochondrial particles: Effect of potassium chloride and intermolecular interactions with redox partners. J. Biochemistry 109, 594–599.

Okamura, T., Kagimoto, M., Simpson, E.R., & Waterman, M.R. (1987). Multiple species of bovine adrenodoxin mRNA. Occurrence of two different mitochondrial precursor sequences associated with the same mature sequence. J. Biol. Chem. 262, 10335–10338.

Omura, T. (1993a). In: Cytochrome P-450, Second edition (Omura, T., Ishimura, Y., & Fujii-Kuriyama, Y., eds.), pp. 1–15, Kodansha, Tokyo.

Omura, T. (1993b). Localization of cytochrome P450 in membranes: Mitochondria. Handb. Expl. Pharmacol. 105, 61–69.

Oprian, D.D., & Coon, M.J. (1982). Oxidation-reduction states of FMN and FAD in NADPH-cytochrome P-450 reductase during reduction by NADPH. J. Biol. Chem. 257, 8935–8944.

Orme-Johnson, N.R. (1990). Distinctive properties of adrenal cortex mitochondria. Biochim. Biophys. Acta 1020, 213–231.

Oster, T., Boddupalli, S.S., & Peterson, J.A. (1991). Expression, purification, and properties of the flavoprotein domain of cytochrome P-450BM-3. Evidence for the importance of the amino-terminal region for FMN binding. J. Biol. Chem. 266, 22718–22725.

Pernecky, S.J., Larson, J.R., Philpot, R.M., & Coon, M.J. (1993). Expression of truncated forms of liver microsomal P450 cytochromes 2B4 and 2E1 in Escherichia coli: Influence of NH2 terminal region on localization in cytosol and membranes. Proc. Natl. Acad. Sci. USA 90, 2651–2655.

Peterson, J.A., Ebel, R.E., O'Keeffe, D.H., Matsubara, T., & Estabrook, R.W. (1976). Temperature dependence of cytochrome P-450 reduction: A model for NADPH-cytochrome P-450 reductase interaction. J. Biol. Chem. 251, 4010–4016.

Peterson, J.A., Lorence, M.C., & Amarneh, B. (1990). Putidaredoxin reductase and putidaredoxin: Cloning, sequence determination, and heterologous expression of the proteins. J. Biol. Chem. 265, 6066–6073.

Peyronneau, M.A., Renaud, J.P., Truan, G., Urban, P., Pompon, D., & Mansuy, D. (1992). Optimization of yeast-expressed human liver cytochrome-P450 3A4 catalytic activities by coexpressing NADPH-cytochrome P450 reductase and cytochrome b_5. Eur. J. Biochem. 207, 109–116.

Porter, T.D., & Kasper, C.B. (1986). NADPH-cytochrome P-450 oxidoreductase: Flavin mononucleotide and dinucleotide domains evolved from different flavoproteins. Biochemistry 25, 1682–1687.

Porter, T.D., Wilson, T.E., & Kasper, C.B. (1987). Expression of a functional 78,000 dalton mammalian flavoprotein, NADPH-cytochrome P-450 oxidoreductase in *Escherichia coli*. Arch. Biochem. Biophys. 254, 353–367.

Porter, T.D., Beck, T.W., & Kasper, C.B. (1990). NADPH-cytochrome P-450 oxidoreductase gene organization correlates with structural domains of the protein. Biochemistry 29, 9814–9818.

Rapoport, R., Sklan, D., & Hanukoglu, I. (1995). Electron leakage from the adrenal cortex mitochondrial P450scc and P450c11 systems: NADPH and steroid dependence. Arch. Biochem. Biophys. 317, 412–416.

Ravichandran, K.G., Boddupalli, S.S., Hasemann, C.A., Peterson, J.A., & Deisenhofer, J. (1993). Crystal structure of hemoprotein domain of P450BM-3, a prototype for microsomal P450's. Science 261, 731–736.

Roome, P.W., Peterson, J.A., & Philley, J.C. (1983). Purification and properties of putidaredoxin reductase. J. Biol. Chem. 258, 2593–2598.

Rossman, M.G., Liljas, A., Branden, C.I., & Banaszak, L.J. (1975). Evolutionary and structural relationships among dehydrogenases. Enzymes 9, 61–102.

Ruettinger, R.T., Wen, L.-P., & Fulco, A.J. (1989). Coding nucleotide, 5′ regulatory and deduced amino acid sequences of P-450BM-3, a single peptide cytochrome P-450:NADPH-P-450 reductase from *Bacillus megaterium*. J. Biol. Chem. 264, 10987–10995.

Sagara, Y., Takata, Y., Miyata, T., Hara, T., & Horiuchi, T. (1987). Cloning and sequence analysis of adrenodoxin reductase cDNA from bovine adrenal cortex. J. Biochem. 102, 1333–1336.

Sagara, Y., Sawae, H., Kimura, A., Sagara-Nakano, Y., Morohashi, K., Miyoshi, K., & Horiuchi, T. (1990). Structural organization of the bovine adrenodoxin gene. J. Biochem. 107, 77–83.

Sagara, Y., Hara, T., Ariyasu, Y., Ando, F., Tokunaga, N., & Horiuchi, T. (1992). Direct expression in *Escherichia coli* and characterization of bovine adrenodoxins with modified amino-terminal regions. FEBS Lett. 3000, 208–212.

Sagara, Y., Wada, A., Takata, Y., Waterman, M.R., Sekimizu, K., & Horiuchi, T. (1993). Direct expression of adrenodoxin reductase in *Escherichia coli* and the functional characterization. Biol. Pharm. Bull. 16, 627–630.

Sakaguchi, M., & Omura, T. (1993). Topology and biogenesis of microsomal cytochrome P-450s. Frontiers Biotransformation 8, 60–73.

Sakaki, T., Shibata, M., Yabusaki, Y., Murakami, H., & Ohkawa, H. (1990). Expression of bovine cytochrome P450c21 and its fused enzymes with yeast NADPH-cytochrome P450 reductase in Saccharomyces cerevisiae. DNA Cell Biol. 9, 603–614.

Sakamoto, H., Ichikawa, Y., Yamano, T., & Takagi, T. (1981). Circular dichroic studies on the interaction between reduced nicotinamide adenine dinucleotide phosphate-adrenodoxin reductase and adrenodoxin. J. Biochem. 90, 1445–1452.

Sakihama, N., Hiwatashi, A., Miyatake, A., Shin, M., & Ichikawa, Y. (1988). Isolation and purification of mature bovine adrenocortical ferredoxin with an elongated carboxyl end. Heterogeneity of adrenocortical ferredoxin. Arch. Biochem. Biophys. 264, 23–29.

Sarkissova, Y.G., Mardanian, S.S., & Haroutunian, A.V. (1990). The role of tryptophanyl residues in electron transfer from NADPH-adrenodoxin reductase to adrenodoxin. Biochem. Int. 22, 977–982.

Sawada, M., Kitamura, R., Ohgiya, S., & Kamataki, T. (1993). Stable expression of mouse NADPH-cytochrome P450 reductase and monkey P4501A1 cDNAs in Chinese hamster cells: Establishment of cell lines highly sensitive to aflatoxin B_1. Arch. Biochem. Biophys. 300, 164–168.

Schenkman, J.B. (1993). Protein-protein interactions. Handbook Expl. Pharmacol. 105, 527–545.

Schenkman, J.B., & Greim, H. (1993). Cytochrome P450. (Handbook Expl. Pharmacol. vol. 105), Springer-Verlag, Berlin.

Scrutton, N.S., Berry, A., & Perham, R.N. (1990). Redesign of the coenzyme specificity of a dehydrogenase by protein engineering. Nature 343, 38–43.

Sem, D.S., & Kasper, C.B. (1992). Geometric relationship between the nicotinamide and isoalloxazine rings in NADPH-cytochrome P-450 oxidoreductase: Implications for the classification of evolutionarily and functionally related flavoproteins. Biochemistry 31, 3391–3398.

Sem, D.S., & Kasper, C.B. (1995). Effect of ionic strength on the kinetic mechanism and relative rate limitation of steps in the model NADPH-cytochrome P450 oxidoreductase reaction with cytochrome c. Biochem. 34, 12768–12774.

Shen, A.L., Porter, T.D., Wilson, T.E., & Kasper, C.B. (1989). Structural analysis of the FMN binding domain of NADPH-cytochrome P-450 oxidoreductase by site-directed mutagenesis. J. Biol. Chem. 264, 7584–7589.

Shen, A.L., Christensen, M.J., & Kasper, C.B. (1991). NADPH-cytochrome P-450 oxidoreductase. The role of cysteine 566 in catalysis and cofactor binding. J. Biol. Chem. 266, 19976–19980.

Shen, A.L., & Kasper, C.B. (1995). Role of acidic residues in the interaction of NADPH-cytochrome P450 oxidoreductase with cytochrome P450 and cytochrome c. J. Biol. Chem. 270. 27475–27480.

Shen, A.L., & Kasper, C.B. (1996). Ser^{457} in NADPH-cytochrome P450 reductase plays a main role in hydride transfer from NADPH to FAD. Biochemistry (In press).

Shen, S., & Strobel, H.W. (1993). Role of lysine and arginine residues of cytochrome P450 in the interaction between cytochrome P4502B1 and NADPH-cytochrome P450 reductase. Arch. Biochem. Biophys. 304, 257–265.

Shephard, E.A., Bayney, R.M., Phillips, I.R., Pike, S.F., & Rabin, B.R. (1983). Quantification of NADPH - cytochrome P-450 reductase in liver microsomes by a specific radioimmunoassay technique. Biochem. J. 211, 333–340.

Shephard, E.A., Palmer, C.N., Segall, H.J., & Phillips, I.R. (1992). Quantification of cytochrome P450 reductase gene expression in human tissues. Arch. Biochem. Biophys. 294, 168–172.

Shibata, M., Sakaki, T., Yabusaki, Y., Murakami, H., & Ohkawa, H. (1990). Genetically engineered P450 monooxygenases: Construction of bovine P450c17/yeast reductase fused enzymes. DNA Cell Biol. 9, 27–36.

Shimizu, T., Tateishi, T., Hatano, M., & Fujii-Kuriyama Y. (1991). Probing the role of lysines and arginines in the catalytic function of cytochrome pfd by site-directed mutagenesis. J. Biol. Chem. 266, 3372–3375.

Shiraki, H., & Guengerich, F.P. (1984). Turnover of membrane proteins: Kinetics of induction and degradation of seven forms of rat liver microsomal cytochrome P-450, NADPH-cytochrome P-450 reductase, and epoxide hydrolase. Arch. Biochem. Biophys. 235, 86–96.

Slepneva, I.A., & Weiner, L.M. (1988). Affinity modification of NADPH-cytochrome P450 reductase. Biochem. Biophys. Res. Commun. 155, 1026–1032.

Solish, S.B., Picado-Leonard, J., Morel, Y., Kuhn, R.W., Mohandas, T.K., Hanukoglu, I., & Miller, W.L. (1988). Human adrenodoxin reductase—Two mRNAs encoded by a single gene on chromosome 17cen-q25 are expressed in steroidogenic tissues. Proc. Natl. Acad. Sci. U.S.A. 85, 7104–7108.

Strobel, H.W., Dignam, J.D., & Gum, J.R. (1980). NADPH cytochrome P-450 reductase and its role in the mixed function oxidase reaction. Pharmac. Ther. 8, 525–537.

Suhara, K., Nakayama, K., Takikawa, O., & Katagiri, M. (1982). Two forms of adrenodoxin reductase from mitochondria of bovine adrenal cortex. Eur. J. Biochem. 125, 659–664.

Ta, D.T., & Vickery, L.E. (1992). Cloning, sequencing, and overexpression of a [2Fe-2S] ferredoxin gene from *Escherichia coli*. J. Biol. Chem. 267, 11120–11125.

Takemori, S., & Kominami, S. (1991). Adrenal microsomal cytochrome P-450 dependent reactions in steroidogenesis and biochemical properties of the enzymes involved therein. Frontiers Biotransformation 3, 153–203.

Takemori, S., Yamazaki, T., & Ikushiro, S.-I. (1993). Cytochrome P-450-linked electron transport system in monooxygenase reaction. In: Cytochrome P-450, Second Edition (Omura, T., Ishimura, Y., & Fujii-Kuriyama, Y., eds.), pp. 44–63, Kodansha, Tokyo.

Tamburini, P.P., White, R.E., & Schenkman, J.B. (1985). Chemical characterization of protein-protein interactions between cytochrome P-450 and cytochrome b_5. J. Biol. Chem. 260, 4007–4015.

Tamburini, P.P., & Schenkman, J.B. (1987). Purification to homogeneity and enzymological characterization of a functional complex composed of cytochromes P450 isozymes 2 and b_5 from rabbit liver. Proc. Natl. Acad. Sci. USA 84, 11–15.

Tamura, S., Korzekwa, K.R., Kimura, S., Gelboin, H.V., & Gonzalez, F.J. (1992). Baculovirus-mediated expression and functional characterization of human NADPH-P450 oxidoreductase. Arch. Biochem. Biophys. 293, 219–223.

Tang, C., & Henry, H.L. (1993). Overexpression in *Escherichia coli* and affinity purification of chick kidney ferredoxin. J. Biol. Chem. 268, 5069–5076.

Taniguchi, H., Imai, Y., Iyanagi, T., & Sato, R. (1979). Interaction between NADPH-cytochrome P-450 reductase and cytochrome P-450 in the membrane of phosphatidylcholine vesicles. Biochim. Biophys. Acta 550, 341–356.

Tashiro, Y., Masaki, R., & Yamamoto, A. (1993). Cytochrome P-450 in the endoplasmic reticulum: Biosynthesis, distribution, induction, and degradation. Subcell. Biochem. 21, 287–311.

Truan, G., Cullin, C., Reisdorf, P., Urban, P., & Pompon, D. (1993). Enhanced in vivo monooxygenase activities of mammalian P450s in engineered yeast cells producing high levels of NADPH-P450 reductase and human cytochrome b_5. Gene 125, 49–55.

Tuls, J., Geren, L., Lambeth, J.D., & Millett, F. (1987). The use of a specific fluorescence probe to study the interaction of adrenodoxin with adrenodoxin reductase and cytochrome P-450scc. J. Biol. Chem. 262, 10020–10025.

Turko, I.V., Adamovich, T.B., Kirillova, N.M., Usanov, S.A., Chashchin, V.L. (1989). Cross-linking studies of the cholesterol hydroxylation system from bovine adrenocortical mitochondria. Biochim. Biophys. Acta. 996, 37–42.

Uhlmann, H., Beckert, V., Schwarz, D., & Bernhardt, R. (1992). Expression of bovine adrenodoxin in *E. Coli* and site-directed mutagenesis of /2Fe-2S/ cluster ligands. Biochem. Biophys. Res. Commun. 188, 1131–1138.

Urban, P. Breakfield, X.O., & Pompon D. (1990). Expression of membrane-bound flavoenzymes in yeast. In: Flavins and Flavoproteins 1990 (Curti, B., Ronchi, S., & Zanetti, G., eds.). pp. 869–872, Walter de Gruyter, Berlin.

Vermilion, J.L., Ballou, D.P., Massey, V., & Coon, M.J. (1981). Separate roles for FMN and FAD in catalysis by liver microsomal NADPH-cytochrome P-450 reductase. J. Biol. Chem. 256, 266–277.

Vogel, F., & Lumper, L. (1986). Complete structure of the hydrophilic domain in the porcine NADPH-cytochrome P-450 reductase. Biochem. J. 236, 871–878.

Voznesensky, A.I., & Schenkman, J.B. (1994). Quantitative analyses of electrostatic interactions between NADPH-cytochrome P450 reductase and cytochrome P450 enzymes. J. Biol. Chem. in press.

Waki, N., Hiwatashi, A., & Ichikawa, Y. (1986). Purification and biochemical characterization of hepatic ferredoxin (hepatoredoxin) from bovine liver mitochondria. FEBS Lett. 195, 87–91.

Warburton, R.J., & Seybert, D.W. (1988). Limited proteolysis of bovine adrenodoxin reductase: Evidence for a domain structure. Biochem. Biophys. Res. Commun. 152, 177–183.

Warburton, R.J., & Seybert, D.W. (1995). Structural and functional characterization of bovine adrenodoxin reductase by limited proteolysis. Biochim. Biophys. Acta 1246, 39–46.

Waxman, D.J. (1992). Modes of regulation of liver-specific steroid metabolizing cytochromes P450: Cholesterol 7α-hydroxylase, bile acid 6β-hydroxylase, and growth hormone-responsive steroid hormone hydroxylases. J. Steroid Biochem. Mol. Biol. 43, 1055–1072.

White, R.E., & Coon, M.J. (1980). Oxygen activation by cytochrome P-450. Ann. Rev. Biochem. 49, 315–356.

Wierenga, R.K., DeMaeyer, M.C.H., & Hol, W.G.J. (1985). Interaction of pyrophosphate moieties with α-helixes in dinucleotide binding proteins. Biochemistry 24, 1346–1357.

Yabusaki, Y., Murakami, H., & Ohkawa, H. (1988). Primary structure of Saccharomyces cerevisiae NADPH-cytochrome P450 reductase deduced from nucleotide sequence of its cloned gene. J. Biochem. 103, 1004–1010.

Yamazaki, M., & Ichikawa, Y. (1990). Crystallization and comparative characterization of reduced nicotinamide adenine dinucleotide phosphate-ferredoxin reductase from sheep adrenocortical mitochondria. Comp. Biochem. Physiol. 96B, 93–100.

Yamazaki, M., Ohnishi, T., & Ichikawa, Y. (1992). Selective chemical modification of amino acid residues in the flavin adenine dinucleotide binding site of NADPH-ferredoxin reductase. Int. J. Biochem. 24, 223–228.

Yamazaki, T., McNamara, B.C., & Jefcoate, C.R. (1993). Competition for electron transfer between cytochromes P450scc and P45011β in rat adrenal mitochondria. Mol. Cell. Endocr. 95, 1–11.

Yang, C.S. (1977). The organization and interaction of monooxygenase enzymes in the microsomal membrane. Life Sci. 21, 1047–1058.

Zylstra, G.J., & Gibson, D.T. (1989). Toluene degradation by Pseudomonas putida F1: Nucleotide sequence of the todc1c2BADE genes and their expression in *Escherichia coli*. J. Biol. Chem. 264, 14940–14946.

THE MOLECULAR STRUCTURE OF P450S:

THE CONSERVED AND THE VARIABLE ELEMENTS

Sandra E. Graham-Lorence and Julian A. Peterson

I. INTRODUCTION

Cytochromes P450 comprise a gene superfamily of hemoproteins whose fifth ligand of the heme iron is a cysteinyl sulfur and which on binding carbon monoxide

Advances in Molecular and Cell Biology
Volume 14, pages 57–79.
Copyright © 1996 by JAI Press Inc.
All rights of reproduction in any form reserved.
ISBN: 0-7623-0113-9

produces a characteristic absorbance maximum at 450 nm (Garfinkel, 1958; Klingenberg, 1958; Estabrook et al., 1963; Omura and Sato, 1962). P450s are most frequently monooxygenases whose substrates are hydrophobic organic compounds. They are present in bacteria, fungi, plants, and animals, and catalyze both regio- and stereo-specific reactions in metabolic pathways (e.g., steroid hormones and prostaglandins) and less specific reactions to remove/metabolize chemical toxins such as carcinogens, phytoalexins, barbiturates, herbicides, and small alcohols.

P450s can be divided into three classes based on their proximal electron donor. Generally, *Class I* and *Class II* proteins are monooxygenases, while *Class III* proteins act on organic peroxides. *Class I* P450s require both an iron-sulfur protein and an FAD-containing reductase, and are found generally in bacteria and in the mitochondria of eukaryotes, for example, the bacterial camphor 5-exo-hydroxylase $P450_{cam}$ (CYP101), and the eukaryotic cholesterol side chain cleavage enzyme— $P450_{scc}$ (CYP11A) and 11β-hydroxylase—$P450_{11\beta}$ (CYP11B) (Nelson et al., 1993). *Class II* P450s require an FAD/FMN-containing NADPH-P450-reductase and are generally found in the endoplasmic reticulum of eukaryotic organisms, for example, P450s involved in steroid metabolism such as 21α-hydroxylase (CYP21), 17α-hydroxylase—$P450_{c17}$ (CYP17), and aromatase—$P450_{arom}$ (CYP19), or P450s of drug (CYP2), carcinogen (CYP1), and fatty acid (CYP4) metabolism (Nelson et al., 1993). Finally, *Class III* P450s do not require electron donors, for example, thromboxane synthase (CYP5) which causes a carbon rearrangement of the endoperoxide PGH_2 to thromboxane A_2, or the plant P450 allene oxide synthase (CYP74) which rearranges a peroxy fatty acid to an allene oxide (Ullrich and Hecker, 1990; Song, Funk, and Brash, 1993).

$P450_{cam}$ has been the most widely studied of all P450s because of the ease of purification of this soluble, *Class I*, bacterial enzyme (Katagiri et al., 1968; Peterson, 1971; Gunsalus and Wagner, 1978). Detailed mechanistic characterization of this enzyme system (e.g., Griffin and Peterson, 1972; Brewer and Peterson, 1988; Sligar and Murray, 1986), which includes $P450_{cam}$ along with its cognate iron-sulfur protein and NADH-reductase, has permitted groups to compare the results obtained with distantly related P450 sub-families to those from $P450_{cam}$. The structure of $P450_{cam}$ has been determined at atomic resolution by Poulos et al. (1986, 1987) and has served as a structural model for other P450s. The structure of a second soluble, *Class I*, bacterial α-terpineol monooxygenase, $P450_{terp}$ (CYP108), has been reported recently (Peterson et al., 1992; Hasemann et al., 1994). Since the primary structure (i.e., the peptide sequence) of $P450_{terp}$ indicates that it is evolutionarily the closest relative to $P450_{cam}$, the determination of its structure enables us to better understand the limits of computer-generated sequence alignments and structure predictions. Finally, $P450_{BM-P}$ (CYP102), a soluble, *Class II*, bacterial fatty acid monooxygenase from *Bacillus megaterium*, was crystalized and its three-dimensional structure determined (Ravichandran et al., 1993). This P450 in its native state as $P450_{BM-3}$, is a fusion protein with its FAD/FMN-containing redox partner (Nahri and Fulco, 1986 and 1987; and Boddupalli et al., 1990, 1992a, and 1992b), and therefore, is both a good structural and functional model for eukaryotic *Class II* P450s.

In any consideration of the structure and function of a P450, one should first have a good understanding of the structurally-known P450s—their similarities and differences. Then, using this knowledge, generate sequence alignments with the known P450s identifying potential structural elements and critical residues. With this analysis, residues which may be important in P450 structure, in protein-protein interactions, and in substrate-protein interactions may be studied.

II. COMPUTER ALIGNMENTS OF P450S AND ANALYSIS OF THE ALIGNMENTS

Using sequence alignments and hydropathy plots, Nelson and Strobel (1987 and 1988) recognized a number of years ago that almost all of the proteins which had an absorbance maximum at 450 nm as a result of a carbon monoxide-iron complex belonged to a gene superfamily of proteins. Nelson and his colleagues have continued to provide a service to the P450 community by accepting and assigning family numbers to new P450s as they are identified (Nelson et al., 1993). In the assignments, Nelson has aligned each of the submitted sequences with his P450 database, which as of April, 1996, contained 401 public sequences which had been approved for distribution by their authors. Nelson uses a computer program to roughly align the new sequence(s) with the P450 database sequences. Subsequently, he manually realigns the sequence to improve gapping and to take advantage of several sequences which he refers to as landmarks (Nelson, 1996).

Gotoh has developed an algorithm which he has used to align some of the members of the CYP2 subfamily (Gotoh and Fujii-Kuriyama, 1989; Gotoh, 1992). Based on this alignment and knowledge of changes in structure/function with respect to substrate specificity in this gene subfamily, Gotoh has identified six Substrate Recognition Sites (SRSs) which may be involved in substrate specificity of this protein family. One of the surprising outcomes of Gotoh's analysis is that SRSs cover approximately 16% of the linear sequence of these proteins—which implies that approximately 80–90 amino acid residues are involved in some way in substrate specificity. In $P450_{BM-P}$, we have identified those sequences important in substrate recognition, access, and binding, and they are all within SRSs identified by Gotoh.

One readily available program for DNA and protein sequence analysis is the University of Wisconsin GCG suite of programs. A pairwise alignment generated with the GAP program of the GCG suite between $P450_{cam}$ and $P450_{BM-P}$ indicates that there is 20% identity and 47% similarity. $P450_{cam}$ and $P450_{terp}$ when compared in a pairwise alignment seem more closely related than $P450_{cam}$ and $P450_{BM-P}$ with 25% identity and 45% similarity. In contrast to the GAP program, the PILEUP program of the GCG suite uses a more complex algorithm which first clusters the places the proteins into a sequential order based on this relatedness; and finally, it aligns the proteins in the order of relatedness using the Needlman-Wunsch algorithm both in the forward and reverse direction (Needleman and Wunsch, 1970). Figure 1 shows a computer-generated alignment of P450$_{cam}$, P450$_{terp}$, and P450$_{BM-P}$ using PILEUP in which there is 21% identity and 44% similarity between P450$_{cam}$ and P450$_{terp}$, and 10% identity and 29% similarity between P450$_{cam}$ and P450$_{BM-P}$. An examination of the alignment illustrates certain important aspects of this gene superfamily: there are relatively few similar or identical residues among the different gene family proteins; and the similarities that do exist are not evenly distributed throughout the linear sequence of the molecules, rather, they are clustered in regions which may be important in the structure/function of P450s. When more than 100 sequences with representatives from each gene family are included in the PILEUP analysis, one can conclude from the graph or tree of

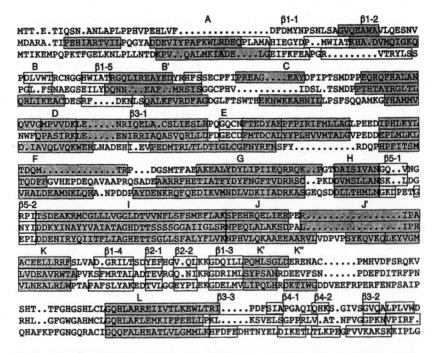

Figure 1. A PILEUP sequence alignment of P450$_{cam}$, P450$_{terp}$, and P450$_{BM-P}$. The PILEUP program of the GCC suite of programs was used to align these proteins. The secondary structural elements as determined from the atomic structures are indicated in the figure. The α-helical elements are shown by the shaded boxes while the strands of β-sheets are shown by the open boxes. The positions of each of these elements in P450$_{BM-P}$ are indicated above the alignment. The first line in each triplet is P450$_{cam}$, the second is P450$_{terp}$, and the third is P450$_{BM-P}$.

relatedness generated by PILEUP that P450$_{cam}$ and P450$_{terp}$ are closely related while P450$_{BM-P}$ is more distantly related.

With the availability of the atomic structure of three P450s—P450$_{cam}$, P450$_{terp}$, and P450$_{BM-P}$, we are now in a position to assess the effectiveness of the various algorithms either in aligning P450s or at predicting the secondary structural elements of P450s. Figure 2 shows the "structural alignment" along with the structural elements generated from the crystallographic data of the three-dimensional structures of P450$_{cam}$, P450$_{terp}$, and P450$_{BM-P}$. Table I contrasts the alignments done using the GAP program, the PILEUP program, and from the structural alignment among the three crystalized P450s. In comparing the computer-generated alignment from Figure 1 with the structural alignment in Figure 2, one finds that the C-terminal ends of the three P450s are aligned with relatively good accuracy using the PILEUP program in that conserved residues as well as the structural elements are correctly aligned. Unfortunately, the alignments of the N-terminal half of the molecules are not structurally relevant, i.e., helices and sheets are not correctly aligned because of the diversity of the sequences. These regions which

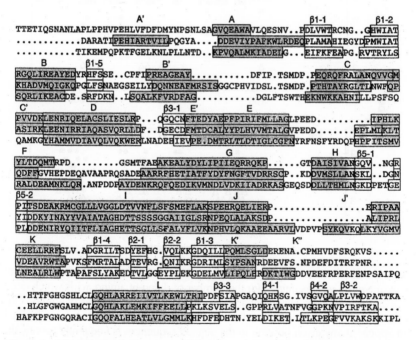

Figure 2. The structural alignment of P450$_{cam}$, P450$_{terp}$, and P450$_{BM-P}$. The alignment of the linear sequences was taken from the overlay of the three dimensional structures of these proteins. The order of the sequences and the designation of the elements is given in Figure 1.

Table 1.

	$P450_{cam}$	$P450_{terp}$	$P450_{BM-P}$
$P450_{cam}$	414	1.9 A/319[a]	2.3 A/227
$P450_{terp}$	21%/44%[b]	428	1.9 A/302
$P450_{BM-P}$	10%/28%	22%/43%	472

Source: [a] R.M.S. deviation/Number of residues compared
             [b] Identical/similar residues

are involved in substrate specificity should be evaluated carefully and/or manually adjusted as Nelson has done.

To further analyze the eukaryotic P450s, Figure 3 contains a compilation of sequence similarity and identity from 213 P450s extracted from the alignment of Nelson (1994). (Note: In this figure, the numbers refer to the alignment numbers, not the sequence numbers, e.g., the amino terminus of $P450_{BM-P}$ is at residue

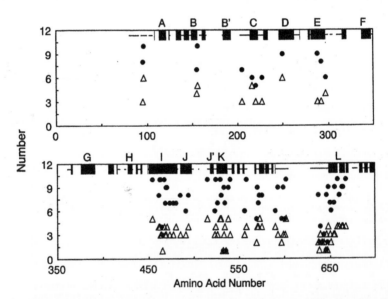

Figure 3. Similarities and identities in the compiled sequences of P450s. The alignment of Nelson was examined for the number of different amino acids which occur at each position in the alignment and the number of different types of amino acids. The total number is shown by the filled circles. Because number of amino acids which can appear at any given position in the aligned sequences is frequently twenty, this figure is limited to those positions which are represented by fewer than ten amino acids. Finally, the number of different types of amino acids is shown by the open triangles. For reference, the secondary structural elements of P450BM-P is shown by a line and dark bars in the top of the figure.

number 75—not at 1.) In this figure, we have indicated the total number of conserved amino acids which appear at any position of the aligned sequence, as well as, the number of different groups of similar amino acids (e.g., hydrophobic, aromatic, charged). The secondary structural elements of P450$_{BM-P}$ are included in the figure for reference. What is most striking about this figure is that there are relatively few conserved amino acids in P450s. In the I-helix, there are four highly conserved residues, (A/G)Gx(D/E)T; and in the K-helix, there are two absolutely conserved residues, ExxR. Finally, there is the cluster of seven conserved residues such as, (A/G)Gx(D/E)T in the heme binding loop immediately N-terminal to the L-helix which includes the invariant cysteine—the fifth coordinating ligand of the heme iron. The seven conserved residues and the heme-binding region in the spatial orientation indicated are hallmarks for cytochromes P450.

An alternative way to examine the alignments is to compile the net charge at each aligned position in the sequence as shown in Figure 4. In comparing Figures 3 and 4, it should be noted that even though many different amino acids can appear in the N-terminal 100 residues of this alignment, almost without exception there are no conserved residues which are charged. This charge analysis brings out additional conserved residues which were not obvious from Figure 3. For example, in the

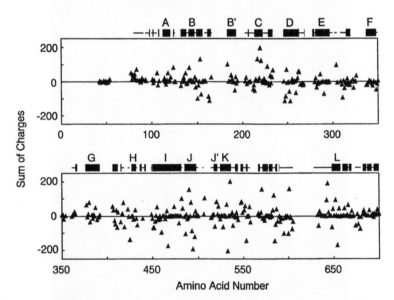

Figure 4. Net charge on individual residues in the aligned sequences of P450s. The sequence alignment of Nelson was examined for the net charge at each position. Arginines and lysines were assigned a charge of 1. Glutamic acid and aspartic acid were assigned a charge of –1. Histidine was assigned a charge of 0.5. The sum of the charges at each position is shown by a filled triangle. For reference, the secondary structural elements of P450$_{BM-P}$ are shown.

middle of the I-helix in most P450s there is a carboxy amino acid (Asp or Glu) immediately N-terminal of the conserved Thr, i.e., (A/G)Gx(D/E)T. Other conserved charged residues noted in this figure will be discussed in the structural analysis section of this chapter.

Thus, although the generation of alignments is relatively easy with modern computers and programs, the generation of *good* sequence alignments to use for structural prediction and model building of eukaryotic P450s is not a trivial task and requires additional analysis.

III. THE STRUCTURES OF P450CAM, P450TERP, AND P450BM-P

There are three key features when comparing and contrasting P450s: (1) the overall domain structures and their conserved structural elements, (2) the substrate binding regions and how substrate specificity is conferred, and (3) the redox-partner binding regions and what differentiates the three classes of P450s.

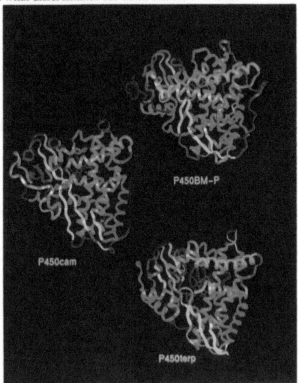

Figure 5. A ribbon diagram of the distal faces of P450cam, P450terp, and P450BM-P. The α-helices are in blue, the β-sheets in orange, the loops or random coils in purple, and the heme in red. All three p450s are shown in the same orientation.

A.   The Overall Structure and the Conserved Elements

Generally, in comparing the three-dimensional structures of $P450_{cam}$, $P450_{terp}$, and $P450_{BM-P}$, one should first approach the problem at a gross level to identify the features in common before focusing on the differences which make each P450 unique. On doing this as shown in Figure 5, one finds that the three P450s appear to have the same structure composed of two domains: a portion that is predominantly α-helices, accounting for approximately 70% of the protein, and one that is predominantly β-sheets, accounting for 22% of the protein (Poulos et al., 1987; Ravichandran et al., 1993; Hasemann et al., 1994). The α-helical domain contains helices B′ through K, helix L, and sheets β3, β4, and β5. The β-sheet domain contains sheets, β1 and β2, and helices A, B, and K′. The regions of highest structural similarity in these proteins are a four-helix bundle in the α-helical domain, helices J and K, sheets β1 and β2, the heme-binding cysteinyl loop, and the "meander"—a 14 or 15 residue stretch immediately C-terminal to the K′ helix referred to by some authors as the "aromatic region" (e.g., Goto and Fujii-Kuriyama, 1989). The regions of greatest structural difference are those proposed to be involved in substrate recognition and binding, and those involved in redox-partner binding. Thus, the structures of the three P450s determined to date are similar, yet they are sufficiently different in the fine details that any one can not be used as a template to build in precise detail any other one since the lengths and positions of the α-helices and β-sheet strands may differ as can be seen in Figure 5.

The α-Helical Domain

In the α-helical domain, the four-helix bundle along with helices J and K are part of the core structure comprising almost 20% of the Cα backbone of these molecules. Figure 6 shows the highly conserved structural elements of these P450s. The four-helix bundle is composed of three parallel helices and one anti-parallel helix: helix I, the long central helix which bounds one side of the heme pocket, helix L forming a portion of the heme-binding region, and helix D, while helix E is anti-parallel. This conserved structure is bordered at the top by five α-helices and the 3_{10} helices, and at the bottom by three α-helices and two β-sheets.

The other elements of the α-helical domain are structurally less well conserved. The region between helices B and B′ is quite different in the three crystalized P450s. In $P450_{cam}$ and $P450_{terp}$, the B-B′ loop has residues protruding into the active site, but in $P450_{terp}$ there is a five amino acid insert; and in $P450_{BM-P}$, this region is predominantly a β-sheet with its carbon backbone lining the heme pocket. In each of these P450s, the B′ helix is not superimposable. The B′-C loop also invaginates into the active site/heme pocket and is variable as well. In each case, however, the residues protruding from the B′-C loop into the heme pocket have an effect on substrate binding and orientation, as discussed later. The two antiparallel helices F and G vary structurally between $P450_{cam}$, $P450_{terp}$, and $P450_{BM-P}$ in both length and position relative to the four-helix bundle (Fig. 4). The F-G loop joining them has

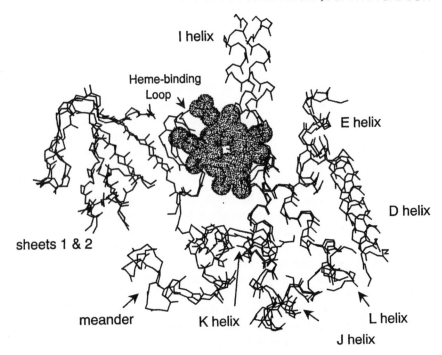

Figure 6. The core structures of P450cam and P450BM-P superimposed. The core structure includes the four-helix bundle of helices D, E, I, and L, helices J and K, β-sheets 1 and 2, the heme-binding region, and the meander.

the highest crystallographic B Factors of any part of these molecules. This implies that in the crystal, this loop does not occupy a single conformation but is highly flexible, and in fact, this loop is completely disordered in $P450_{terp}$ and was not included in the final published crystal structure (Hasemann et al., 1994). On examination of the three structures, the residues in this loop appear to act as "gate-keepers" for the substrate-access channel.

On the proximal face of the P450 molecules (i.e., on the side of the molecule where the cysteinyl loop binds the heme iron), the J, K, B, and C helices bound the putative redox-partner binding regions. Helix K contains two residues, a Glu and an Arg (corresponding to E320 and R323 in $P450_{BM-3}$) which are conserved in *all* P450s. The Arg of this pair was presumed to be involved in electrostatic interaction with potential redox partners by some groups; however, in each of the P450s for which the structure is known, these residues are electrostatically coupled and buried on the hydrophobic interior of the protein. In $P450_{BM-P}$ there is an additional helix, the J' helix, as compared to $P450_{cam}$ and $P450_{terp}$. There also is a small insert N-terminal of the heme-binding region which appears to be present in all eukaryotic

P450s. Amino acid residues present in these two insertions may be involved in redox-partner binding.

There are three β-sheets in the α-helical domain. Sheets β3 and β4 are close together with β4 (called β-5 in P450$_{cam}$) forming a portion of the heme pocket. The other β-sheet is between helices H and I and has the appearance of an appendix atop the molecules.

The β-Sheet Domain

The three-dimensional orientation of the β-sheets in the β-sheet domains of the three crystalized proteins are very similar, i.e., β1 and β2 are simultaneously superimposable (Fig 6). There are also three helices in this domain: A, B, and K'. P450$_{terp}$ has an additional helix A' which is amino terminal to helix A. The A helix on the distal face of the molecules varies in size, length, and position; whereas, the B helix on the proximal face is similar in sequence and position among the three P450s. Helix K' is a very short, newly-defined helix which is present in each of these structures and is just carboxy-terminal to the β-3 strand. It is not clear whether the primary function of these helices is the stabilization of the β-sheet domain or for interaction with the redox partner or both.

B. The Substrate-Binding Regions

The second key feature in the structures of P450s is the substrate binding regions. Although the structures of substrate-free and substrate-bound P450$_{cam}$ have been reported (Poulos et al., 1986 and 1987), it is not immediately obvious how the substrate enters the active site from bulk solvent and the product exits. With the availability of the structures of additional P450s, the process of substrate binding may be envisioned to be more complex than in proteins whose substrates are hydrophilic and whose active sites are exposed directly to solvent. That is, it may involve substrate recognition on the surface of the molecule, substrate access to the active site down a channel, and substrate orientation in the active site or heme pocket.

The structural elements and the residues which are important in substrate recognition and binding differ depending on the P450 and the substrate. These differences are exemplified when comparing P450$_{cam}$ to P450$_{BM-3}$ and their corresponding substrates—camphor, a small bicyclic monoterpene for P450$_{cam}$, and long-chain fatty acids and eicosanoids for P450$_{BM-3}$. To accommodate the large substrates in P450$_{BM-P}$, the access channel to the heme active site is cavernous; whereas, in P450$_{cam}$, the access channel is almost nonexistent.

There are interesting contrasts in the specificity of the hydroxylations catalyzed by P450$_{terp}$, P450$_{cam}$, and P450$_{BM-3}$. P450$_{cam}$ and P450$_{terp}$ are regio- and stereo-specific in their sites of hydroxylation respectively forming 5-exohydroxy-camphor from camphor (Atkins and Sligar, 1988 and 1989) and 7-hydroxy-terpineol from α-terpineol (J.A. Frutel and P. Ortiz de Montellano). P450$_{BM-3}$, on the other hand, metabolizes many different compounds, and is capable of monooxygenation at several positions e.g., hydroxylation at ω-1, ω-2 or ω-3 of palmitate (Miura and Fulco, 1975; Boddupalli et al., 1992). With these things in mind, one should not be surprised to find that there are differences in the lengths of the helices involved in substrate binding and their orientation in three-dimensional space. Thus, when comparing these P450s to other P450s, one must consider the size and shape of the substrate, the specificity of the P450 for its substrate, as well as, the stereo- and regio-specificity of hydroxylation.

The Mouth or Substrate Recognition Region

Several years ago, Fulco noted that trypsin treatment of the holoenzyme of P450$_{BM-3}$ plus or minus the substrate myristate gave slightly different results (Narhi and Fulco, 1987). In the presence of myristate, only the single trypsin sensitive site at K472 connecting the P450 and reductase domains was cleaved. In the absence of myristate, two additional cleavage sites were observed in the N-terminus of the protein at K10 and K15. They also made the interesting observation that the hemoprotein domain could be converted partially from the low spin state to the high spin state of the iron on the addition of myristate to enzyme which had been trypsin treated in the presence of the fatty acid; however, the enzyme which was treated in the absence of myristate would not undergo this spin state change. Interpretation of these results in the absence of a three dimensional structure of this protein was difficult. Now, we realize that several of the removed residues are located at the mouth of the access channel. This also correlates with the substrate-binding region of P450$_{scc}$ identified by Tsujita and Ichikawa (1993). When they digested [^{14}C]methoxychlor-bound P450$_{scc}$ with trypsin, they found the bound fragment was at the N-terminus within residues 8–28. The results of Nahri and Fulco and of Tsujita and Ichikawa implies that substrate binds initially at the surface of the molecule on this hydrophobic patch at a considerable distance from the heme, in a region we refer to as the substrate recognition region or docking region.

The mouth of the substrate access channel and the access channel itself in P450$_{BM-P}$ are composed of the N-terminal residues Phe11, Leu14, Leu17, Pro18, and Leu19, the F-G loop and portions of helices F and G, helices B' and C, and β-sheet 1. A part of the aligned sequences which has received considerable attention because of the work of Negishi is the loop between the F and G helices (Lindberg and Negishi, 1989). Negishi identified residue F209 which we believe corresponds to Thr185 in P450$_{cam}$ as being located in the F-G loop of Cyp2A5 and as conferring substrate specificity such that when it was mutated to a Leu the preferred substrate changed from coumarin to 11-deoxysteroids (e.g., testosterone). The F-G loop appears to act as a "gatekeeper," i.e., it is highly flexible and may act as a "flap" to cover the mouth of the access channel. In the three crystal structures, both helices F and G and their loop vary in length and orientation. The sheet β1, however, is structurally very conserved in these P450s. In P450$_{BM-P}$, Arg47 which is at the mouth of the access channel in strand β1-2 is approximately 25Å from the heme iron, yet it plays an important role in substrate recognition and binding. When this residue is mutated to Glu, arachidonate is bound and metabolized poorly by P450BM-P (manuscript in preparation). In contrast, the hydrophobic depression in $P450_{terp}$, which has a much smaller substrate, may initially bind α-terpineol adjacent to the substrate binding pocket above the heme iron at a distance approximately two thirds that of the $P450_{BM-P}$ channel.

The Access Channel

From the mouth of the protein, the substrate must travel down a channel to the active site believed to be driven by the hydrophobic effect. In $P450_{BM-P}$, this channel is composed predominantly of backbone carbon atoms, of hydrophobic residues found in the strand $\beta1$-5, and of portions of helices F and B'. This channel is exceptionally long in $P450_{BM-P}$ (> 25Å) to accommodate the long-chain fatty acids, with the length of this channel being partially determined by the length of helix F and its juxtaposed helix G. In P450s with small substrates (e.g., $P450_{terp}$ and $P450_{cam}$), this channel may be almost nonexistent.

The Active Site or Heme Pocket

In each of these P450s, the active site is composed of residues from the B-B' loop ($\beta1$-5 in $P450_{BM-P}$), the B'-C loop, $\beta1$-4, $\beta4$, and a portion of the I-helix. Figure 7 shows the active site of $P450_{BM-P}$ as an example. Residues from the B-B' and the B'-C loops are located over the C and D pyrrole rings of the heme and are discussed below. Residues from $\beta4$ are above or near pyrrole ring A of the heme while the conserved Thr from the I helix which is believed to be involved in oxygen activation in *Class I* and *Class II* P450s (Imai et al., 1989), actually blocks ring A from substrate. Residues in the region of the conserved Ala/Gly-Gly pair from the I-helix cover pyrrole ring B. Strand $\beta1$-4 cuts across the substrate binding site between pyrrole rings D and A. From this strand, the sidechains of residues V295 and D297 of $P450_{cam}$ and of F317 of $P450_{terp}$ form part of the walls of their respective substrate binding sites; whereas, in the corresponding region of $P450_{BM-P}$, the backbone atoms rather than the amino acid sidechain atoms line the pocket.

The lengths and positions in three-dimensional space of the B-B' loop/strand and the B'-C loop are quite different in the three P450s; however, they are both important in substrate binding and orientation. The loop between helices B and B' in $P450_{cam}$ and $P450_{terp}$ is predominantly sheet in $P450_{BM-P}$. In $P450_{BM-P}$, the B-B' region is located in a similar region to that of $P450_{cam}$, but in $P450_{BM-P}$ as a sheet, this region is extended, and thus, accommodates the long-chain fatty acid in the access channel. Alternatively, the residues of this strand do not protrude into the channel or pocket as they do in $P450_{cam}$ and $P450_{terp}$. In $P450_{terp}$, there are two residues from this loop in the heme pocket: E76 and I77. In the $P450_{cam}$ loop, the aromatic ring of F87 inserts above the heme between pyrrole ring C and D and from computer dynamics studies F87 appears to have high mobility (Paulsen et al., 1991). This residue in $P450_{cam}$ may act in an analogous fashion to R47 from $\beta1$-2 in

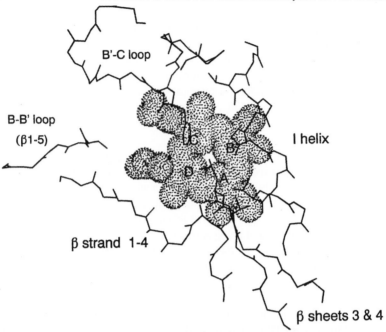

Figure 7. The active site of P450$_{BM-P}$ showing the protein backbone structures and the heme. These structural elements are also present in P450$_{cam}$ and P450$_{terp}$, and all but helix I are different in length and position among P450s.

P450$_{BM-P}$ which is at the mouth of the access channel and appears to be involved in substrate access to the pocket and fatty acid binding. In contrast, in P450$_{cam}$ and P450$_{terp}$ which have smaller substrates, residues critical for substrate access and binding may be located closer to the heme pocket e.g., F87 in P450$_{cam}$.

The other important loop in the heme pocket is the B'-C loop which snakes its way in and out of the active site. In P450$_{cam}$, three residues from this loop are located in the active site: Y96, F98, and T101. Y96 is located between the C and D rings and hydrogen bonds with the 2-keto group of camphor to orient it in the active site. Changing this residue to a Phe (Atkins and Sligar, 1988 and 1989) or a His (unpublished data S. Graham-Lorence and P. Ortiz de Montellano) allows for altered regiospecificity of hydroxylation. T101 which is just above the C ring of the heme when mutated to Ala may slightly tilt the plane of the heme ring (unpublished data, S. Graham-Lorence and P. Ortiz de Montellano). While in P450$_{cam}$ there are three residues protruding into the active site, there are four residues from this loop in P450$_{terp}$ with three of them being consecutive, i.e., I98, E101, T102, and S103, and in P450$_{BM-P}$, only the sidechain of residue F87 is present in the pocket. This F87 in P450$_{BM-P}$ (different than F87 from P450$_{cam}$) is perpendicular to the plane of the heme over the C pyrrole ring. From preliminary results, the pattern of hydroxylation of palmitate is changed when F87 is mutated to Val indicating that F87 probably serves to channel long-chain fatty acids into their final orientation in the substrate pocket through steric interactions.

E.F. Johnson's group has altered hydroxylation activity in several of the CYP2 family proteins by altering the B-B'-C helix regions. They have formed a chimeric protein by substituting most of the B to C helix regions in $P450_{2C1}$ (CYP2C1) for $P450_{2c5}$ (CYP2C5), and thus, conferring progesterone 21-hydroxylase activity on the new protein (Kronbach et al., 1991). They have also made a series of cassette mutations in the B' to C loop in $P450_{2C2}$ (CYP2C2) using the predictions of Gotoh (1992) to successfully alter the lauric acid hydroxylase activity (Straub et al., 1993a and 1993b). Thus, these regions in the mammalian P450s which align with the B to C helix region of the crystalized P450s appear to alter substrate affinity possibly by changing the shape of the active site.

C. The Redox-Partner Binding Regions

The third key feature in the structural analysis of P450s is the redoxpartner binding region. The shape of the docking region for the redox partner is similar in $P450_{cam}$ and $P450_{terp}$, but is different for $P450_{BM-P}$. This is as one might expect since $P450_{cam}$ and $P450_{terp}$ are *Class I* proteins which bind small, iron-sulfur proteins (approximately 11 kD), while $P450_{BM-P}$ is a *Class II* P450 which binds the 23 kD FMN domain of the FAD/FMN-containing NADPH-dependent P450 reductase (approximately 65 kD; Sevrioukova et al., 1996). In the former case, the iron-sulfur protein is believed to shuttle back and forth between the P450 and the reductase delivering one electron at a time to the P450; whereas, in $P450_{BM-3}$, the reductase binds directly to the P450 domain donating its electrons.

The shape of the putative redox-partner binding region of P450cam and $P450_{terp}$ is relatively flat with four positively charged residues clustered near the center of the proximal face of each protein (Figure 8). In $P450_{cam}$, these basic residues have been mutated to uncharged residues, thus decreasing the electron transfer from its redox partner putidaredoxin (Stayton and Sligar, 1990). In contrast, the proximal face of $P450_{BM-P}$ is like a bowl with the central region composed almost exclusively of hydrophobic and aromatic residues, while the rim has numerous negatively and positively charged residues. In the past, several authors have discussed the "aromatic regions" in microsomal P450s which we have referred to as the "meander" just C-terminal of the K' helix. It appears that most of those aromatic residues, as suggested by other authors, as well as aromatic residues from other regions, cluster together forming an aromatic patch at the reductase-binding site. The added depth to the $P450_{BM-P}$ binding region is a result of two insertions in $P450_{BM-P}$: the J' helix and 3_{10} helix f downstream of the K' helix. Sequence alignments indicate that these insertions are present in both mitochondrial and microsomal P450s of eukaryotes, and may also produce a bowl shaped redox-partner binding region in them.

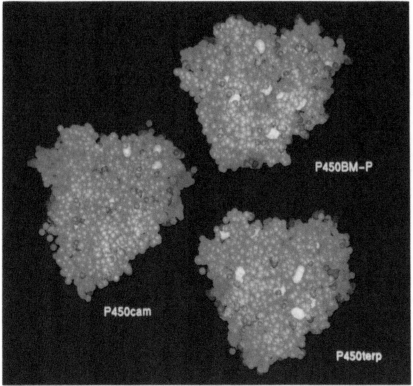

Figure 8. CPK models of the proximal face of P450$_{cam}$, P450$_{terp}$, and P450$_{BM-P}$ showing the difference in charge distribution between the *Class II* P450, P450$_{BM-P}$, and the *Class I* P450s, P450$_{terp}$ and P450$_{cam}$. Basic charges are in red, acidic charges are in dark blue, and aromatic residues are in white, with the heme in orange.

In Figure 4, a composite charge distribution for mitochondrial and microsomal P450s was compiled from 187 full-length sequences in the alignment of Nelson (1994) to determine whether there are unique charge distribution patterns for mitochondrial versus microsomal P450s, and to compare these patterns to the composite structures of the known P450s. In order to identify the pattern of residues critical for redox-partner binding in *Class I* versus *Class II* P450s, Figure 9 focuses on the charge differences in mitochondrial and microsomal P450s on what is believed to be the proximal face as determined from alignments with the three known structures. In helix B, we identified a positively charged surface residue which is negatively charged in almost all mitochondrial P450s. This residue is one of the lysine residues identified by Shen and Strobel (1993) as being involved in reductase interaction in CYP2B1. In helix C, there is a positive residue which is also present in both mitochondrial and microsomal P450s and which is coupled to one of the heme propionates; however, in microsomal P450s, there is an additional

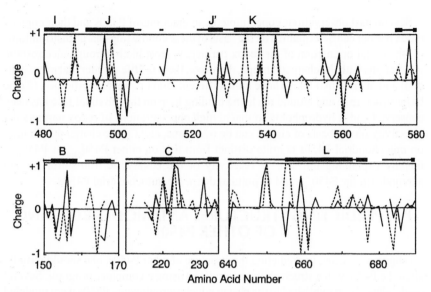

Figure 9. Charges on selected structural elements in microsomal and mitochondrial P450s. The sequences in the alignment of Nelson were divided into three groups: those which are clearly microsomal, those which are clearly mitochondrial, and all others. The charges on each of these groups of sequences was determined as described in Figure 4 and the total charge at any position was divided by the number of sequences being examined. Mitochondrial P450s are shown in this figure by the dashed line while microsomal P450s are shown by the solid line.

positive residue adjacent to the first producing a double charge. There is a group of positive charges in $P450_{BM-P}$ in the insertion associated with the J' helix, but because of the variability in this region between mitochondrial and microsomal P450s, the alignment is probably not very good. At the amino terminus of the K helix on the proximal face, mitochondrial P450s have a positively-charged residue while almost all microsomal P450s are negatively charged. In $P450_{scc}$, Wada and Waterman (1992) mutated these residues and showed that adrenodoxin interaction was dramatically changed. Finally, among the most striking switches in charge are in the L helix where the sequence for all mitochondrial P450s sequenced to date begins with RRhAEuE (where "h" is an L, I, or V and "u" uncharged). In microsomal P450s, however, the pattern is almost the inverse where frequently the helix begins with a negative residue, followed by two uncharged residues, a positive residue, an uncharged residue, and finally, a negative residue i.e., -uu+-. Thus, while the proximal redox-partner binding face may be bowl-shaped for both mitochondrial and microsomal P450s, the charge distribution across these faces may reflect the charge distribution on the redox partner.

Several past observations concerning the interactions of P450s and their cognate redox partner can be explained on the basis of these contrasting charge distributions: (1) the initial orientation of the redox partners is most likely through electrostatic interactions with the charged groups; therefore, there is little wonder that mitochondrial and microsomal P450s have quite different redox partner specificities, (2) the basic structure in and around the heme binding loop on the proximal face has been conserved and is hydrophobic so that at high concentrations of redox partners the selectivity on the basis of charge can be overcome, and (3) the charge distribution on mitochondrial P450s is quite distinct from that on either $P450_{cam}$ or $P450_{terp}$, thereby explaining the lack of reactivity of the eukaryotic iron-sulfur protein adrenodoxin for $P450_{cam}$ and of putidaredoxin for mitochondrial P450s.

IV. PREDICTING STRUCTURES AND MODEL BUILDING OF OTHER P450S

As suggested earlier, in predicting the structure of other P450s, one should initially align the P450 with a structurally known P450 using a computer program. Several models have been published in the past in which steroidogenic P450s were modeled using $P450_{cam}$ as a template. (Laughton et al., 1990 and 1993). However, because of the differences in the proximal faces between $P450_{BM-P}$, and $P450_{cam}$ and $P450_{terp}$, i.e., the inserts of the J' helix and the region between the K' helix and the heme-binding region which are involved in redox-partner binding, the authors believe that $P450_{BM-P}$ is the more appropriate prototype for eukaryotic P450s. Thus, initial computer alignments should be done with $P450_{BM-P}$, followed by hand alignments of the eukaryotic P450 sequence with the structural alignments of $P450_{cam}$, $P450_{terp}$, and $P450_{BM-P}$, using certain motifs as bench marks such as the positive charge in the C helix, the cluster of positively charged residues at the end of the G helix, or the i + 3 or i + 4 nature of the helices and their amphipathic properties to align the more hypervariable N-terminal region. Using the crystallographic structures as the basis for alignment, one obtains 14% identity and 36% similarity between $P450_{cam}$ and $P450_{BM-P}$. With the type of care taken as discussed above, $P450_{arom}$ was aligned with $P450_{BM-P}$ giving an 18% identity and a 30% similarity as shown in Figure 10 (Amarneh et al., 1993; Graham-Lorence et al., 1995).

With this alignment, the four-helix bundle of the helix domain (i.e., helices D, E, I, and L), helices J and K, and β-sheets 1 and 2 of the sheet domain which appear to be conserved structural elements may be useful in the model building of the core structure. Unfortunately, the remainder of the molecule is not as structurally conserved because of the variable nature of the substrate recognition, access, and binding regions to accommodate different substrates as well as simple evolutionary diversity. That is, models of P450s can not be built by simple molecular replacement. For example, during the refinement of the $P450_{terp}$ structure, attempts were made to use molecular replacement of $P450_{cam}$ with $P450_{terp}$ (i.e., substitution of

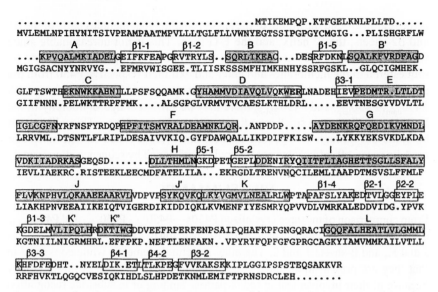

```
.........................................MTIKEMPQP.KTFGELKNLPLLTD.....
MVLEMLNPIHYNITSIVPEAMPAATMPVLLLTGLFLLVWNYEGTSSIPGPGYCMGIG...PLISHGRFLW
          A        β1-1      β1-2          B           β1-5         B'
....KPVQALMKIADELGEIFKFEAPGRVTRYLS..SQRLIKEAC...DESRFDKNLSQALKFVRDFAGD
MGIGSACNYYNRVYG...EFMRVWISGEE.TLIISKSSSMFHIMKHNHYSSRFGSKL..GLQCIGMHEK.
          C                          D                       β3-1      E
GLFTSWTHEKNWKKAHNILLPSFSQQAMK.GYHAMMVDIAVQLVQKWERLNADEHIEVPEDMTR.LTLDT
GIIFNNN.PELWKTTRPFFMK....ALSGPGLVRMVTVCAESLKTHLDRL.....EEVTNESGYVDVLTL
          F                                    G
IGLCGFNYRFNSFYRDQPHPFITSMVRALDEAMNKLQR..ANPDDP.....AYDENKRQFQEDIKVMNDL
LRRVML.DTSNTLFLRIPLDESAIVVKIQ.GYFDAWQALLIKPDIFFKISW....LYKKYEKSVKDLKDA
                        H    β5-1    β5-2              I
VDKIIADRKASGEQSD.......DLLTHMLNGKDPETGEPLDDENIRYQIITFLIAGHETTSGLLSFALY
IEVLIAEKRC.RISTEEKLEECMDFATELILA...EKRGDLTRENVNQCILEMLIAAPDTMSVSLFFMLF
          J            J'         K           β1-4      β2-1   β2-2
FLVKNPHVLQKAABEAARVLVDPVPSYKQVKQLKYVGMVLNEALRLWPTAPAFSLYAKEDTVLGGEYPLE
LIAKHPNVEEAIIKEIQTVIGERDIKIDDIQKLKVMENFIYESMRYQPVVDLVMRKALEDDVIDG.YPVK
     β1-3    K'     K"                                          L
KGDELMVLIPQLHRDKTIWGDDVEEFRPERFENPSAIPQHAFKPFGNGQRACIGQQFALHEATLVLGMMI
KGTNIILNIGRMHRL.EFFPKP.NEFTLENFAKN..VPYRYFQPFGFGPRGCAGKYIAMVMMKAILVTLL
   β3-3          β4-1     β4-2    β3-2
KHFDFEDHT..NYELDIK.ETLTLKPEGFVVKAKSKKIPLGGIPSPSTEQSAKKVR
RRFHVKTLQGQCVESIQKIHDLSLHPDETKNMLEMIFTPRNSDRCLEH........
```

Figure 10. The alignment between P450$_{arom}$ and P450$_{BM-P}$ designating and showing the structural elements of P450$_{BM-P}$.

the amino acids of one for the other) to determine the phase of the crystal lattice, but there was not enough structural similarity between the two to build and refine the model. With these limitations in mind, we have built a model of P450$_{arom}$ by first replacing residues in the highly conserved regions—the four-bundle helix, helix J and K, sheets β1 and β2, and the heme-binding region. Then, we have constructed the remaining helices and sheets, placing them in such a way as to increase hydrophobic contacts. Finally, we constructed the loops which joined the structural elements. We then ran molecular dynamics simulations with substrates and inhibitors in the active site in the hopes that the loops will find a minima compatible with knowledge acquired through inhibitor studies and mutagenesis experiments.

Using the modeled structure identify important tertiary structures involved in protein-protein interactions and in substrate-protein recognition. For example, while the most hydrophobic stretch of amino acids and the most probable location for the membrane-spanning region is the amino-terminal portion (Ahn et al., 1993; Szsesna-Skorupa and Kemper, 1989; Szczesna-Skorupa, E., Browne, N., Mead, D., and Kemper, B. 1988), deletion of this region does not result in a soluble protein. Rather, there appears to be at least one other region which causes a loose association with the membrane as demonstrated by the inability to solubilize, without denaturation, truncated eukaryotic P450 proteins on purification. From the P450$_{arom}$ model, it appears that the surface of the protein which is the most hydrophobic is the region around the mouth of the access channel that includes parts of the $\beta 1$ and $\beta 2$ sheets, the B′ helix, and the F-G loop. This is not unreasonable since substrates of the P450s are generally hydrophobic compounds that may be sequestered in the membranes and enter and exit their respective P450 from the membrane. Additionally, from the alignments presented here of P450$_{arom}$ with P450$_{cam}$, P450$_{terp}$, and P450$_{BM-P}$, there may be sufficient hydrophobicity in the F-G loop to allow for a small interdigitation into the membrane. This will have to be borne out through mutagenesis studies; however, as stated earlier, molecular modeling may afford insights not otherwise envisioned by sequence alignments.

VII. SUMMARY

From comparison of the three crystalized P450s—P450$_{cam}$, P450$_{terp}$, and P450$_{BM-P}$, a core structure for P450s can be identified consisting of a four-helix bundle, helices J and K, and β-sheets 1 and 2. The other structural elements, while conserved in that they are present in all three structures, vary in length and position, largely because of their involvement in redox-partner binding or in substrate recognition and binding. From the structural alignments of the three crystalized P450 along with the careful alignment of Nelson (1996), one can identify those residues present in all P450s which can be used as hallmarks: the (A/G)Gx(E/D)T in the I helix, the absolutely conserved ExxR in the K helix, and the heme-binding region containing the conserved cysteine. On the other hand, computer generated alignments of the three P450s only give a reasonable alignment in the C-terminal half of the molecules, but they do not give a true picture of the structures at the N-terminal half of the molecules primarily because those regions confer substrate specificity.

In comparing the three-dimensional structures of P450$_{cam}$, P450$_{terp}$, and P450$_{BM-P}$, one realizes that there are three regions of importance in the substrate-associated regions: the substrate recognition or docking region, the substrate access channel, and the active site. These regions are discernable not only from the crystal structures, but from experiments which identify substrate specificity, and stereo- and regio-specificity of hydroxylation, e.g., mutagenesis experiments and inhibitor studies.

When studying the redox-partner binding sites of P450$_{cam}$, P450$_{terp}$, and P450$_{BM-P}$, one discovers that in the Class II P450, i.e., P450$_{BM-P}$, there are two insertions present as compared to the Class I P450s, P450$_{cam}$ and P450$_{terp}$. From multiple sequence alignments, these insertions also seem to be present in both eukaryotic mitochondrial and microsomal P450s, whether or not they are Class I or Class II P450s. These insertions in P450$_{BM-P}$ cause the proximal face to appear as a bowl with the center devoid of charged residues, rather than sloped as is the case for P450$_{cam}$ and P450$_{terp}$ with basic charges towards the center of the proximal face.

Using P450$_{BM-P}$ as the structural model for eukaryotic Class I and Class II P450s, one can identify a charge signature for mitochondrial versus microsomal P450s contained in the elements on the proximal face which might be overlayed on a conserved structure to permit appropriate redox-partner binding.

REFERENCES

Ahn, K., Szczesna-Skorupa, E., & Kemper, B. (1993). The amino-terminal 29 amino acids of cytochrome P450 2C1 are sufficient for retention in the endoplasmic reticulum. J. Biol. Chem. 268, 18726–18733.

Amarneh, B., Corbin, J.C., Peterson, J.A., Simpson, E.R., & Graham-Lorence, S.E. (1993). Functional domains of human aromatase cytochrome P450 characterized by linear alignment and site-directed mutagenesis. Mol. Endo. 7, 1617–1624.

Atkins, W.M., & Sligar, S.G. (1988). The roles of active site hydrogen bonding in cytochrome P-450$_{cam}$ as revealed by site-directed mutagenesis. J. Biol. Chem. 263, 18842–18849.

Atkins, W.M., & Sligar, S.G. (1989). Molecular recognition in cytochrome P-450: Alteration of regioselective alkane hydroxylation via protein engineering. J. Am. Chem. Soc. 111, 2715–2717.

Boddupalli, S.S., Estabrook, R.W., & Peterson, J.A. (1990). Fatty acid monooxygentaion by cytochrome P-450$_{BM-3}$. J. Biol. Chem. 265, 4233–4239.

Boddupalli, S.S., Oster, T., Estabrook, R.W., & Peterson, J.A. (1992a). Reconstitution of the fatty acid hydroxylation function of cytochrome P450$_{BM-3}$ utilizing its individual recombinant hemo- and flavoprotein domains. J. Biol. Chem. 267, 10375–10380.

Boddupalli, S.S., Pramanik, B.C., Slaughter, C.A., Estabrook, R.W., & Peterson, J.A. (1992b). Fatty acid monooxygenation of P450$_{BM-3}$: Product identification and proposed mechanisms for the sequential hydroxylation reactions. Arch. Biochem. Biophys. 292, 20–28.

Brewer, C.B., & Peterson, J.A. (1988). Single turnover kinetics of the reaction between oxycytochrome P-450$_{cam}$ and reduced putidaredoxin. J. Biol. Chem. 263, 791–798.

Estabrook, R.W., Cooper, D.Y., & Rosenthal, O. (1963). The light reversible carbon monoxide inhibition of the steroid c21-hydroxylase system of the adrenal cortex. Biochemische Zeitschrift 338, 741–755.

Garfinkel, D. (1958). Studies on pig liver microsomes. I. Enzymic and pigment composition of different microsomal fractions. Arch. Biochem. Biophys. 77, 493–509.

Gotoh, O. (1992). Substrate recognition sites in cytochrome P450 family 2 (CYP2) proteins inferred from comparative analyses of amino acid and coding nucleotide sequences. J. Biol. Chem. 267, 83–90.

Gotoh, O., & Fujii-Kuriyama, Y. (1989). Evolution, structure, and gene regulation of cytochrome P-450. In: Frontiers in Biotransformation (Ruchpaul, K., ed.), pp. 195–243, Taylor and Francis, London.

Graham-Lorence, S., Amarneh, B., White, R.E., Peterson, J.A., & Simpson, E.R. (1995). A three-dimensional model of aromatase cytochrome P450. Prot. Sci. 4, 1065–1080.

Griffin, B.W., & Peterson, J.A. (1972). Camphor binding by *Pseudomonas putida* cytochrome P-450. Kinetics and thermodynamics of the reaction. Biochemistry 11, 4740–4746.

Gunsalus, I.C., & Wagner, G.C. (1978). Bacterial P-450$_{cam}$ methylene monooxygenase components: Cytochrome M, putidaredoxin, and pituidaredoxin reductase. Methods Enzymol. 52, 166–188.

Hasemann, C.A., Ravichandran, K.G., Peterson, J.A., & Deisenhofer, J. (1994). Crystal structure and refinement of cytochrome P450$_{terp}$ at 2.3 A resolution. J. Mol. Biol. 236, 1169–1185.

Imai, M., Shimada, H., Watanabe, Y., Matsushima-Hibiya, Y., Makino, R., Koga, H., Horiuchi, T., & Ishimura, Y. (1989). Uncoupling of the cytochrome P-450$_{cam}$ monooxygenase reaction by a single mutation, threonine-252 to alanine or valine: A possible role of the hydroxy amino acid in oxygen activation. Proc. Natl. Acad. Sci. USA 86, 7823–7827.

Katagiri, M., Ganguli, B.N., & Gunsalus, I.C. (1968). A soluble cytochrome P450 functional in methylene hydroxylation. J. Biol. Chem. 243, 3543–3546.

Klingenberg, M. (1958). Pigments of rat liver microsomes. Arch. Biochem. Biophys. 75, 376–386.

Kronbach, T., Kemper, B., & Johnson, E.F. (1991). A hypervariable region of P450IIC5 confers progesterone 21-hydroxylase activity to P450IIC1. Biochemistry 30, 6097–6102.

Laughton, C.A., Neidle, S., Zvelebil, M.J.J.M., & Sternberg, M.J.E. (1990). A molecular model for the enzyme cytochrome P450-17α, a major target for the chemotherapy of prostatic cancer. Biochem. Biophys. Res. Commun. 171, 1160–1167.

Laughton, C.A., Zvelebil, M.J.J.M., & Neidle, S. (1993). A detailed molecular model for human aromatase. J. Steroid Biochem. Molec. Biol. 44, 399–407.

Lindberg, R.L.P., & Negishi, M. (1989). Alteration of mouse cytochrome P450coh substrate specificity by mutation of a single amino-acid residue. Nature 339, 632–634.

Miura, Y., & Fulco, A.J. (1975). ω-1, ω-2 or ω-3 hydroxylation of long-chain fatty acids, amides and alcohols by a soluble enzyme system from *Bacillus megaterium*. Biochim. Biophys. Acta 487, 487–494.

Narhi, L.O., & Fulco, A.J. (1986). Characterization of a catalytically self-sufficient 119,000-Dalton cytochrome P-450 monooxygenase induced by barbituates in *Bacillus megaterium*. J. Biol. Chem. *261*, 7160–7169.

Narhi, L.O., & Fulco, A.J. (1987). Identification and characterization of two functional domains in cytochrome P-450$_{BM-3}$, a catalytically self-sufficient monooygenase induced by barbituates in *Bacillus megaterium*. J. Biol. Chem. 262, 6683–6690.

Needleman, S.B., & Wunsch, C.D. (1970). A general method applicable to the search for similarities in the amino acid sequence of two proteins. J. Mol. Biol. 48, 443–453.

Nelson, D.R. (1996). INTERNET via http://DRNELSON.UTMEM.EDU/NELSONHOMEPAGE.HTML.

Nelson, D.R., & Strobel, H.W. (1987). Evolution of cytochrome P-450 proteins. Mol. Biol. Evol. 4, 572–593. Erratum (1988). Mol. Biol. Evol. 5, 199.

Nelson, D.R., & Strobel, H.W. (1988). On the membrane topology of vertebrate cytochrome P-450 proteins. J. Biol. Chem. 263, 6038–6050.

Nelson, D.R., Kamataki, T., Waxman, D.J., Guengerich, P., Estabrook, R.W., Feyereisen, R., Gonzalez, F.J., Coon, M.J., Gunsalus, I.C., Gotoh, O., Kuda, K., & Nebert, D.W. (1993). The P450 superfamily: Update on new sequences, gene mapping, accession numbers, early trivial names of enzymes, and nomenclature. DNA Cell Biol. 12, 1–51.

Omura, T., & Sato, R. (1962). A new cytochrome in liver microsomes. J. Biol. Chem. 237, 1375–1376.

Paulsen, M.D., Bass, M.D., & Ornstein, R.L. (1991). Analysis of active site motions from a 175 picosecond molecular dynamics simulation of camphor-bound cytochrome P450$_{cam}$. J. Biomol. Struct. Dyn. 9, 187–203.

Peterson, J.A. (1971). Camphor binding by *Pseudomonas putida* cytochrome P-450. Arch. Biochem. Biophys. 144, 678–693.

Peterson, J.A., Lu, J-Y, Geisselsoder, J., Graham-Lorence, S., Carmona, C., Witney, F., & Lorence, M.C. (1992). Cytochrome P450$_{terp}$. J. Biol. Chem. 267, 14193–14203.

Poulos, T.L., Finzell, B.C., & Howard, A.J. (1986). Crystal structure of substrate-free *Pseudomonas putida* cytochrome P-450. Biochemistry 25, 5314–5322.

Poulos, T.L., Finzell, B.C., & Howard, A.J. (1987). High-resolution crystal structure of cytochrome P450$_{cam}$. J. Mol. Biol. 195, 687–700.

Ravichandran, K.G., Boddupalli, S.S., Haseman, C.A., Peterson, J.A., & Deisenhofer, J. (1993). Crystal structure of hemoprotein domain of P450$_{BM-3}$, a prototype for microsomal P450's. Science 261, 731–736.

Sevrioukova, I., Truan, G., & Peterson, J.A. (1996). The flavoprotein domain of P450BM-3: Expression, purification, and properties of the FAD- and FMN-binding sub-domains. Biochemistry (in press).

Shen, S., & Strobel, H.W. (1993). Role of lysine and arginine residues of cytochrome P450 in the interaction between cytochrome P4502B1 and NADPH-cytochrome P450 reductase. Arch. Biochem. Biophys. 304, 257–265.

Sligar, S.G., & Murray, R.I. (1986). Cytochrome P-450$_{cam}$ and other bacterial P-450 enzymes. In: Cytochrome P-450: Structure, Mechanism, and Biochemistry (Ortiz de Montellano, P., ed.), pp. 429–502, Plenum Press, New York.

Song, W.C., Funk, C.D., & Brash, A.R. (1993). Molecular cloning of an allene oxide synthase: A cytochrome P450 specialized for the metabolism of fatty acid hydroperoxides. Proc. Natl. Acad. Sci. USA 90, 8519–8523.

Stayton, P.S., & Sligar, S.G. (1990). The cytochrome P-450$_{cam}$ binding surface as defined by site-directed mutagenesis and electrostatic modeling. Biochemistry 29, 7381–7386.

Straub, P., Johnson, E.F., & Kemper, B. (1993a). Hydrophobic side chain requirements for lauric acid and progesterone hydroxylation at amino acid 113 in cytochrome P450 2C2, a potential determinant of substrate specificity. Arch. Biochem. Biophys. 306, 521–527.

Straub, P., Lloyd, M., Johnson, E.F., & Kemper, B. (1993b). Cassette mutagenesis of a potential substrate recognition region of cytochrome P450 2C2. J. Biol. Chem. 268, 21997–22003.

Szczesna-Skorupa, E., Browne, N., Mead, D., & Kemper, B. (1988). Positive charges at the NH2 terminus convert the membrane-anchor signal peptide of cytochrome P-450 to a secretory signal peptide. Proc. Natl. Acad. Sci. USA 85, 738–742.

Szczesna-Skorupa, E., & Kemper, B. (1989). NH$_2$-terminal substitutions of basic amino acids induce translocation across the microsomal membrane and glycosylation of rabbit cytochrome P450IIC2. J. Cell. Biol. 108, 1237–1243.

Tsujita, M., & Ichikawa, Y. (1993). Substrate-binding region of cytochrome P-450$_{scc}$ (P-450 XIa1). Identification and primary structure of the cholesterol binding region in cytochrome P-450$_{scc}$. Biochim. Biophys. Acta 1161, 124–130.

Ullrich, V., & Hecker, M. (1990). A concept for the mechanism of prostacyclin and thromboxane A2 biosynthesis. In: Advances In Prostaglandin, Thromboxane, and Leukotriene Research (Samuelsson, B., et al., eds.), vol. 20, pp. 95–101, Raven Press, Ltd., New York.

Wada, A., & Waterman, M.R. (1992). Identification by site-directed mutagenesis of two lysine residues in cholesterol side chain cleavage cytochrome P450 that are essential for adrenodoxin binding. J. Biol. Chem. 267, 22877–22882.

DIVERSE MOLECULAR MECHANISMS REGULATE THE EXPRESSION OF STEROID HYDROXYLASE GENES REQUIRED FOR PRODUCTION OF LIGANDS FOR NUCLEAR RECEPTORS

Michael R. Waterman and Diane S. Keeney

Advances in Molecular and Cell Biology
Volume 14, pages 81–102.
Copyright © 1996 by JAI Press Inc.
All rights of reproduction in any form reserved.
ISBN: 0-7623-0113-9

I. INTRODUCTION

A major advance in biology during the 1980s has been the elucidation of biochemical and physiological mechanisms underlying the function of cellular receptors. Members of one group of receptors, those belonging to the steroid/thyroid/vitamin nuclear receptor superfamily, play essential roles in regulating the transcription of many genes involved in cellular growth and differentiation of the embryo, tissue-specific differentiated functions, and maintenance of homeostasis. These receptors are DNA-binding proteins of the zinc finger type and regulate gene transcription in a ligand-dependent manner. Table 1 catalogs genes that are transcriptionally regulated by steroid hormone ligands bound to their specific nuclear receptors. In a few instances, the biochemical details of how steroid hormones regulate gene transcription are being elucidated, and a general picture of steroid hormone action is emerging. Most recently, it has been demonstrated that ligand-activated nuclear receptors may not always function alone but in some cases may interact with other DNA-binding proteins to regulate gene transcription (Lucas and Granner, 1992; Imai et al., 1993). Surely, many additional discoveries will be made during the present decade that will describe a spectrum of unique biochemical mechanisms by which steroid hormones regulate transcription of specific genes. Thus we can expect that while activation of glucocorticoid receptors by glucocorticoids is a common event in most cells of most species, the specific biochemical processes by which the activated receptors regulate transcription may vary from one gene to another. Such complexity and diversity at the transcriptional level must be an important contributor to tissue-specific and species-specific gene expression mediated by steroid hormones.

An additional factor contributing to the complexity and diversity of steroid hormone action lies in the process of steroidogenesis itself and in those mechanisms regulating the availability of ligands for steroid hormone receptors. The controlled production of steroidal ligands, at levels required to regulate the vast array of genes in Table 1, must involve active and highly regulated processes. Evolution has led to a remarkable diversity among species in the capacity for steroid hormone biosynthesis and in endogenous steroid hormone profiles throughout development and adulthood. General schemes can be presented for the production of steroid hormones such as the generic biosynthetic pathways for the adrenal gland depicted in Figure 1, yet exceptions to such schemes are well known. The biochemical basis for variant pathways among species and tissues is poorly understood. Given that such variations exist as a result of evolutionary and/or physiological pressures, can we reasonably hope to find common mechanisms by which the biosynthesis of key steroidal ligands is regulated? As described in this chapter, the answer seems to be yes and no.

When considering steroid hormone biosynthesis, two broad aspects must be examined, the catalytic activities of steroidogenic enzymes and their levels of expression in different tissues and species. On one hand, steroidogenic enzymes

Table 1. Examples of Genes Regulated by Steroid Hormone Ligands and Their Nuclear Receptors

Ligand	Gene	Tissue (Species)
Glucocorticoid	synapsin	hippocampus (rat)
	MMTV	mouse mammary tumor virus
	lysozyme	oviduct (avian)
	α2μ-globulin	liver (rat)
	tryptophan oxygenase	liver (rat)
	tyrosine aminotransferase	liver (rat)
	metallothionein	liver (human)
	growth hormone	adenohypophysis (human)
	proopiomelanocortin	adenohypophysis (rat)
	prolactin	adenohypophysis
	adrenocorticotropin	adenohypophysis
	CRF	hypothalamus
	vasopressin	hypothalamus
	glucocorticoid receptor	
	casein genes	mammary gland (mouse, rat)
	whey acidic protein	mammary gland
	α-lactalbumin	mammary gland
	cyclooxygenase	
	lipocortin	
Mineralocorticoid	Na^+,K^+-ATPase	kidney (rat)
	citrate synthase	kidney (rat)
Progesterone	ovalbumin	oviduct (avian)
	ovomucoid	oviduct (avian)
	lysozyme	oviduct (avian)
	c-myc	oviduct (avian)
	uteroglobin	uterus (rabbit)
	casein genes	mammary gland (mouse, rat)
	α-lactalbumin	mammary gland
Estrogen	casein genes	mammary gland
	pS2	MCF-7 breast cancer cells
	albumin	liver (avian)
	apo-VLDL III	liver (avian)
	vitellogenin	liver (frog, avian)
	ovalbumin	oviduct (avian)
	ovomucoid	oviduct (avian)
	lysozyme	oviduct (avian)
	c-myc	breast cancer cells
	n-myc	uterus (rat)
	lactoferrin	uterus (mouse)
	prolactin	adenohypophysis

(continued)

Table 1. (Continued)

Ligand	Gene	Tissue (Species)
Androgen	major urinary proteins	liver (mouse)
	α2μ-globulin	liver (rat)
	ovalbumin	oviduct (avian)
	ovomucoid	oviduct (avian)
	lysozyme	oviduct (avian)
	β-glucuronidase	kidney (mouse)
	ornithine decarboxylase	kidney (mouse)
	alcohol dehydrogenase	kidney (mouse)
	steroid binding protein	prostate

such as the steroid hydroxylases catalyze specific functional activities which are regulated largely by the immediately available levels of steroidal substrates. This aspect of steroid hormone biosynthesis is discussed in detail elsewhere in this volume. This chapter, however, will focus on the other aspect, the regulation of steroid hydroxylase gene expression which will be seen to be multifactorial in

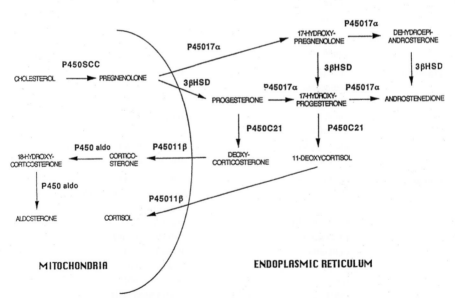

Figure 1. Steroidogenic pathways in the adrenal cortex of most species. P450scc, cholesterol side chain cleavage cytochrome P450; P45017α, 17α-hydroxylase cytochrome P450; 3βHSD, 3β-hydroxysteroid dehydrogenase/$\Delta^{5\rightarrow4}$ isomerase; P450c21, 21-hydroxylase cytochrome P450; P45011β, 11β-hydroxylase cytochrome P450; P450aldo, aldosterone synthase cytochrome P450. Modified from Zanger et al. (1992).

nature. Diversity in the levels of expression of these steroidogenic enzymes among tissues and species results in remarkable differences in the capacity for steroid hormone biosynthesis. Although the biochemical and physiological basis of such diversity is not yet explainable, attempts at understanding it are fundamental to understanding steroid hormone action. A number of investigators have contributed to our understanding of the regulation of steroid hydroxylase gene expression. Different laboratories have investigated the same steroid hydroxylase genes in different steroidogenic tissues and species, and a detailed chapter citing all of these results would be quite lengthy and boring if written by us. Thus, in this chapter we cite results that provide a general view of the complexity associated with steroid hydroxylase gene expression, with no intention to be comprehensive in the citation of important results.

II. STEROID HYDROXYLASE GENES

The steroid hydroxylase genes are members of the cytochrome P450 superfamily of genes (Nelson et al., 1996). Schematic representations of these genes and their CYP designations are given in Figure 2. Of the three gene families encoding microsomal steroid hydroxylases, *CYP17*, *CYP19*, and *CYP21*, each is known to produce a single steroid hydroxylase. The *CYP19* gene encoding aromatase cytochrome P450 (P450arom) is by far the largest known *P450* gene, and it exhibits a very complex regulatory pattern (Means et al., 1989, 1991). P450arom is also one of the most ancient eukaryotic *CYP* genes. The *CYP17* gene which encodes 17α-hydroxylase cytochrome P450 (P450c17) and the *CYP21* gene which encodes 21-hydroxylase cytochrome P450 (P450c21) have evolved more recently and are imagined to have arisen from a common progenitor *CYP* gene (Nebert et al., 1991). The microsomal steroid hydroxylases, like other microsomal forms of cytochrome P450, are inserted into the endoplasmic reticulum (ER) of steroidogenic cells by the signal recognition particle pathway (Sakaguchi et al., 1987). These proteins contain an amino-terminal hydrophobic signal anchor sequence that is required for proper insertion of the P450 into the ER membrane. This anchoring at the amino terminus during synthesis of the steroid hydroxylase is an important step in the normal folding pathway leading to a functional P450 (Clark and Waterman, 1992). Other hydrophobic regions within the protein may ultimately play a key role in proper association of these mixed-function oxidases with the ER membrane, but the most important association involves the signal anchor sequences. Of course, the activities of all microsomal steroid hydroxylases are supported by the ubiquitous ER flavoprotein, NADPH cytochrome P450 reductase.

The genes encoding mitochondrial steroid hydroxylases are all members of the CYP11 family which contains two known subfamilies. CYP11A encodes cholesterol side chain cleavage cytochrome P450 (P450scc), and CYP11B encodes 11β-hydroxylase cytochrome P450 (P450011β). An interesting recent development in the field of steroidogenesis is the discovery that in some species (including rats

MICHAEL R. WATERMAN and DIANE S. KEENEY

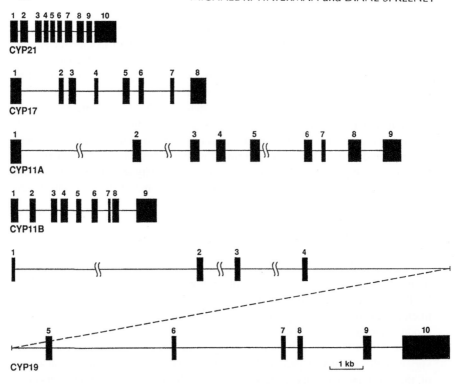

Figure 2. Structure of the steroid hydroxylase genes encoding CYP21, CYP17, CYP11A, CYP11B and CYP19 in humans. Exons are numbered and depicted by solid boxes. Introns are depicted by solid lines. Taken from Keeney and Waterman (1993).

and humans), there are two members of the CYP11B subfamily (Ogishima et al., 1989). One, the product of CYP11B1, is the traditional P45011β required for cortisol biosynthesis. The other, the product of CYP11B2, is expressed in the zona glomerulosa (Domalik et al., 1991) and is required for C18-hydroxylation of corticosterone and aldosterone biosynthesis. This enzyme has been named aldosterone synthase cytochrome P450 (P450aldo). Mitochondrial P450s including those described here are synthesized as higher molecular weight precursor proteins and processed proteolytically upon uptake into mitochondria. Thus, they contain amino-terminal extensions that are removed by mitochondrial proteases before they associate with the inner membrane of the mitochondrion where mature P450s reside as integral membrane proteins. The steps involved in translocation from the outer to the inner mitochondrial membranes are not yet well characterized. Unlike P450s localized in the ER, mitochondrial P450s do not contain amino-terminal signal anchor sequences. In fact, they do not contain any discernable anchor sequences, even though they apparently are integral proteins at the inner aspect of the inner mitochondrial membrane, facing the matrix compartment. An electron transport chain located in the matrix consisting of a flavoprotein (ferredoxin reductase) and an iron-sulfur protein (ferredoxin) transfers reducing equivalents from NADPH to P450scc, P45011β or P450aldo.

III. OVERVIEW OF STEROID HYDROXYLASE GENE EXPRESSION

Steroid hydroxylase genes are traditional eukaryotic genes having exons and introns (Fig. 2) and regulatory elements such as enhancers 5′ to (upstream from) the promoter and transcriptional start site. Mitochondrial steroid hydroxylases are encoded by nuclear genes, not by the mitochondrial genome. Transcriptional initiation start sites and chromosomal localizations have been reported for several steroid hydroxylase genes in different species. In the human genome, these genes are not clustered together: CYP17 is found on chromosome 10 (Matteson et al., 1986); CYP21 on chromosome 6 (White et al., 1985); CYP19 on chromosome 15 (Chen et al., 1988); CYP11A on chromosome 15 (Chung et al., 1986); and the CYP11B subfamily on chromosome 8 (Chua et al., 1987).

The cellular sites and patterns of expression of steroidogenic enzymes are complex both among tissues within a given species and among species. The most well studied example of steroid hydroxylases that function in different cell types is the adrenal cortex which produces principally glucocorticoids and mineralocorticoids. Steroidogenic cells within this gland secrete steroid hormones into the adrenal vascular sinusoids, and hence into the systemic circulation for delivery to a large number of target tissues where they serve as ligands for nuclear receptors regulating gene transcription (Table 1). The adrenal cortex also secretes androgens; however, the primary source of this and other sex steroid hormones such as estrogens and progestins is the gonads. The gonads are endocrine organs that secrete sex steroid hormones into the systemic circulation for delivery to peripheral target tissues, but, in addition, high intragonadal concentrations of steroid hormones are required for normal gonadal function (Gore-Langton and Armstrong, 1988; Sharpe et al., 1992). Testosterone is necessary for spermatogenesis in the testis and estrogen for follicular development in the ovary. Estrogen production which requires P450arom occurs in a variety of extragonadal sites as well, most notably in adipose tissue. The placenta is also a highly steroidogenic organ that is important for maintenance of pregnancy and, in some species, for parturition as well. Steroid hydroxylases are also known to exist at several sites in the brain (Stromstedt & Waterman, 1995), although their roles in the biochemistry in the brain have not yet been elucidated. More recently steroid hydroxylases have been localized to intestine (Keeney et al., 1995), stomach, and other nonsteroidogenic tissues, again without known functions.

The primary focus of studies on the regulation of steroidogenic enzymes has been the adrenal gland and gonads where the mechanisms involved in regulating the levels of steroid hydroxylases are multifactorial. It is possible to catalog four distinct levels or types of regulatory processes involved in transcription of steroid hydroxylase genes: developmental, tissue-specific, cAMP-dependent and cAMP-independent. While we may consider each of these independently, there is obviously a great deal of interaction among these different levels of regulation. Furthermore, unlike metabolic processes such as glycolysis, the tricarboxylic acid cycle and heme biosynthesis which occur in virtually every cell and may appear to share common regulatory events among species, we find that the transcriptional regulation of key genes in steroidogenesis is very species-dependent. While it is not immediately obvious why the same steroidogenic gene in different species has different biochemical systems regulating its expression, this diversity at the biochemical level must be related to the wide variation in physiology among mammalian species. Accordingly, we can expect that interactions among the four levels of transcriptional regulation noted above will also be species-dependent. For convenience we will discuss individually the different levels of transcriptional regulation, providing comments on their interplay where appropriate. Species variations will be presented where appropriate to broaden the scope; however, no attempt is made to be inclusive.

IV. DEVELOPMENTAL REGULATION

Detailed biochemical analysis of the developmental regulation of steroid hormone biosynthesis and function has not yet been carried out. Perhaps the clearest example of the role of steroid hormones in development is their requirement for the male phenotype. Development of the embryonic testis in eutherians is dependent on a gene product named SRY encoded on the Y chromosome (Goodfellow et al., 1992). It is the production of testosterone by Leydig cells of the fetal testis, however, that is required for full development of the male phenotype, in particular the secondary sex characteristics of adult males. At a critical time during embryogenesis, the onset of expression of several genes encoding steroidogenic enzymes must occur, and these include the CYP11A and *CYP17* genes along with others encoding non-P450 enzymes necessary for testosterone and dihydrotestosterone biosynthesis (Waterman and Keeney, 1992). Although the biochemical basis of these events is unknown, the timely onset of expression of these genes permits the male to achieve full reproductive capacity in adulthood. Nothing is known to date concerning the requirements for initiation of transcription of the steroid hydroxylase genes during embryonic development; however, the consequences of inadequate or complete failure to express one of these genes is exemplified by 46XY individuals having certain mutations in the *CYP17* gene. Such affected individuals do not exhibit the male phenotype but rather female secondary sex characteristics (Yanase et al., 1991). Even though the biochemical details of developmental regulation of steroid hydroxylase gene expression are not known, we have a glimpse of the biochemical events associated with fetal expression of these genes. As will be described in more detail in the next section, a novel DNA-binding protein called steroidogenic factor-1 (SF-1) has been localized in steroidogenic tissues, and binding sites for this protein have been identified in all steroid hydroxylase genes (Parker and Schimmer, 1993). SF-1 binding is thought to be essential for the expression of steroid hydroxylase genes in steroidogenic tissues and thus somehow involved in mediating tissue-specific expression of these genes. Quite interesting, however, is the observation that SF-1 transcripts are expressed in steroidogenic tissues of the mouse embryo prior to the appearance of steroid hydroxylase transcripts (Ikeda et al., 1993). Thus, this apparent overlap of tissue-specific and developmental aspects of the regulation of steroid hydroxylase gene expression may involve common factors like SF-1. Disruption of the SF-1 gene leads to absence of steroidogenic glands (adrenal and gonads) establishing a fundamental role for SF-1 in organogenesis in addition to potential roles in expression of genes encoding steroid hydroxylases.

One anecdotal observation concerning fetal expression of steroid hydroxylases comes from studies of bovine and ovine fetal adrenal glands. During early and late gestation, all of the expected steroid hydroxylases required for glucocorticoid biosynthesis are expressed in the fetal adrenal glands. During the middle third of gestation, however, P450c17 is absent, resulting in the absence of fetal cortisol biosynthesis during this period, a fact of unknown physiological consequence (Lund et al., 1988; Tangalakis et al., 1989). This transient disappearance of P450c17 in the fetal adrenal correlates with the absence of ACTH in fetal blood. As will be considered in detail in a subsequent section, ACTH plays a key role in transcriptional regulation of the steroid hydroxylase genes. Consequently, in the case of the bovine fetal adrenal gland, we see an interplay between developmental and cAMP-dependent mechanisms regulating steroid hydroxylase gene expression. Interestingly, P450c17 is present in the bovine fetal testis throughout development, including the gestational period during which it is absent from the fetal adrenal gland (Lund et al., 1988). This species-specific event is not observed in the human fetal adrenal gland (Voutilainen and Miller, 1986).

While not developmental in itself, maternal steroid hormone biosynthesis is essential for maintenance of pregnancy and thus normal development of the fetus. Examples include the conversion of the ovarian follicle from an estrogen-producing to a progesterone-producing tissue following ovulation. This results from the process of luteinization and involves the differentiation of ovarian glomerulosa and thecal cells into luteal cells. While levels of P450scc increase substantially during this transition from follicle to corpus luteum, those of P450c17 decrease and essentially disappear, further enhancing progesterone production (Rodgers et al., 1986, 1987). The production of large quantities of progesterone from the corpus luteum during pregnancy seems to be important for implantation and subsequent development of the fertilized egg. In short, we know little about the biochemical basis of developmental regulation of steroid hydroxylase gene expression. We do know, however, that the timely production of steroid hormones during development is of great importance, and detailed analyses of the developmental events necessary for steroid hormone biosynthesis will be extremely interesting.

V. TISSUE-SPECIFIC REGULATION

Tissue-specific expression of steroid hydroxylase genes leads to the presence of different steroidogenic pathways in different cell types. As indicated previously, the adrenal cortex is the source of both glucocorticoids and mineralocorticoids. By examining Fig. 1, it is obvious that there are specific steroid hydroxylases required for production of these two classes of steroids that must be expressed in this tissue. Notably, microsomal P450c21 and mitochondrial P45011β and P450aldo are found nearly exclusively in the adrenal cortex, although a recent report of P45011β in rat brain (Ozaki et al., 1991) suggests a somewhat broader tissue distribution. It can be seen in Fig. 1 that the key enzyme at the point of divergence of the glucocorticoid and mineralocorticoid biosynthetic pathways is P450c17. This unique P450 catalyzes two distinct reactions, 17α-hydroxylation which is required for glucocorticoid (but not mineralocorticoid) production and 17,20-lyase activity which converts C21 steroids to C19 steroids, a key step in production of androgenic and estrogenic sex steroid hormones. Thus P450c17 must be expressed not only in the adrenal cortex, but also in the gonads. In the adrenal cortex, the cell-specific expression of this enzyme is a key determinant regulating the production of different steroid hormones (glucocorticoids and mineralocorticoids) in the inner and outer cortical zones. The inner zones, the zona fasciculata and zona reticularis, express P450c17 and produce primarily glucocorticoids. The outer zone or zona glomerulosa does not express detectable levels of P450c17 and produces primarily aldosterone. Some species, including rat and humans, require P450aldo for aldosterone biosynthesis which is expressed exclusively in the zona glomerulosa (Domalik et al., 1991). Other species, for example bovine, apparently use one form of P45011β for both mineralocorticoid (aldosterone) and glucocorticoid (cortisol) biosynthesis (Morohashi et al., 1990; Mathew et al., 1990). Thus, a complex cell-specific pattern of expression of steroid hydroxylases exists within the adrenal cortex, superimposed on a species-specific component. Another example of species-specific variation of steroid hydroxylase gene expression is P450c17 in rats and mice where P450c17 is not expressed in the adrenals but is expressed in the gonads. The adrenal glands of these species thus cannot produce cortisol, and corticosterone is the major secreted glucocorticoid.

The testis is less complex than the adrenal in that only one cell type, the Leydig cells, is responsible for steroid hormone biosynthesis. Leydig cells of all species studied express both P450scc and P450c17 as required for testosterone biosynthesis; however, it is well known that species specificity is associated with the substrate specificity of P450c17. In all species studied, this enzyme can 17α-hydroxylate both the $\triangle$5-C21 steroid pregnenolone and the $\triangle$4-C21 steroid progesterone, but the ability of $\triangle$4- and $\triangle$5-17α-hydroxylated C21 intermediates to serve as substrate for the lyase reaction is species-dependent. In humans, cows, and sheep, P450c17 can convert only $\triangle$5-C21 steroids to androgens (17α-hydroxypregnenolone $\rightarrow$ dehydroepiandrosterone), but in pigs, rats, chickens, and fish (trout) both $\triangle$5- and $\triangle$4-17α-hydroxylated C21 steroids serve as substrates for androgen biosynthesis. Presumably the determinant of this species variation lies within the P450c17 primary sequence since only a single *CYP17* gene is known.

Ovarian follicular cells also display a complex cell-specific pattern of steroid hydroxylase gene expression, with two steroidogenic cell types showing different patterns of expression. Granulosa cells express P450scc and P450arom, while thecal cells express P450scc and P450c17. As noted previously, following ovulation the granulosa and thecal cells differentiate into luteal cells, and the resultant corpus luteum produces large quantities of progesterone. The shift from estrogen to progesterone production, coincident with luteinization, can be explained by the profound increase in the levels of expression of P450scc and the absolute disappearance of P450c17.

It is apparent, then, that the biochemical mechanisms involved in tissue-specific expression of steroid hydroxylases in different species are complex. Details of the biochemical requirements for expression of a given steroid hydroxylase gene in one cell type, but not in another, are not yet established; however, there have been some recent, interesting developments. During the past three years a DNA-binding protein that is expressed nearly exclusively in steroidogenic tissues has been found to be commonly associated with all steroid hydroxylase genes. Investigators working independently in two different laboratories have made this discovery in the course of their studies of steroid hydroxylase genes in different species. In the laboratories of Omura and Morohashi, several nuclear protein binding sites were identified within the 5'-flanking sequence of the bovine *CYP11B* gene (Honda et al., 1990). One of these sequences, designated Ad4, was also found in other bovine steroid hydroxylase genes (Morohashi et al., 1992). This led to the suggestion that the protein that binds this site, Ad4 binding protein (Ad4BP), must play an important role in expression of steroid hydroxylases. Independently, in their studies of the transcriptional regulation of murine steroid hydroxylases, Parker and Schimmer found one or more copies of the same conserved sequence (consensus DNA sequence AGGTCA) in all steroid hydroxylase genes (Rice et al., 1991). They named the DNA-binding protein associated with this site steroidogenic factor-1 (SF-1). Subsequently they cloned SF-1 and discovered that the encoded protein is a zinc finger orphan receptor (unknown ligand), a member of the same superfamily as the nuclear receptors for steroid and thyroid hormones and vitamins A and D (Lala et al., 1992). The primary sequence of SF-1 indicates that it is closely related to other orphan receptors and most closely related to a protein named FTZ-FA that regulates the *fushi tarazu* homeobox gene during development in *Drosophila* (Lala et al., 1992). In certain instances, SF-1 is found to interact with other closely related transcription factors, perhaps even by forming heterodimers required for transcription (Parker and Schimmer, 1993). In the 5'-flanking region of the murine *Cyp21* gene, two different SF-1 binding sites overlap with those of other transcription factors, a COUP-TF binding site and a NGFI-B binding site (Wilson et al., 1993). Ad4BP has also been cloned and found to be the bovine complement of SF-1 (Honda et al., 1993). In the bovine *CYP11B* gene, the Ad4 binding site interacts at a distance with an almost consensus cAMP response element (CRE) to maximize cAMP-responsive transcriptional activation of this gene (Hashimoto et al., 1992). At present it is believed that all steroid hydroxylase genes (and perhaps other steroidogenic cell-specific genes) from all species contain SF-1/Ad4 binding sites. This common regulatory element is thought to be important in the tissue-specific expression of steroidogenic genes, and its binding protein SF-1/Ad4BP may in some instances interact with other DNA-binding proteins (Parker and Schimmer, 1993). Thus, the action of SF-1 appears to be complex, perhaps overlapping all levels of the multifactorial regulation of steroid hydroxylase genes.

VI. cAMP-DEPENDENT REGULATION

The earliest studies on the regulation of steroidogenesis were those by Purvis and colleagues (1973a, 1973b) and Kimura (1969) who demonstrated that hypophysectomy of rats leads to a decline in the activities of steroid hydroxylases and that treatment of hypophysectomized animals with pharmacological doses of peptide hormones from the anterior pituitary gland can restore enzymatic activities. We now know that peptide hormones (ACTH in the adrenal and LH/hCG in the testis) activate adenylate cyclase leading to elevated intracellular levels of cAMP. One of the actions of cAMP is to enhance the transcription of the genes encoding the steroid hydroxylases. This appears to be the primary mechanism whereby optimal levels of steroid hydroxylases are maintained in steroidogenic cells. These cells are thus primed for the synthesis of steroid hormones on demand. Rather than thinking of this process as induction as is the case for *CYP* genes encoding drug metabolizing P450s, in steroidogenic tissues this is more correctly described as maintenance. Optimal steroidogenic capacity is maintained in steroidogenic cells by a cAMP-dependent process. While developmental and tissue-specific mechanisms provide for expression of steroid hydroxylases at the proper time and in the appropriate cellular sites, cAMP-dependent mechanisms sustain appropriate levels of expression of these genes throughout life.

The action of cAMP on steroid hydroxylase genes is at the transcriptional level (John et al., 1986), but it does not involve a consensus CRE (TGACGTCA) in most cases. The absence of consensus CREs in these genes was predicted for two reasons. First, the transcriptional response of steroid hydroxylase genes to cAMP takes several hours to be manifest, much slower than that observed for genes containing a consensus CRE which binds the transcription factor known as CREB. Second, cAMP-dependent transcription of the steroid hydroxylase genes, at least in the cAMP, but it is unaffected by protein synthesis inhibitors (Roesler et al., 1988). Thus, once steroid hydroxylase genes from different species were cloned and the search began for cAMP response elements in their 5′-flanking sequences, the expectation was that novel transcriptional response elements would be found that mediate the effects of cAMP in these genes. Because the transcriptional activation of different steroid hydroxylase genes by cAMP appeared to occur coordinately (John et al., 1986), it was also expected that a common, novel cAMP-responsive sequence (CRS) would be found in all of the steroid hydroxylase genes, just as the SF-1 binding sequence is found in all of these genes. At the time when "promoter-bashing" studies of the steroid hydroxylase genes began, the following working hypothesis was defined:

$$ACTH \xrightarrow[\text{cyclase}]{\text{adenylate}} cAMP \xrightarrow[\text{kinase A}]{\text{protein}} X{\sim}P \xrightarrow{?} \rightarrow \rightarrow \text{Steroid Hydroxylase Gene Transcription}$$

We now know that cAMP-dependent transcription of these genes involves the action of cAMP-dependent protein kinase (kinase A), which represents a common point for cAMP-dependent regulation of all steroid hydroxylase genes. The working hypothesis at points distal to kinase A as outlined above is incorrect, although we do not yet know to what extent. As will be described next, each gene seems to have its own unique CRS. At present, we still do not know how many steps exist between the action of kinase A and transcriptional activation through the different CRS sequences. Moreover, the same gene in different species sometimes uses different CRS elements. The degree of complexity of cAMP-dependent regulation for the steroid hydroxylase genes of different species remains to be determined.

Most of the studies of cAMP-responsiveness of *CYP* genes have focused on steroid hydroxylases from two species, mice and cows; incomplete data from studies of the human and rat genes are also available. The CRS elements of the steroid hydroxylase genes have been localized by coupling 5′-flanking regions of different steroid hydroxylase genes with reporter genes (e.g., chloramphenicol acetyltransferase, globin, growth hormone) and then transfecting these reporter gene constructs into cultured cells. The Y1 mouse adrenal tumor cell line has been used extensively for studies in which the CRS elements for several bovine genes, *CYP11A*, *CYP11B*, *CYP17*, and *CYP21*, have been localized. Surprisingly, unique CRS elements have been found for each of these genes.

The biochemical details of cAMP-dependent transcription of members of the *CYP11* gene family are not yet clear for many different species, but it is evident that in the bovine members of the two CYP11 subfamilies utilize very distinct cAMP-responsive systems. Bovine CYP11B contains an almost consensus CRE very near its promoter (Honda et al., 1990). It is assumed that this CRE-like element binds CREB leading to cAMP-dependent transcriptional activation; however, this activation is enhanced by interaction with the Ad4 binding site in this gene and its

CREB leading to cAMP-dependent transcriptional activation; however, this activation is enhanced by interaction with the Ad4 binding site in this gene and its binding protein, Ad4BP/SF-1. The biochemical basis of this interaction is not yet understood. Bovine CYP11A utilizes a very different CRS system, by comparison with CYP11B (Ahlgren et al., 1990). Evidence strongly suggests that binding of the ubiquitous transcription factor Sp1 to a sequence in the 5′-flanking region (−118 to −100 bp) of CYP11A is necessary for cAMP-responsiveness of this gene (Momoi et al., 1992; Venepally and Waterman, 1995). This finding is novel because no previous reports can be found in which Sp1 has been shown to be involved in cAMP-dependent transcriptional activation. Sp1 is a well characterized zinc finger DNA-binding protein (Kadonaga et al., 1988). It can be phosphorylated by a DNA-dependent protein kinase (not kinase A), but the role of phosphorylation in transcriptional activation has not been determined (Jackson et al., 1990). For a ubiquitous transcription factor like Sp1 to play such a role in cAMP-mediated transcription of bovine CYP11A, an accessory protein must be involved. This could be a steroidogenic-specific kinase or a protein that interacts with Sp1 and is itself phosphorylated, perhaps by kinase A. If an accessory protein is involved, it apparently is not a DNA-binding protein as none is detectable by gel shift analysis (Momoi et al., 1992; Venepally and Waterman, 1995). It is notable that in humans, the *CYP11A* gene contains a sequence very similar to that of the CRS in bovine CYP11A, but this sequence has not been shown to be involved in cAMP-responsiveness. The human CYP11A does, however, contain a CRE far upstream (−1697 to −1523 bp) that seems to be required for cAMP-responsiveness (Inoue et al., 1991; Moore et al., 1990).

The bovine *CYP17* gene contains two CRS elements (CRSI, −243 to −225 bp; CRSII, −80 to −40 bp) (Lund et al., 1990). Sequence homology between these two elements is not apparent and they do not compete nuclear protein binding to each other. Neither CRS contains a consensus CRE indicating that CREB does not bind to either CRS element. The CRSI and its binding proteins have been studied in some detail. Results of *in vitro* transcription assays have established that CRSI will enhance reporter gene transcription in steroidogenic cells but not in nonsteroidogenic cells (Zanger et al., 1991). Gel shift analysis, however, reveals the same gel shift pattern for CRSI in both steroidogenic and nonsteroidogenic cells (Zanger et al., 1992), indicating the presence of a binding protein(s) in both types of cells. Furthermore, addition of partially purified CRSI-binding protein from steroidogenic cells activates CRSI-dependent transcriptional activation of reporter genes in nonsteroidogenic cells (Zanger et al., 1991). Thus, the CRSI-binding protein from steroidogenic cells is somehow activated, enabling it to enhance CRSI-dependent transcription, while that from nonsteroidogenic cells is inactive. CRSI binding proteins have recently been purified and found to include Pbx1a and Pbx1b, which are homeodomain proteins (Kagawa et al., 1994). The present hypothesis is that Pbx1b forms a heterodimer with a yet to be characterized 60 kDa protein to mediate cAMP-dependent transcription through CRS1. CRSII binds SF-1 and another

```
                         -129              -115
          CYP21 CRS      CTGTTTTGTGGGCGG

                         -243                   -225
          CYP17 CRSI     TTGATGGACAGTGAGCAAG

                         -76                                          -40
          CYP17 CRSII    AGCATTAACATAAAGTCAAGGAGAAGGTCAGGGG

                         -118                                   -83
          CYP11A CRS     ACTGAGTCTGGGAGGAGCTGTGTGGGCTGGAGTCAG

                         -67        -59
          CYP11B CRS (Ad1)  TGACGTGAT
```

Figure 3. Comparison of nucleotide sequences of CRS elements in the four bovine steroid hydroxylase genes.

orphan receptor COUP-TF although their roles in cAMP dependent transcription remain to be elucidated (Bakke and Lund, 1995). In humans, CYP17 contains sequences similar to those of bovine CRSI and CRSII; however, deletion analyses of the promoter region of the human gene suggest that a functional CRS lies in a region (–184 to –104 bp) having no sequence homology to either CRSI or CRSII (Brentano et al., 1990).

The human *CYP21* gene contains a single CRS element within the region –129 to –96 bp (Kagawa and Waterman, 1990). This sequence contains two overlapping binding sites, one for a nuclear protein originally thought to be an adrenal-specific protein (ASP) and one for the ubiquitous transcription factor Sp1 (Kagawa and Waterman, 1991). ASP, found almost exclusively in adrenal cells, has been purified from nuclear extracts of mouse adrenal Y1 tumor cells by DNA affinity chromatography using the binding site –126 to –113 bp which does not bind to Sp1. It is 78 kDa protein and is present in very low abundance (Kagawa and Waterman, 1992). Addition of ASP to Y1 cells in an *in vitro* transcription system enhances transcription mediated by the CYP21 CRS. This effect is inhibited by an antibody prepared against ASP, further confirming a role for ASP in transcription of CYP21. The bovine CYP21 gene contains a CRS that is very closely related to that in the human gene, based on sequence homology and biochemical properties (Kagawa and Waterman, 1991).

A comparison of the nucleotide sequences of the CRS elements in the four bovine steroid hydroxylase genes (Fig. 3) reveals little similarity. For example, CRS elements from the bovine CYP17 and CYP21 genes are unrelated, even though these two genes are thought to have evolved from a common progenitor gene and thus were predicted to share common CRS elements (Zanger et al., 1992). Likewise, CYP11A and CYP11B exhibit no obvious relationship to each other by comparison of their CRS elements, even though they are members of the same gene family. Surprisingly, a significant degree of relatedness is observed between the CRS elements of CYP21 and CYP11A. Both bind ASP and Sp1, although the orientation of the two binding sites is reversed in the two genes (Fig. 4). In CYP21, the ASP

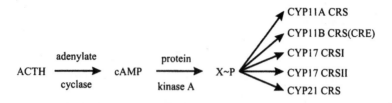

binding site is at the 5' end and that for Sp1 at the 3' end of the CRS, and the ASP site is required for cAMP-responsiveness. In CYP11A, the Sp1 binding site is at the 5' end and that for ASP is at the 3' end of the CRS, and the Sp1 site is required for cAMP responsiveness. The degree of overlap of the two sites is less extensive in CYP11A than in CYP21. At present, the significance of the relatedness between these two CRS elements from very divergent CYP genes is not clear.

Based on the results obtained from investigations of the 5'-flanking regions of the bovine steroid hydroxylase genes, the original working hypothesis has been revised dramatically. Results of recent studies indicate that the bovine gene encoding adrenodoxin (ferredoxin), the iron-sulfur protein required for activity of mitochondrial P450s, also utilizes a novel CRS for cAMP-responsiveness (Chen and Waterman, 1992). At present it is not possible to explain why this apparent degree of complexity exists among CRS elements of different steroid hydroxylase genes which respond in a coordinate manner to ACTH in the bovine adrenal cortex. These genes do represent different gene families that have arisen from different evolutionary paths. Nevertheless, it seems surprising that common biochemical systems are not involved in the chronic action of ACTH to coordinately regulate the transcription of all of the steroid hydroxylase genes, thereby maintaining optimal steroidogenic capacity in the adrenal cortex. The most distal, common point in the pathway from ACTH to activation of the steroid hydroxylase genes seems to be kinase A. In the gonads, we suspect that CYP11A and CYP17 are regulated through the same biochemical systems as in the adrenal gland, the only difference being the peptide hormone effectors.

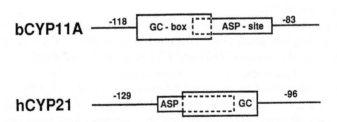

Figure 4. Schematic representation of the location and orientation of binding sites for Sp1 and ASP within the CRS elements of genes encoding CYP11A and CYP21B. Taken from Waterman et al. (1992).

There not only is an absence of common biochemical systems among the steroid hydroxylase genes in a given species but for the same gene among species as well. This is evident in comparing the results outlined above from studies of bovine steroid hydroxylase genes with the following results for murine genes. The mouse *Cyp11A* gene does not contain a CRS that can be assigned to a single 5'-flanking region. Rather, it requires a combination of elements which also play a role in constitutive expression (Rice et al., 1990). It does have a sequence closely related to the CRS in bovine CYP11A (Momoi et al., 1992), but it has not been determined that this sequence, by itself, can mediate cAMP-responsiveness. The mouse, unlike the cow, has two *Cyp11B* genes, one encoding P450011β and one encoding P450aldo. The gene encoding P450aldo utilizes a CRS that includes a CRE-like sequence at −56 to −49 bp (Rice et al., 1989). The transcriptional response of this gene to cAMP is relatively rapid, whereas that encoding P450011β, which lacks a CRE-like element, responds more slowly to stimulation by cAMP (Domalik et al., 1991). Mouse Cyp21 also appears to contain multiple elements involved in cAMP-responsiveness, one around −65 bp in the 5'-flanking region and one around −170 bp (Parissenti et al., 1993). The sequence at -65 bp binds both SF-1 and NGFI-B (Wilson et al., 1993), while the binding proteins at the -170 bp region remain to be elucidated. Neither of the regions correlate with the ASP binding site in bovine CYP21.

One way to rationalize the complexity of the biochemistry of cAMP-responsiveness of steroid hydroxylase gene expression is to recall the extent of species variation in endogenous steroid hormone profiles. This variation is most certainly a result of physiological differences that exist among species. Remembering that peptide hormone (cAMP-mediated) regulation of the steroid hydroxylases is the basis of maintenance of optimal steroidogenic capacity, perhaps unique regulatory mechanisms are necessary to generate the different steroid hormone profiles found in different species.

VII. cAMP-INDEPENDENT REGULATION

It is expected that this level of regulation provides a basal level of transcription on which cAMP-responsive transcriptional regulation is superimposed. The unique physiology associated with different species and steroidogenic tissues may be an important contributor to the variation observed in cAMP-independent regulation. There are numerous reports of regulation of steroid hydroxylase protein and mRNA levels by factors other than cAMP; insulin-like growth factor, epidermal growth factor, interferons, calcium, angiotensin II, phorbol esters, salt, androgens, and transforming growth factor-β are examples. The biochemical details of how these effectors control the expression of steroid hydroxylase genes are not yet known. However, it may not be surprising to find intra- and inter-species variation for cAMP-independent regulation as has been described for cAMP-dependent regulatory mechanisms. An example is found in the human and bovine *CYP17* genes.

Analysis of human CYP17 suggests that the cAMP-responsive sequence lies at −184 to −104 bp and is unrelated to the CRS elements (CRSI and CRSII) in bovine CYP17 (Brentano et al., 1990). Human CYP17 also contains a negative regulatory sequence element responsive to phorbol esters which has been localized at a site (−310 to −184 bp) distinct from the cAMP response element. In bovine CYP17, the negative regulatory element for phorbol esters is located within CRSI (Bakke and Lund, 1992). Interestingly, the sequence in the human gene corresponding to bovine CRSI lies at −310 to −184 bp. Thus, it appears that CYP17 in these two species may share a common negative response element for phorbol esters, but they do not share common CRS elements. Evidence from different experimental approaches suggests that ACTH can also influence the levels of steroid hydroxylases independently of cAMP, although the biochemical basis of this effect has not been established (Hanukoglu et al., 1990; Enyeart et al., 1993).

VIII. CONCLUSIONS

The question is posed in the introduction as to whether we can find common biochemical mechanisms underlying the multifactorial regulation of steroid hydroxylase genes. With respect to tissue-specific expression, common mechanisms involving SF-1 seem to be important for both the expression of different steroid hydroxylase genes within a species and for expression of a given steroid hydroxylase gene across species. For cAMP-dependent gene expression on the other hand, distinct CRS sequences associated with each of the steroid hydroxylase genes are responsible for maintenance of optimal steroidogenic capacity by the action of peptide hormones. Furthermore, the same gene in different species seems to utilize different cAMP-responsive biochemical systems. Thus, the answer to the question is both yes and no. The diversity of transcriptional regulatory mechanisms is most likely related to the different physiological requirements for steroid hormone biosynthesis in different species. It can be expected that the apparent complexity and diversity in regulatory mechanisms will be greatly amplified when considering the possible interactions among the different levels or types of transcriptional regulation. Because the characteristics of steroid hormone production do vary among species, perhaps we should expect variation in the biochemical mechanisms regulating steroidogenesis.

The genes represented in Table 1 are all regulated by steroid hormones through their respective nuclear receptors. There are many more members of the steroid hormone receptor superfamily than those that bind steroid hormones. Many others are orphan receptors and some of these likely bind derivatives of vitamins A and D as their endogenous ligands. Like those that bind steroid hormones, these other nuclear receptors and their endogenous ligands apparently also play important roles in development and homeostasis. It will not be surprising if members of P450 gene families other than the steroid hydroxylases described in Figure 2 will also be found to play key roles in metabolism (biosynthesis or inactivation) of important bioactive ligands.

ACKNOWLEDGMENTS

Studies carried out by the authors and reported in this chapter were supported by NIH Grant DK28350.

REFERENCES

Ahlgren, R., Simpson, E.R., Waterman, M.R., & Lund, J. (1990). Characterization of the promoter/regulatory region of the bovine *CYP11A (P450scc)* gene: Basal and cAMP-dependent expression. J. Biol. Chem. 265, 3313–3319.

Bakke, M., & Lund, J. (1992). A Novel 3',5' -cyclic adenosine monophosphate-responsive sequence in the bovine *CYP17* gene is a target of negative regulation by protein kinase C. Mol. Endocrinol. 6, 1323–1331.

Brentano, S.T., Picado-Leonard, J., Mellon, S.H., Moore, C.C.D., & Miller, W.L. (1990). Tissue-specific, cyclic adenosine 3',5'-monophosphate-induced, and phorbol ester-repressed transcription from the human P450c17 promoter in mouse cells. Mol. Endocrinol. 4, 1972–1979.

Chen, S., Besman, M.J., Sparkes, R.S., Zollman, S., Klisak, I., Mohandas, T., Hall, P.F., & Shively, J.E. (1988). Human aromatase: cDNA cloning, Southern blot analysis, and assignment of the gene to chromosome 15. DNA 7, 27–38.

Chen, J.-Y., & Waterman, M.R. (1992). Two promoters in the bovine adrenodoxin gene and the role of associated, unique cAMP-responsive sequences. Biochemistry 31, 2400–2407.

Chua, S.C., Szabo, P., Vitek, A., Grzeschik, K.-H., John, M., & White, P.C. (1987). Cloning of cDNA encoding steroid 11β-hydroxylase (P450c11). Proc. Natl. Acad. Sci. USA 84, 7193–7197.

Chung, B.-C., Matteson, K.J., Voutilainen, R., Mohandas, T.K., & Miller, W.L. (1986). Human cholesterol side-chain cleavage enzyme, P450scc: cDNA cloning, assignment of the gene to chromosome 15, and expression in the placenta. Proc. Natl. Acad. Sci. USA 83, 8962–8966.

Clark, B.J., & Waterman, M.R. (1992). Functional expression of bovine 17α-hydroxylase in COS 1 cells is dependent upon the presence of an amino-terminal signal anchor sequence. J. Biol. Chem. 34, 24568–24574.

Domalik, L.J., Chaplin, D.D., Kirkman, M.S., Wu, R.C., Liu, W.W., Howard, T.A., Seldin, M.F., & Parker, K.L. (1991). Different isozymes of mouse 11β-hydroxylase produce mineralocorticoids and glucocorticoids. Mol. Endocrinol. 5, 1853–1861.

Enyeart, J.J., Mlinar, B., & Enyeart, J.A. (1993). T-Type Ca^{2+} channels are required for adrenocorticotropin-stimulated cortisol production by bovine adrenal zona fasciculata cells. Mol. Endocrinol. 7, 1031–1040.

Goascogne, C.L., Sananes, N., Eychenne, B., Gouezou, M., Baulieu, E-E., & Robel, P. (1995). Androgen biosynthesis in the stomach: Expression of cytochrome P450 17α-hydroxylase/17,20-lyase messenger ribonucleic acid and protein, and metabolism of pregnenolone and progesterone by parietal cells of the rat gastric mucosa. Endocrinology 136, 1744–1752.

Goodfellow, P.N., Berkovitz, G., Hawkins, J.R., Harley, V.R., & Lovell-Badge, R. (1992). Sex Determination, Sex-reversal and SRY: A review. In: Gonadal Development and Function (Hillier, S. G., ed.), pp. 5–15. Serono Symposia Publications, volume 94, Raven Press, New York.

Gore-Langton, R.E., & Armstrong, D.T. (1988). Follicular steroidogenesis and its control. In: The Physiology of Reproduction (Knobil, E., & Neill, J., eds.), pp. 331–385, Raven Press, New York.

Hanukoglu, I., Feuchtwanger, R., & Hanukoglu, A. (1990). Mechanism of corticotropin and cAMP induction of mitochondrial cytochrome P450 system enzymes in adrenal cortex cells. J. Biol. Chem. 265, 20602–20608.

Hashimoto, T., Morohashi, K.-I., Takayama, K., Honda, S.-I., Wada, T., Handa, H., & Omura, T. (1992). Cooperative transcription activation between Ad1, CRE-like element, and other elements in *CYP11B* gene promoter. J. Biochem. 112, 573–575.

Honda, S., Morohashi, K., Nomura, M., Takeya, M., Kitajimi, M., & Omura, T. (1993). Ad4BP regulating steroidogenic P-450 genes is a member of steroid hormone receptor superfamily. J. Biol. Chem. 268, 7479–7502.

Honda, S., Morohashi, K., & Omura, T. (1990). Novel cAMP regulatory elements in the promoter region of bovine P45011β gene. J. Biochem. 108, 1042–1049.

Ikeda, Y., Lala, D.S., Luo, X., Kim, E., Moisan, M.-P., & Parker, K.L. (1993). Characterization of the mouse FTZ-F1 gene, which encodes a key regulator of steroid hydroxylase gene expression. Mol. Endocrinol. 7, 852–860.

Imai, E., Miner, J.N., Mitchell, J.A., Yamamoto, K.R., & Granner, D.K. (1993). Glucocorticoid receptor-cAMP response element-binding protein interaction and the response of the phosphoenolpyruvate carboxykinase gene to glucocorticoids. J. Biol. Chem. 268, 5353–5356.

Inoue, H., Watanabe, N., Higashi, Y., & Fujii-Kuriyama, Y. (1991). Structures of regulatory regions in the human cytochrome $P-450_{scc}$ (desmolase) gene. Eur. J. Biochem. 195, 563–569.

Jackson, S.P., MacDonald, J.J., Lees-Miller, S., & Tjian, R. (1990). GC box binding induces phosphorylation of Sp1 by a DNA-dependent protein kinase. Cell 63, 155–165.

John, M.E., John, M.C., Boggaram, V., Simpson, E.R., & Waterman, M.R. (1986). Transcriptional regulation of steroid hydroxylase genes by corticotropin. Proc. Natl. Acad. Sci. USA 83, 4715–4719.

Kadonaga, J.T., Courey, A.J., Ladika, J., & Tjian, R. (1988). Distinct regions of Sp1 modulate DNA binding and transcriptional activation. Science 242, 1566–1570.

Kagawa, N., & Waterman, M.R. (1990). cAMP-dependent transcription of the human CYP21B (P-450_{c21}) gene requires a cis-regulatory element distinct from the consensus cAMP-regulatory element. J. Biol. Chem. 19, 11299–11305.

Kagawa, N., & Waterman, M.R. (1991). Evidence that an adrenal-specific nuclear protein regulates cAMP responsiveness of the human CYP21B (P450c21) gene. J. Biol. Chem. 266, 11199–11204.

Kagawa, N., & Waterman, M.R. (1992). Purification and characterization of a transcription factor which appears to regulate cAMP responsiveness of the human CYP21B gene. J. Biol. Chem. 267, 25213–21219.

Kagawa, N., Ogo, A., Takahashi, Y., Iwamatsu, A., & Waterman, M.R. (1994). a cAMP-responsive sequence (CRS1) of CYP17 is a cellular target for the homeodomain protein Pbx1. J. Biol. Chem. 269, 18716–18719.

Keeney, D.S., Ikeda, Y., Waterman, M.R., & Parker, K. (1995). Cholesterol side-chain cleavage cytochrome P450 gene expression in primitive gut of the mouse embryo does not require steroidogenic factor-1. Mol. Endocrinol. 19, 1091–1098.

Keeney, D.S., & Waterman, M.R. (1993). Regulation of steroid hydroxylase gene expression: Importance to physiology and disease. Pharmacol. Ther. (in press).

Kimura, T. (1969). Effects of hypophysectomy and ACTH administration on the level of adrenal cholesterol side-chain desmolase. Endocrinology 85, 492–499.

Lala, D.S., Rice, D.A., & Parker, K.L. (1992). Steroidogenic factor 1, a key regulator of steroidogenic enzyme expression, is the mouse homolog of fushi tarazu-factor 1. Mol. Endocrinol. 6, 1249–1258.

Lucas, P.C., & Granner, D.K. (1992). Hormone response domains in gene transcription. Annu. Rev. Biochem. 1131–1173.

Lund, J., Faucher, D.J., Ford, S.P., Porter, J.C., Waterman, M.R., & Mason, J.I. (1988). Developmental expression of bovine adrenocortical steroid hydroxylases: Regulation of P45017α expression leads to episodic fetal cortisol production. J. Biol. Chem. 263, 16195–16201.

Lund, J., Ahlgren, R., Wu, D., Kagimoto, M., Simpson, E.R., & Waterman, M.R. (1990). Transcriptional regulation of the bovine CYP17 (P-$450_{17\alpha}$) gene. J. Biol. Chem. 265, 3304–3312.

Luo, X., Ikeda, Y., & Parker, K.L. (1994). A cell-specific nuclear receptor is essential for adrenal and gonadal development and sexual differentiation. Cell 77, 481–490.

Mathew, P.A., Mason, J.I., Trant, J.M., Sanders, D., & Waterman, M.R. (1990). Amino acid substitutions Phe[66] → Leu and Ser[126] → Pro abolish cortisol and aldosterone synthesis by bovine cytochrome P450$_{11\beta}$. J. Biol. Chem. 265, 20228–20233.

Matteson, K.J., Picado-Leonard, J., Chung, B.-C., Mohandas, T.K., & Miller, W.L. (1986). Assignment of the gene for adrenal P450c17 (steroid 17α-hydroxylase/17,20-lyase) to human chromosome 10. J. Clin. Endocrinol. Metab. 63, 789–791.

Means, G.D., Mahendroo, M.S., Corbin, C.J., Mathis, J.M., Powell, F.E., Mendelson, C.R., & Simpson, E.R. (1989). Structural analysis of the gene encoding human aromatase cytochrome P450, the enzyme responsible for estrogen biosynthesis. J. Biol. Chem. 264, 19385–19391.

Means, G.D., Kilgore, M.W., Mahendroo, M.S., Mendelson, C.R., & Simpson, E.R. (1991). Tissue-specific promoters regulate aromatase cytochrome *P450* gene expression in human ovary and fetal tissues. Mol. Endocrinol. 5, 2005–2013.

Momoi, K., Waterman, M.R., Simpson, E.R., & Zanger, U.M. (1992). 3',5'-Cyclic adenosine monophosphate-dependent transcription of the *CYP11A* (cholesterol side chain cleavage cytochrome P450) gene involves a DNA response element containing a putative binding site for transcription factor Sp1. Mol. Endocrinol. 6, 1682–1690.

Moore, C.C.D., Brentano, S.T., & Miller, W.L. (1990). Human *P450scc* gene transcription is induced by cAMP and repressed by 12-O-tetradecanoylphorbol-13-acetate and A23187 through independent *cis*-elements. Mol. Cell. Biol. 10, 6013–6023.

Morohashi, K., Nonaka, Y., Kirita, S., Hatano, O., Takakusu, A., Okamoto, M., & Omura, T. (1990). Enzymatic activities of P-450(11β)s expressed by two cDNAs in COS-7 cells. J. Biochem. 107, 635–640.

Morohashi, K., Honda, S., Inomata, Y., Handa, H., & Omura, T. (1992). A common trans-acting factor, Ad4BP, to the promoters of steroidogenic P450s. J. Biol. Chem. 267, 17913–17919.

Nebert, D.W., Nelson, D.R., Coon, M.J., Estabrook, R.W., Feyereisen, R., Fujii-Kuriyama, Y., Gonzalez, F.J., Guengerich, F.P., Gunsalus, I.C., Johnson, E.F., Loper, J.C., Sato, R., Waterman, M.R., & Waxman, D.J. (1991). The P450 superfamily: Update on new sequences, gene mapping, and recommended nomenclature. DNA Cell Biol. 10, 1–14.

Nelson, D.R., Koymans, L., Kamataki, T., Stegeman, J.J., Feyereisen, R., Waxman, D.J., Waterman, M.R., Gotoh, O., Coon, M.J., Estabrook, R.W., Gunsalus, I.C., & Nebert, D.W. (1996). P450 superfamily: Update on new sequences, gene mapping, accession numbers, and nomenclature. Pharmacogenetics 6, 1–42.

Ogishima, T., Mitani, F., & Ishimura, Y. (1989). Isolation of aldosterone synthase cytochrome P450 from zona glomerulosa mitochondria of rat adrenal cortex. J. Biol. Chem. 264, 10935–10938.

Ozaki, H.S., Iwahashi, K., Tsubaki, M., Fukui, Y., Ichikawa, Y., & Takeuchi, Y. (1991). Cytochrome P45011β in rat brain. J. Neurosci. Res. 28, 518–524.

Parissenti, A., Parker, K.L., & Schimmer, B.P. (1993). Identification of promoter elements in the mouse 21-hydroxylase (*Cyp21*) gene that require a functional cAMP-dependent protein kinase. Mol. Endocrinol. 7, 283–290.

Parker, K.L., & Schimmer, B.P. (1993). Transcriptional regulation of the adrenal steroidogenic enzymes. Trends Endocrinol. Metab. 4, 46–50.

Purvis, J.L., Canick, J.A., Latif, S.A., Rosenbaum, J.H., Hologgitas, J., & Menard, R.H. (1973a). Lifetime of microsomal cytochrome P450 and steroidogenic enzymes in rat testis as influenced by human chorionic gonadotropin. Arch. Biochem. Biophys. 159, 39–49.

Purvis, J.L., Canick, J.A., Mason, J.I., Estabrook, R.W., & McCarthy, J.L. (1973b). Lifetime of adrenal cytochrome P450 as influenced by ACTH. Ann. N. Y. Acad. Sci. 212, 319–342.

Rice, D.A., Aitken, L.D., Vandenbark, G.R., Mouw, A.R., Franklin, A., Schimmer, B.P., & Parker, K.L. (1989). A cAMP-responsive element regulates expression of the mouse steroid 11β-hydroxylase gene. J. Biol. Chem. 264, 14011–14015.

Rice, D.A., Kirkman, M.S., Aitken, L.D., Mouw, A.R., Schimmer, B.P., & Parker, K.L. (1990). Analysis of the promoter region of the gene encoding mouse cholesterol side chain cleavage enzyme. J. Biol. Chem. 265, 11713–11720.

Rice, D.A., Mouw, A.R., Bogerd, A., & Parker, K.L. (1991). A shared promoter element regulates the expression of three steroidogenic enzymes. Mol. Endocrinol. 5, 1552–1561.

Rodgers, R.J., Waterman, M.R., & Simpson, E.R. (1986). Cytochromes P450scc, P45017α, adrenodoxin and reduced nicotinamide adenine dinucleotide phosphate-cytochrome P450 reductase in bovine follicles and corpora lutea. Changes in specific contents during the ovarian cycle. Endocrinology 118, 1366–1374.

Rodgers, R.J., Waterman, M.R., & Simpson, E.R. (1987). Levels of messenger ribonucleic acid encoding cholesterol side chain cleavage cytochromes P450, 17α-hydroxylase cytochrome P450, adrenodoxin, and low density lipoprotein receptor in bovine follicles and corpora lutea throughout the ovarian cycle. Mol. Endocrinol. 1, 274–279.

Roesler, W.J., Vandenbar, G.R., & Hanson, R.W. (1988). Cyclic AMP and the induction of eucaryotic gene transcription. J. Biol. Chem. 263, 9063–9066.

Sakaguchi, M., Mihara, K., & Sato, R. (1987). A short amino-terminal segment of microsomal cytochrome P-450 functions both as an insertion signal and as a stop-transfer sequence. EMBO J. 6, 2425–2431.

Sharpe, R.M., McKinnell, C., Millar, M., Maddocks, S., Kerr, J.B., Hargreave, T.B., & Saunders, P.T.K. (1992). Intratesticular androgen action. In: Gonadal Development and Function (Hillier, S.G., ed.), pp. 127–138. Serono Symposia Publications, volume 94, Raven Press, New York.

Strömstedt, M., & Waterman, M.R. (1995). Messenger RNAs encoding steroidogenic enzymes are expressed in rodent brain. Mol. Brain Res. 34, 75–88.

Tangalakis, K., Coghlan, J.R., Connell, J., Crawford, R., Darling, P., Hammond, V.E., Haralambidis, J., Penschow, J., & Wintour, E.M. (1989). Tissue distribution and levels of gene expression of three steroid hydroxylases in ovine fetal adrenal glands. Acta Endocrinol. 120, 225–232.

Venepally, P., & Waterman, M.R. (1995). Two Sp1-binding sites mediate cAMP-induced transcription of the bovine CYP11A gene through the protein kinase A signaling pathway. J. Biol. Chem. 270, 25402–25410.

Voutilainen, R., & Miller, W.L. (1986). Developmental expression of genes for the steroidogenic enzymes P450scc (20,20-desmolase), P450c17 (1α-hydroxylase/17,20-lyase), and P450c21 (21-hydroxylase) in the human fetus. J. Clin. Endocrinol. Metab. 63, 1145–1150.

Waterman, M.R., Kagawa, N., Zanger, U.M., Momi, K., Lund, J., & Simpson, E.R. (1992). Comparison of cAMP-responsive DNA sequences and their binding proteins associated with expression of the bovine CYP17 and CYP11A and human CYP21 genes. J. Steroid Biochem. Mol. Biol. 43, 931–935.

Waterman, M.R., & Keeney, D.S. (1992). Genes involved in androgen biosynthesis and the male phenotype. Horm. Res. 38, 217–221.

White, P.C., Grossberger, D., Onufer, B.J., Chaplin, D.D., New, M.L., Dupont, B., & Strominger, J.L. (1985). Two genes encoding steroid 21-hydroxylase are located near the genes encoding the fourth component of complement in man. Proc. Natl. Acad. Sci. USA 82, 1089–1093.

Wilson, T., Mouw, A.R., Weaver, C.A., Milbrandt, J., & Parker, K.L. (1993). The orphan nuclear receptor NGFI-B regulates steroid 21-hydroxylase gene expression. Mol. Cell Biol. 13, 861–868.

Yanase, T., Simpson, E.R., & Waterman, M.R. (1991). 17α-hydroxylase/17,20-lyase deficiency: From clinical investigation to molecular definition. Endocrine Rev. 12, 91–108.

Zanger, U.M., Lund, J., Simpson, E.R., & Waterman, M.R. (1991). Activation of transcription in cell-free extracts by a novel cAMP-responsive sequence from the bovine CYP17 gene. J. Biol. Chem. 266, 11417–11420.

Zanger, U.M., Kagawa, N., Lund, J., & Waterman, M.R. (1992). Distinct biochemical mechanisms for cAMP-dependent transcription of CYP17 and CYP21. FASEB J. 6, 719–713.

CYTOCHROME P450scc AND REGULATION OF CHOLESTEROL CONVERSION TO STEROID HORMONES

Colin R. Jefcoate

Advances in Molecular and Cell Biology
Volume 14, pages 103–148.
Copyright © 1996 by JAI Press Inc.
All rights of reproduction in any form reserved.
ISBN: 0-7623-0113-9

1. INTRODUCTION

Steroids formed from cholesterol are synthesized in many animal cell types each with specific regulatory functions and each under distinct mechanisms of endocrine control. In each tissue the rate of conversion of cholesterol to pregnenolone through cytochrome $P450_{scc}$ (CYP11Al) determines the total formation of steroids, while the final steroid product is determined by the extent of expression of the individual genes in the steroidogenic pathway. The most extensive synthesis of steroids per cell occurs in the fasciculata cells of the adrenal cortex which produce glucocorticoids within a few minutes after physiological or environmental stress. This rapid response to stress mediated by elevation of ACTH is met by high levels of steroidogenic enzymes in both mitochondria and endoplasmic reticulum and a great facility to mobilize serum cholesterol (Jefcoate et al., 1987; 1991). The adrenal glomerulosa cells also have high levels of $P450_{scc}$ but these cells are specialized to produce aldosterone which serves as one of several regulators of blood pressure. These cells also respond to additional stimulants that are responsive to renal cardiovascular functions (Angiotensin, K^+). The distinct biosynthetic demands are met by differences in downstream enzymes (CYP11B2 but no CYP17).

Other tissues which express $P450_{scc}$ do not show such rapid fluctuations in steroid synthesis. The ovary produces estrogens and progesterone while individual follicles undergo a progressive program of differentiation over the course of several days under partial control of these steroids (Keyes and Wiltbank, 1987; Richards, 1994). The testis produces testosterone in the Leydig cells without sudden fluctuations in demand, although exhibiting a diurnal fluctuation in hormonal stimulation (Dufau, 1988). The placenta has a single program of differentiation culminating in birth and involving a slow progressive increase in estrogen production (Gibori et al., 1988). The fetal adrenal produces changes in glucocorticoid synthesis which also regulate fetal development and parturition in a species-specific manner (Rogler and Pintar, 1993). The role of the placenta in steroid production differs greatly between species: in primates there is substantial steroid synthesis from cholesterol and expression of $P450_{scc}$ while in rodents there is only metabolism of progesterone provided by the corpus luteum (ovary). Specific neuronal tissues in the brain also convert cholesterol to pregnenolone and progesterone in very low amounts and this may be associated with the regulation of GABA receptors by 5a-reduced progesterone (Robel et al., 1987).

II. BIOCHEMISTRY OF P450scc

Cytochrome $P450_{scc}$ is an unusual enzyme in many respects. The substrate cholesterol has a structural role in mitochondrial membranes while the reaction involves multiple mono-oxygenase reactions and is also dependent on reducing equivalents generated within the mitochondrial matrix. The overall activity may therefore be susceptible to many influences in addition to the level of expression of the enzyme.

A. Protein

$P450_{scc}$ shares with $P450_{AROM}$ the function of catalyzing multiple consecutive oxygenase steps. This means that established intermediates 22R-hydroxy cholesterol (22R HOC) and 20S,20R-dihydroxy cholesterol are metabolized more rapidly than they dissociate from the protein. Each intermediate indeed binds with high affinity and is metabolized more rapidly than cholesterol thus avoiding build-up of these intermediates (Heyl et al., 1986). Other side chain hydroxy-cholesterol derivatives are also good substrates and provide the advantage that, unlike cholesterol, they are not restrained by tight binding to lipids and insolubility in water. Unlike cholesterol, their metabolism is not enhanced by ACTH (Jefcoate et al., 1974), a major piece of evidence indicating that substrate availability rather than cytochrome $P450_{scc}$ turnover limits pregnenolone formation.

As with other biosynthetic P450 cytochromes the substrate binds with a conformational change which converts the heme from a low spin Fe^{3+} state to an almost completely high spin state (Jefcoate et al., 1976). Consistent with changes observed in the X-ray structures of $P450_{CAM}$ (Poulos, 1991) this low spin state probably corresponds to a water-Fe^{3+} complex. When cholesterol tightly occupies the substrate binding pocket this water is excluded converting the heme to a pentacoordinate thiolate complex. Although the product pregnenolone forms a low spin complex which is competitive with cholesterol there is also evidence for a weaker secondary binding by pregnenolone in which cholesterol is not displaced (Jefcoate, 1982). One possibility is that pregnenolone changes the spin state through binding to a docking site similar to that described for P450 BM3 in Chapter 3. The various hydroxycholesterol analogs form mixed spin states either through interaction of the side chain hydroxyl with Fe^{3+} (22 or 20 hydroxyls) or through a less exact fit in the substrate site (25-hydroxyl) which then allows entry of H_2O. Cholesterol preferentially binds with the reduced form of $P450_{scc}$ thus raising the redox potential and facilitating reduction (Heyl et al., 1986). Interestingly hydroxylated reaction intermediates bind more readily to the oxidized form. Cholesterol has only a small effect on the binding of O_2 or CO to the reduced heme of $P450_{scc}$ (Tuckey and Kamin,

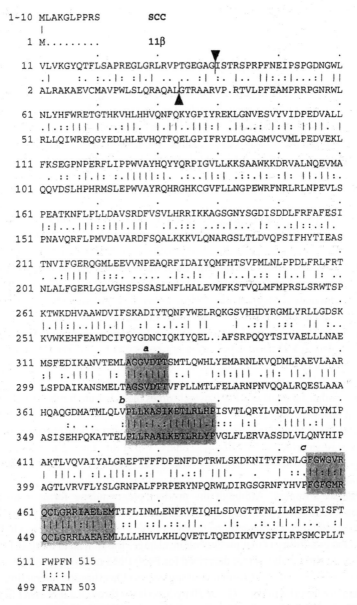

```
1-10   MLAKGLPPRS          SCC
         |
  1    M........           11β

 11    VLVKGYQTFLSAPREGLGRLRVPTGEGAGISTRSPRPFNEIPSPGDNGWL
        .|      :: :..|: :|.|  ..  .. :|  :.  |..  ||:.:|....:| ||
  2    ALRAKAEVCMAVPWLSLQRAQALGTRAARVP.RTVLPFEAMPRRPGNRWL

 61    NLYHFWRETGTHKVHLHHVQNFQKYGPIYREKLGNVESVYVIDPEDVALL
        .|.::|||  |  ..:||.  |.||..|||:| .||..:  |:|: |||||. |
 51    RLLQIWREQGYEDLHLEVHQTFQELGPIFRYDLGGAGMVCVMLPEDVEKL

111    FKSEGPNPERFLIPPWVAYHQYYQRPIGVLLKKSAAWKKDRVALNQEVMA
        . ::  :|.|: :.|||||:|. .:..||:| .::.|: :|: ||.||:.
101    QQVDSLHPHRMSLEPWVAYRQHRGHKCGVFLLNGPEWRFNRLRLNPEVLS

161    PEATKNFLPLLDAVSRDFVSVLHRRIKKAGSGNYSGDISDDLFRFAFESI
        |:|...|||::|||.||| .|.::: ...|... |: ..:|::::|.
151    PNAVQRFLPMVDAVARDFSQALKKKVLQNARGSLTLDVQPSIFHYTIEAS

211    TNVIFGERQGMLEEVVNPEAQRFIDAIYQMFHTSVPMLNLPPDLFRLFRT
        . .:|||| |:::. .....  .|:.|:  ||...|.:: :|..| |: ..
201    NLALFGERLGLVGHSPSSASLNFLHALEVMFKSTVQLMFMPRSLSRWTSP

261    KTWKDHVAAWDVIFSKADIYTQNFYWELRQKGSVHHDYRGMLYRLLGDSK
        |.||:|..|||.|| :| :.|.:| ||  .::| ::: || :.
251    KVWKEHFEAWDCIFQYGDNCIQKIYQEL..AFSRPQQYTSIVAELLLNAE

                        a
311    MSFEDIKANVTEMLAGGVDTTSMTLQWHLYEMARNLKVQDMLRAEVLAAR
        :| :.|||| |: ||:||||  :.| |:|:||| .||: ||.| |||
299    LSPDAIKANSMELTAGSVDTTVFPLLMTLFELARNPNVQQALRQESLAAA

                      b
361    HQAQGDMATMLQLVPLLKASIKETLRLHPISVTLQRYLVNDLVLRDYMIP
        :. ..    :|||:|.:||||||-|:::  |:|   .||||.:| ||
349    ASISEHPQKATTELPLLRAALKETLRLYPVGLFLERVASSDLVLQNYHIP

                                                     c
411    AKTLVQVAIYALGREPTFFFDPENFDPTRWLSKDKNITYFRNLGFGWGVR
        |  |||.| :|.|||:|.:| ||.::| |||.   .   | :: .||:|:|
399    AGTLVRVFLYSLGRNPALFPRPERYNPQRWLDIRGSGRNFYHVPFGFGMR

461    QCLGRRIAELEMTIFLINMLENFRVEIQHLSDVGTTFNLILMPEKPISFT
        |||||:||.|| || ::|  ::|.::.||.    .|:  .:.:||.|...  :|
449    QCLGRRLAEAEMLLLLHHVLKHLQVETLTQEDIKMVYSFILRPSMCPLLT

511    FWPFN 515
        |:::|
499    FRAIN 503
```

Figure 1. Sequence comparison for human P450scc (upper) with human P45011β1 (lower). Proteolytic cleavage sites are indicated by arrows. Boxes a, b and c designate conserved sequences associated with respectively O_2 binding/proton transfer, electron transfer and heme binding.

1983). However hydroxylcholesterol reaction intermediates strongly inhibit these complexes implicating the presence of the 22R-hydroxyl close to the heme.

The reductant adrenodoxin binds with high affinity producing a small increase in high spin state but more importantly enhances the affinity for cholesterol. Affinity for cholesterol is also increased by a decrease in pH (Jefcoate, 1982) and by binding of cardiolipin or more hydrophilic analogs that retain the same head-group. Interestingly when the smaller pregnenolone occupies the binding site there is no effect of pH on the interaction and binding of ADX is inhibited. The pH effect corresponds to a pk_a of about 7.0 and analysis indicates formation of a high pH, largely low spin, cholesterol complex where presumably the active site again opens to let in H_2O. It remains possible that the effects of low pH, cardiolipin and ADX all operate through changes in a single unique protein domain.

The amino acid sequence is 77 percent identical between human and rat sequences and is 39 percent identical to $P450_{11\beta}$ (Miller, 1988) (Fig. 1). Protein modification (Tuls et al., 1989) and site-directed mutagenesis (Wada and Waterman, 1992) implicate positively charged amino acids (lysines) at position 377 and 381 in the binding of ADX. Site-directed mutagenesis of human ADX has identified aspartates 76 and 79 as critical to this interaction (Coghland and Vickery, 1991, 14b).This alignment of the two proteins for electron transfer is shown in Fig. 2. Fluorescence energy transfer calculations indicate a distance of about 26-A between the ADX binding site and the heme of $P450_{scc}$ (Tuls et al., 1989). The surrounding

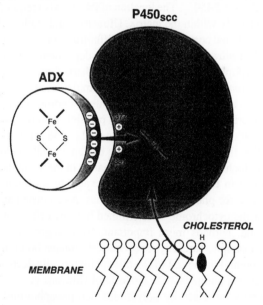

Figure 2. Alignment of negatively charged amino acids on adrendoxin with positively charged amino acids on P450scc.

domain of 12 amino acids is conserved with only a single conservative substitution in $P450_{11\beta}$ (CYP11B1) and CYP27. (Shown in Fig. 1) The sequence is also conserved between species. While $P450_{scc}$ exhibits only very low similarity to $P450_{CAM}$, tyrosine 93 of $P450_{scc}$ is located in an equivalent position to tyrosine 96 of the bacterial cytochrome where it forms a hydrogen bond with camphor (Pikuleva et al., 1994). Mutagenesis of this amino acid in $P450_{scc}$ effects a 5-fold increase in K_d for cholesterol suggesting a similar role in $P450_{scc}$. The heme binding sequence of $P450_{scc}$ is identical in $P450_{11\beta}$ and retains high identity even when extended to 25 amino acids (Fig. 1, sequence c). This region contains additional positively charged amino acids that are probably involved in binding adrenodoxin. The threonine residue found in the center of the I-helices of $P450_{CAM}$ and $P450_{BM3}$ (see Chapter 3) which is involved in O_2-binding is also part of a sequence which is highly conserved with $P450_{11\beta}$ (Fig. 1, sequence a).

B. Membrane Interaction

The membrane location and disposition of P450 cytochromes is critical to their function. Cytochrome $P450_{scc}$, for example, has been localized to the matrix face of the inner membrane of adrenal cortex and ovarian mitochondria (Mitani et al., 1982; Farkash et al., 1985). The location of cytochrome $P450_{scc}$ has an important bearing on the availability of the substrate cholesterol which must enter the mitochondria from the cytosol and transfer across two membranes prior to combination with cytochrome $P450_{scc}$. The location of $P450_{scc}$ also requires that NADPH is generated exclusively within the matrix. Ultrastructural studies show that mitochondria from the adrenal cortex and ovaries differ from those of non-steroidogenic tissues and that their ultrastructure is hormonally dependent. Immature rat ovaries or adrenals after hypophysectomy exhibit small spherical mitochondria containing elongated tubular cristae. Prolonged hormonal stimulation of either rat adrenals or rat ovaries produces a progressive increase in mitochondrial size and proliferation of small vesicular cristae within the matrix in parallel to increases in cytochrome $P450_{scc}$ content (Farkash et al., 1986; Goldring et al., 1986; Zlotkin et al., 1986). It seems possible that the two phenomena may be functionally associated; that is, higher levels of cytochrome $P450_{scc}$ direct a change in the vesicular structure.

$P450_{scc}$ is unique in the superfamily with respect to the high affinity of the substrate cholesterol for phospholipid membranes. As a consequence the affinity of cholesterol for the cytochrome is determined both by effects of phospholipids on the protein and, perhaps more important, by changes in cholesterol lipid interactions. For example, cardiolipin probably stimulates binding through interaction with the protein (Lambeth, 1981) while sphingomyelin decreases the binding by virtue of its strong complex with cholesterol (Stevens et al., 1986). Enzyme turnover is generally increased by polyunsaturated phospholipids and by acidic head groups (Kowluru et al, 1983). Photolysis of CO. $P450_{scc}$ complexes has been used to measure the rotational mobility of the cytochrome in submitochondrial particles (Ohta et al. 1991). Changes in rotational mobility are consistent with other observations in $P450_{scc}$-induced vesicle aggregation (Dhariwal et al. 1991). ADX substantially increases mobility even though forming a larger complex with $P450_{scc}$ consistent with decreases in secondary vesicle interactions. This photolysis technique promises to provide the means for direct measurement of $P450_{scc}$ mobility within intact mitochondria.

The interaction of purified $P450_{scc}$ with phospholipid vesicles is complex. The insertion of solubilized $P450_{scc}$ into phospholipid vesicles and cholesterol transfer to the cytochrome may be limited by dissociation of oligomeric $P450_{scc}$ on the membrane surface (Dhariwal and Jefcoate, unpublished). This step is much faster for unsaturated phospholipids which favor insertion and for head groups which may facilitate oligomeric dissociation. Even though fully bound by the membrane, $P450_{scc}$ readily exchanges between phospholipid vesicles (Dhariwal et al, 1991). Transfer of $P450_{scc}$ from adrenal mitochondrial particles to phospholipid vesicles also demonstrates this mobility (Kominami et al, 1995). Acidic and unsaturated phospholipids slow this exchange just as they enhance insertion. $P450_{scc}$ enhances vesicle-vesicle aggregation preferentially with these lipids through a weak secondary interaction of basic amino acids from the ADX binding site with phospholipid head groups of additional vesicles. This mobility of $P450_{scc}$ may improve access to immobile pools of cholesterol within mitochondrial membranes. Interestingly in reconstituted vesicles maximum rates are achieved when even a small fraction of the cholesterol is mobilized by low concentrations of detergent (Dhariwal and Jefcoate, 1989). $P450_{scc}$ is located on the outside of matrix vesicles where it may modulate intervesicle interactions. The distribution between vesicles also is heterogeneous leaving regions deficient in $P450_{scc}$ where cholesterol metabolism may be slow.

This labile interaction with the membrane is consistent with the absence of the hydrophobic aminoterminal domain which is typical of microsomal P450 proteins. However, $P450_{11\beta}$ which also lacks this sequence is far more firmly integrated into phospholipid membranes while this interaction additionally stabilizes the protein (Seybert, 1990). Evidently other sequences contribute to these membrane interactions, possibly via an insertion of polar head groups into the core P450 structure. When $P450_{scc}$ is membrane bound trypsin cleavage is slowed indicating protection of the cleavage site which has been localized in the region containing residues 247–257 (Chaschin et al., 1986). Interestingly the effects of phospholipids on $P450_{11\beta}$ are exactly opposite. Saturated phospholipids such as dipalmitoyl stabilize the enzyme and increase activity while cardiolipin is potently inhibitory (Seybert, 1990). The potential regulatory effect of phospholipids is further demonstrated by the *stimulatory* effect on the conversion of corticosterone to aldosterone by $P450_{11\beta}$. Phospholipid mixtures that stimulate $P450_{scc}$ inhibit 11β-hydroxylation by $P450_{11b}$.

C. Enzymatic Activity

The turnover of $P450_{scc}$ like that of $P450_{11\beta}$ (Lambeth et al., 1979) is proportional to the concentration of free ADX (Hanukoglu and Jefcoate, 1980). This is indicative of a shuttling of ADX between ADX-reductase and $P450_{scc}$ rather than a ternary complex of reductase, ADX and $P450_{scc}$. The striking characteristic of this reaction is that rates are *decreased* by excess reductase as ADX preferentially forms complexes with reductase rather than $P450_{scc}$ (see Chapter 2). Significantly the ratio of reductase to P450 in adrenal mitochondria is roughly 1:1 while ADX and $P450_{scc}$ are typically near stoichiometric (Hanukoglu and Hanukoglu, 1986). The electron transfer and cholesterol metabolism are optimally tightly coupled with a stoichiometry of 3NADPH consumed per pregnenolone formed (Shikita and Hall, 1974). Ca^{++}, however, increases this ratio while slowing the reaction (Hanukoglu et al., 1993). Ca^{++} stimulates electron transfer from reduced ADX to oxygen which competes with reduction of $P450_{scc}$. Within the mitochondria it is unlikely that the low proportion of reductase distributes evenly in relation to $P450_{scc}$. Regions of unusually low reductase may account for the incapacity of weaker reducing conditions to maintain metabolism of more than about half the mitochondrial cholesterol that can be metabolized under strong reducing conditions (McNamarra and Jefcoate, 1990 a,b).

Although ADX greatly potentiates the binding of cholesterol and vice versa, higher cholesterol concentrations require higher ADX concentrations to sustain the maximum rates (Hanukoglu and Jefcoate, 1980). Not only does the K_m for ADX *increase* as the cholesterol concentration rises but, except at very low concentrations, this K_m is far in excess of the K_d. The same trend applies to the K_m for cholesterol. The K_m of either ADX or cholesterol increases from levels equal to high affinity binding in the ternary complex up to levels higher than are seen for binary complex formation. At high reactant levels complex formation is evidently not rate limiting. One possibility is that dissociation of pregnenolone may be rate limiting and facilitated by high concentrations of ADX and cholesterol.

Relatively little is known about these reactions in mitochondria. Recent work shows that under most conditions rates are limited both by electron transfer and by cholesterol availability. However when mitochondrial reductant processes are near a peak, cholesterol transport becomes the sole limiting factor (Yamazaki et al., 1993). Strong evidence has also been presented that $P450_{scc}$ and $P450_{11\beta}$ form 1:1 hetero oligomeric complexes which affect the capacity of the latter to metabolize corticosterone to aldosterone (Ikushiro et al., 1992). Electron transfer in mitochondria from ADX to $P450_{scc}$ seems to be about 5-fold favored relative to electron transfer to $P450_{11b}$ (Yamazaki et al., 1993).

III. EXPRESSION OF P450scc IN INDIVIDUAL TISSUES

In each tissue the expression of $P450_{scc}$ is cell-type specific and each cell type that contains $P450_{scc}$ exhibits a unique endocrine control involving usually a primary peptide hormone and secondary regulation by local modulators (cytokines, prostaglandins, endothelins, insulin growth factors, etc.).

A. Adrenal Cortex

The adrenal cortex is divided into three zones each with steroidogenic cells with distinct phenotypes both in terms of cellular characteristics and steroid production (Roskelley and Auersperg, 1990). Each cell type expresses $P450_{scc}$ and shows selective expression of other steroidogenic enzymes, as will be discussed in other chapters. The small glomerulosa cells that form a very narrow zone immediately under the capsule are distinguished from the much more abundant adjacent fasciculata cells by the unique expression of CYP11B2 and absence of CYP17. Reticularis cells predominate over fasciculata cells in the inner region of the cortex and are characterized by lower CYP11B1 in relation to both $P450_{scc}$ and CYP17. This change causes a shift in the distribution of cholesterol metabolism products towards adrenal androgens. This variability is also partially reproduced as a function of the time that adrenal cells are maintained in culture (Hornsby et al. 1992). $P450_{scc}$ is relatively persistent and retains responsiveness to cAMP while other forms become sequentially suppressed; first CYP11B forms followed by CYP21 and then CYP17.

In vivo adrenal steroid secretion exhibits a diurnal pattern of secretion that parallels ACTH secretion (James, 1991). This corresponds to changes in the activity of $P450_{scc}$ that are largely determined by cholesterol availability rather than the expression of the protein. Following hypophysectomy of rats all steroidogenic enzymes levels decline and $P450_{scc}$ typically exhibits a half-time for this response of about 5 days (Purvis et al., 1973). Following stimulation by ACTH $P450_{scc}$ protein levels increase substantially after the peak in mRNA is attained (delay 6–12h). These protein levels then remain elevated after mRNA levels decline. Both features are consistent with a much slower turnover of protein relative to mRNA. These changes in $P450_{scc}$ mRNA are also delayed relative to rapid changes in c-fos, c-Jun, Jun-B and cMyc mRNA which potentially may mediate delayed ACTH responses (Miyamoto et al., 1992; Hall et al., 1991). In bovine adrenal cells ACTH also selectively stabilizes $P450_{scc}$ mRNA about 5-fold (Boggaram et al., 1989). Each of these different levels of regulation may contribute to the distinct time courses of induction of steroidogenic genes in adrenal cells by cAMP. Notably, there are distinct delays for induction of each gene except ADX which responds rapidly (Hanukoglu et al. 1990).

The fetal adrenal plays a key role in determining the development of several fetal organs including the lung, liver, brain, retina and gastrointestinal tract. Cells of the adrenal cortex arise early in development from the splanchic mesoderm and exhibit the three cell types seen in adults (Rogler and Pintar, 1993). $P450_{scc}$ and ADX can be detected in the rat embryos as early as day 12 but increase over 3-fold between days 14 and 16. This period corresponds to a surge in secretion of ACTH from the fetal pituitary.

B. Testis

P450$_{scc}$ is expressed in the interstitial or Leydig cells of the testis at levels which correlate in inbred strains of mice with the variable levels of testosterone secretion (Perkins et al., 1988). As in the adrenal, this activity shows diurnal fluctuations which parallel the levels of LH, the primary stimulant for the Leydig cells. This gonadotrophin probably largely functions through elevation of cAMP. Recent studies of the development of the rat testis have identified three distinct stages of Leydig cell development in which regulation of P450$_{scc}$ differs (Shan et al., 1993). These stages are characterized by the relative activities of P450$_{scc}$, CYP17 and 3β-hydroxysteroid dehydrogenase together with differing morphological appearances and are best defined at respectively day 21 (progenitor cells), day 35 (immature cells) and day 90 (adult). When 20,22R dihydroxycholesterol is the substrate pregnenolone formation is dependent only on the level and activity of P450$_{scc}$. Then testosterone production in purified isolated Leydig cells progressively increased by 40- and 1000-fold at respectively day 35 and day 90. Serum cholesterol combined with LH stimulation was progressively less effective relative to dihydroxycholesterol as the gland matured—suggesting that cholesterol availability was rate limiting. Again diurnal changes reflect cholesterol availability rather than P450 expression even though in the testis cholesterol synthesis is a more important determinant than lipoprotein uptake (Gwynne and Strauss, 1982). Recent studies of the rat testis have shown regulation of steroidogenesis from seminiferous tubules. A complex of two proteins released by Sertol: cells (metalloproteinase inhibitor-I and procathepsin L) also potently stimulates cholesterol metabolism in Leydig cells (Boujrad et al., 1995).

Like the adrenal, inhibition of P450$_{scc}$ (e.g., with aminogluthethimide) leads to decreased steroid production. This causes a lowering of feedback suppression in the brain and pituitary gland with consequent stimulation of the adrenal and, in males, the testis. This involves secretion from the hypothalamus of releasing hormones (CRH and LRH) and of the subsequent stimulation of the corresponding pituitary hormones (ACTH and LH). The increase in LH stimulates testis Leydig cells to produce more P450$_{scc}$, mitochondria and endoplasmic reticulum while also increasing numbers of this cell type in the organ.

C. Ovary

Steroid regulation and P450$_{scc}$ activity in the ovary presents a much more complex situation. Here steroidogenesis fluctuates over the course of the relatively prolonged period of the female reproductive cycle during which extensive changes occur in the growth and differentiation of multiple cell types, each of which express P450$_{scc}$ (Zlotkin et al., 1986; Richards, 1944). This cycle comprises separate stages of follicle development, ovulation and corpus luteum formation and regression. Each stage involves concerted regulation between gonadotrophins (LH/FSH) and steroids. A rapid increase in estradiol precedes and triggers the surge of LH which then effects ovulation. The post-ovulatory rise in progesterone secretion results from proliferation of cells which form the corpus luteum, from increased expression of $P450_{scc}$ and also from factors determining cholesterol availability (See Section V).

Cholesterol metabolism in the ovary occurs in three cell types found in follicles and in luteal cells of the corpus luteum that develops from ovulating follicles. The follicular basement membrane separates granulosa cells (inside) from 2–4 layers of elongated theca cells. On the inner side of the granulosa zone cumulus cells surround the oocyte. The highest concentrations of $P450_{scc}$ and lipid droplets are seen in granulosa cells are similar to those seen in adrenal cells. In the immature ovary granulosa cells do not express $P450_{scc}$ while theca cells exhibit significant levels. LH initially stimulates a rise in $P450_{scc}$ in theca and interstitial cells while granulosa cells continue to be without $P450_{scc}$. LH also effects a selective expression of CYP17 and $P450_{AROM}$ in, respectively, theca and granulosa cells which therefore cooperate in the formation of estradiol for the ovulatory surge. Subsequent to ovulation $P450_{scc}$ levels rise nearly 100-fold in conjunction with elevations in the electron transfer proteins and 3β-hydroxysterol dehydrogenase. The increase starts in theca cells at about the time of the LH surge then eventually spreads to granulosa cells and finally reaches cumulus cells immediately adjacent to the oocyte. The increase in cumulus cells occurs just prior to expulsion of the oocyte (Farkash et al., 1986; Zlotkin et al., 1987). $P450_{scc}$ however is not detectable in follicles that do not ultimately ovulate (Zlotkin et al., 1986). $P450_{scc}$ in granulosa cells is elevated additionally by estradiol (Goldring et al., 1986). Significantly LH receptors necessary to mediate this response are seen in large preovulatory follicles but not in small follicles (Meduri et al., 1992). The theca and granulosa cells then act as precursors for development of the corpus luteum (Alila et al., 1988). The complexity of these regulatory changes suggests the involvement of multiple factors. Potential contributors include inhibin, activin, insulin-like growth factors and FSH which may selectively regulate $P450_{scc}$ in granulosa cells (Urban et al., 1991; Hanukoglu, 1992).

During the estrous cycle theca and granulosa cells increase in number, change progressively and then after ovulation differentiate into small and large luteal cells that constitute the corpus luteum. As both follicular cell types undergo luteinization following the LH surge there is a decline in CYP17, and cytochrome $P450_{scc}$ levels rise substantially in the resulting corpus luteum (Goldring et al., 1986; Richards and Hedin, 1988; Rodgers et al., 1986; Richards, 1994). This change is accompanied by an increase in mitochondria with vesicular cristae typical of those containing $P450_{scc}$ (Farkash et al., 1986; Goldring et al., 1986; Rodgers et al., 1986). Steroid production is maintained in the corpus luteum by pituitary LH which, through elevation of cAMP, both stimulates cholesterol availability for metabolism and gene expression of the steroidogenic P450s (Strauss et al., 1983; Ghosh et al. 1987). Peak expression of $P450_{scc}$ in the luteal cells of the corpus luteum also matches levels seen in the adrenal cortex.

D. Ovary-Placenta Unit

During pregnancy, steroids are produced both by the placenta and by the corpus luteum although the distribution of these functions is very different in rodents and in primates. Progesterone and estradiol are necessary for maintenance of pregnancy and fetal development. In rodents, the ovary is the source of progesterone and estradiol which are maintained throughout pregnancy in parallel with the consistently differentiated state of the luteal cells. This differentiation is controlled by the combined effects of LH, prolactin, prolactin-like hormones (produced by the pituitary and placenta), (Khan et al., 1987; Strauss et al., 1983) and also by estradiol (Gibori et al., 1988). The effect of LH on steroidogenesis in luteal cells of pregnant rats is primarily to maintain and increase cholesterol supply and the levels of cytochromes $P450_{scc}$ and CYP17 (Khan et al., 1987). Prolactin-like hormones act largely through maintenance of LH receptors and potentiation of the response to estradiol (Gibori et al., 1979, 1981). Estradiol additionally enhances $P450_{scc}$-dependent cholesterol metabolism through enhanced availability of substrate cholesterol (Gibori et al., 1988) which is no longer under the control of gonadotropins (Richards, 1994).

The prolactin-like hormones that are produced by the placenta (placental lactogens) and steroid hormones each arise from trophoblast giant cells. Recently, progress has been made in studying rat placental differentiation *in vitro* by use of a rat trophoblast cell line (Rcho-1) that has been established from a transplantable choriocarcinoma (Yamamoto et al., 1994). The trophoblast giant cells of the rat placenta express $P450_{scc}$ while Rcho-1 cells also express $P450_{scc}$ as they differentiate to giant cells. $P450_{scc}$ increases to high levels in the placenta by day 12 of pregnancy, well before $P450_{17\alpha}$ (days 16–18). However, $P450_{scc}$ mRNA is detectable even earlier (days 4–6) in both maternal cells of the decidua and in trophoblast giant cells (Schiff et al., 1993). The rat placenta, unlike the human placenta, has no aromatase and therefore estrogens are provided entirely by the ovary.

In primates, the placenta coordinates with the ovary via placental choriogonadotropin (CG) which has a high degree of homology to LH (Ryan et al., 1988;) and acts at the LH receptor. CG is produced shortly after implantation and stimulates the luteal cholesterol metabolism, P450 synthesis and steroidogenesis. As the level of CG rises during pregnancy, desensitization of the corpus luteum occurs leaving the gland quiescent throughout the rest of pregnancy. As luteal function fades steroidogenesis is taken over by the placenta which then secretes large amounts of progesterone for the remainder of the pregnancy. The human placenta contains mitochondrial $P450_{scc}$ in association with adrenodoxin and adrenodoxin reductase and $P450_{AROM}$ in the endoplasmic reticulum (Mason and Boyd, 1973; Simpson and Miller, 1978). The activities of $P450_{scc}$ in the human placenta are maintained by placentally derived CG. Since CG does not induce CYP17 the human placenta cannot metabolize progesterone which is therefore secreted in abundance. There is little influence from the corpus luteum where steroid metabolism is low. Instead a complex steroidogenic relationship exists between the placenta and the fetal steroid producing tissues (Aromatase; See Chapter 8).

E. Brain and Thymus

Specific cells of the brain produce steroid hormones from metabolism of cholesterol in the brain or possibly from adrenal steroids that transfer from the general circulation to the brain. The principle steroidogenic reactions that have been detected in the brain are cholesterol side chain cleavage (LeGoascogne et al., 1987), aromatase (Roselli et al., 1985), estradiol 2-hydroxylation (Theron et al., 1985) and, most recently hydroxylation of 5α-reduced androgens (Warner et al., 1989). Aromatization has also been observed in hypothalamus, amygdala and hippocampus (LeGoascogne et al., 1987). CYP17 or associated activities have not been detected. Although the level of these activities in the brain is low (10–200 pmole/h/g tissue) this may reflect activities that are relatively high in specialized cells. Side chain cleavage and aromatase reactions may be associated with the potent agonist and antagonist activity of steroids on Type A γ-aminobutyric acid (GABA) receptors. 5α-reduced steroids (e.g., 3α, 5α-tetrahydroprogesterone) have a depressant effect by enhancing the effect of GABA. Pregnenolone sulfate is excitatory by inhibiting the effect of GABA (Robel and Baulien, 1994). Thus the net effect of localized cholesterol metabolism depends on the activities of enzymes subsequently metabolizing pregnenolone.

The conversion of cholesterol to pregnenolone and cytochrome $P450_{scc}$ are each exclusively located in glial cells of human and rat brain (LeGoascogne et al., 1987; Robel and Baulieu, 1994). This activity has also been detected in primary cultures of newborn rat glial cells (Robel and Baulieu, 1994). The onset of this activity correlates with the differentiation and maturation of the oligodendrocytes (Robel and Baulieu, 1994). These neurosteroid activities can therefore be seen as a cooperation between synthesis in the glial cells and receptor activity in adjacent neurons. These active steroids can be removed by a P450-dependent hydroxylase (Warner et al., 1989) which seems to be distributed throughout the rat CNS and other target cells for dihydrotestosterone (e.g., hypothalamus, pituitary, prostate).

While oligodendrocytes were initially identified as the major glial cells producing progesterone (Robel and Baulieu, 1994), the activity has been seen in cultured astrocytes (Akwa et al., 1993). Recent analyses of primary cultures of rat glial cells indicate that astrocyte $P450_{scc}$ can account for most of this activity (Mellon and Deschepper, 1993). Primary glial cultures and the C6 glioma cell line each express $P450_{scc}$ mRNA at about $1/10^4$ of adrenal levels while the protein is one percent as abundant. This difference suggests very different turnover of $P450_{scc}$ mRNA and protein in glia as compared to adrenal.

In rodent embryos, $P450_{scc}$ appears in the developing nervous system well prior to appearance of adrenal cells (day 12). Thus, $P450_{scc}$ appears in the neural crest at day 9.5 and by days 12.5–13.5 is visible in spinal cord and dorsal root ganglia

(Compagnone et al., 1995). The sites of expression are located mostly in sensory structures during embryogenesis. In the adrenal gland and gonads $P450_{scc}$ expression is dependent on the nuclear regulatory factor SF-1 (see Section IV) but absence of this protein from the nervous system suggests a different type of regulation.

$P450_{scc}$ expression is also clearly measurable in the retina. Here cAMP stimulation of pregnenolone formation is detectable (Guarneri et al., 1994). Immunocytochemical staining shows localized expression in the retinal ganglial cell layer. Pregnenolone can be further metabolized to several other steroid products including sulfate esters and those resulting from activity of $P450_{17\alpha}$.

During late embryogenesis, $P450_{scc}$ also appears at substantial levels in the thymus (Compagnone et al., 1995). The thymus also produces both CRF and ACTH. This pregnenolone formation seems to arise primarily from thymic epithelial cells, is limited by availability of cholesterol and is elevated twofold by ACTH (Vacchio et al., 1994). Peak activity is seen in the late fetus and then declines after birth. The thymus also stains for $P450_{11\beta}$ suggesting that a localized production of glucocorticoids may regulate apoptotic deletion of thymocytes. In support of this hypothesis, inhibition of $P450_{scc}$ in fetal thymic organ cultures greatly affects thymocyte apoptotic losses.

IV. REGULATION OF P450scc PROTEIN LEVELS

In all tissues the activity of cholesterol side chain cleavage is determined both by the levels of $P450_{scc}$ and associated electron transfer proteins, and the availability of cholesterol to the cytochrome. Hormonal stimulation of each of these steroidogenic proteins involves enhanced gene transcription as evidenced by increased levels of the corresponding mRNA (Miller, 1988; Waterman and Simpson, 1989; Lauber et al., 1991; Waterman, 1994). Mechanisms for these regulatory processes are also discussed in more detail in Chapter 4. The tissue levels of $P450_{scc}$ are highly variable but this in part was due to variation in the proportion of cells expressing $P450_{scc}$. Regulation of the expression of $P450_{scc}$ will be discussed in the next sections.

A. Transcriptional Control

The $P450_{scc}$ (CYP11A) gene encoding $P450_{scc}$ is located on chromosome 15 in humans (Sparkes et al., 1991) and comprises 9 exons over 20 kb (Morohashi et al., 1987). Exon 1 encodes a short 5'-untranslated sequence, and the mitochondrial signal sequence in addition to the aminoterminal sequence of the mature protein. The terminal exon encodes a relatively short 3'-untranslated sequence following on the C-terminal sequence. Transcription of the $P450_{scc}$ gene like the other steroidogenic genes is stimulated by cAMP-dependent processes whether stimulated by ACTH in the adrenal, FSH in granulosa cells or LH in Leydig cells (Golos et al., 1987; Miller, 1988). Analysis of the 5'-flanking region of the $P450_{scc}$ gene when attached to reporter genes has allowed identification of cAMP-dependent regulatory domains (Waterman and Simpson, 1989; Begeot et al., 1993; Parker and Schimmer, 1993). (See Chapter 4.) Deletion analysis of the 5'-upstream region indicates that cAMP stimulation is mediated by elements distinct from the typical CREB binding element. This, among responsive steroidogenic genes, only appears to regulate CYP11B transcripts (Waterman et al., 1992; Rice et al., 1989). cAMP stimulation of rat $P450_{scc}$ is effected through two SF 1 (also called Ad4) elements that are bound by an orphan Zn-finger protein of the steroid receptor family that has also been called Ad4BP (Morohashi et al., 1992). This protein functions in all steroidogenic genes and probably contributes to tissue specificity of expression. Deletion of the gene encoding SF1/Ad4BP results in a complete loss of development of steroidogenic organs, thus indicating a broader role for the protein in gene expression (Luo et al., 1994). This SF1 element appears to be less critical for bovine $P450_{scc}$ transcription. Here, cooperation between elements binding an SP1 factor and an adrenal specific protein mediate the cAMP response (Waterman, 1994; see Chapter 4). Other factors may also modulate $P450_{scc}$ synthesis, presumably through further regulatory sequences in the gene. Thus, IGF-I, VIP and EGF each act synergistically with cAMP in increasing $P450_{scc}$ synthesis in pig granulosa cells (Trzeciak et al., 1986, 1987a; Veldhuis et al., 1986; Voutilainen et al., 1988). In rat granulosa cells, the large rise in $P450_{scc}$ expression that is seen during ovulation and with hCG treatment does not correspond to an increase in SF1 protein (Richards, 1994). A second region that is 1.5–1.7 kb upstream of the start site also mediates cAMP responsiveness. This region contains elements that interact with, respectively, SF1 and cAMP response element binding protein (CREB (Watanabe et al., 1994). Interestingly, the placenta of mice deficient in SF1/Ad4BP nevertheless express normal levels of $P450_{scc}$ in the placenta (Sadovsky et al., 1995). In C6 Glioma cells, regulation of $P450_{scc}$ promoter constructs and gel shift assays also indicate that a factor other than SF1 mediates transcription (Zhang et al., 1995). Placenta and brain therefore appear to be sites of alternative regulatory mechanisms for $P450_{scc}$.

In adrenal cells hormone and cAMP stimulation of transcription is totally inhibited by protein synthesis inhibitors such as cycloheximide and this has led to the suggestion that transcription requires labile steroid-hydroxylase inducing proteins (SHIP's) (John et al., 1986; Golos et al., 1987; Waterman and Simpson, 1989; Voutilainen and Miller, 1987, 1988). While the 5'-flanking regions of CYP11B, $P450_{scc}$ and adrenodoxin (ADX) genes have been sequenced no obvious consensus domain has been identified corresponding to a binding site for a trans-acting SHIP. When several steroidogenic tissues are compared it is apparent that SHIP mechanisms are very selective. For example, an absence of cycloheximide-sensitivity suggests that SHIP regulation is not functional in either human granulosa cells (Picado-Leonard et al., 1988) or a Leydig cell tumor (Mellon and Vaisse, 1989). In a transformed cytotrophoblast line (JEG-3) SHIP regulation is seen for $P450_{scc}$ transcription but cycloheximide causes superinduction thus implicating a labile repressor (Picado-Leonard et al., 1988). A similar cycloheximide sensitive repression has been observed in other genes not involved in steroid synthesis (Nebert and Gonzales, 1987).

The expression of adrenodoxin and adrenodoxin-reductase appear to be coordinately regulated with $P450_{scc}$ (Waterman and Simpson, 1989). Thus, following hypophysectomy of male rats the levels of all three protein decrease in both adrenals and testis and can be restored by addition of, respectively, ACTH or hCG (Simpson and Waterman, 1988; Waterman and Simpson, 1989). Each protein is synthesized on cytosolic ribosomes as a larger precursor that has the appropriate polar aminoterminal sequence for transfer into the mitochondria. After transfer thought the peptide channels into the matrix this polar sequence is processed by a Mn-dependent mitochondrial protease to form the mature protein (Matocha and Waterman, 1986; Ogishima et al., 1985; Waterman and Simpson, 1989).

Protein kinase C has opposing effects on $P450_{scc}$ transcription; mediating transcription through phosphorylation of JUN proteins but also causing down regulation of $P450_{scc}$. In human adrenal cells phorbol esters effect a weak stimulation of $P450_{scc}$ transcription but antagonize the much greater stimulation by ACTH (Ilvesmaki and Vontilainen, 1991). In addition, Y1-adrenal tumor cells exhibit down regulation of $P450_{scc}$ stimulated by apoprotein E which is apparently linked in some way to PKC activity (Reyland et al., 1992). Activation of protein kinase C by phorbol esters also results in suppression of $P450_{scc}$ apparently by opposing the cAMP activation in a regulatory element at -118/-101 of the upstream region of the gene (Begeot et al, 1993). $P450_{scc}$, ADX and ADX-reductase naturally exhibit broader expression than the other steroidogenic enzymes which determine a cell specific steroid product (Hanukoglu and Hanukoglu, 1986). In adrenal cells recent evidence suggests that transcriptional activation by ACTH is not clearly mediated exclusively by cAMP (Hanukoglu et al., 1990).

While glucocorticoids effect a feedback regulation by suppressing ACTH secretion they also suppress expression of steroidogenic enzymes, most notably $P450_{scc}$. Addition of dexamethasone to cultured bovine adrenal cells causes a 3-fold suppression of the level of $P450_{scc}$ mRNA (Trzeciak et al., 1993). Orphan receptors related to steroid receptors appear to have a broad role in the ACTH responses and are themselves inducible by the hormone thus contributing a delayed response (Wilson et al, 1993). The active concentration of the synthetic glucocorticoid (0.1 μM) is consistent with mediation by the glucocorticoid receptor and with levels of glucocorticoids likely to be present in the adrenal gland during stimulation *in vivo*. However, this is not unique to adrenal cells since rat Leydig cells show even stronger glucocorticoid suppression of $P450_{scc}$ mRNA (Payne et al., 1992). This down regulation is also seen with $P450_{scc}$ promoter constructs that contain the elements necessary for cAMP stimulation suggesting a direct repression of this effect. A similar down regulation by glucocorticoids is seen in these cells with CYP 17 transcription. This regulatory process, however, is slower in producing down regulation than suppression of ACTH secretion from the pituitary. This suppression by glucocorticoids contrasts with the stimulation produced for transcription of

$P450_{AROM}$ (Mendelson et al., 1982) and has been seen for several other genes but without identification of a conserved regulatory element.

B. Suppression of Cholesterol Metabolism

Perhaps the best characterized inhibitory modulator is $TGF_{\beta1}$, which partially inhibits steroidogenesis in most steroidogenic cells (Hotta and Baird, 1987; Perrin et al., 1991). $TGF_{\beta1}$ down-regulates several steroidogenic enzymes at the transcriptional level, most notably CYP17, but also decreases cholesterol mobilization. $TGF_{\beta1}$ also strongly inhibits aldosterone synthesis both at the level of cholesterol mobilization and at the conversion of corticosterone to aldosterone (Gupta et al. 1993). In bovine adrenal cells, this effect in part involves decreased cholesterol uptake through the LDL pathway. Interestingly, α_2-macroglobulin, a protein which strongly sequesters $TGF_{\beta1}$, is stimulated in adrenal cells by $TGF_{\beta1}$ and suppressed by ACTH (Shi et al. 1990). This suggests that this represents a physiological regulatory mechanism. Interleukins 1α and 1_β suppress steroidogenesis in ovarian and testis cells, most obviously by modulating expression of P450 cytochromes, but changes in cholesterol transfer mechanisms have typically not been excluded (Hurwitz et al., 1991; Lin et al., 1991). However, in the adrenal gland *in vivo* interleukins stimulate steroid secretion largely through elevation of ACTH (Besedovsky et al. 1991). Atrial natriuretic peptide (ANP) decreases steroid synthesis in adrenal glomerulosa cells (Elliott and Goodfriend, 1986). This occurs at least in part through decreased cholesterol availability to $P450_{scc}$, probably mediated by cGMP and antagonism of the Ca^{++}-mediated phosphorylation step (Elliott and Goodfriend, 1986).

C. Genetic Deficiency in P450scc

Unlike most other steroidogenic P450 cytochromes no genetic deficiency for $P450_{scc}$ has yet been identified in humans. Loss of pregnenolone synthesis is a lethally recessive condition. Examination of one of the few surviving infants with a deficiency in pregnenolone formation failed to find any deficiency or alteration in the $P450_{scc}$ sequence (Matteson et al., 1986). It has recently been established that this involves a block in cholesterol delivery mediated by processes described in section V. It remains to be seen whether partially effective $P450_{scc}$ mutants exist in the human population. Such defects could easily be compensated by elevated basal ACTH but may exhibit chemical problems due to this chronic overexpression.

A congenital adrenal hyperplasia which can be inherited and seen in newborn rabbits is due to absence of $P450_{scc}$ expression (Yang et al., 1993) that was not even detectable by rtPCR. Analysis of genome DNA showed normal $P450_{scc}$ sequence for 85 percent of the coding region suggesting a major deletion elsewhere in the gene.

V. ACUTE HORMONAL REGULATION OF P450$_{scc}$

For each steroidogenic cell type, cholesterol metabolism consists of three distinct processes, each of which involves multiple steps and each of which can potentially be rate limiting: cholesterol uptake, intracellular transfer, and reaction at P450$_{scc}$. Cholesterol is an essential component of cell membranes, and consequently steroidogenesis depends on a sufficient supply of cell cholesterol that exceeds constitutive cellular requirements. This cholesterol must be transferred to cytochrome P450$_{scc}$ on the inner membranes of the mitochondria (Churchill and Kimura, 1979) prior to metabolic cleavage to pregnenolone. Significantly, expression of the full complement of ADX-reductase, ADX and P450$_{scc}$ in COS-cells by transient transfection failed to elevate cholesterol metabolism to activities seen in steroidogenic tissues with comparable enzyme levels (Zuber et al., 1988). 25-hydroxycholesterol showed much higher activity indicating a shortage of available cholesterol, which probably also limits activity in the human deficiency condition.

Hormonal stimulation of cholesterol metabolism may occur within minutes but may also change over the course of days. Acute stimulation inevitably involves activation of metabolism without increased gene expression and depends exclusively on enhanced cholesterol transfer to P450$_{scc}$. This cholesterol can be supplied from serum as various cholesterol esters, by transfer from either low or high density lipoproteins (LDL or HDL) (Kovanen et al., 1980; Gwynne and Mahaffee, 1989), or cholesterol may be provided by enhanced *de novo* synthesis of cholesterol. The various processes for internal transfer of cholesterol may additionally depend on the source of the cholesterol.

In ACTH-stimulated rat adrenal glands, mitochondrial ADX has been measured in a nearly fully reduced state, even though these conditions provide the highest rate of pregnenolone formation (Williams-Smith and Cammack, 1977). During *in vivo* turnover in the rat adrenal, the P450$_{scc}$ is nearly fully depleted of cholesterol, consistent with rate-limiting cholesterol P450$_{scc}$ (Jefcoate and Orme-Johnson, 1975). Although this suggests that the supply of reducing equivalents is unlikely to be a limiting factor in adrenal mitochondria *in vivo*, rates of cholesterol metabolism in isolated mitochondria are substantially dependent on the reductant even when cholesterol availability is low (McNamara and Jefcoate, 1990 a,b). Cellular conditions may therefore frequently be such that reductant transfer processes affect the rate of cholesterol metabolism.

A.   Intracellular Signaling and Acute Stimulation

Steroidogenic cells clearly share many mechanisms of cholesterol regulation that can be controlled externally by a variety of cell-specific hormonal processes (e.g., ACTH for adrenal fasciculata cells, LH for testis Leydig cells). The intracellular processes that stimulate steroid synthesis can be activated either by hormonal activation of cAMP (ACTH/adrenal fasciculata) or through elevation of cytosolic

Ca^{++} (angiotensin/adrenal glomerulosa). Both mechanisms may also operate in the same cells. Thus, cholesterol metabolism in cultured bovine adrenal fasciculata cells can be stimulated effectively by both ACTH and angiotensin (Walker et al., 1991). In human fetal adrenal glands the stimulation of pregnenolone formation by ACTH and by cAMP was equally sensitive to the calmodulin inhibitor calmidazolin, indicating a role of Ca^{++}/calmodulin complexes (Carr et al., 1987). This inhibition was effective for metabolism of constitutive cholesterol but not hydroxycholesterols that are readily available to $P450_{scc}$ thus pointing to a role for Ca^{++} in cholesterol mobilization. In bovine adrenal cells, cholesterol is mobilized within 5 minutes by ACTH and cAMP through an overall process that shows very similar characteristics to those of rat adrenals *in vivo* (DiBartolomeis and Jefcoate, 1984).

The capacity of a hormone to stimulate a particular cell type is determined by the expression of the corresponding cell surface receptors and on the coupling of these receptors to signaling enzymes through heterotrimeric G proteins. The expression of receptors and G protein, together with their mutual interactions, may change substantially when cells are cultured, depending on many factors, including the culture medium and time in culture (Langlois et al., 1987; Hornsby, 1991). While a hormone such as ACTH may act entirely through cAMP and protein kinase A in cultured adrenal cells (Wong and Schimmer, 1989), this does not exclude additional regulatory factors (e.g., Ca-dependent steps). Irrespective of whether ACTH signaling is mediated by cytosolic cAMP or Ca^{++}, a sequence of protein phosphory-

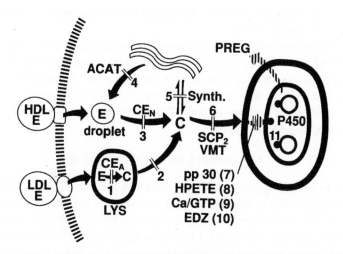

Figure 3. Steps involved in the transfer of cholesterol to mitochondrial cytochrome $P450_{scc}$ in steroidogenic tissues. The figure shows transfer between lysosomes, endoplasmic reticulum, and mitochondria. Numbers 1–11 refer to specific inhibitors listed in Table 1. E represents cholesterol esters, and MT refers to microfilaments and microtubules. Other abbreviations are referred to in the text.

Table 1. Selective Inhibitors of Cholesterol Metabolism

Process	
1. lysosomal cholesterol esterase	chloroquine
2. transfer of cholesterol from lysosome to cytosol	imipramine progesterone
3. neutral cholesterol esterase	diethyl phosphate
4. ACAT	Sandoz 58035
5. cholesterol synthesis	lovastatin
6. cytoskeletal transfer of cholesterol to mitochondria	cytochalasins (microfilaments) vinblastine (microtubules) acrylamide (intermediate filaments)
7. pp30	cycloheximide
8. 5-HPETE	nordihydroguaretic acid
9. Ca/GTP	GTPγS
10. endozepine peripheral benzodiazepine receptor	flunitrazepam
11. cytochrome P450$_{scc}$	aminoglutethimide

Note: *Steps 1–11 are shown in Figure 3.

lation steps presumably mediates the stimulation of adrenal steroidogenesis (Koroscil and Gallant, 1981).

One of the major challenges in understanding the activation of steroidogenesis is the identification of key targets for these protein phosphorylation steps. The following sections outline several steps in the regulation of cholesterol transfer where phosphorylation may modulate activity. Figure 3 and Table 1 provides a summary of the steps involved in intracellular cholesterol transfer to P450$_{scc}$ and lists selective inhibitors of these steps.

B. Lipoprotein Uptake and Processing

Current evidence suggests that, for the most part, the availability of cholesterol in steroidogenic cells is determined by processes that are equivalent to those analyzed in non-steroidogenic cells (Gwynne and Strauss, 1982; Goldstein and Brown, 1990; Johnson et al., 1991; Liscum and Underwood, 1995). Cholesterol is transferred into the cytosolic pool either by uptake of LDL (most species) or HDL (rats) via distinct receptors. The rat adrenal has recently been identified as the most abundant site of the recently characterized HDL receptor (Acton et al., 1996), while adrenal cells from most species provide the most plentiful source of LDL receptors. When these lipoproteins are depleted or the pathway is blocked, cholesterol synthesis is activated (Faust et al., 1977; Pittman et al., 1987). In adrenal cortex cells or ovarian granulosa cells lipoproteins provide the major source of steroidogenic cholesterol and this is stored as lipid droplets in the cytosol which comprise mostly of cholesterol esters (Azhar et al., 1981). However, in the testis cholesterol biosynthesis provides the major source and this process may be rate limiting (Strauss and Gwynne, 1982). Lipoproteins nevertheless increase testosterone production in hormonally stimulated Leydig cells in culture (Quinn et al., 1981). LDL cholesterol reaches the cytosol via lysosomes where an acidic esterase hydrolyzes the predominating esters. The lysosomal processing of cholesterol esters is blocked not only by agents such as chloroquine which raise lysosomal pH (Poole and Ohkuma, 1981) but also by imipramine (Rodriguez-Lafrasse et al., 1990) and several steroids including progesterone (Butler et al., 1992) which cause cholesterol accumulation in the lysosomes by preventing transport out of the organelle. A similar change is seen in the Niemann-Pick Type C syndrome (Blanchette-Mackie et al., 1989).

In the cytosol equilibration between free cholesterol and cholesterol esters is determined by a neutral esterase (Nishikawa et al., 1981) which is activated by cAMP-kinase (ser 563) and inhibited by Ca/Calmodulin kinase or AMP kinase (ser 565). α-MSH, a peptide which is formed from the same precursor-protein as ACTH, may stimulate cholesterol availability by increasing esterase activity (Pedersen et al., 1980). ACTH also induces an inhibitory phosphorylation of acyl CoA acyl transferase (ACAT) which esterifies cholesterol (Yeaman, 1990). However inhibition of this enzyme with the specific inhibitor Sandoz 58035 does not stimulate cholesterol metabolism in rat adrenal cells (Jamal et al., 1985; Gwynne and Mahaffee, 1989) suggesting that this reaction does not limit cholesterol availability to $P450_{scc}$. When cytosolic cholesterol is depleted by any of the above deficiencies, ACTH stimulates cholesterol synthesis via activation of HMG CoA reductase (Lehoux et al., 1990; Rainey et al., 1986). Organophosphates, particularly triaryl phosphates, potently inhibit the neutral esterase causing cytosolic accumulation of cholesterol esters (Harrison et al., 1990). In isolated cells transfer to $P450_{scc}$ seems to be more sensitive to these inhibitors than the neutral esterase (Gocze and Freeman, 1992). Transport is complicated by vesicular cholesterol transfer to the plasma membrane which may act as a source both for export of cholesterol from the cell or transfer to mitochondria (Freeman, 1987, 1989).

ACTH stimulation of adrenal glands is associated with rapid depletion of cytosolic cholesterol esters from the gland (Garren et al., 1971). Apoprotein E (Apo E) mediates binding to LDL receptors and has been associated with reverse cholesterol transfer to plasma HDL possibly by targeting intracellular cholesterol/phospholipid particles for exocytosis (Johnson et al., 1991). Apo E is expressed at high levels in steroidogenic cells (Reyland and Williams, 1991). Although the synthesis of this protein in macrophage is triggered by elevated free cytosolic free cholesterol (Dory, 1989), this homeostatic mechanism for cellular cholesterol has not been established in adrenal cells. Although elevation of Apo E in transfected Y-1 adrenal cells results in suppression of cholesterol transport to mitochondria (Reyland and Williams, 1991), this is accompanied by suppression of cAMP-dependent activation of the transcription of several steroidogenic genes including $P450_{scc}$. Elevation of PKC in these cells probably accounts for suppression of basal levels of $P450_{scc}$ but not the diminished effects of cAMP (Reyland et al., 1992). Apo E, either derived from lipoproteins or constitutive expression may have effects on steroidogenesis well beyond cholesterol transport and may even affect steroid products (see Reyland et al., 1992).

C. Cholesterol Movement and the Cytoskeleton

The cytoskeleton and cell shape plays an important role in the regulation of cholesterol movement to adrenal mitochondria. In Y-1 adrenal tumor cells, disruption of actin stress fibers with cytochalasins (Mrotek and Hall, 1977; Hall et al., 1979), microtubules with colchicine (Crivello and Jefcoate, 1978), or intermediate filaments with acrylamide (Shiver et al., 1992) each cause small stimulations of basal steroidogenesis. On the other hand, microfilament and microtubule disruptions inhibit ACTH-stimulated cholesterol metabolism while the interaction of acrylamide with intermediate filaments is additive with ACTH-stimulation. Taxol which stabilizes microtubules inhibits all of these effects (Rainey et al., 1985). However the detailed relationship between cytoskeleton and cholesterol transfer seems to depend on cell type. In bovine adrenal cells under certain conditions there is no effect of cytochalasin or taxol on cholesterol metabolism (Rainey et al., 1984; Rainey et al., 1985) although serum free conditions in these experiments may place a higher reliance on a rate limiting synthesis of cholesterol.

Cell shape also modulates basal cholesterol metabolism as evidenced by the effects of the adhering surface. A charged polylysine surface enhances cell spreading, and diminishes basal cholesterol metabolism (Hall and Almabobi, 1992). However although ACTH stimulates cell rounding through changes in the cytoskeleton in bovine adrenal cells this rounding seems to be due to activation of urokinase, an enzyme which disrupts extracellular matrix (Hornsby et al., 1989). Since inhibition of this enzyme prevents rounding but not activation of cholesterol metabolism it was suggested that these changes are not directly linked.

The intermediate filament protein vimentin binds both lipid droplets and mitochondria (Hall and Almabobi, 1992) and vimentin-deficient cells fail to move cholesterol out of lysosomes into the cytosolic pool (Sarria et al., 1992). Hall has suggested that phosphorylation of vimentin may play an important role in hormonal control of cholesterol movement (Hall, 1995). Vimentin deficiency is superficially very similar to changes associated with Type C Niemann-Pick disease and lysosomal inhibition induced by progesterone and various drugs (see previous section). This mechanism may account for the affect of acrylamide which also interacts with intermediate filaments. These various cytoskeletal elements possibly keep lipid droplets and mitochondria apart except during hormonal stimulation. Although relaxation of the cytoskeleton can provide a small stimulation, this structural change needs to be coupled to additional cAMP-stimulated processes particularly directed to the mitochondria to effect a major activation of cholesterol metabolism.

D. Sterol Carrier Protein-2 (SCP₂)

Sterol carrier protein₂ (SCP_2), a 13 kd protein that has been purified from liver, stimulates the activity of several microsomal enzymes involved in the synthesis and utilization of cholesterol. For example, cholesterol esterification in microsomal membranes by ACAT is stimulated by SCP_2 (Gavey et al., 1981), probably through enhanced delivery of the substrate. SCP_2 enhances transfer of cholesterol from lipid droplets to mitochondria and also weakly stimulates mitochondrial steroidogenesis (Vahouny et al., 1983). SCP_2 is synthesized from a 60 kd precursor protein, and both precursor and mature forms are primarily located in peroxisomes (van Amerongen et al., 1989). Treatment of rat adrenal cells for 24 h with ACTH elevates SCP_2 several fold (Trzeciak et al., 1987). Likewise, in rat corpus luteum, estradiol elevates SCP_2 while increasing the availability of cholesterol to $P450_{scc}$ (McClean et al., 1989). In Leydig cells, hormonal stimulation is associated with redistribution of SCP_2 (Van Noort et al., 1988). In rat adrenal cells, steroidogenesis is partially suppressed by intracellular administration of anti-SCP_2 IgG (Chanderbahn et al., 1986). Overexpression of SCP_2 in COS cells also enhances steroidogenesis, but the site of this effect has not been defined and overall rates of steroid production remain very low (Yamamoto et al., 1991). Depletion of SCP_2 in fibroblasts by use of anti-sense oligonucleotides results in much slower intracellular cholesterol transport that then is much more dependent on an intact cytoskeleton (Puglielli et al., 1995)

SCP_2 has also been characterized as a non-specific phospholipid transfer protein that facilitates the transfer of several types of phospholipids between cell membranes (Westerman and Wirtz, 1985), although this protein binds cholesterol more weakly than phospholipids. This affinity may be sufficient to enhance cholesterol transfer from lipid droplets where the competition from phospholipids is low. SCP_2 is however completely inactive in cholesterol transfer from phospholipid-cholesterol membranes. SCP_2 may work in concert with vimentin and other cytoskeletal elements. There is also evidence suggesting that SCP_2 is activated in adrenal cytosol relative to purified liver SCP_2 (McNamara and Jefcoate, 1989).

E. Intramitochondrial Cholesterol Transfer

The rate of cholesterol transfer to mitochondria in ACTH-stimulated bovine adrenal cells equals the rate of pregnenolone formation at $P450_{scc}$ and there is very little elevation of mitochondrial cholesterol (DiBartolomeis and Jefcoate, 1984). Thus mitochondrial processes are not rate limiting. However, as will be discussed in the next section, ACTH also stimulates intramitochondrial cholesterol transfer. Impairment of this activation causes intramitochondrial movement of cholesterol to become rate limiting for metabolism at P450$_{scc}$. The next sections discuss several contributors to an apparently complex process.

1. Cycloheximide-Sensitive Factors

ACTH-treatment *in vivo* enhances the activity of cholesterol metabolism in isolated adrenal mitochondria in parallel with changes in cholesterol-P450$_{scc}$ complex formation (Simpson et al., 1972; Jefcoate et al., 1974). Following ACTH stimulation *in vivo*, rat adrenal mitochondria retain a pool of reactive cholesterol and exhibit enhanced cholesterol P450$_{scc}$-cholesterol complex formation, as indicated by optical and EPR spectroscopy (Simpson et al., 1972; Jefcoate et al., 1974; Brownie et al., 1973). When inhibitors of protein synthesis block this cholesterol transfer to P450$_{scc}$, as evidenced by low metabolism and complex formation. This inhibition has no effect on the metabolism of more soluble hydroxy-cholesterol analogs, which readily transfer across the intermembrane space (Jefcoate et al., 1974). Similar effects of cycloheximide have been seen in mitochondria from the ovaries of luteinized rats (Arthur and Boyd, 1976) and Leydig cells (Cooke et al., 1975). Complex formation between cholesterol and P450$_{scc}$ in isolated rat adrenal mitochondria parallels the size of the pool of reactive cholesterol. A comparison of these parameters following different adrenal excision procedures indicates that most of this cholesterol accumulates as the gland becomes deprived of oxygen during removal from the animal (Jefcoate and Orme-Johnson, 1975). This change corresponds in part to the accumulation seen after inhibition of P450$_{scc}$ with aminoglutethimide (Mahatee et al., 1974; Simpson et al., 1979), suggesting that cholesterol transfer continues in the gland after cessation of metabolism at P450$_{scc}$ through lack of oxygen. When these mitochondria are disrupted by sonication, these effects of pretreatment are lost, presumably because of a loss of the intermembrane barrier (Jefcoate et al., 1974).

When *in vivo* protein synthesis is inhibited, cholesterol mobilized by ACTH accumulates in the outer membrane of the mitochondria (Privalle et al., 1983; Ohno et al., 1983; Cheng et al., 1985). This and other experiments have led to the concept that ACTH increases the rate of cholesterol transfer between outer and inner mitochondrial membranes and that this process is mediated not only by cAMP but also by a labile protein (Privalle et al., 1983). The association of cholesterol with P450$_{scc}$ in mitochondria of rat corpus luteum is similarly stimulated by gonadotrophins and blocked by cycloheximide (Ghosh et al., 1987). A variety of protein synthesis inhibitors reverse the stimulation of cholesterol side chain cleavage initiated by cAMP with a half-time of 3–5 mins. (Shulster et al., 1970; Crivello and Jefcoate, 1978), a rate equal to the rate after removal of ACTH. The first candidate for this labile protein was a 30 amino acid peptide which was found in steroidogenic cells after hormonal activation. This has turned out to be identical to the C-terminus of glucose regulatory protein (GRP94), a common stress protein (Pedersen and Brownie, 1987). While appearance of this peptide parallels hormonal activation (Mertz and Pedersen, 1989) most laboratories find the peptide has only a weak capacity to directly stimulate mitochondrial cholesterol metabolism in isolated adrenal mitochondria (Xu et al., 1991).

A more probable candidate that explains both hormonal stimulation and cyclo-heximide sensitivity is a set of approximately 30 kD proteins which rapidly appear in all steroidogenic mitochondria following hormonal stimulation (Krueger and Orme-Johnson, 1983; Alberta et al., 1989). These proteins are formed from a 37 kD precursor which is co-translationally phosphorylated after cAMP or ACTH stimulation. This step is a prerequisite for entry into the mitochondria. Subsequent intramitochondrial processing produces a set of 28 and 30 kd membrane proteins evidently formed by phosphorylation and other post translational changes to the same core protein (Epstein and Orme-Johnson, 1991). After treatment with cyclo-heximide some of these proteins disappear roughly in parallel to the loss of cholesterol metabolism. Comparable changes occur in Leydig cells (Stocco and Soderman, 1991; Stocco and Chen, 1991) and in both bovine adrenal glomerulosa and fasciculata cells (Elliott et al., 1993). Significantly the changes in p30 and p28 proteins are qualitatively similar after cAMP, angiotensin or K^+ stimulation of glomerulosa cells but nevertheless increase for each stimulant in proportion to cholesterol metabolism (cAMP angiotensin K^+). The latter stimulations occur via elevation of cytosolic Ca^{++} (Braley et al., 1986) suggesting that cAMP- and Ca^{++}-dependent kinases can activate the same stimulatory processes. Atrial naturetic peptides inhibited p30 formation in proportion to inhibition of cholesterol metabolism (Elliott et al., 1993). Threonine phosphorylation has been implicated by use of threonine analogs which effect parallel decreases p30 phosphorylation and cholesterol metabolism (Stocco and Clarke, 1993).

A cDNA corresponding to a p30 protein from MA 10 Leydig tumor cells has recently been cloned (Clark et al., 1994). Expression of this cDNA in non-steroi-dogenic COS 1 cells that contain transfected cytochrome $P450_{scc}$ and adrenodoxin resulted in a substantial elevation of mitochondrial cholesterol metabolism (King et al., 1995). This unique gene product which shows little or no homology with other known protein sequences has been named the Steroidogenic Acute Regulatory Protein (StAR). Stimulation of MA 10 mouse Leydig tumor cells with cAMP results in a steady increase in 3.4, 2.7 and 1.6 kb transcripts during a period of 1–12 hrs (Clark et al., 1995). Interestingly, cAMP stimulates cholesterol metabolism even when StAR mRNA is extremely low. Similarly, there is a maximum rise in cholesterol metabolism in cultured rat adrenal cells prior to any increase in mRNA, consistent with a previous understanding that cAMP-induced translation and mito-chondrial insertion is the key process (Epstein and Orme-Johnson, 1991). Indeed, while cAMP elevates StAR mRNA in both rat fasciculata cells and glomerulosa cells, this does not cause an increase in cholesterol metabolism over the early stimulated activity (after 30 mins.) produced from basal StAR mRNA levels. These transcription products are formed from a gene that encodes the translated protein sequence across 7 closely spaced exons (Clark et al., 1995). We find that the seventh exon is actually substantially extended to over 3kb and contains multiple polyadenylation signals consistent wit the multiple transcripts (Kim et al., 1996). The functional role of these longer transcripts, which are universally seen wherever StAR is expressed, remains to be determined.

The sequence of the human StAR cDNA is highly conserved (87 percent identical to mouse). Expression is restricted to adrenal testis, ovaries and possible kidney, but significantly is not found in the placenta where cytochrome $P450_{scc}$ is nevertheless active (Sugawara et al., 1995). The gene resides on chromosome 8 while a pseudogene has been detected on chromosome 13. The deficiency of metabolism of cholesterol to pregnenolone that is found occasionally (mostly in Japanese and Korean populations) is due to insertion of premature stop codons in this gene (Lin et al., 1995). These proteins were truncated by 28 or 93 amino acids when expressed in CoS-1 cells and were completely inactive.

Recent work shows that cholesterol metabolism in MA 10 cells is highly sensitive to organophosphate compounds (umbelliferone phosphate) that inhibit cholesterol esterase (Choi et al., 1995). Interestingly, this treatment also blocks the entry of StAR into the mitochondria. It is not clear whether this implicates a new esterase or a requirement for active cholesterol esterase/free cholesterol for this StAR transfer. A presumably different type of inhibition of adrenal cholesterol metabolism is effected by DMSO but again seems to involve a lowering of StAR (Stocco et al., 1995).

2. Contact Sites and Peripheral Benzodiazepine Receptors

The rapid transfer of cholesterol between the outer and inner mitochondrial membranes that is seen in activated adrenal mitochondria probably involves contact between the two membranes. Cholesterol exchange between separate membranes is typically very slow ($t_{1/2}$ 1–2 h) (Bar et al., 1987). Such contact sites, which may represent as much as 10 percent of the mitochondrial surface, provide the location for the import of mitochondrial proteins (Pfanner et al. , 1990; Hwang et al., 1991) and phospholipids (Volker, 1989; Van Venetie and Verkleij, 1982). They also transfer mitochondrial energy to the cytoplasm via the ADP/ATP translocator and creatine kinase (CK) (Kusnetsov, 1989; Rojo et al., 1991). Cardiolipin enhances both the affinity of $P450_{scc}$ for cholesterol (Lambeth, 1981) and intermembrane fusion (Van Venetie and Verkleis, 1982). Stress activation of rat adrenal cells appears to decrease the intermembrane space (Stevens et al., 1985), which may also increase the frequency of intermembrane contacts. This suggests the possibility of facilitated cholesterol transfer to $P450_{scc}$ at cardiolipin-rich contact sites. Contact sites are also associated with porin, a 30 kd outer membrane protein that permits entry of small molecules into the intermembrane space. The channel formed by oligomeric porin is voltage dependent; at voltages more positive than 30 mV, large anions (ADP, ATP, and creatine phosphate) are partially excluded while cation transport is retained (Brdiczka, 1991).

Recently, evidence has been presented for interaction between peripheral benzodiazepine receptors (PBR) and porins in adrenal mitochondria (Ganier et al.,

1993) suggesting that the receptor regulates the channel in some way. The high concentration of peripheral benzodiazepine receptors (PBR) in adrenal mitochondrial outer membranes may be associated with regulation of specific cholesterol transferring contact sites. PBR are 18 kd proteins (Papadopoulos et al., 1990) which are activated by a range of benzodiazepines and isoquinolines which, however, bind to separate sites. Isoquinolines bind directly to purified PBR while benzodiazepines require the further interaction with porin (Garnier et al., 1994). The expression of these receptors is increased in steroidogenic tissues by hormonal stimulation, presumably through enhanced transcription (Amsterdam and Suh, 1991). This receptor is also present in many non-steroidogenic cell types where benzodiazepines are active (pituitary, salivary gland, nasal epithelium, lung, kidney, uterus). In many cases, secretory functions are implicated, suggesting some more general regulation of ion fluxes possibly through coordination of cytosolic and mitochondrial processes. PBR regulated voltage-differences between the mitochondrial intramembrane space and the cytosol possibly coordinate these ion fluxes in and out of the mitochondria and the frequency of intermembrane contacts.

Benzodiazepine agonists stimulate steroidogenesis in adrenal, Leydig, and granulosa cells and are also effective when added to isolated mitochondria from adrenal or testis tissue (Amsterdam and Suh, 1991; Muhkin et al., 1989; Papadopoulos et al., 1990). This stimulation of isolated mitochondria partially overcomes the block caused by treatment *in vivo* with cycloheximide and results in enhanced cholesterol transfer from outer membrane to inner membrane (Krueger and Papadopoulos, 1990; Papadopoulos et al., 1991). Active isoquinolines and benzodiazepines provide a relatively modest stimulation of cholesterol metabolism in steroidogenic cells (3- to 4-fold). Co-addition of those agonists with ACTH does not increase the sensitivity to the hormone, suggesting that stimulation of the benzodiazepine receptor is not a limiting factor in the cellular process. Hormone-stimulated steroidogenesis, but not basal activity, is, however, inhibited 50% by the partial agonist, flunitrazepam implying a role for the receptor in the activation of these mitochondria by cAMP (Papadopoulos et al., 1991).

Endozepine, a natural 10 KD polypeptide agonist for both central and peripheral receptors through different domains (Guidotti, 1990) is present in adrenal cells at high levels and stimulates mitochondrial cholesterol metabolism (Besman et al., 1989). This protein is also identical to the acyl CoA binding protein which has been purified from liver (Knudsen and Nielsen, 1990) and which stimulates the availability of acyl CoA to the inner mitochondria (Rasmussen et al., 1994). A second 10 kDa protein that is not acyl CoA binding protein/endozepine is also completed to PBR in all tissues (Blahos et al., 1995) and is likely to be critical to this process. Hormonal stimulation of steroid synthesis is not paralleled by elevated endozepine peptide (Brown et al., 1992), again suggesting a permissive role in the activation process. Fatty acids which are elevated after cholesterol ester hydrolysis may also increase acyl CoA and affect binding of endozepine to the receptor. Antisense suppression of endozepine synthesis is associated with suppression of steroidogenesis (Boujrad et al., 1993; Garnier et al., 1994), suggesting that the PBR-endozepine complex is critical to mitochondrial cholesterol transfer. However, this protein is also critical for fatty acid utilization for β-oxidation and glycerolipid synthesis (Rasmussen et al., 1994).

PBR also undergoes cAMP-dependent phosphorylation, which is associated with enhanced benzodiazepine binding and enhanced entry of cholesterol into the mitochondria (Boujrad et al., 1994; Papadopoulos and Brown, 1995). This activation also seems to be associated with a clustering of PBR polypeptides. It has been suggested that these clusters may form a cholesterol channel (Papadopoulos et al., 1994). Interestingly, a possible larger form of endozepine may be present in the mitochondria based on immunoblotting experiments and appears to be decreased by cAMP-dependent activity (Papadopoulos and Brown, 1995). Activated PBR and porins in the outer mitochondrial membrane may associate with pp30 (StAR) in the inner membrane to facilitate cholesterol transfer (Fig. 4). The activation of PBR also parallels the appearance of pp30 (StAR) in the mitochondria.

3. Lipoxygenase Metabolites

Several experiments implicate 5-lipoxygenase metabolites in the mitochondrial transfer of cholesterol (Mikami et al., 1990; Jones et al., 1987). However, indomethacin, which inhibits prostaglandin synthesis synthase, has no effect on steroidogenesis. Specific lipoxygenase inhibitors which inhibit formation of 5-HETE and LTB_4 in rat adrenal cells decrease hormonally stimulated steroidogenesis. Dibutyryl cAMP (db cAMP) has no effect on 5-lipoxygenase activities, indicating that this is a permissive effect rather than a mechanism of hormonal

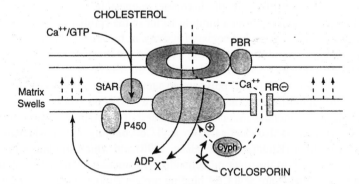

Figure 4. A model for cholesterol translocation into mitochondria. The peripheral benzodiazepine receptor (PBR) modulates ion movement through the adjacent porin channel and contacts with the inner membrane ADP/ATP translocator. StAR (inner membrane location uncertain) mediates uptake of cholesterol by the inner membrane. Ca^{++} modifies inner membrane permeability through a cyclophilin (CYPH) sensitive opening of the translocator to multiple ions with accompanying matrix swelling.

activation. 5-hydroperoxyeicosatrienoic acid (5-HPETE), the initial product of the 5-lipoxygenase, reversed such inhibition, suggesting that this may be the active product. A similar involvement of lipoxygenase products has been deduced from inhibitor effects on LH-stimulated steroidogenesis in rat Leydig cells (Dix et al., 1984, 1985).

Lipoxygenase products probably affect the transfer of cholesterol to P450$_{scc}$. There was no effect of lipoxygenase inhibitors on 25-hydroxycholesterol metabolism to pregnenolone or on steroid synthesis from pregnenolone. In other cells, 5-HPETE has been associated with calcium-dependent processes, while cholesterol metabolism in adrenal mitochondria is particularly Ca-sensitive (Przylipiak et al, 1990). Several other arachidonic acid metabolites (14,15-epoxide, 12- and 15-lipoxygenase products) have been reported in glomerulosa cells following angiotensin-II stimulation, and 12-lipoxygenase products have been implicated in aldosterone synthesis (Campbell et al., 1991).

4. GTP and Ca^{++}

When isolated rat adrenal mitochondria are incubated with GTP, metabolism of exogenously added cholesterol is stimulated (Xu et al., 1989, 1991). This stimulation reflects enhanced transfer of cholesterol from outer to inner mitochondrial membranes. This effect was additive with stimulation by SCP$_2$ in a manner consistent with cholesterol transfer to the mitochondria through a complex with SCP$_2$. GTPγS is inactive but inhibits the activity of GTP, suggesting that hydrolysis of the nucleotide at a mitochondrial G protein may be necessary to mediate intermembrane cholesterol transfer. Ca^{++} also stimulates cholesterol metabolism in adrenal mitochondria (Arthur et al., 1976). Comparable stimulations to those produced by GTP are also seen with low concentrations of buffered Ca^{++} [<1 μM] over a wide range of conditions. Addition of Ca^{++} and GTP produces only a small additional effect (Kowluru et al., 1995). The effect of Ca^{++} is unaffected by GTP$_\gamma$S and by Ruthenium Red, an inhibitor of the inner membrane Ca^{++} entry channel. Ca^{++} therefore stimulates cholesterol transfer by a mechanism that is independent of GTP and also of Ca^{++} uptake into the matrix.

When cholesterol accumulates in the outer membrane as a result of deficient *in vivo* intermembrane cholesterol transport (e.g., cycloheximide depletion of pp30) low levels of Ca^{++}, but not GTP, stimulate cholesterol transfer to P450$_{scc}$. This clearly distinct effect of Ca^{++} is blocked by Ruthenium Red indicating a site of action in the matrix. This suggests that cholesterol may slowly enter a second outer membrane pool that is regulated by Ca^{++} and pp30 and which does not participate in the *in vitro* GTP-sensitive transfer process. Interestingly cholesterol sulfate potently inhibits intermembrane cholesterol transfer possibly at the level of the Ca^{++}/GTP regulated process (Lambeth et al., 1987).

5. Changes in Adrenal Lipids

ACTH stimulates a very rapid net synthesis of phospholipids, apparently secondary to *de novo* synthesis of phosphatidic acid (Cozza et al., 1990; Igarashi and Kimura, 1984). This increase in lipid synthesis results in an elevation of phosphatidyl inositol (PI) and phosphatidyl ethanolamine (PE) in the mitochondrial outer membrane (Privalle et al., 1987). Other studies indicate an ACTH-induced enrichment of the PE fraction with the polyunsaturated (22:6) adrenic acid (Igarashi and Kimura, 1986). Adrenic acid-substituted PE, like cardiolipin, directly stimulates mitochondrial steroidogenesis. This increase, taken together with the enrichment of cardiolipin in contact sites (Ardail et al., 1990), suggests that cholesterol may actually transfer through hexagonal phase structures in which the normal bilayer inverts. This membrane inversion is favored by both polyunsaturated PE and cardiolipin (Killian and de Kruijff, 1986) and has been observed in mitochondria at suspected contact sites (Van Venetie and Verkleij, 1982). In one possible mechanism (Fig. 4), activation of PBR/porin complexes in the outer membrane changes intermembrane ion concentrations, notably Ca^{++}, resulting in formation of hexogonal cholesterol-lipid structures around these protein complexes. The p30 StAR proteins provide a cAMP-sensitive mechanism for subsequent uptake by the inner membrane.

6. Heterogeneity of the Intramitochondrial Cholesterol Pool

Studies with rat adrenal mitochondria point towards additional requirements for metabolism of cholesterol within the inner mitochondria (McNamara and Jefcoate, 1990 a and b). Succinate generates NADPH through an energy dependent NADH/NADP transhydrogenase that is linked to succinate dehydrogenase (Robinson and Stevenson, 1972). NADPH generation from succinate is less effective than from isocitrate, as evidenced by lower activities for metabolism of 20α-hydroxycholesterol at $P450_{scc}$ and deoxycorticosterone at $P450_{11\beta}$.

Metabolism supported by succinate is almost completely lost after even a 2 min preincubation prior to addition of reductant. NADPH generation, as evidenced by metabolism of 20α-hydroxycholesterol or deoxycorticosterone, is only slightly decreased. This preincubation-induced transition corresponds to a loss of succinate-dependent ATP generation and probably membrane potential. These same processes are highly sensitive to inhibition by Ca^{++} and are each equally blocked by EGTA and Ruthenium Red and partially prevented by cyclosporin A. The latter binds to immunophilins, which may therefore mediate Ca-sensitive opening of low specificity inner membrane channels (Connern and Halestrap, 1994). The effects of these inhibitors indicate uptake of Ca^{++} into the matrix and inhibition through opening of a low permeability channel (Yamazaki et al., 1995). This can in turn affect cholesterol transfer by causing matrix swelling and increased contacts (Fig. 4).

Cholesterol availability to $P450_{scc}$, at least when succinate is the reductant, is evidently uniquely sensitive to this transition, possibly through changes in mem-

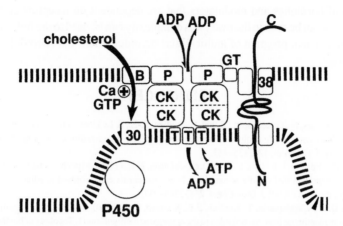

Figure 5. A model for cholesterol transfer and metabolism in adrenal mitochondria. Cholesterol is transferred from outer to inner membrane mediated by various factors at contact sites and then moves to intramatrix vesicles (B). In this model, only metabolism in the inner membrane (Pool A) is mediated by succinate dehydrogenase (SDH) and NADH/NADP transhydrogenase (TH). NADPH generated by NADP/isocitrate dehydrogenase mediates metabolism at all P450$_{scc}$.

brane fluidity that are linked to the mitochondrial energy state (Kusnetzov et al., 1989). In rat adrenal mitochondria we envisage two pools of reactive cholesterol (Fig. 5) that differ in their dependence on mitochondrial membrane potential. Cholesterol is transferred first to an inner membrane pool that remains mobile only when the membrane potential is maintained. Much of the cholesterol then transfers to matrix vesicles which are a clear characteristic of highly active steroidogenic mitochondria (Farkash et al., 1986). This second pool should be available to cytochrome P450$_{scc}$ independent of mitochondrial energy state. This model is dependent on mitochondrial metabolism pathways. In bovine adrenal mitochondria, very high levels of malate enzyme act downstream from succinate to generate NADPH (Mandella and Sauer, 1975), avoiding reliance on the transhydrogenase.

VI. CONCLUSION

Cholesterol metabolism in steroidogenic cells is evidently dependent not only on the level of expression of cytochrome P450$_{scc}$ and its associated electron transfer proteins but also on proteins and energy sources involved in reductant transfer and on the levels, modification and organization of many proteins involved in the cholesterol transfer process. Hormones affect these proteins both through fast acting processes, usually kinases, and slow processes typically involving transcriptional activation. Each of these processes will be highly cell specific involving the actions of hormones and modulators that are dependent on receptors on the cell surface. As can be seen in the ovary this complexity can be used to the full to provide a time-dependent program of multicellular steroidogenesis and steroid-dependent physiology.

REFERENCES

Akwa, Y., Sananès, N., Gouèzou, M., Robel, P., Baulieu, E.-E., & Le Goascogne, C. (1993). Astrocytes and Neurosteroids: Metabolism of Pregnenolone and Dehydroepiandrosterone. Regulation by Cell Density. J. Cell Biol. 121, 135–143.

Alberta, J.A., Epstein, L.F., Pon, L.A., & Orme-Johnson, N.R. (1989). Mitochondrial localization of a phosphoprotein that rapidly accumulates in adrenal cortex cells exposed to adrenocorticotropic hormone or to cAMP. J. Biol. Chem. 264, 2368–2372.

Allmann, D.W., Wakabayashi, T., Korman, E.F., & Green, D.E. (1970). Studies on the transition of the cristae membrane from the orthodox to the aggregated configuration I: Topology of bovine adrenal cortex mitochondria in the orthodox configuration. Bioenergetics 1, 73–86.

Amsterdam, A., & Suh, B.S. (1991). An inducible functional peripheral benzodiazepine receptor in mitochondria of steroidogenic granulosa cells. Endocrinology 129, 503–510.

Ardail, D., Privat, J.-P., Egret-Charlier, M., Levrat, C., Lerme, F., & Louisot, P. (1990). Mitochondrial contact sites: Lipid composition & dynamics. J. Biol. Chem. 265, 18797–18802.

Arthur, J.R., & Boyd, G.S. (1976). The effect of inhibitors of protein synthesis on cholesterol side-chain cleavage in the mitochondria of luteinized rat ovaries. Eur. J. Biochem. 49, 117–127.

Arthur, J.R., Mason, J.I., & Boyd, G.S. (1976). Effect of calcium ions on metabolism of exogenous cholesterol by rat adrenal mitochondria. FEBS Lett. 66, 206–209.

Azhar, S., Menon, M., & Menon, K.M.J. (1981). Receptor-mediated gonadotropin action in the ovary. Demonstration of acute dependence of rat luteal cells on exogenously supplied sterols. Biochim. Biophys. Acta 665, 362–375.

Bar, L.K., Barenholz, Y., & Thompson, T.E. (1987). Dependence on phospholipid composition of the fraction or cholesterol undergoing spontaneous exchange between small unilamellar vesicles. Biochemistry 26, 5460–5465.

Begeot, M., Shetty, U., Kilgore, M., Waterman, M., & Simpson, E. (1993). Regulation of expression of the CYP11A (P450$_{scc}$) gene in bovine ovarian luteal cells by forskolin and phorbol esters. J. Biol. Chem. 268, 17317–17325.

Benz, R., Kottke, M., & Brdiczka, D. (1990). The cationically selective state of the mitochondrial outer membrane pore: A study with intact mitochondria and reconstituted mitochondrial porin. Biochim. Biophys. Acta 1022, 311–318.

Besedovsky, H.O., Sel Rey, A., Klusman, I., Furukawa, H., Monge Arditi, G., & Kabiersch, A. (1991). Cytokines as modulators of the hypothalamus-pituitary-adrenal axis. J. Biochem. Molec. Biol. 40, 613–618.

Besman, M.J., Yanagibashi, K., Lee, T.D., Kawamura, M., Hall, P.F., & Shively, J.E. (1989). Identification of des-(Gly-Ile)-endozepine as an effector of corticotropin-dependent adrenal steroidogenesis: Stimulation of cholesterol delivery is mediated by the peripheral benzodiazepine receptor. Proc. Natl. Acad. Sci. USA 86, 4897–4901.

Biermans, W., Bakker, A., & Jacob, W. (1990). Contact sites between inner and outer mitochondrial membrane: A dynamic microcompartment for creatine kinase activity. Biochim. Biophys. Acta 1018, 225–228.

Blahos, J., Whalin, M.E., & Krueger (1995). Identification and purification of a 10kDa-protein associated with mitochondrial benzodiazepine receptors. J. Biol. Chem. 270, 20285–20291.

Blanchette Mackie, E.J., Dwyer, N.K., Vanier, M.T., Solol, J., Merrick, H.F., Comly, M.E., Argoff, C.E., & Pentchev, P.G. (1989). Type C Niemann-Pick disease: Dimethyl sulfoxide moderates abnormal LDL-cholesterol processing in mutant fibroblasts. Biochem. Biophys. Acta. 1006, 219–226.

Boujrad, N., Gaillard, J-L., Garnier, M., & Papadopoulos, V. (1994). Acute action of choriogonadotropin on Leydig tumor cells: Induction of a high affinity benzodiazepine-binding site related to steroid biosynthesis. Endocrinology 1576–1583.

Boujrad, N., Ogwuegba, S.O., Garnier, M., Lee, C.-H., Martin, B.M., & Papadopoulos, V. (1995). Identification of a stimulator of steroid hormone synthesis isolated from testis. Science 268, 1609–1612.

Boggaram, V., John, M.E., Simpson, E.R., & Waterman, M.R. (1989). Effect of ACTH on the stability of mRNAs encoding bovine adrenocortical P-450scc, P-45011β, P-45017α, P-450C21 and adrenodoxin. Biochem. Biophyus. Res. Commun. 160, 1227–1232.

Boujrad, N., Hudson, J.R., Jr., & Papadopoulos, V. (1993). Inhibition of hormone-stimulated steroidogenesis in cultured leydig tumor cells by a cholesterol-linked phosphorothioate oligodeoxynucleotide antisense to diazepam binding inhibitor. Proc. Natl. Acad. Sci. USA 90, 5728–5733.

Braley, L.M., Menachery, A.I., Brown, E.M., & Williams, G.H. (1986). Comparative effect of angiotensin II, potassium, adrenocorticotropin, and cyclic adenosine 3',5'-monophosphate on cytosolic calcium in rat adrenal cells. Endocrinology 119, 1010–1019.

Brdiczka, D. (1991). Contact sites between mitochondrial envelope membranes, structure, and function in energy- and protein-transfer. Biochim. Biophys. Acta 1071, 291–312.

Brdiczka, D., Knoll, G., Riesinger, I., Weiler, U., Klug, G., Benz, R., & Krause, J. (1986). Microcompartmentation at the mitochondrial surface: Its function in metabolic regulation. In: Myocardial and Muscle Bioenergetics (Brantbar, N., ed.), pp. 55–70. Plenum Press, New York.

Brown, A.S., Hall, P.F., Shoyab, M., & Papadopoulos, V. (1992). Endozepine/diazepam binding inhibitor in adrenocortical and Leydig cell lines: Absence of hormonal regulation. Mol. Cell. Endocrinol. 83, 1–9.

Brown, M.S., & Goldstein, J.L. (1986). A Receptor-mediated pathway for cholesterol homeostasis. Science 232, 34–48.

Brownie, A.C., Alfano, J., Jefcoate, C.R., Beinert, H., & Orme-Johnson, W.H. (1973). Effect of ACTH on adrenal mitochondrial cytochrome P450 in rat. Ann. N.Y. Acad. Sci. 212, 344–360.

Butler, J.D., Blanchette-Mackie, J., Goldin, E., O'Neil, R.R., Carstea, G., Roff, C.F., Patterson, M.C., Patel, S., Comly, M.E., Cooney, A., Vanier, M.T., Brady, R.O., & Pentchev, P.G. (1992). Progesterone blocks cholesterol translocation from lysosomes. J. Biol. Chem. 267, 23797–23805.

Cali, J.J., & Russell, D.W. (1991). Characterization of human sterol 27-hydroxylase. J. Biol. Chem. 266, 7774–7778.

Campbell, W.B., Brady, M.T., Rosolowsky, L.J., & Falck, J.R. (1991). Metabolism of arachidonic acid by rat adrenal glomerulosa cells: Synthesis of hydroxyeicosatetraenoic acids and epoxyeicosatrienoic acids. Endocrinology 128, 2183–2194.

Carr, B. R., Rainey, W. E., & Mason, J. I. (1987). The role of calmodulin antagonists on steroidogenesis by fetal zone cells of the human fetal adrenal gland. Endocrinology 120, 995–999.

Chanderbhan, R.F., Kharroubi, A.T., Noland, B.J., Scallen, T.J., & Vahouny, G.V. (1986). SCP$_2$: Further evidence for its role in adrenal steroidogenesis. Endocrine Res. 12, 351–360.

Chanderbhan, R., Noland, B.J., Scallen, T.J., & Vahouny, G.V. (1982). Sterol carrier protein$_2$: Delivery of cholesterol from adrenal lipid droplets to mitochondria for pregnenolone synthesis. J. Biol. Chem. 257, 8928–8934.

Chashchin, V.L., Lapko, V.N., Adamovich, T.B., Lapko, A.G., Kuprina, N.S., & Akhrem, A.A. (1986). Primary structure of the cholesterol side-chain cleavage cytochrome P-450 from bovine adrenocortical mitochondria and some aspects of its functioning on a structural level. Biochim. Biophys. Acta 871, 217–223.

Cheng, B., Hsu, D.K., & Kimura, T. (1985). Utilization of mitochondrial membrane cholesterol by cytochrome P450-dependent cholesterol side-chain cleavage reaction in bovine adrenocortical mitochondria: Steroidogenic and non-steroidogenic pools of cholesterol in the mitochondrial inner membranes. Mol. Cell. Endocrinol. 40, 233–243.

Choi, Y.-S., Stocco, D.M., & Freeman, D.A. (1995). Diethylumbelliferyl phosphate inhibits steroidogenesis by interfering with a long-lived factor acting between protein kinase A activation and induction of the steroidogenic acute regulatory protein (STAR). Eur. J. Biochem. 234, 680–685.

Churchill, P.F., & Kimura, T. (1979). Topological studies of cytochromes $P450_{scc}$ and $P450_{11b}$ in bovine adrenocortical inner mitochondrial membranes. J. Biol. Chem. 254, 10443–10448.

Clark, B.J., Ikeda, Y., Parker, K.L., & Stocco, D.M. (1995). Hormonal and developmental regulation of the steroidogenic acute regulatory (STAR) protein. Molec. Endocrinol. 9, 1346–1355.

Clark, B.J., Wells, J., King, S.R., & Stocco, D.M. (1994). The purification, cloning, and expression of a novel luteinizing hormone-induced mitochondrial protein in MA-10 mouse Leydig tumor cells: Characterization of the steroidogenic acute regulatory protein (StAR). J. Biol. Chem., 269, 28314–28322.

Coghlan, V. M., & Vickery, L. E. (1991). Site-specific mutations in human ferredoxin that affect binding to ferredoxin reductase and cytochrome $P450_{scc}$. J. Biol. Chem. 266, 18606–18612.

Compagnone, N.A., Bulfone, A., Rubenstein, J.L.R., & Mellon, S.H. (1995). Expression of the steroidogenic enzyme P450 scc in the central and peripheral nervous systems during rodent embryogenesis. Endocrinology 136, 2689–2696.

Cooke, B.A., Janszen, F.H.A., Clotscher, W.F., & van der Molen, H.J. (1975). Effect of protein synthesis inhibitors on testosterone production in rat testis interstitial tissue and Leydig cell preparations. Biochem. J. 150, 413–418.

Corbin, C.J., Graham-Lorence, S., McPhaul, M., Mason, J.I., Mendelson, C.R., & Simpson, E.R. (1988). Proc. Natl. Acad. Sci. USA 85, 8948–8952.

Cozza, E.N., del Carmen Vila, M., Acevedo-Duncan, M., Gomez-Sanchez, C.E., & Farese, R.V. (1990). ACTH increases *de novo* synthesis of diacylglycerol and translocates protein kinase C in primary cultures of calf adrenal glomerulosa cells. J. Steroid Biochem. 35, 343–351.

Crivello, J.F., & Jefcoate, C.R. (1978). Mechanism of corticotropin action in rat adrenal cells. 1. Effect of inhibitors of protein synthesis and of microfilament formation on corticosterone synthesis. Biochim. Biophys. Acta 542, 315–329.

Dalziel, K. (1980). Isocitrate dehydrogenase and related oxidative decarboxylases. FEBS Lett. 117, K45–K55.

Dhariwal, M. S., & Jefcoate, C. R. (1989). Cholesterol metabolism by purified cytochrome $P-450_{scc}$ is highly stimulated by octyl glucoside and stearic acid exclusively in large unilamellar phospholipid vesicles. Biochemistry 28, 8397–8402.

Dhariwal, M. S., Kowluru, R. A., & Jefcoate, C. R. (1991). Cytochrome $P-450_{scc}$ induces vesicle aggregation through a secondary interaction at the adrenodoxin binding sites (in competition with protein exchange). Biochemistry 30, 4940–4949.

DiBartolomeis, M.J., & Jefcoate, C.R. (1984). Characterization of the acute stimulation of steroidogenesis in primary bovine adrenal cortical cell cultures. J. Biol. Chem. 259, 10159–10167.

Dix, C.J., Habberfield, A.D., Sullivan, M.H.F., & Cooke, B.A. (1984). Inhibition of steroid production in Leydig cells by non-steroidal anti-inflammatory and related compounds: Evidence for the involvement of lipoxygenase products in steroidogenesis. Biochem. J. 219, 529–537.

Dix, C.J., Habberfield, A.D., Sullivan, M.H., & Cooke, B.A. (1985). Evidence for the involvement of lipoxygenase products in steroidogenesis. Biochem. Soc. Trans. B. 60–63.

Dory, L. (1989). Synthesis and secretion of apoE in thioglycolate-elicited mouse peritoneal macrophages: Effect of cholesterol efflux. J. Lipid Res. 30, 809–816.

Dufau, M.L. (1988). Endocrine Regulation and Communicating Functions in the Leydig Cell. Ann. Rev. Physiol. 50, 483–508.

Dyer, C.A., & Curtiss, L.K. (1988). Apoprotein E-rich high density lipoproteins inhibit ovarian androgen synthesis. J. Biol. Chem. 263, 10965–10973.

Elliott, M.E., & Goodfriend, T.L. (1986). Inhibition of aldosterone synthesis by atrial natriuretic factor. Fed. Proc. 45, 2376–2381.

Elliott, M. E., Goodfriend, T. L., & Jefcoate, C. R. (1993). Bovine adrenal glomerulosa and fasciculata cells exhibit 28.5-kilodalton proteins sensitive to angiotensin, other agonists, and atrial natriuretic peptide. Endocrinology 133, 1669–1677.

Epstein, L.F., & Orme-Johnson, N.R. (1991). Regulation of steroid hormone biosynthesis: Identification of precursors of a phosphoprotein targeted to the mitochondrion in stimulated rat adrenal cortex cells. J. Biol. Chem. 266, 19739–19745.

Fahien, L.A., MacDonald, M.J., Teller, J.K., Fibich, B., & Fahien, C.M. (1989). Kinetic advantages of hetero-enzyme complexes with glutamate dehydrogenase and the a-ketoglutarate dehydrogenase complex. J. Biol. Chem. 264, 12303–12312.

Farese, R.V., & Sabir, A.M. (1980). Polyphosphoinositides: Stimulator of mitochondrial cholesterol side chain cleavage and possible identification as an adrenocorticotropin-induced, cycloheximide-sensitive, cytsolic, steroidogenic factor. Endocrinology 106, 1869–1979.

Farkash, Y., Timberg, R., & Orly, J. (1986). Preparation of antiserum to rat cytochrome P-450 cholesterol side chain cleavage, and its use for ultrastructural localization of the immunoreactive enzyme by protein A-gold technique. Endocrinology 118, 1353–1365.

Faust, J.R., Goldstein, J.L., & Brown, M.S. (1977). Receptor-mediated uptake of LDL and utilization of its cholesterol for steroid synthesis in cultured mouse adrenal cells. J. Biol. Chem. 252, 4861–4871.

Fowler, S.D., & Brown, W.J. (1984). Lysosomal acid lipase in lipases. In: Lipases (Bergstrom, B., and Brockman, H.L., eds.), pp. 329–364. Elsevier Science Publishing Co., Inc., New York.

Freeman, D.A. (1987). cAMP mediated modification of cholesterol traffic in Leydig tumor cells. J. Biol. Chem. 262, 13061–13068.

Freeman, D.A. (1989). Plasma membrane cholesterol: Removal and insertion into the membrane and utilization as a substrate for steroidogenesis. Endocrinology 124, 2527–2535.

Frustaci, J., Mertz, L.M., & Pedersen, R.C. (1989). Steroidogenesis activator polypeptide (SAP) in the guinea pig adrenal cortex. Mol. Cell. Endocrinol. 64, 137–143.

Garnier, M., Boujrad, N., Ogwuegbu, S.O., Hudson, Jr., J.R., & Papdopoulos, V. The polypeptide diazepam-binding inhibitor and a higher affinity mitochondrial peripheral-type benzodiazepine receptor sustain constitutive steroidogenesis in the R2C Leydig tumor cell line. J. Biol. Chem., 269, 22105–22112.

Garnier, M., Dimchev, A.B., Boujrad, N., Price, J.M., Musto, N.A., & Papadopoulos, V. (1994). *In vitro* reconstitution of a functional peripheral-type benzodiazepine receptor from mouse leydig tumor cells. Mol. Pharmacol. 45, 201–211.

Garren, L.D., Gill, G.N., Masui, H., & Walton, G.M. (1971). On the mechanism of action of ACTH. Rec. Prog. Horm. Res. 27, 433–478.

Gavey, K.L., Noland, B.J., & Scallen, T.J. (1981). The participation of sterol carrier protein$_2$ in the conversion of cholesterol to cholesterol ester rat liver microsomes. J. Biol. Chem. 256, 2993–2999.

Gibori, G., Basuray, R., & McReynolds, B. (1981). Placental-derived regulators and the complex control of luteal cell function. Endocrinology 108, 2060–2066.

Gibori, G., Khan, I., Warshaw, M.L., McLean, M.P., Puryear, T.K., Nelson, S., Durkee, T.J., Azhar, S., Steinschneider, A., & Rao, M.C. Placental-derived regulators and the complex control of luteal cell function. Recent Progress in Hormone Research, 44, 377–425.

Gibori, G., Richards, J.S., & Keyes, P.L. (1979). Differential action of decidual luteotropin on luteal and follicular production of testosterone and estradiol. Biol. Reprod. 21, 419–423.

Gocze, P. M., & Freeman, D. A. (1992). A cholesteryl ester hydrolase inhibitor blocks cholesterol translocation into the mitochondria of MA-10 leydig tumor cells. Endocrinology 131, 2972–2978.

Ghosh, D. K., Dunham, W. R., Sands, R. H., & Menon, K. M. J. (1987). Regulation of cholesterol side-chain cleavage enzyme activity by gonadotropin in rat corpus luteum. Endocrinology 121, 21–27.

Goldring, N.B., Farkash, Y., Goldschmidt, D., & Orly, J. (1986). Immunofluorescent Probing of the Mitochondrial Cytochrome P450$_{scc}$ in Differentiating Granulosa Cells in Culture. Endocrinology 119, 2821–2832.

Goldstein, J.L., & Brown, M.S. (1990). Regulation of the mevalonate pathway. Nature 343, 425–430.

Golos, T.G., Strauss, J.F., & Miller, W.L. (1987a). Regulation of low density lipoprotein receptor and Cytochrome P450$_{scc}$ mRNA levels in human granulosa cells. J. Steroid Biochem. 27, 767–773.

Guidotti, A., Berkovich, A., Muhkin, A., & Costa, E. (1990). Diazepam-binding inhibitor: Response to Knudsen and Nielsen. Biochem. J. 265, 928–929.

Gupta, P., Franco–Saenz, R., & Mulrow, P.J. (1993). Transforming Growth Factor-β1 Inhibits Aldosterone Biosynthesis in Cultured Bovine Zona Glomerulosa Cells. Endocrinology 132, 1184–1188.

Gwynne, J.T., & Mahaffee, D.D. (1989). Rat adrenal uptake and metabolism of high density lipoprotein cholesteryl ester. J. Biol. Chem. 264, 8141–8150.

Gwynne, J.T., & Strauss, J.F. III. (1982). The role of lipoproteins in steroidogenesis and cholesterol metabolism in steroidogenic glands. Endocrine Rev. 3, 299–329.

Hall, P.F. (1995). The roles of microfilaments and intermediate filaments in the regulation of steroid synthesis. J. Steroid Biochem. Molec. Biol. 55, 601–605.

Hall, P.F., & Almabobi, G. (1992). The role of cytoskeleton in the regulation of steroidogenesis. J. Steroid Biochem. Mol. Biol. 43, 769–779.

Hall, P.F., Charponnier, C., Nakamura, M., & Gabbiani, G. (1979). The role of microfilaments in the response of adrenal tumor cells to ACTH. J. Biol. Chem. 254, 9080–9084.

Hall, S. H., Berthelon, M.-C., Avallet, O., & Saez, J. M. (1991). Regulation of c-*fos*, c-*jun*, *jun*-B, and c-*myc* messenger ribonucleic acids by gonadotropin and growth factors in cultured pig leydig cell. Endocrinology 129, 1243–1249.

Hanukoglu, I. (1992). Steroidogenic enzymes: Structure, function and role in regulation of steroid hormone biosynthesis. J. Steroid Biochem. Molec. Biol. 43, 779–804.

Hanukoglu, I., Feuchtwanger, R., & Hanukoglu, A. (1990). Mechanism of corticotropin and cAMP induction of mitochondrial cytochrome P450 system enzymes in adrenal cortex. J. Biol. Chem. 265, 20602–20608.

Hanukoglu, I., & Hanukoglu, Z. (1986). Stoichiometry of mitochondrial's, cytochromes P450, adrenodoxin and adrenodoxin reductase in adrenal cortex and corpus luteum. Eur. J. Biochem. 157, 27–31.

Hanukoglu, I., & Jefcoate, C.R. (1980). Adrenal mitochondrial cytochrome P450$_{scc}$: Mechanism of electron transport by adrendoxin. J. Biol. Chem. 255, 3057–3061.

Hanukoglu, I., Rapoport, R., Weiner, L., & Sklan, D. (1993). Electron leakage from the mitochondrial NADPH-adrenodoxin reductase-adrenodoxin-P450$_{scc}$ (cholesterol side chain cleavage) system. Arch. Biochem. Biophys. 305, 489–498.

Harrison, E.H., Bernard, D.W., Scholm, P., Quinn, D.M., Rothblat, G.H., & Glick, J.M. (1990). Inhibitors of neutral cholesteryl ester hydrolase. J. Lipid Res. 31, 2187–2193.

Heyl, B. L., Tyrrell, D. J., & Lambeth, J. D. (1986). Cytochrome P-450$_{scc}$-substrate interactions: Roles of the 3β- and side chain hydroxyls in binding to oxidized and reduced forms of the enzyme. J. Biol. Chem. 261, 2743–2749.

Hornsby, P.J. (1988). In: Hormones and Their Actions, Part II. (Cooke, B.A., King, R.J.B., & Van der Molen, H.J., eds.), pp. 193–210. Elsevier Science Publishers, Amsterdam-New York-Oxford.

Hornsby, P.J. (1991). Regulation of steroid hydroxylases in normal and SV40 T antigen-transfected bovine adrenocortical cells in long-term culture. Endocrine Res. 17, 109–134.

Hornsby, P.J., Cheng, C.Y., Lala, D.S., Maghsoudlou, S.S., Raju, S.G., & Yang, L. (1992). Changes in gene expression during senescence of adrenocortical cells in culture. J. Steroid Biochem. Molec. Biol. 43, 951–960.

Hornsby, P.J., & Gill, G.N. (1981). Regulation of responsiveness of cultured adrenal cells to adrenocorticotropin and prostaglandin E$_1$: Cell density, cell division, and inhibitors of protein synthesis. Endocrinology 108, 183–188.

Hornsby, P.J., Sepehr, S.M., Cheng, V., & Cheng, C.Y. (1989). cAMP-mediated cytoskeletal effects in adrenal cells are modified by serum, insulin, insulin-like growth factor-J and an antibody against urokinase plasminogen activator. Mol. Cell. Endocrinol. 67, 185–193.

Hotta, M., & Baird, A. (1987). The inhibition of low density lipoprotein metabolism by transforming growth factor-b mediates its effects on steroidogenesis in bovine adrenocortical cells *in vitro*. Endocrinology 121, 150–159.

Hurwitz, A., Payne, D.W., Packmann, J.N., Andreani, C.L., Resnick, C.E., Hernandez, E.R., & Adashi, E.Y. (1991). Cytokine-mediated regulation of ovarian function. Endocrinology 129, 1250–1256.

Hwang, S.T., Wachter, C., & Schatz, G. (1991). Protein import into the yeast mitochondrial matrix: A new translocation intermediate between the two mitochondrial membranes. J. Biol. Chem. 266, 21083–21089.

Igarashi, Y., & Kimura, T. (1984). Adrenocorticotropic hormone-mediated changes in rat adrenal mitochondrial phospholipids. J. Biol. Chem. 259, 10745–10753.

Igarashi, Y., & Kimura, T. (1986). Adrenic acid content in rat adrenal mitochondrial phosphatidylethanolamine and its relation to ACTH-mediated stimulation of cholesterol side chain cleavage reaction. J. Biol. Chem. 261, 14118–14124.

Ikushiro, S.-I., Kominami, S., & Takemori, S. (1992). Adrenal P-450$_{scc}$ modulates activity of P-450$_{11b}$ in liposomal and mitochondrial membranes: Implication of P-450$_{scc}$ in zone specificity of aldosterone biosynthesis in bovine adrenal. J. Biol. Chem. 267, 1464–1469.

Ilvesmäki, V., & Voutilainen, R. (1991). Interaction of phorbol ester and adrenocorticotropin in the regulation of steroidogenic P450 genes in human fetal and adult adrenal cell cultures. Endocrinology 128, 1450–1458.

Jamal, Z., Suffolk, R.A., Boyd, G.S., & Suckling, K.E. (1985). Metabolism of cholesteryl ester in monolayers of bovine adrenal cortical cells. Effect of an inhibitor of acyl-CoA: Cholesterol acyltransferase. Biochim. Biophys. Acta 834, 230–237.

James, V.H.T. (1991) Adrenal steroid endocrinology — some unsolved problems. J. Steroid Biochem. Molec. Biol., 39, 867–821.

Jefcoate, C. R. (1982). pH modulation of ligand binding to adrenal mitochondrial cytochrome P-450$_{scc}$. J. Biol. Chem. 257, 4731–4737.

Jefcoate, C.R., DiBartolomeis, M.J., Williams, C.A., & McNamara, B.C. (1987). ACTH regulation of cholesterol movement in isolated adrenal cells. J. Steroid Biochem. 27, 721–729.

Jefcoate, C.R., McNamara, B.C., & Artemenko, J., & Yamazaki, T. (1992). Regulation of cholesterol movement to mitochondrial cytochrome P450$_{scc}$ in steroid hormone synthesis. J. Steroid Biochem. Molec. Biol. 43, 751–767.

Jefcoate, C.R., & Orme-Johnson, W.H. (1975). Cytochrome P450 of adrenal mitochondria: *In vitro* and *in vivo* changes in spin state. J. Biol. Chem. 250, 4671–4677.

Jefcoate, C. R., Orme-Johnson, W. H., & Beinert, H. (1976). Cytochrome P-450 of bovine adrenal mitochondria: Ligand binding to two forms resolved by EPR spectroscopy. J. Biol. Chem. 251, 3706–3715.

Jefcoate, C.R., Simpson, E.R., & Boyd, G.S. (1974). Spectral properties of rat adrenal mitochondrial cytochrome P450. Eur. J. Biochem. 42, 539–551.

John, M.E., John, M.C., Boggaram, U., Simpson, E.R., & Waterman, M.R. (1986). Transcriptional regulation of steroid hydroxylase genes by corticotropin. Proc. Natl. Acad. Sci. USA 83, 4715–4719.

Johnson, W.J., Chacko, G.K., Phillips, M.C., & Rothblat, G.H. (1990). Efflux of lysosomal cholesterol from cells. J. Biol. Chem. 265, 5546–5553.

Johnson, W.J., Mahlberg, F.H., Rothblat, G.H., & Phillips, M.C. (1991). Cholesterol transport between cells and high-density lipoproteins. Biochim. Biophys. Acta 1085, 273–298.

Jones, D.B., Marante, D., Williams, B.C., & Edwards, C.R.W. (1987). Adrenal synthesis of corticosterone in response to ACTH in rats is influenced by leukotriene A$_4$ and by lipoxygenase intermediates. J. Endocrinol. 112, 253–258.

Karaboyas, G.C., & Koritz, S.B. (1954). Identity of the site of action of cAMP and ACTH in corticosteroidogenesis in rat adrenal and beef adrenal cortex slices. Biochemistry 4, 462–468.

Keyes, P.L., & Wiltbank, M.C. (1988). Endocrine regulation of the corpus luteum. Ann. Rev. Physiol. 50, 465–482.

Khan, I., Sridaran, R., Johnson, D.C., & Gibori, G. (1987). Selective stimulation of luteal androgen biosynthesis by luteinizing hormone: Comparison of hormonal regulation of P450$_{17a}$ activity in corposa lutea and follicles. Endocrinology 121, 1312–1319.

Killian, J.A., & deKruijff, B. (1986). The influence of proteins and peptides on the phase properties of lipids. Chem. Phys. Lipids 40, 259–284.

Kim, Y., Aryoshi, N., Artimenko, I., Elliott, M., & Jeffcoate, C.R. (1996). Regulation of adrenal steroid synthesis. Steroids (In press).

Knudsen, J., & Nielsen, M. (1990). Diazepam-binding inhibitor: A neuropeptide and/or an acyl-CoA ester binding protein? Biochem. J. 265, 927–928.

Kojima, I., Kojima, K., & Rasmussen, H. (1985). Role of calcium and cAMP in the action of ACTH on aldosterone secretion. J. Biol. Chem. 260, 4248–4256.

Kominami, S., Onizuka, M., Tasaka-Marumoto, C., & Takemori, S. (1995). Cytochrome P450 transfer from adrenocortical submitochondrial particles to liposome membranes. Biochemistry 34, 4839–4845.

Koroscil, T.M., & Gallant, S. (1981). The phosphorylation of adrenal proteins in response to adreno-corticotropic hormone. J. Biol. Chem. 256, 6700–6707.

Kottke, M., Adam, V., Riesinger, I., Bremm, G., Bosch, W., Brdiczka, D., Sandri, G., & Panfili, E. (1988). Mitochondrial boundary membrane contact sites in brain: Points of hexokinase and creatine kinase location, and control of Ca^{2+} transport. Biochim. Biophys. Acta 935, 87–102.

Kovanen, P.T., Goldstein, J.L., Chappell, D.A., & Brown, M.S. (1980). Regulation of low density lipoprotein receptors by adrenocorticotropin in the adrenal gland of mice and rats in vivo. J. Biol. Chem. 255, 5591–5598.

Kowluru, R.A., George, R., & Jefcoate, C.R. (1983). Polyphosphoinositide activation of cholesterol side chain cleavage with purified cytochrome P-450$_{scc}$. J. Biol. Chem. 258, 8053–8059.

Kowluru, R., Yamazaki, T., McNamara, B.C., & Jefcoate, C.R. (1995). Metabolism of exogenous cholesterol by rat adrenal mitochondria is stimulated equally by physiological levels of free Ca^{2+} and by GTP. Mol. & Cell. Endocrinology, 107, 1–8.

Kramer, R.E., Ramey, W.E., Funkenstein, B., Dee, A., Simpson, E.R., & Waterman, M.R. (1984). Induction of synthesis of mitochondrial steroidogenic enzymes of bovine adrenal cortical cells by analogs of cAMP. J. Biol. Chem. 259, 707–715.

Krause, J., Hay, R., Kowollik, C., & Brdiczka, D. (1986). Cross-linking analysis of yeast mitochondrial outer membrane. Biochim. Biophys. Acta 860, 690–698.

Kreuger, R.J., & Orme-Johnson, N.R. (1983). Acute adronocortocotropic hormone stimulation of adrenal corticosteroidogenesis. Discovery of a rapidly induced protein. J. Biol. Chem. 258, 10159–10167.

Kruger, K.E., & Papadopolous, V. (1990). Peripheral-type benzodiazepine receptors mediate transloca-tion of cholesterol from outer to inner mitochondrial membranes in adrenocortical cells. J. Biol. Chem. 265, 15015–15022.

Kuznetsov, A.V., Khuchua, Z.A., Vassil'eva, E.V., Medved'eva, N.V., & Saks, V.A. (1989). Heart mitochondrial creatine kinase revisited: The outer mitochondrial membrane is not important for coupling of phosphocreatine production to oxidative phosphorylation. Arch. Biochem. Biophys. 268, 176–190.

Lambeth, J.D. (1981). Cytochrome P450$_{scc}$: Cardiolipin as an effector of activity of a mitochondrial cytochrome P450. J. Biol. Chem. 256, 4757–4762.

Lambeth, J.D., Seybert, D.W., & Kamin, H. (1979). Ionic effects on adrenal steroidogenic electron transport. The role of adrenodoxin as an electron shuttle. J. Biol. Chem. 254, 7255–7264.

Lambeth, J.D., Xu, X.X., & Glover, M. (1987). Cholesterol sulfate inhibits adrenal mitochondrial cholesterol side chain cleavage at a site distinct from cytochrome P-450$_{scc}$: Evidence for an intramitochondrial cholesterol translocator. J. Biol. Chem. 262, 9181–9187.

Langlois, D., Saez, J.M., & Begeot, P (1987). The potenting effects of phorbol ester on ACTH cholera toxin- and forskolin-induced cAMP production in cultured bovine adrenal cells is not mediated by inactivation of a-subunit of G$_i$ protein. Biochem. Biophys. Res. Commun. 146, 517–523.

Lauber, M.E., Bengtson, T., Waterman, M.R., & Simpson, E.R. (1991). Regulation of *CYP11A* (P450$_{scc}$) and *CYP17* (P450$_{17a}$) gene expression in bovine luteal cells in primary culture. J. Biol. Chem. 266, 11170–11175.

Le Goascogne, C., Robel, P., Gouezou, M., Sananes, N., Baulieu, E.E., & Waterman, M. (1987). Neurosteroids and Cytochrome P450$_{scc}$ in rat brain. Science 237, 1212–1215.

Lehoux, J.-G., Lefebvre, A., Belisle, S., & Bellabarba, D. (1990). Effect of ACTH suppression on adrenal 3-hydroxy-3-methylglutaryl coenzyme A reductase mRNA in 4-aminopyrazolopyrimidine-treated rats. Mol. Cell. Endocrinol. 69, 41–49.

Lin, D., Sugawara, T., Strauss, J.F., Clark, B.J., Stocco, D.M., Saenger, P., Rogol, A., & Miller, W.L. (1995). Role of steroidogenic acute regulatory protein in adrenal and gonadal steroidogenesis. Science 267, 1828–1831.

Lin, T., Wang, D. Nagpal, M.L., Calkins, J.H., Chang, W., & Chi, R. (1991). Interleukin-1 inhibits cytochrome P450$_{scc}$ expression in primary cultures of Leydig cells. Endocrinology 129, 1305–1311.

Liscum, L., & Underwood, K.W. (1995). Intracellular cholesterol transport and compartmentation. J. Biol. Chem. 270, 15443–15446.

Luo, X., Ikeda, Y., & Parker, K.L. (1994). A cell specific nuclear receptor is essential for adrenal and gonadal development and sexual differentiation. Cell 77, 481–490.

Mahaffee, D., Reitz, R.C., & Ney, R.L. (1974). Mechanism of action of ACTH-role of mitochondrial cholesterol accumulation in regulation of steroidogenesis. J. Biol. Chem. 249, 227–233.

Majewska, M.D., Harrison, N.L., Schwartz, R.D., Barker, J.L., & Paul, S.M. (1986). Steroid hormone metabolites are barbiturate-like modulators of the GABA receptor. Science 232, 1004–1007.

Mandella, R.D., & Sauer, L.A. (1975). The mitochondrial malic enzymes. I. Submitochondrial localization and purification and properties of the NAD(P)$^+$-dependent enzyme from adrenal cortex. J. Biol. Chem. 250, 5877–5884.

Mason, J.I., & Boyd, G.S. (1973). Mitochondrial cholesterol side-chain cleavage system in mitochondria of human term placenta. Eur. J. Biochem. 21, 308–321.

Mathew, P.A., Mason, J.I., Trant, J.M., & Waterman, M.R. (1990). Incorporation of steroidogenic pathways which produce cortisol and aldosterone from cholesterol into nonsteroidogenic cells. Mol. Cell. Endocrinol. 73, 73–80.

Matocha, M.F., & Waterman, M.R. (1986). Import and processing of P450$_{scc}$ and P450$_{11b}$ precursors by corpus luteal mitochondria: A processing pathway recognizing homologous and heterologous precursors. Arch. Biochem. Biophys. 250, 456–460.

Matteson, K.J., Chung, B.-C., Urdea, M.S., & Miller, W.L. (1986). Study of cholesterol side-chain cleavage (20, 22 desmolase) deficiency causing congenital lipoid adrenal hyperplasia using bovine-sequence P450$_{scc}$ oligodeoxyribonucleotide probes. Endocrinology 118, 1296–1305.

McLean, M.P., Puryear, T.K., Khan, I., Azhar, S., Billheimer, J.T., Orly, J., & Gibori, G. (1989). Estradiol regulation of sterol carrier protein-2 independent of cytochrome P450 side-chain cleavage expression in the rat corpus luteum. Endocrinology 125, 1337–1344.

McNamara, B.C., & Jefcoate, C.R. (1988). Synergistic stimulation of pregnenolone synthesis in rat adrenal mitochondria by n-hexane and cardiolipin. Arch. Biochem. Biophys. 260, 780–788.

McNamara, B.C., & Jefcoate, C.R. (1989). The role of SCP$_2$ in stimulation of steroidogenesis in rat adrenal mitochondria by adrenal cytosol. Arch. Biochem. Biophys. 275, 53–62.

McNamara, B.C., & Jefcoate, C.R. (1990a). Heterogeneous pools of cholesterol side-chain cleavage activity in adrenal mitochondria from ACTH-treated rats: Differential responses to different reducing precursors. Mol. Cell. Endocrinol. 73, 123–134.

McNamara, B.C., & Jefcoate, C.R. (1990b). Heterogenous pools of cholesterol side-chain cleavage activity in adrenal mitochondria from adrenocorticotropic hormone-treated rats: Reconstitution of the isocitrate response with succinate and low concentrations of isocitrate. Arch. Biochem. Biophys. 283, 464–471.

Meduri, G., Vuhai-Luuthi, T., Jolivet, A., & Milgrom, E. (1992). New functional zonation in the ovary as shown by immunohistochemistry of luteinizing hormone receptor. Endocrinology 131, 366–373.

Mellon, S.H., & Vaiss, E.C. (1989). cAMP regulates $P450_{scc}$ gene expression by a cycloheximide-insensitive mechanism in cultured mouse Leydig MA-10 cells. Proc. Natl. Acad. Sci. USA 86, 7775–7779.

Mendelson, C.R., Cleland, W.H., Smith, M.E., & Simpson, E.R. (1982). Regulation of aromatase activity of stromal cells derived from human adipose tissue. Endocrinology 111, 1077–1084.

Mertz, L.M., & Pedersen, R.C. (1989). The kinetics of steroidogenesis activator polypeptide in the rat adrenal cortex. J. Biol. Chem. 264, 15274–15279.

Mikami, K., Omura, M., Tamura, Y., & Yoshida, S. (1990). Possible site of action of 5-hydroperoxyeicosatetraenoic acid derived from arachidonic acid in ACTH-stimulated steroidogenesis in rat adrenal glands. J. Endocrinol. 125, 89–96.

Miller, W.L. (1988). Molecular biology of steroid hormone synthesis. Endocrin. Rev. 9, 295–318.

Mitani, F., Iizuka, T., Ueno, R., Ishimura, Y., Kimura, T., Izumi, S., Komatsu, N., & Watanabe, K. (1982). Regulation of cytochrome P450 activities in adrenocortical mitochondria from normal rats and human neoplastic tissues. Adv. Enz. Reg. 20, 213–231.

Miyamoto, N., Seo, H., Kanda, K., Hidaka, H., & Matsui, N. (1992). A 3′,5′-cyclic adenosine monophosphate-dependent pathway is responsible for a rapid increase in c-fos messenger ribonucleic acid by adrenocorticotropin. Endocrinology 130, 3231–3236.

Morohashi, K., Yoshioka, H., Gotoh, O., Okada, T., Yamamoto, K., Miyata, T., Sogawa, K., Fujii-Kuriyama, Y., & Omura, T. (1987). Molecular cloning and nucleotide sequence of DNA of mitochondrial cytochrome P-450(11b) of bovine adrenal cortex. J. Biochem., 102, 559–568.

Morohashi, K.I., Honda, S.-I., Inomata, Y., Handa, H., & Omura, T. (1992). A common trans-acting factor, Ad4-binding protein, to the promoters of steroidogenic P-450s. J. Biol. Chem. 267, 17913–17919.

Morohashi, K.I., Sogawa, K., Omura, T., & Fujii-Kuriyama, Y. (1987). Gene structure of human cytochrome P-450(SCC), cholesterol desmolase. J. Biochem. 101, 879–887.

Mrotek, J., & Hall, P.F. (1977). Response of adrenal tumor cells to ACTH: Site of inhibition by cytochalasin B. Biochemistry 16, 3177–3181.

Mukhin, A.G., Papadopoulos, V., Costa, E., & Krueger, K.E. (1989). Mitochondrial benzodiazepine receptors regulate steroid biosynthesis. Proc. Natl. Acad. Sci. USA 86, 9813–9816.

Muller, M., Moser, R., Cheneval, D., & Carafoli, E. (1985). Cardiolipin is the membrane receptor for mitochondrial creatine phosphokinase. J. Biol. Chem. 260, 3839–3843.

Nebert, D.W., & Feyereisen, R. (1994). Evolutionary argument for a connection between drug metabolism and signal transduction. In: Cytochrome P450, 8th International Conference (Lechner, M.C., ed.), pp. 3–13, John Libbey Eurotext, Paris.

Nebert, D.W., & Gonzales, F.J. (1987). P450 genes: Structure, evolution and regulation. Ann. Rev. Biochem. 56, 945–993.

Nishikawa, T., Mikami, K., Saito, Y., Tamura, Y., & Kumagai, A. (1981). Studies on cholesterol esterase in the rat adrenal. Endocrinology 108, 932–936.

Ogishima, T., Okada, Y., & Omura, T. (1985). Import and processing of the precursor of cytochrome P-450(SCC) by bovine adrenal cortex mitochondria. J. Biochem. 98, 781–791.

Ohno, Y., Yanagibashi, K., Yonezawa, Y., Ishiwatari, S., & Matsuba, M. (1983). A possible role of "steroidogenic factor" in the corticoidogenic response to ACTH; Effect of ACTH, cycloheximide and aminoglutethimide on the content of cholesterol in the outer and inner mitochondrial membrane of rat adrenal cortex. Endocrinol. Japan 30, 335–338.

Ohta, Y., Yanagibashi, K., Hara, T., Kawamura, M., & Kawato, S. (1991). Protein rotation study of cytochrome P-450 in submitochondrial particles: Effect of KCl and intermolecular interactions with redox partners. J. Biochem. 109, 594–599.

Papadopoulos, V., Boujrad, N., Ikonomovic, M.D., Ferrara, P., & Vidic, B. (1994). Topography of the Leydig cell mitochondrial peripheral-type benzodiazepine receptor. Mol. Cell Endocrinol. 104, R5–R9.

Papadopoulos, V., & Brown, A.S. (1995). Role of the peripheral-type benzodiazepine receptor and the polypeptide diazepam binding inhibitor in steroidogenesis, J. Steroid Biochem. Molec. Biol. 53, 103–110.

Papadopoulos, V., Mukhin, A.G., Costa, E., & Krueger, K.E. (1990). The peripheral-type benzodiazepine receptor is functionally-linked to Leydig cell steroidogenesis. J. Biol. Chem. 265, 3772–3779.

Papadopoulos, V., Nowzari, F.B., & Krueger, V.E. (1991). Hormone-stimulated steroidogenesis is coupled to mitochondrial benzodiazepine receptors. J. Biol. Chem. 266, 3682–3687.

Parker, K.L., & Schimmer, B.P. (1993). Transcriptional regulation of the adrenal steroidogenic enzymes. Trends Endocrinol. Metab. 4, 46–50.

Payne, A.H., Youngblood, G.L., Sha, L., Burgos-Trinidad, M., & Hammond, S.H. (1992). Hormonal regulation of steroidogenic enzyme gene expression in leydig cells. J. Steroid Biochem. Molec. Biol. 43, 895–906.

Pedersen, R.C., Brownie, A.C., & Ling, N. (1980). Pro-adrenocorticotropin/ endorphin-derived peptides: Coordinate action on adrenal steroidogenesis. Science 208, 1044–1046.

Pedersen, R.C., & Brownie, A.C. (1982). Immunoreactive g-melanotropin in rat pituitary and plasma. A partial characterization. Endocrinology 110, 825–834.

Pedersen, R.C., & Brownie, A.C. (1983). Cholesterol side-chain cleavage in the rat adrenal cortex: Isolation of a cycloheximide-sensitive activator peptide. Proc. Natl. Acad. Sci. U.S.A. 80, 1882–1886.

Pedersen, R.C., & Brownie, A.C. (1987). Steroidogenesis-activator polypeptide isolated from a rat Leydig cell tumor. Science 236, 188–190.

Perkins, L.M., & Payne, A.H. (1988). Quantification of $P450_{scc}$, $P450_{17a}$ and iron sulfur protein reductase in leydig cells and adrenals of inbred strains of mice. Endocrinology 123, 2675–2682.

Perrin, A., Pascal, O., Defaye, G., Feige, J.-J., & Chambaz, E.M. (1991). Transforming growth factor b_1 is a negative regulator of steroid 17a-hydroxylase expression in bovine adrenocortical cells. Endocrinology 128, 357–362.

Pfanner, N., Rassow, J., Wienhues, U., Hergersberg, C., Sollner, T., Becker, K., & Neupert, W. (1990). Contact sites between inner and outer membranes: Structure and role in protein translocation into the mitochondria. Biochim. Biophys. Acta 1018, 239–242.

Picado-Leonard, J., Voutilainen, R., Kao, L.C., Chung, B.C., Strauss, J.F., & Miller, W.R. (1988). Human adrenodoxin and cloning of 3 cDNA's and cycloheximide enhancement in JEG-3 cells. J. Biol. Chem. 263, 3240–3244.

Pikuleva, I., Kagawa, N., & Waterman, M. R. (1994). Site directed mutagenesis of tyrosines 93 and 94 of bovine cholesterol side chain cleavage cytochrome $P450_{scc}$. Symposium on Microsomes & Drug Oxidations, Toronto, Canada.

Pittman, R.C., Knecht, T.P., Rosenbaum, M.S., & Taylor, C.A., Jr. (1987). A nonendocytotic mechanism for the selective uptake of high density lipoprotein-associated cholesterol esters. J. Biol. Chem. 262, 2443–2450.

Pon, L., Moll, T., Vestweber, D., Marshallsay, B., & Schatz, G. (1989). Protein import into mitochondria ATP-dependent protein translocation activity in submitochondrial fraction enriched in contact sites and specific proteins. J. Cell Biol. 109, 2603–2616.

Poole, B., & Ohkuma, S. (1981). Effect of weak bases on the intralysosomal pH in mouse peritoneal macrophages. J. Cell. Biol. 90, 665–669.

Poulos, T.L. (1991). In: Methods in Enzymology (Waterman, M. R., and Johnson, E.F., eds.), pp. 11–20, Academic Press, New York.

Privalle, C.T., Crivello, J.F., & Jefcoate, C.R. (1983). Regulation of intramitochondrial cholesterol transfer to side-chain cleavage cytochrome P450$_{scc}$ in rat adrenal gland. Proc. Natl. Acad. Sci. USA 80, 702–706.

Privalle, C.T., McNamara, B.C., Dhariwal, M.S., & Jefcoate, C.R. (1987). ACTH control of cholesterol side-chain cleavage at adrenal mitochondrial cytochrome P450$_{scc}$. Regulation of intramitochondrial cholesterol transfer. Mol. Cell. Endocrinol. 53, 87–101.

Przylipiak, A., Kiesel, L., Habenicht, A.J.R., Przylipiak, M., & Runnebaum, B. (1990). Exogenous action of 5-lipoxygenase by its metabolites on luteinizing hormone release in rat pituitary cells. Mol. Cell. Endocrinol. 69, 33–39.

Pugielli, L., Rigotti, A., Greco, A.V., Santos, M.J., & Nervi, F. (1995). Sterol carrier protein-2 is involved in cholesterol transfer from endoplasmic reticulum to the plasma membrane in human fibroblasts. J. Biol. Chem. 270, 18723–18726.

Purvis, J.L., Canick, J.A., Mason, J.I., Estabrook, R.W., & McCarthy, J.L. (1973). Lifetime of adrenal cytochrome P-450 as influenced by ACTH. Ann. NY Acad. Sci., 212, 319–343.

Quinn, P.G., Dombransky, L.J., Chen, Y.J., & Payne, A.H. (1981). Serum lipoproteins increase testosterone production in HCG-sensitized Leydig cells. Endocrinology 109, 1790–1792.

Rainey, W.E., Kramer, R.E., Mason, J.I., & Shay, J.W. (1985). The effects of taxol, a microtubule-stabilizing drug, on steroidogenic cells. J. Cell. Physiol. 123, 17–24.

Rainey, W.E., Shay, J.W., & Mason, J.J. (1984). The effect of cytochalasin D on steroid production and stress fiber organization in cultured bovine adrenal cortical cells. Mol. Cell. Endocrinol. 35, 189–197.

Rainey, W.E., Shay, J.W., & Mason, J.I. (1986). ACTH induction of 3-hydroxy-3-methylglutaryl coenzyme A reductase, cholesterol biosynthesis, and steroidogenesis in primary cultures of bovine adrenocortical cells. J. Biol. Chem. 261, 7322–7326.

Rasmussen, J.T., Faergeman, N.J., Kristiansen, K., & Knudsen, J. (1994). Acyl CoA-binding protein can mediate intermembrane acyl-CoA transport and donate acyl-CoA for B-oxidation and glycerolipid synthesis. Biochem. J. 299, 165–170.

Ray, P., & Strott, C.A. (1981). Cytosol stimulation of pregnenolone synthesis by isolated adrenal mitochondria. Life Sci. 28, 1529–1534.

Reyland, M. E., Prack, M. M., & Williams, D. L. (1992). Elevated levels of protein kinase C in Y1 cells which express apolipoprotein E decrease basal steroidogenesis by inhibiting expression of P450-cholesterol side chain cleavage mRNA. J. Biol. Chem. 267, 17933–17938.

Reyland, M.E., & Williams, D.L. (1991). Suppression of cAMP-mediated signal transduction in mouse adrenocortical cells which express apolipoprotein E. J. Biol. Chem. 266, 21099–21200.

Rice, D.A., Aitken, L.D., Vandenbark, G.R., Mouw, A.R., Franklin, A., Schimmer, B.P., & Parker, K.L. (1989). A cAMP-responsive element regulates expression of the mouse 11β-hydroxylase gene. J. Biol. Chem. 264, 14011–14015.

Richards, J.S. (1994). Hormonal control of gene expression in the ovary. Endocr. Rev. 15, 725–751.

Richards, J.S., & Hedin, L. (1988). Molecular aspects of hormone action in ovarian follicular development, ovulation and luteinization. Ann. Rev. Physiol. 50, 441–463.

Richardson, M.C., & Schulster, D. (1973). The role of protein kinase activation in the control of steroidogenesis in the adrenal cortex. Biochem. J. 136, 993–998.

Robel, P., & Baulieu, E.E. (1994). Neurosteroids, biosynthesis, and function. Trends in Endocrinol. Metab. 5, 1–8.

Robel, P., Bourreau, E., Corpechot, L., Dang, D.C., Halberg, F., Clarke, C., Haug, M., Schlegel, M.L., Synguelakis, M., Vourch, C., & Baulieu, E.E. (1987). Neurosteroids and 3b-D5-derivatives in rat and monkey brain. J. Steroid Biochem. 27, 649–655.

Robinson, J., & Stevenson, P.M. (1972). The energy-linked reduction of $NADP^+$ by succinate, and its relationship to cholesterol side-chain cleavage. FEBS Lett. 23, 327–331.

Rodgers, R.J., Waterman, M.R., & Simpson, E.R. (1986). $P450_{scc}$, $P450_{17a}$, adrenodoxin and NADPH-P450 reductase in bovine follicles and corpora lutea. Changes in specific contents during ovarian cycle. Cytochromes Endocrinology 118, 1366–1374.

Rodriguez-Lafrasse, C., Rousson, R., Bonnet, J., Pentchev, P.G., Louisot, P., & Vanier, M.T. (1990). Abnormal cholesterol metabolism in imipramine-treated fibroblast cultures. Similarities with Niemann-Pick type C disease. Biochim. Biophys. Acta 1043, 123–128.

Rogler, L. E., & Pintar, J. E. (1993). Expression of the P450 side-chain cleavage and adrenodoxin genes begins during early stages of adrenal cortex development. Mol. Endocrinol. 7, 453–461.

Rojo, M., Hovius, R., Demel, R.A., Nicolay, K., & Wallimann, T. (1991). Mitochondrial creatine kinase mediates contact formation between mitochondrial membranes. J. Biol. Chem. 266, 20290–20295.

Roselli, C.E., Morton, L.E., & Resko, J.A. (1985). Distribution and regulation of aromatase activity in the rat hypothalamus and limbic system. Endocrinology 117, 2471–2477.

Roskelley, C. D., & Auersperg, N. (1990). Density separation of rat adrenocortical cells: Morphology, steroidogenesis, and $P-450_{SCC}$ expression in primary culture. In Vitro Cell. Dev. Biol. 25, 493–501.

Ryan, R.J., Charlesworth, M.C., McCormick, D.J., Milius, R.P., & Keutmann, H.T. (1988). The glycoprotein hormones: Recent studies of structure-function relationships. FASEB J. 2, 2661–2669.

Sadovsky, Y., Tourtelott, L.M., Simburgerm K., Crawford, P.A., Woodson, K.G., Polish, J.A., Clements, M.A., & Milbrandt, J. (1995). Mice deficient in the orphan receptor SF-1 lack adrenal glands placenta. Proc. Natl. Acad. Sci. 92, 10939–10943.

Sarria, A.J., Panini, S.R., & Evens, R.M. (1992). A functional role for vimentin intermediate filaments in the metabolism of lipoprotein-derived cholesterol in human SW-13 cells. J. Biol. Chem. 267, 19455–19465.

Schiff, R., Arensburg, J., Ahuva, I., Keshnet, E., & Orly, J. (1993). Expression and cellular localisation of uterine side chain cleavage P450 messenger RNA during dearly pregnancy in mice. Endocrinology 133, 529–537.

Schlegel, J., Zurbriggen, B., Wegmann, G., Wyss, M., Eppenberger, H.M., & Wallimann, T. (1988). Native mitochondrial creatine kinase forms octameric structures. I. Isolation of two interconvertible mitochondrial creatine kinase forms, dimeric and octameric mitochondrial creatine kinase characterization, localization and structure-function relationships. J. Biol. Chem. 263, 16942–16962.

Schulster, D., Burstein, S., & Cooke, B.A. (1976). Molecular Endocrinology of the Steroid Hormones. John Wiley and Sons, London.

Seybert, D. W. (1990). Lipid regulation of bovine cytochrome $P450_{11\beta}$ activity. Arch. Biochem. Biophys. 279, 188–194.

Shan, L.-X., Phillips, D.M., Bardin, C.W., & Hardy, M.P. (1993). Differential regulation of steroidogenic enzymes during differentiation optimizes testosterone production by adult rat leydig cells. Endocrinology 133, 2277–2283.

Shi, D.L., Savona, C., Gagnon, J., Cochet, C., Chambaz, E.M., & Feige, J.-J. (1990). Transforming growth factor-β stimulates the expression of a2-macroglobulin by cultured bovine adrenocortical cells. J. Biol. Chem. 265, 2881–2887.

Shikita, M., & Hall, P. (1974). Stoichiometry of cholesterol side chain cleavage by adrenal cytochrome $P450_{scc}$. Proc. Natl. Acad. Sci. USA 71, 1441–1445.

Shiver, I.M., Sackett, D.L., Knipling, L., & Wolff, J. (1992) Intermediate filaments and steroidogenesis in adrenal Y-1 cells: Acrylamide stimulation of steroid production. Endocrinology 131, 201–207.

Simpson, E.R., McCarthy, J.L., & Peterson, J.A. (1979). Evidence that the cycloheximide-sensitive site of ACTH action is in the mitochondrion. J. Biol. Chem. 253, 3135–3139.

Simpson, E.R., & Miller, D.A. (1978). Cytochrome P450$_{scc}$ and iron-sulfur protein in human placental mitochondria. Arch. Biochem. Biophys. 190, 800–808.

Simpson, E.R., & Waterman, M.R. (1988). Regulation of the synthesis of steroidogenic enzymes in adrenal cortical cells by ACTH. Ann. Rev. Physiol. 50, 427–440.

Simpson, E.R., Jefcoate, C.R., Brownie, A.C., & Boyd, G.S. (1972). The effect of either stress anaesthesia on cholesterol side-chain cleavage and cytochrome P450 in rat adrenal mitochondria. Eur. J. Biochem. 28, 443–450.

Slotte, J.P., & Bierman, E.L. (1988). Depletion of plasma membrane sphingomyelin rapidly alters the distribution of cholesterol between plasma membranes and intracellular cholesterol pools in cultured fibroblasts. Biochem. J. 250, 653–658.

Sprengel, R., Werner, P., & Seeburg, P.H. (1989). Molecular cloning and expression of cDNA encoding a peripheral-type benzodiazepine receptor. J. Biol. Chem. 264, 20415–20421.

Stevens, V.L., Tribble, D.L., & Lambeth, J.D. (1985). Regulation of mitochondrial compartment volumes in rat adrenal cortex by ether stress. Arch. Biochem. Biophys. 242, 324–327.

Stevens, V.L., Aw, T.Y., Jones, D.P., & Lambeth, J.D. (1984). Oxygen dependence of adrenal cortex cholesterol side chain cleavage. J. Biol. Chem. 259, 1174–1179.

Stocco, D.M., & Chen, W. (1991). Presence of identical mitochondrial proteins in unstimulated constitutive steroid-producing R$_2$C rat Leydig tumor and stimulated nonconstitutive steroid-producing MA-10 mouse Leydig tumor cells. Endocrinology 128, 1918–1926.

Stocco, D.M., & Clark, B.J. (1993). The requirement of phosphorylation on a threonine residue in the acute regulation of steroidogenesis in MA-10 mouse Leydig tumor cells. J. Steroid Biochem. and Mol. Biol. 46, 337–347.

Stocco, D.M., & Sodeman, T.C. (1991). The 30-kDa mitochondrial proteins induced by hormone stimulation in MA-10 mouse Leydig tumor cells are processed from larger precursors. J. Biol. Chem. 266, 19731–19738.

Stone, D., & Hechter, O. (1954). Studies on ACTH action in perfused bovine adrenals: Site of action of ACTH in corticosteroidogenesis. Arch. Biochem. Biophys. 51, 457–469.

Strauss, J.F., Schuler, L.A., Rosenblum, M.F., & Tanaka, T. (1983). Cholesterol metabolism by ovarian tissue. Adv. Lipid Res. 18, 99–157.

Sugawara, T., Holt, J.A., Driscoll, D., Strauss, J.F., III, Lin, D., Miller, W.L., Patterson, D., Clancy, K.P., Hart, I.M., Clark, B.J., & Stocco, D.M. (1995). Cloning of the cDNA encoding human homolog of steroidogenic acute regulatory protein: Tissue specific expression of the cognate mRNA and mapping of the gene to Sp11.2. Proc. Natl. Acad. Sci. USA 92, 4778–4782.

Taylor, F.R., Kandutsch, A.A., Anzalone, L., Phirwa, S., & Spencer, T.A. (1988). Photoaffinity labeling of the oxysterol receptor. J. Biol. Chem. 263, 2264–2269.

Theron, C.N., Russel, V.A., & Taljaard, J.J.F. (1985). Evidence that estradiol 2/4-hydroxylase activities in rat hypothalamus and hippocampus differ qualitatively and involve multiple forms of P450: Ontogenetic and inhibition studies. J. Steroid Biochem. 23, 919–927.

Trzeciak, W.H., Ahmed, C.E., Simpson, E.R., & Ojeda, S.R. (1986). Vasoactive intestinal peptide induces synthesis of cytochrome P450$_{scc}$ in cultured rat ovarian granulosa cells. Proc. Natl. Acad. Sci. USA 83, 7490–7494.

Trzeciak, W.H., Duda, T., Waterman, M.R., & Simpson, E.R. (1987). Effects of EGF on the synthesis of Cytochrome P450$_{scc}$ in rat granulosa cells in primary culture. Mol. Cell. Endocrinol. 52, 43–50.

Trzeciak, W. H., LeHoux, J.G., Waterman, M. R., & Simpson, E. R. (1993). Dexamethasone inhibits corticotropin-induced accumulation of CYP11A and CYP17 messenger RNAs in bovine adrenocortical cells. Mol. Endocrinol. 7, 206–213.

Trzeciak, W.H., Simpson, E.R., Scallen, T.J., Vahouny, G.V., & Waterman, M.R. (1987). Studies on the synthesis of SCP-2 in rat adrenal cortical cells in monolayer culture. J. Biol. Chem. 262, 3713–3720.

Tuckey, R.C., & Kamin, H. (1983). Kinetics of O_2 and CO binding to adrenal cytochrome P-450$_{scc}$. Effect of cholesterol, intermediates, and phosphatidylcholine vesicles. J. Biol. Chem. 258, 4232–4237.

Tuls, J., Geren, L., Lambeth, J.D., & Millett, F. (1987). The use of a specific fluorescence probe to study the interaction of adrenodoxin with adrenodoxin reductase and cytochrome P450$_{scc}$. J. Biol. Chem. 262, 10020–10025.

Tuls, J., Geren, L., & Millett, F. (1989). Fluorescein isothiocyanate specifically modifies lysine 338 of cytochrome P-450$_{scc}$ and inhibits adrenodoxin binding. J. Biol. Chem. 264, 16421–16425.

Vahouny, G.V., Chanderbhan, R., Noland, B.J., Irwin, D., Dennis, P., Lambeth, J.D., & Scallen, T.J. (1983). Sterol carrier protein$_2$: Identification of adrenal sterol carrier protein$_2$ and site of action for mitochondrial cholesterol utilization. J. Biol. Chem. 258, 11731–11737.

Vahouny, G.V., Dennis, P., Chanderbhan, R., Fiskum, G., Noland, B.J., & Scallen, T.J. (1984). Sterol carrier protein$_2$ (SCP$_2$)-mediated transfer of cholesterol to mitochondrial inner membranes. Biochem. Biophys. Res. Comm. 122, 509–515.

van Amerongen, A., van Noort, M., van Beckhoven, J.R.C.M., Rommerts, F.F.G., Orly, J., & Wirtz, K.W.A. (1989). The subcellular distribution of the nonspecific lipid transfer protein (sterol carrier protein 2) in rat liver and adrenal gland. Biochim. Biophys. Acta 1001, 243–148.

van Meer, G. (1989). Lipid traffic in animal cells. Annu. Rev. Cell Biol. 5, 247–275.

Van Noort, M., Rommerts, F.F.G., Van Amerongen, A., & Wirtz, K.W.A. (1988). Regulation of SCP$_2$ in the soluble fraction of rat Leydig cells. Kinetics and possible role of calcium influx. Mol. Cell. Endocrinol. 56, 133–138.

Van Venetie, R., & Verkleij, A.J. (1982). Possible role of non-bilayer lipids in the structure of mitochondria: A freeze-fraction electron microscopy study. Biochim. Biophys. Acta 692, 397–405.

Verma, A., & Snyder, S.H. (1989). Peripheral type benzodiazepine receptors. Annu. Rev. Pharmacol. Toxicol. 29, 307–322.

Voelker, D.R. (1989). Reconstitution of phosphatidylserine import into rat liver mitochondria. J. Biol. Chem. 264, 8019–8025.

Wada, A., & Waterman, M.R. (1992). Identification by site-directed mutagenesis of two lysine residues in cholesterol side chain cleavage cytochrome P450 that are essential for adrenodoxin binding. J. Biol. Chem. 267, 22877–22882.

Walker, S.W., Lightly, E.R.T., Clyne, C., Williams, B.C., & Bird, I.M. (1991). Adrenergic and cholinergic regulation of cortisol secretion from the zona fasciculata/reticularis of bovine adrenal cortex. Endocrine Res. 17, 237–265.

Warner, M., Stromstedt, M., Moller, L., & Gustafsson, J.A. (1989). Endocrinology 124, 2699–2706.

Watanabe, N., Inoie, H., & Fujii-Kuriyama, Y. (1994). Regulatory mechanisms of cAMP-dependent and cell-specific expression of human steroidogenic cytochrome P450scc gene. Eur. J. Biochem. 222, 825–834.

Waterman, M.R., Kagawa, N., Zanger, U.M., Momoi, K., Lund, J., & Simpson, E.R. (1992). Comparison of cAMP-responsive DNA sequences and their binding proteins associated with expression of the bovine CYP17 and CYP11A and human CYP21B genes. J. Steroid Biochem. Molec. Biol. 43, 931–935.

Waterman, M.R. (1994). Biochemical diversity of cAMP-dependent transcription of steroid hydroxylase genes in the adrenal cortex. J. Biol. Chem. 269, 27783–27786.

Waterman, M.R., & Simpson, E.R. (1989). Regulation of steroid hydroxylase gene expression is multifactorial in nature. Rec. Prog. Horm. Res. 45, 533–566.

Westerman, J., & Wirtz, K.W.A. (1985). The primary structure of the nonspecific lipid transfer protein (sterol carrier protein 2) from bovine liver. Biochem. Biophys. Res. Comm. 127, 333–338.

Williams-Smith, D.L., & Cammack, R. (1977). Oxidation-reduction potentials of cytochromes P450 and ferredoxin in the bovine adrenal. Their modification by substrates and inhibitors. Biochim. Biophys. Acta 499, 432–442.

Wilson, T.E., Mouw, A.R., Weaver, C.A., Milbrandt, J., & Parker, K.L. (1993). The orphan nuclear receptor NGFI-B regulates expression of the gene encoding steroid 21-hydroxylase. Mol. Cell. Biol. 13, 861–868.

Wong, M., & Schimmer, B.P. (1989). Recovery of responsiveness to ACTH and cAMP in a protein kinase-defective adrenal cell mutant following transfection with a protein kinase A gene. Endocrine Res. 15, 49–65.

Xu, X., Xu, T., Robertson, D.G., & Lambeth, D.J. (1989). GTP stimulates pregnenolone generation in isolated rat adrenal mitochondria. J. Biol. Chem. 264, 17674–17680.

Xu, T., Bowman, E.P., Glass, D.V., & Lambeth, J.D. (1991). Stimulation of adrenal mitochondrial cholesterol side-chain cleavage by GTP, steroidogenesis activator polypeptide (SAP), and sterol carrier protein. J.Biol. Chem. 266, 6801–6807.

Yamamoto, R., Kallen, C.B., Babalola, G.O., Rennert, H., Billheimer, J.T., & Strauss, J.F. III. (1991). Cloning and expression of a cDNA encoding human sterol carrier protein 2. Proc. Natl. Acad. Sci. USA 88, 463–467.

Yamamoto, T., Roby, K.F., Kwok, S.C.M., & Soures, M.J. (1994). Transcriptional activation of cytochrome $P450_{scc}$ expression during trophoblast cell differentiation. J. Biol. Chem. 269, 6517–6523.

Yamazaki, T., Kowluru, R., & Jefcoate, C.R. (1995). Cholesterol availability and electron transport are independently sensitive in rat adrenal mitochondria to elevation of matrix Ca^{++}.

Yamazaki, T., Kowluru, R., McNamara, B.C., & Jefcoate, C.R. (1995). $P450_{scc}$-dependent cholesterol metabolism in rat adrenal mitochondria is inhibited by low concentrations of Matrix Ca^{++}. Arch. Biochem. Biophys. (In press.)

Yamazaki, T., McNamara, B.C., & Jefcoate, C.R. (1993). Competition for electron transfer between cytochromes $P450_{scc}$ and $P450_{11b}$ in rat adrenal mitochondria. Mol. and Cell. Endo. 95, 1–11.

Yanagibashi, K., Ohno, Y., Nakamichi, N., Matsui, T., Hayashida, K., Takamura, M., Yamada, K., Tou, S., & Kawamura, M. (1989). Peripheral-type benzodiazepine receptors are involved in the regulation of cholesterol side chain cleavage in adrenocortical mitochondria. J. Biochem. 106, 1026–1029.

Yang, X., Iwamoto, K., Wang, M., Artwohl, J., Mason, J.I., & Pang, S. (1993). Inherited congenital adrenal hyperplasia in the rabbit is caused by a deletion in the gene encoding cytochrome P450 cholesterol side-chain cleavage enzyme. Endocrinology 132, 1977–1982.

Yeaman, S.J. (1990). Hormone-sensitive lipase — A multi-purpose enzyme in lipid metabolism. Biochim. Biophys. Acta 1052, 128–152.

Zlotkin, T., Farkash, Y., & Orly, J. (1986). Cell-specific expression of immunoreactive cytochrome $P450_{scc}$ during follicular development in the rat ovary. Endocrinology 119, 2809–2820.

THE REGULATION OF THE FORMATION OF GLUCOCORTICOIDS AND MINERALOCORTICOIDS *IN VIVO*

Jean-Guy LeHoux, Hugues Bernard, Lyne Ducharme,
Andrée Lefebvre, Dennis Shapcott, André Tremblay,
and Steeve Véronneau

Advances in Molecular and Cell Biology
Volume 14, pages 149–201.
Copyright © 1996 by JAI Press Inc.
All rights of reproduction in any form reserved.
ISBN: 0-7623-0113-9

149

I. INTRODUCTION

Although structurally related, glucocorticoids and mineralocorticoids have contrasting functions. Indeed, glucocorticoids are a class of corticosteroids whose
function lies in the regulation of intermediary metabolism, whereas mineralocorticoids regulate ion transport across epithelial cell membranes resulting in the
retention of sodium and excretion of potassium and the maintenance of the volume
of extracellular fluid. Most of the actions of these two classes of steroids are
mediated by their specific receptors in target tissues. As stated in the introductory
chapter, mineralocorticoid receptors bind cortisol and aldosterone with similar
affinity, but those tissues with mineralocorticoid receptors possess high 11β-
hydroxysteroid dehydrogenase (11β-HSD) activity which converts cortisol to
cortisone, its inactive metabolite. Aldosterone can therefore exert its specific
activity despite circulating levels much lower than those of cortisol.

Corticosteroids are synthesized by every species of vertebrate so far studied
(Sandor and Mehdi, 1979). In most mammalian species cortisol is the principal
glucocorticoid together with varying amounts of corticosterone and cortisone.
While the guinea pig and hamster synthesize cortisol as their principal glucocorticoid, small rodents such as the rat and the mouse differ from this general pattern of
adrenal steroid excretion in that corticosterone is their major glucocorticoid.
Corticosterone is the sole glucocorticoid produced by birds and most reptiles and
amphibia, it is also the principal glucocorticoid of bony fish, whereas cortisol is
found in cartilaginous fish and in cyclostomes.

The conservation of water was not a problem for the earliest animals in their
uniquely aquatic environment, but the emerging terrestrial animals needed a means

to prevent dehydration. In invertebrates the impermeable integument was adequate, but for the purely terrestrial vertebrates the development of the glomerular-tubular kidney and the hormonal regulation of cations became necessary. Higher vertebrates including reptiles, birds, and mammals synthesize a novel corticosteroid, aldosterone, for this purpose, which appears to be restricted to terrestrial vertebrates.

The mammalian adrenal gland contains two types of tissue, the medulla which is responsible for the synthesis of catecholamines, while corticosteroids are synthesized in the cortex which is subdivided histologically into two regions, the zona glomerulosa (ZG) and the zona fasciculata-reticularis (ZF-R). The gross anatomy of the adrenal in lower vertebrates differs considerably from that of mammals. Indeed, in cyclostomes for example, the gland is merely a clump of secreting cells, whereas in fishes the interrenal tissues assume a more homogeneous structure, although they are separate from the chromaffin cells. In birds, the chromaffin and corticosteroid producing cells are intimately intermixed. These secretary cells, whether interrenals or adrenal glands are, however, very similar in function.

A. The Biosynthesis of the Corticosteroids

Glucocorticoids and mineralocorticoids are formed by the excision of the lateral chain of the 27 carbon sterol, cholesterol. The cholesterol used for corticosteroid formation comes from circulating cholesterol, from cholesterol stored in adrenal lipid droplets or from *de novo* cholesterol synthesis within the adrenal (Lehoux, 1979). Cholesterol is converted to pregnenolone by a mitochondrial cytochrome $P450_{scc}$ ($P450_{scc}$) the product of *CYP11A1* gene (Morohashi et al., 1984; John et al., 1984) while pregnenolone is metabolized into progesterone by a microsomal 3β-hydroxysteroid dehydrogenase (3β-HSD) system (type II) (Rhéaume et al., 1991). Progesterone is further metabolized into deoxycorticosterone by the addition of an hydroxyl group at position C21 by the microsomal cytochrome $P450_{C21}$, the product of the *CYP21* gene (Higashi et al., 1986; Chung et al., 1986). The transformation of deoxycorticosterone to corticosterone is catalyzed by the mitochondrial cytochrome $P450_{11\beta}$ ($P450_{11\beta}$) the product of *CYP11B1* gene. In bovines, $P450_{11\beta}$ appears to be responsible for both 11β-hydroxylase and aldosterone synthase activities (Yanagibashi et al., 1986). In the human, rat, mouse and hamster adrenal ZG, however, corticosterone is transformed to aldosterone by cytochrome P450 aldosterone synthase ($P450_{C18}$) the product of *CYP11B2*. The involvement of $P450_{C18}$ in the conversion of corticosterone to aldosterone in these species will be discussed in details in the following sections of this chapter. *CYP11B2* is repressed in ZF-R and therefore aldosterone cannot be formed in this zone.

In cortisol producing mammalian species, cytochrome $P450_{17\alpha}$ ($P450_{17\alpha}$), the product of the *CYP17* gene which is expressed in the ZF-R (Nakajin et al., 1984; Zuber et al., 1986, 1986a), contributes to divert metabolism away from mineralocorticoids to glucocorticoids. $P450_{17\alpha}$ uses either pregnenolone or progesterone as substrate to yield 17α-hydroxypregnenolone or 17α-hydroxyprogesterone. These intermediates are subsequently transformed into cortisol by $P450_{C21}$ and $P450_{11\beta}$. The *CYP17* gene is repressed in the ZG and consequently cortisol is not formed in this zone. Furthermore, in species producing androgens in their adrenals, $P450_{17\alpha}$ is also responsible for the transformation of C21-hydroxylated steroids into C19-androgenic steroids.

The genes coding for adrenal P450s contain intronic and exonic sequences as well as regulatory elements upstream from the transcriptional start site. The highly lipophilic $P450_{scc}$, $P450_{11\beta}$ and $P450_{C18}$ are synthesized as precursor proteins with an extension peptide which is proteolytically cleaved upon transportation into the mitochondrial inner membrane to generate the mature form of the enzyme. $P450_{17\alpha}$ and $P450_{C21}$ are anchored in the endoplasmic reticulum by a specific hydrophobic sequence at the amino acid terminus. Adrenal P450s are haemoproteins containing the protophophyrin IX as prosthetic group which is attached to a highly conserved peptide sequence present in all P450 protein molecules. To be hydroxylated, the steroid substrate binds to the oxidized form of P450. Upon substrate binding, the P450 heme-iron changes in most cases from a low spin-state to a high-spin state. The P450 heme-iron is more readily reduced in the high-spin state, suggesting a link between spin state and enzyme function. The enzyme-substrate complex is then reduced, which allows the formation of a ternary complex by the addition of molecular oxygen. A second reduction then occurs and chemical rearrangements take place to finally yield one molecule of hydroxylated steroid and one molecule of H_2O. Mitochondrial P450s are reduced by adrenodoxin whereas microsomal P450s require the ubiquitous enzyme NADPH cytochrome P450 reductase.

B.   Regulation of Adrenal Steroidogenesis

The adrenal cortex is a target organ under the influence of many regulatory factors. The formation of mineralocorticoids is regulated, among other factors, by angiotensin-II (AII), sodium and potassium ions, adrenocorticotropin (ACTH), and the atrial natriuretic peptide (ANP), whereas the synthesis of glucocorticoids is mainly regulated by ACTH.

Two major levels of regulation have been documented for the formation of mineralocorticoids, namely the control of cholesterol metabolism, including *de novo* cholesterol synthesis and uptake by low density lipoprotein receptors, and the *CYP11B2* gene expression. AII binds to specific membrane receptors to activate the phosphoinositide transduction pathway, resulting to the mobilization of intracellular calcium and the opening of calcium channels to increase intracellular calcium. Calcium and diacylglycerol formed during this process activate protein kinase C (PKC) known to modify the transcription of many genes. A low sodium intake acts through AII and its receptors to increase aldosterone secretion. Potassium ions can stimulate per se the adrenal cortex steroidogenesis, although they also act through AII. The mechanism of action of ACTH differs from above mentioned mechanisms as after binding to membrane receptors the second messenger involved is c-AMP and protein kinase A (PKA). Within minutes, ACTH activates mechanisms making cholesterol available to $P450_{scc}$ to produce pregnenolone, an intermediary in corticosteroid formation. In ZF-R of cortisol secreting species, ACTH regulates also the expression of the *CYP17A* gene. There is also a long term action of ACTH that maintains the integrity of adrenocortical cells. For its action, ANP binds to specific cell membrane receptors to activate the formation of c-GMP. ANP has a negative regulatory action on adrenocortical functions which is probably localized at the level of the conversion of cholesterol to pregnenolone, although mechanisms of action of ANP are yet to be fully elucidated.

In this review discussion will be confined to the effects of the above mentioned factors on adrenocortical steroidogenic enzymes in mammals. The presence of one enzyme $P450_{11\beta}$ in bovine adrenal and of two different enzymes, $P450_{11\beta}$ and $P450_{C18}$, in human, mouse, rat and hamster adrenals will be discussed in relation to glucocorticoid and mineralocorticoid formation. A comparison of $P450_{11\beta}$ and $P450_{C18}$ between mammalian species will be done. 11β-hydroxylase deficiency, corticosterone methyloxidase II deficiency and glucocorticoid-suppressible aldosteronism will be discussed in relation to mutations in *CYP11B1* and *CYP11B2* genes. We will comment on the effect of high potassium and low sodium intake on the expression of adrenocortical steroidogenic enzyme genes. We will show that $P450_{scc}$ and $P450_{C18}$ gene expression is stimulated by these two intakes and that AII and AII receptors are involved in their mechanism of action. We will show that ACTH affect mainly glucocorticoid formation at early steps of steroidogenesis, at the level of cholesterol metabolism, but that the control of the $P450_{C17}$ enzyme is also physiologically important. The inhibitory effect of ANP on adrenocortical functions will be discussed in relation to the effect of ACTH and AII. Furthermore, the AII and ACTH post-receptors events will be examined in relation to activation by trans-acting factors at the steroidogenic genes level.

II. EFFECTS OF POTASSIUM AND SODIUM INTAKE ON MINERALOCORTICOID FORMATION AND P450 SYSTEMS

Earlier studies have well established that potassium functions as a major regulator of aldosterone secretion *in vivo*. Variations in plasma potassium within the physiological range provoked significant changes in aldosterone secretion in man (Gann et al., 1964; Dluhy et al 1972., Brown et al., 1972) and in experimental animals (Funder et al., 1969; Boyd et al., 1971). The width of the rat adrenal ZG decreased during restriction of dietary potassium and increased after potassium supplementation (Deane et al., 1948). Potassium appears to have a stimulatory effect at an early stage in the corticosteroid biosynthetic pathway, before the formation of pregnenolone, and also during the final steps of aldosterone formation (Müller, 1971). Indeed, it has been demonstrated that potassium can act directly on the ZG *in vitro* to increase the conversion of cholesterol to pregnenolone and of corticosterone to aldosterone (Kaplan, 1965; Haning, 1970).

Earlier works also showed that changes in sodium intake greatly influence the plasma aldosterone level (Luetscher and Alxelrad, 1954; Blair-West et al., 1963; Lehoux, 1979; Aguilera and Catt, 1983). A low sodium intake stimulates the production of aldosterone by the adrenal; there is hypertrophy of the ZG (Deane et al., 1948) and an increased capacity to subsequently synthesize aldosterone *in vitro* (Marusic and Mulrow, 1967; Lehoux et al., 1974; Kramer et al., 1979). We found that adrenal homogenates of rats maintained on a low sodium intake metabolized more cholesterol to corticosterone and aldosterone than did homogenates from rats maintained on a normal diet (Lehoux et al., 1974). Kramer et al., (1979) reported that sodium depletion enhances aldosterone synthesis in the rat adrenal ZG by increasing the rate of cholesterol side chain cleavage and the rate of conversion of corticosterone to aldosterone.

It does not appear that sodium ions, per se, can influence the secretion of aldosterone *in vivo*. Moreover, changes in the sodium concentration of the incubation media had no effect upon the output of aldosterone in ZG cells *in vitro*, (Schneider et al., 1985; Schneider and Kramer, 1986). Increased aldosterone secretion under low sodium intake in the rat (Debreceni and Csete, 1975) was not inhibited by a low potassium intake, indicating that sodium depletion affects aldosterone secretion through mechanisms other than by an increased level of potassium.

A. The Effects of High Potassium and Low Sodium Intake on Adrenal Steroidogenic Enzymes mRNA and Protein

We recently reevaluated the effects of high potassium and low sodium intake on aldosterone synthesis by analysis of the mRNA and protein of the adrenal steroidogenic enzymes involved, and we found that both intake influence aldosterone formation mainly at the level of $P450_{scc}$ and $P450_{C18}$ (Tremblay and Lehoux, 1989, 1989a, 1989b; Tremblay et al., 1991). Plasma aldosterone levels of rats maintained on a high potassium or a low sodium intake for up to seven days were significantly increased after one day and remained elevated thereafter. In the ZG, under either regimen, the levels of mRNA and protein of $P450_{scc}$ and $P450_{11\beta}$-$P450_{C18}$ (using a cDNA and an antibody that recognized both $P450_{11\beta}$ and $P450_{C18}$), gradually increased although to varying degree. In potassium supplemented rats, the mRNA levels for $P450_{scc}$ and $P450_{11\beta}$-$P450_{C18}$ each increased by 2-fold after seven days of diet. The greatest induction of mRNA was observed at day seven in sodium depleted rats for $P450_{11\beta}$-$P450_{C18}$ with a 4-fold increase followed by a 2.7-fold increase for $P450_{scc}$. In subsequent experiments, an 8.2-fold increase was found in the level of $P450_{11\beta}$-$P450_{C18}$ mRNA in the ZG of rats maintained on a low sodium intake for 21 days (Lehoux and Tremblay, 1992). Similar increases were obtained in the level of $P450_{11\beta}$-$P450_{C18}$ and $P450_{scc}$ proteins, indicating that the mRNAs were effectively translated into their respective protein. In contrast, no changes were found in the ZG-F with either ionic intake.

B. The Effects of High Potassium and Low Sodium Intake on the Transcription Rate of Rat Adrenal Steroidogenic Enzymes

Our studies (Tremblay and Lehoux, 1993) to assess the transcription rate of steroidogenic enzymes following a high potassium or low sodium intake showed similar stimulation of nascent RNA for $P450_{scc}$ and $P450_{11\beta}$-$P450_{C18}$ for each treatment (5 to 6 fold) in nuclei isolated from the rat adrenal ZG. A much lower incorporation of labeled nucleotides was found in preparations of the ZF-R, this remained unaltered following changes in the intake of potassium or sodium. This indicates that the initiation of transcription of the genes involved in the early and final steps of the aldosterone biosynthetic pathway in the rat adrenal ZG is up-regulated by high potassium or low sodium intake. In the case of $P450_{C18}$ the magnitude of changes in transcription was probably underestimated because the cDNA probe used, as mentioned earlier, recognized both $P450_{11\beta}$ and $P450_{C18}$ mRNAs (see also the following section). In conclusion, these results indicate that increased gene transcription rates are mainly responsible for the accumulation of $P450_{scc}$ and $P450_{C18}$ mRNA during high potassium or low sodium intake.

C. Regulation of P450s Involved in the Last Steps of Aldosterone Formation

As mentioned earlier, two different enzymes catalyze the final steps of aldosterone formation in the rat, mouse, hamster, and humans but not so in cattle. The differences and similarities in these enzymatic systems for each species will be compared and discussed in the following sections.

Rat Adrenal

It is now well documented that, in rat adrenal, $P450_{11\beta}$ is involved in the transformation of deoxycorticosterone to corticosterone, but not of corticosterone to aldosterone, whereas $P450_{C18}$ catalyzes both reactions. The isolation and characterization of rat adrenal cDNAs encoding $P450_{11\beta}$ (Nonaka et al., 1989) and $P450_{C18}$ (Imai et al., 1990; Matsukawa et al., 1990) showed a strong homology between the two proteins. Furthermore, two genes were isolated from a rat genomic library, one for $P450_{11\beta}$ and the other for $P450_{C18}$, and some structural differences in the 5'-flanking region of the two genes have been described (Mukai et al., 1991). The expression of $P450_{11\beta}$ and $P450_{C18}$ cDNAs in COS-7 cells showed differences in the catalytic activities of the two proteins. $P450_{11\beta}$ catalyzed 11β-, 18- and 19-hydroxylation of 11-deoxycorticosterone but not aldosterone production. $P450_{C18}$, however, catalyzed 11β- and 18-hydroxylations of deoxycorticosterone and the formation of aldosterone but not that of 19-hydroxy-deoxycorticosterone (Matsukawa et al., 1990; Nonaka and Okamoto, 1991). With COS-7 cells transfected with rat $P450_{C18}$ cDNA and having various degrees of expression, the ratios of corticosterone formation from deoxycorticosterone on the production rates of other steroids were found constant (Nonaka and Okamoto, 1991). Similar observations have been made in the case of purified bovine $P450_{11\beta}$ as the relative production rates of corticosterone, 18-hydroxy-deoxycorticosterone, 18-hydroxy-corticosterone and aldosterone from deoxycorticosterone were constant without a lag phase (Ikushiro et al., 1989). An explanation for these observations is that putative intermediaries in the formation of aldosterone such as corticosterone and 18-hydroxy-corticosterone would not leave the catalytic site during the reaction. The catalytic site of the rat $P450_{11\beta}$ might be unable to accommodate corticosterone or 18-hydroxy-corticosterone generated from deoxycorticosterone, which consequently cannot be further hydroxylated. In the case of $P450_{C18}$, its catalytic site would be more flexible and would allow both 11β and 18-hydroxylated intermediaries to go through further hydroxylation (Nonaka and Okamoto, 1991). It is, also, still impossible to determine with certainty if 18-hydroxy-corticosterone is a free intermediate or a reaction by-product of the conversion of corticosterone to aldosterone.

The sequence of the two P450s involved in the last steps of corticosteroidogenesis being known, in subsequent experiments, by northern analysis, we could discriminate between the two forms of rat adrenal $P450_{11\beta}$ and $P450_{C18}$ mRNAs, using oligonucleotides specific to the rat $P450_{11\beta}$ and $P450_{C18}$ respectively (Tremblay et al., 1992). The transcripts detected for the two forms were of similar size but differed in their zonal distribution; $P450_{C18}$ mRNA was detected in the ZG only, whereas $P450_{11\beta}$ was present in both ZG and ZF-R. When rats were fed a high potassium or a low sodium diet for one week, the level of $P450_{C18}$ mRNA was increased by 6- or 5-fold, whereas the level of $P450_{11\beta}$ mRNA was unaffected by such regimens (Figure 1). In agreement with these findings, hybridization studies in situ conclusively demonstrated that the $P450_{C18}$ mRNA was located exclusively in the rat adrenal ZG, whereas $P450_{11\beta}$ mRNA was present in both ZG and ZF-R (Yabu et al., 1991).

As mentioned above, a low sodium or high potassium diet fed to rats increased the levels of a protein which coupled to an antibody to bovine adrenal cytochrome $P450_{11\beta}$. This increase was found in the ZG but not in the ZF-R suggesting that the effects of the diets were directed towards $P450_{C18}$, but not towards $P450_{11\beta}$.

The presence of two distinct $P450_{11\beta}$ proteins in the rat adrenal ZG had been long suspected. Indeed, two proteins immunoreacting against an antibovine adrenal cytochrome $P450_{11\beta}$ antibody were reported to be present in the rat adrenal ZG (Lauber et al., 1987; Onishi et al., 1988; Ogishima et al., 1989; Shibata et al., 1991). These two proteins migrated to the areas of 51 kDa and 49 kDa. The 51 kDa protein band was detected in the ZG and in the zona fasciculata, whereas the 49 kDa band was only present in the ZG preparations. The 49 kDa protein was isolated from the ZG of rats kept on a low-sodium, high-potassium regimen (Lauber and Müller, 1989). The isolated 49 kDa protein was able to transform deoxycorticosterone to

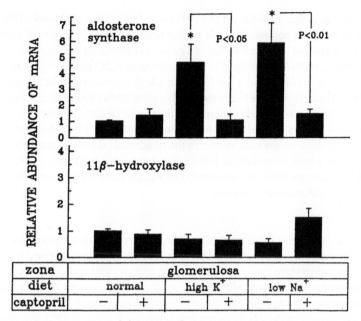

Figure 1. Effects of low sodium or high potassium intake and captopril administration on the expression of P450$_{c11}$ related forms. Rats were maintained for 1 week either on sodium-restricted (low Na$^+$), high-potassium (high K$^+$), or normal diet, with (+) or without (-) captopril. Total RNA samples from ZG were analyzed with [^{32}P]dATP-labeled oligonucleotide probes specific for either (P450C$_{C18}$) aldosterone synthase or (P450$_{11\beta}$) 11β-hydroxylase (Tremblay et al., 1992; Lehoux and Tremblay, 1992).

corticosterone, 18-hydroxy-corticosterone, and aldosterone. More recently, using a specific antibody against synthetic peptides corresponding to a region of P450$_{C18}$ differing from that of P450$_{11\beta}$, it was clearly demonstrated that P450$_{C18}$ is located in the ZG (Ogishima et al., 1992). In contrast to the location of P450$_{11\beta}$ mRNA in both zones of the rat adrenal cortex the P450$_{11\beta}$ protein was found only in the ZF-R using an antibody raised against a peptide containing a specific sequence of P450$_{11\beta}$. Interestingly there was penetration of P450$_{C18}$ positive cells into the intra-cortical regions of sodium depleted rats. When rats were maintained for 10 days under sodium depleted conditions, the ZG cells containing P450$_{C18}$ proliferated from 10 to 15 layers with a concentration of P450$_{C18}$ 5- to 7-fold that of unstimulated rats. Also, the intensity of the staining of P450$_{C18}$ did not differ significantly from cells of sodium-depleted and control rats, suggesting a similar expression of P450$_{C18}$ in individual cells from both groups of animals and that the increase in the level of P450$_{C18}$ protein with sodium depletion was due to an increased number of cells expressing the enzyme. This point, however, is not yet completely elucidated. Shibata et al. (1991) reported an increased level of $P450_{C18}$ protein in the ZG of potassium supplemented rats after 4 days of treatment, although this level decreased to normal values at day 10. As mentioned earlier, we found that both $P450_{C18}$ mRNA and protein levels in the adrenal ZG and also plasma aldosterone were increased in rats after 7 days on a high potassium intake (Tremblay et al., 1991). This difference between the two experiments can be tentatively explained by the fact that in the Shibata experiment, the increased serum potassium concentration after 4 days had returned to control values after 10 days of treatment. Collectively these results indicate that in the rat adrenal (a) $P450_{scc}$ and $P450_{C18}$ are regulated by a high potassium or low sodium intake, but not $P450_{11\beta}$ (b) $P450_{C18}$ and $P450_{11\beta}$ are two different enzymes (c) $P450_{C18}$ is located exclusively in the ZG, at the site of aldosterone synthesis.

Mouse Adrenal

Domalik et al. (1991) also isolated and characterized two isoenzymes designated 11β-hydroxylase and aldosterone synthase in the mouse. When expressed in COS-7 cells, 11β-hydroxylase converted deoxycorticosterone into corticosterone but not to aldosterone. In contrast, aldosterone synthase was capable of transforming deoxycorticosterone to both corticosterone and aldosterone. $P450_{C18}$ mRNA was found exclusively in the ZG by *in situ* hybridization. $P450_{11\beta}$ mRNA was detected in the mouse adrenal ZF-R, but in contrast to the rat no significant expression was observed in the ZG, suggesting that the mouse adrenal cortex exhibits a greater segregation into glucocorticoid-producing and mineralocorticoid-producing zones. In mouse Y1 adrenocortical cells $P450_{11\beta}$ and $P450_{C18}$ mRNAs were detected by northern analysis, and the level of both mRNAs was increased by treatment with ACTH. The increase in $P450_{C18}$ preceded that of $P450_{11\beta}$ with a peak at 2 h and a decline thereafter to below control levels, whereas there was no induction of $P450_{11\beta}$ mRNA until 9 h and then a prolonged peak occurred between 9 to 24 h of ACTH treatment. These results indicate that the two genes are induced by different mechanisms.

Hamster Adrenal

We have established that hamster adrenals synthesize both aldosterone and cortisol and that the synthesis of these two hormones is regulated differently (Lehoux et al., 1992). Indeed, in adrenal cell suspensions the addition of AII into incubation media provoked dose-dependent increases in the output of aldosterone and corticosterone. In contrast, ACTH induced no change in aldosterone output, whereas a dose-dependent increase in glucocorticoid output was found. Furthermore, plasma ACTH and glucocorticoids levels showed a diurnal rhythm which fluctuated in parallel during the day, but a different profile for plasma aldosterone levels was observed, thereby demonstrating that the control of expression of glucocorticoids and mineralocorticoids also differs in this species (Lehoux and Ducharme, 1992).

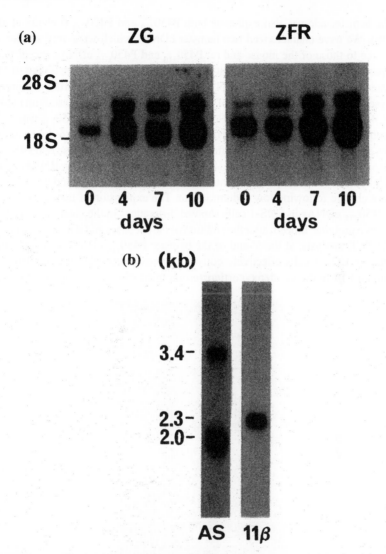

Figure 2. (a) Effects of low sodium intake on P450$_{C18}$ mRNA in hamster adrenals. Groups of hamsters were fed a low sodium diet for 0, 4, and 10 days. Total RNA from zona glomerulosa (ZG) and zona fasciculata-reticularis (ZF-R) were analyzed by northern blotting using a [^{32}P]-labeled cDNA probe specific for hamster adrenal P450$_{C18}$ (Lehoux et al., 1994b). (b) Northern blotting analysis of hamster adrenal zona glomerulosa P450$_{C18}$, and P450$_{11\beta}$. Total RNA samples were analyzed with [^{32}P]-labeled oligonucleotides specific for P450$_{C18}$ (AS) and to P450$_{11\beta}$ (11β) (Lehoux et al., 1994b).

The hamster adrenal also expresses both $P450_{11\beta}$ and $P450_{C18}$ (Lehoux et al., 1994b). We have characterized two hamster cDNAs which were very similar in sequence to those of the mouse and rat $P450_{11\beta}$ and $P450_{C18}$ cDNAs except that, compared to the rat $P450_{11\beta}$, the first six bases were missing in the 5'-end of the hamster $P450_{11\beta}$ cDNA. Two mRNA bands were found by northern blotting analyses performed on the adrenals of hamsters maintained on a low sodium intake for 0, 4, 7, and 10 days, when a $[^{32}P]$-oligonucleotide specific to $P450_{C18}$ was used as probe. The intensity of both bands increased 5-fold after 4 days of low sodium intake and remained elevated thereafter (Figure 2a). The size of these two $P450_{C18}$ mRNAs were estimated at 2.0 kb and 3.4 kb (Figure 2b). A distinct mRNA band of 2.3 kb hybridized with a $[^{32}P]$-oligonucleotide specific to $P450_{11\beta}$ and its intensity did not change following low sodium intake. The expression of hamster $P450_{11\beta}$ and $P450_{C18}$ cDNAs in COS-1 cells showed differences in the catalytic activities of the two proteins. Indeed, after the addition of the six bases ATGGCA, as deduced by RT-PCR analysis, at the 5'-end of the hamster $P450_{11\beta}$ cDNA, its expression product in COS-1 cells effectively converted deoxycorticosterone to corticosterone and 19-hydroxy- deoxycorticosterone in nearly an equimolar ratio, but

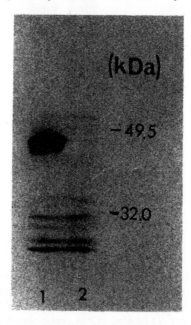

Figure 3. Immunoblotting analysis on hamster adrenal. Tissues were homogenized and solubilized in sodium dodecyl sulfate and then electrophoresed on polyacrylamide gel and transferred onto nitrocellulose membrane. $P450_{C18}$ was detected with an antibody directed against a specific peptide sequence deduced from hamster adrenal $P450_{C18}$. Hamsters were maintained on a low sodium diet for 7 days: lane 1; and on a control diet: lane 2.

produced no detectable aldosterone. In contrast, $P450_{C18}$ converted deoxycorticosterone to aldosterone and 18-hydroxy-corticosterone (Véronneau et al., 1996). With an antibody directed against a specific peptide sequence deduced from the hamster adrenal $P450_{C18}$ cDNA, two protein bands were detected by Western blotting in the area of 45 kDa and 43 kDa in homogenate of whole adrenal (Figure 3). At this stage of our study, however, we cannot rule out that the lower band is not a degradation product of the upper band. A low sodium intake increased the intensity of these two bands compared to similar preparations from hamsters maintained on a normal diet, supporting that $P450_{C18}$ mRNAs could effectively be translated into protein. Using the above mentioned $P450_{C18}$ antibody, we observed by immunocytochemistry that $P450_{C18}$ was localized in ZG and that the intensity of the staining of the $P450_{C18}$ increased in ZG of hamsters kept on a low sodium intake for 11 days (Lehoux et al., 1996). Furthermore $P450_{C18}$ was restricted to mitochondria as determined by electron microscopy (Lehoux et al., 1996). Furthermore $P450_{C18}$ was restricted to mitochondria as determined by electron microscopy (Lehoux et al., 1996). These results indicate that the hamster possesses at least two adrenal isoenzymes involved in the last steps of aldosterone formation. These two isoenzymes have different catalytic activities, $P450_{C18}$ being the only one capable of aldosterone synthesis. Noteworthy is that the hamster $P450_{11\beta}$ possesses a much stronger 19-hydroxylase activity than $P450_{11\beta}$ of other studied species. This high 19-hydroxylase activity makes the hamster an excellent model to study the formation of 19-hydroxy-deoxycorticosterone and its derivative 19-nor-deoxycorticosterone, in relation to hypertension, since elimination of 19-nor-deoxycorticosterone was found in the urine of patients with hypertension due to primary aldosteronism (Griffing et al., 1981).

Bovine Adrenal

The bovine adrenal is an exception to the general rule since $P450_{11\beta}$ is responsible of both 11β-hydroxylation and aldosterone formation, and no gene with unique 11β-hydroxylase activity has yet been found in this species.

A $P450_{11\beta}$ has been isolated from a bovine genomic library (Hashimoto et al., 1989). Two forms of $P450_{11\beta}$ were purified from bovine adrenal (Ogishima et al., 1989a). Four forms of bovine $P450_{11\beta}$ cDNA, however, were cloned [$P450_{11\beta}$-2,3,4, and pB11] (Morohashi et al., 1987; Kirita et al., 1988; Hashimoto et al., 1989; Mathew et al., 1990). Expression studies showed that the $P450_{11\beta}$-4 form had no 11β-hydroxylase activity or aldosterone synthase activity, whereas the products of $P450_{11\beta}$-2 and -3 cDNAs possessed the following catalytic activities: 11β-, 18-, 19-hydroxylations and 18-aldehyde formation, aldosterone production being higher in $P450_{11\beta}$-3 than in $P450_{11\beta}$-2 (Morohashi et al., 1990). In the bovine adrenal $P450_{11\beta}$ is expressed in the ZF-R and also in the ZG, although aldosterone is exclusively produced in ZG cells. It has been postulated that the activity of the bovine $P450_{11\beta}$ is controlled by different mechanism(s) in the ZF-R as compared to the ZG. Indeed, in a reconstituted system, purified mitochondrial $P450_{11\beta}$, isolated from bovine adrenal ZF-R and subsequently incorporated in liposomes, showed catalytic activities not only for the hydroxylation of deoxycorticosterone at the 11β- and 18-positions, but also for the conversion to aldosterone. In the presence of a detergent (0.05% Tween 20), however, no aldosterone was detected, the predominant product of the reconstituted enzyme system being corticosterone. A suppression mechanism for aldosterone synthesizing activity might therefore exist in the bovine adrenal ZG-F, possibly through factors that accelerate the dissociation of intermediates from $P450_{11\beta}$ (Ikushiro et al., 1989). $P450_{11\beta}$ is labile and consequently the solubilized form of the enzyme is difficult to study. However, the presence of the substrate deoxycorticosterone has a stabilizing effect on $P450_{11\beta}$. Moreover, the protein being lipophilic was found more stable when incorporated into liposome membranes. It was observed that when ZF-R mitochondria were solubilized in sodium cholate, they could produce aldosterone, indicating that the suppression of the aldosterone-synthesizing activity probably occurs, in this zone, within the mitochondrial membrane. Furthermore, Ikushiro et al. (1992) found that $P450_{scc}$ is capable of modulating the various activities of $P450_{11\beta}$ in liposomes. Indeed, when corticosterone was used as substrate, corticosterone and 18-hydroxy-corticosterone formation was increased, whereas that of 18-hydroxy-deoxycorticosterone and aldosterone was decreased. These authors conclude that bovine adrenal $P450_{11\beta}$ exists in two forms, one having high 11β-hydroxylase activity (C-form) and low activity for aldosterone formation, and the other having relative low 11β-hydroxylase activity and high aldosterone synthesizing capacity (A-form). It has been suggested that the effect of $P450_{scc}$ on the activity of $P450_{11\beta}$ incorporated into liposomes would be on the change of the A-form to the C-form. Moreover, it was hypothesized that in ZF-R, $P450_{11\beta}$ would be mainly in the C-form whereas in ZG some of the $P450_{11\beta}$ would be in the A-form. More work is needed, however, to study the interactions of $P450_{scc}$ with $P450_{11\beta}$ in mitochondria originating from ZG and ZF-R in order to determine if this mechanism is operational in these organelles.

Human Adrenal

In contradistinction to the animal studies, our information concerning the regulation of corticoid synthesis in humans has been obtained mainly by the study of genetic disorders of the system.

In humans, Mornet et al., (1989) isolated *CYP11B1* and its related *CYP11B2* gene, both genes being located on chromosome 8 at q22 (Chua et al., 1987). Although *CYP11B1* is normally expressed at high levels, (Mornet et al., 1989; Curnow et al., 1991) *CYP11B2* transcripts were not detected in human adrenal mRNA or among cDNA clones (Mornet et al., 1989). However, a cDNA clone encoding $P450_{C18}$ was isolated from a cDNA library derived from adrenal tumor tissue of a patient suffering from primary aldosteronism (Kawamoto et al., 1990). The product expressed by this cDNA catalyzed the conversion of deoxycorticosterone to aldosterone. Using northern blot analysis, these authors found that the *CYP11B2* gene was highly expressed in the adrenal tumor. In accord with these results, a specific aldosterone synthase was isolated and purified from the adrenals of patients with primary aldosteronism (Ogishima et al., 1991). Subsequently, Curnow et al., (1991) reported that the transcript of *CYP11B2* is indeed present at low levels in normal adrenal glands; is increased in cultured ZG cells by AII, but not significantly by ACTH; and converts deoxycorticosterone to aldosterone. In contrast, the same authors reported that $P450_{11\beta}$ is present at high levels in normal adrenal glands, is increased in cultured cells more by ACTH than by AII, and synthesizes corticosterone, but not aldosterone from deoxycorticosterone. These results thus indicate the involvement of two genes contributing to the final steps of aldosterone formation in human adrenal.

The failure to convert 11-deoxycortisol to cortisol is defined as 11β-hydroxylase deficiency, and is the second most common cause of adrenal hyperplasia causing hypertension. The first report on a molecular defect causing 11β-deficiency was on a missense mutation in codon 448 (Arg448Ala) of the *CYP11B1* gene (White et al., 1991) that abolishes 11β-hydroxylase activity. A 2-basepair insertion in codon 394 of *CYP11B1* leading to premature termination of the open reading frame and a presumably non-functional product, was reported by Helmberg et al. (1992). More recently Curnow et al., (1993) described eight additional mutations in *CYP11B1* gene causing 11β-deficiency. Of the eight mutations, four are missense mutations (Thr318Met; Arg374Gln; Arg384Gln; Val441Gly) and COS-1 cells transfected with plasmids containing any of these mutants produced no detectable cortisol when incubated with deoxycortisol, thus explaining the genetic deficiency of these individuals. Three mutations are nonsense mutations (Lys174X; Gln338X; Gln356X) leading to truncated and presumably nonfunctional proteins, and one is a single C deletion in codon 32 (ΔC32) which causes a frame shift and a premature termination of the protein.

Corticosterone methyloxidase II deficiency is an inherited defect of aldosterone synthesis. This pathology is rare among the general population, but its frequency has been reported to be increased among Iranian-Jewish families. In this population the deficiency has been attributed to a recessive trait genetically linked to a unique *Mps* I polymorphysm in the *CYP11B1* gene (Globerman et al., 1988). Pascoe et al. (1992a) characterized the *CYP11B1* and *CYP11B2* genes in thirty seven members of an Iranian-Jewish family in which corticosterone methyloxidase II deficiency was segregating. Twelve members had the disease. The authors found no candidate mutations in *CYP11B1*, although two missense mutations (Arg181Trp; Val386Ala) were detected in *CYP11B2*. When these mutations were individually introduced in CYP11B2 cDNA and expressed in COS-1 cells, the mutation in position 181 reduced 18-hydroxylation and abolished detectable 18-oxidase activities. The mutation in position 386 caused only a small reduction in 18-hydroxycorticosterone production.

Glucocorticoid-suppressible aldosteronism is a primary aldosteronism with hypertension also known variously as dexamethasone-suppressible hyperaldosteronism, glucocorticoid-remediable aldosteronism or familial hyperaldosteronism type I. This pathology is characterized by hypertension and high circulating aldosterone level accompanied by the presence of abnormal adrenal steroids such as 18-hydroxycortisol and 18-oxocortisol. In such cases the regulation of aldosterone synthesis is shifted from the renin-angiotensin system to ACTH. In an autosomal dominant disorder, glucocorticoid-suppressible aldosteronism, Lifton and co-workers (1992) reported a gene duplication arising from unequal crossing over, fusing the 5' regulatory region of *CYP11B1* to the coding sequences of *CYP11B2*. They demonstrated that this chimaeric 11β-hydroxylase/aldosterone synthase gene was at the origin of this disorder in a large kindred characterized by early hypertension and death from stroke before age 45. They have proposed that the product of this gene has aldosterone synthase activity due to the coding sequence of $P450_{C18}$ being aberrantly expressed in the ZF-R by the $P450_{11\beta}$ regulatory sequences. Pascoe et al., (1992) also reported similar findings. In four unrelated patients with glucocorticoid-suppressible hyperaldosteronism each patient had a chromosome which carries three *CYP11B* genes instead of two. Unequal meiotic crossover is believed to be at the origin of the generation of the third gene which is a hybrid containing 5' regulatory and coding regions corresponding to *CYP11B1* and the 3' coding region of *CYP11B2*. Two patients had crossovers in intron 2, whereas the third and fourth patient had crossovers in intron 3, and in exon 4 respectively. *In vitro* studies showed that cells transfected with hybrid cDNAs containing up to the first 3 exons of *CYP11B1* coupled to the remaining 3' exons of CYP11B2 cDNA then synthesized aldosterone to a level comparable to CYP11B2 cDNA, while hybrids containing the first five exons or more of *CYP11B1* had no detectable aldosterone synthase activity. This implies that if other mutations with crossover downstream the fourth exon would occur they would likely lead to the synthesis of a non-active protein. This chimaeric gene would be under the control of ACTH in the ZF-R *in vivo*, and this would explain why it is suppressed by glucocorticoids. These studies demonstrate that glucocorticoid-suppressible hyperaldosteronism can be caused by the expression of a gene which is regulated as *CYP11B1* but expresses an aldosterone synthesizing protein. The discovery of this chimaeric gene is very significant and should contribute to a better understanding of some forms of endocrine hypertension.

$P450_{11\beta}$ and $P450_{C18}$: Comparison between the Hamster and other Mammalian Species

Figure 4 shows the coding sequence of hamster adrenal $P450_{11\beta}$ and $P450_{C18}$ cDNAs. The cDNA coding sequence and the deduced amino acid sequence of these two genes are respectively 90% and 84% homologous indicating that these two cytochromes originated from a common ancestral gene (Lehoux et al., 1994b; Véronneau et al., 1996). The products of these two hamster genes are closely akin to those of mouse, rat and human. The homologies of $P450_{11\beta}$ and $P450_{C18}$ amino acid sequences between the hamster, mouse, rat, human, and bovine are shown in Figure 5 and Figure 6, respectively. The following decreasing order of homology

```
M   A   L   R   A   K   A   D   V   W   L   A   R   P   W    15
ATG GCA CTC AGG GCA AAG GCA GAT GTG TGG CTG GCA AGA CCC TGG   45
--- --- --- --- --- --- --- --- --- --- --- --- --- --- ---

Q   C   L   P   R   T   R   A   L   G   T   T   A   A   L    30
CAG TGC CTG CCC AGG ACG AGG GCA CTG GGC ACC ACA GCA GCA CTG   90
--- --- --- --- --- --- --- --- --- --- --- --- --- --- ---

A   P   N   T   L   R   P   F   E   A   I   P   Q   Y   S    45
GCC CCC AAC ACA CTG CGG CCC TTT GAA GCC ATA CCG CAG TAC TCC  135
--- --- --- --- --- --- --- --- --- --- --- --- --- --- ---

R   N   R   W   L   K   M   L   Q   I   L   R   E   E   G    60
AGA AAC AGG TGG CTG AAG ATG CTA CAG ATC CTG AGG GAG GAG GGC  180
--- --- --- --- --- --- --- --- --- --- --- --- --- --- ---

Q   E   G   L   H   L   E   M   H   E   A   F   R   E   L    75
CAA GAG GGC CTG CAC CTG GAG ATG CAT GAG GCC TTC CGG GAG CTG  225
--- --- --- --- --- --- --- --- --- --- --- --- --- --- ---

G   P   I   F   R   Y   S   M   G   R   T   Q   V   V   S/Y   90
GGG CCC ATT TTC AGG TAC AGC ATG GGA AGA ACA CAG GTT GTG TCT  270
                                                        -A-

V   M   L   P   E   D/V A   E   K   L/V H/F Q   V/A E/D S    105
GTG ATG TTG CCA GAG GAT GCC GAG AAG CTG CAC CAG GTG GAG AGT  315
                    -T-             G-A TT-     -CA --C ---

M/T H/Q P   R/S R   M/T H/L L   E   P   W   V   A   H   R    120
ATG CAC CCT CGT CGG ATG CAC CTG GAA CCT TGG GTA GCC CAC AGA  360
-CA --G --C A-C --C -C- -T- T--

E   H   R   G   L   S   R   G   V   F   L   L   N   G   P    135
GAA CAC CGT GGC CTG AGT CGT GGA GTG TTC TTG CTA AAT GGG CCT  405
--- --- --- --- --- --- --- --- --- --- --- --- --- --- ---

E   W   R   F   N   R   L   R   L/I N   P   H   V/M L   S    150
GAA TGG CGC TTC AAC CGA CTG AGG CTC AAC CCA CAC GTG CTG TCC  450
                            A--         A--

P   K   A   V   Q   K   F   V   P   M   V   D   M   V   A    165
CCA AAG GCC GTT CAG AAG TTT GTC CCC ATG GTG GAC ATG GTA GCA  495
--- --- --- --- --- --- --- --- --- --- --- --- --- --- ---

R   D   F   L   E   S/F L   K   K   K   V   F/L Q/A N   A    180
CGG GAC TTC TTG GAG AGT CTG AAG AAG AAG GTG TTT CAG AAT GCT  540
                    -T-                 CTG GC-     --C

R/H G   S   L   T/S M   D/N V/F Q/Y Q/S S   L/M F   N   Y    195
CGT GGG AGC CTC ACC ATG GAT GTG CAG CAA AGC CTT TTT AAC TAC  585
-A- --A --- T-G T-A --- A-C T-C T-T TCC --T A-G --C --- --T

S/T I   E   A   S   N/H F   V   L   F   G   E   R   L   G    210
AGT ATA GAA GCC AGC AAC TTT GTT CTT TTT GGG GAG CGG CTG GGA  630
-CC             --- C--                                 --C

L   L   G   H/D D   L   S/N P/S A/G S   L   T/K F   I/V H/N  225
CTC CTT GGC CAT GAC CTG AGC CCT GCC AGC CTG ACG TTC ATC CAT  675
--- --- --T G-- --- --- -AT T-- -G- --- --- -A- --- G-- A--

A   L   H/N S   V/I F/M K   T   T   P   Q   L   M   F/L L    240
GCC TTG CAT TCC GTG TTC AAG ACG ACC CCA CAG CTC ATG TTC TTG  720
--- --- A-- --- A-A A-G --- --- --T --- --- --- C-T ---

P   R/S S/G L   T   R   W   T/I S   T   R   V   W   K   E    255
CCC AGG AGC CTG ACT CGC TGG ACA AGC ACC CGG GTG TGG AAA GAG  765
--T --T G-T --- --- --- --- -T- --- --- --- --- --- --- ---

H/N F   E/D A/S W   D   V/F I/V S   E   Y   V   N/T R/K C/N  270
CAT TTT GAG GCC TGG GAT GTC ATC TCT GAG TAT GTC AAC AGA TGT  810
A-C --- --T T-- --- --- T-- G-- --- --- --C --- -CA -A- AA-
```

Figure 4. Primary structures of the hamster adrenal P450$_{C18}$ and P450$_{11\beta}$ cDNAs. Base numbering starts at the initiating methionine. Dashes indicate indentity, and letters positions that differ. P450$_{C18}$ translation is indicated above the nucleotide sequence; where different, the translation of P450$_{11\beta}$ is shown.

```
I/V R/K K/N V   H/Y Q   E   L/V R/Q L/S G   S/G P   H/Q T/S 285
ATC CGG AAG GTG CAC CAG GAG CTC AGA CTT GGC AGC CCT CAC ACC 855
G-- AA- --T --- T-T --A --- G-G CA- AG- --T G-- --A --G T--

Y/W S   G/V I   V/  A/S E/Q L   M/V S/A Q/E G   A   L   P/T 300
TAC AGT GGC ATC GTG GCA GAA CTA ATG TCC CAG GGA GCT TTG CCT 900
-GG --- -T- --A [ ] T-- C-G --G G-A G-A G-- --T --- C-- A-A

L/M D   A   I   R/L A   N   S   I/L E   L   T   A   G   S   315
CTC GAC GCC ATC AGA GCC AAC TCA ATT GAG CTC ACC GCT GGG AGT 945
A-G --T --- --T CTG --- --- --T C-G --A --- --T --- --- ---

V   D   T   T   T/S F/V P   L   V   M   A/T L   F   E   L   330
GTA GAC ACG ACA ACC TTC CCC CTG GTC ATG GCT CTC TTT GAG CTG 990
--G --- --- --- T-A G-G --- --- --- --- A-C --- --- --- ---

A   R   N   P   D   V   Q   Q   A   V   R   Q   E   S   L   345
GCT CGG AAC CCA GAT GTT CAG CAG GCT GTG CGG CAG GAG AGC CTG 1035
--- --- --- --- --- --- --A --- --- --- --- --- --- --- ---

A   A   E   A   S   V   A   A   N   P   Q   R   A   M   S   360
GCA GCT GAG GCC AGC GTG GCT GCA AAT CCC CAG AGG GCT ATG TCG 1080
--- --- --- --- --- --- --- --- --- --- --- --- --- --- ---

D   L   P   L   L   R   A   V   L   K   E   T   L   R   L   375
GAT CTG CCC CTG CTG CGG GCT GTC CTT AAA GAG ACC TTG AGG CTC 1125
--- --- --- --- --- --- --- --- --- --- --- --- --- --- ---

Y   P   V   G/A G/V F   L   E   R   I   L   S   S   D   L   390
TAT CCT GTT GGT GGC TTT TTG GAG AGA ATT CTA AGC TCG GAC TTG 1170
--- --- --- -C- -T- --- --- --- --- --- --- --- --- --- ---

V   L   Q   N   Y   H   V   P   A   G   T   L/I V/L L/H L/M 405
GTG CTT CAG AAC TAC CAC GTC CCT GCT GGG ACA TTG GTC CTA CTT 1215
--- --- --- --- --- --- --- --- --- --- --G A-C C-- -AC A-G

Y/S L   Y   S   M   G   R   N   P   A   V   F   P   R   P   420
TAT CTC TAC TCC ATG GGC CGA AAC CCT GCA GTA TTT CCG AGG CCC 1260
AG- --- --- --- --- --- --- --- --- --- --- --- --- --- ---

E   H   Y   L   P   Q   R   W   L   E   R   N   G   S   F   435
GAG CAC TAC TTG CCC CAG CGC TGG CTG GAG AGG AAT GGG AGT TTC 1305
--- --- --- --- --- --- --- --- --- --- --- --- --- --- ---

Q   H   L   T   F   G   F   G   V   R   Q   C   L   G   K   450
CAG CAC CTG ACC TTC GGC TTT GGG GTG CGC CAG TGC CTG GGG AAG 1350
--- --- --- --- --- --- --- --- --- --- --- --- --- --- ---

R   L   A   Q   V   E   M   L   L   L   L   H   V   L   465
CGC CTG GCT CAG GTG GAG ATG CTC CTC CTG CTG CAC CAT GTG CTG 1395
--- --- --- --- --- --- --- --- --- --- --- --- --- ---

K   S   F   R   V   E   T   Q   E   R   E   D   V   R   M   480
AAA TCC TTC AGG GTG GAG ACG CAG GAG CGA GAG GAT GTG CGG ATG 1440
--- --- --- --- --- --- --- --- --- --- --- --- --- --- ---

V   Y   R   F   V   L   A   P   S   S   S   P   L   L   T   495
GTG TAC CGC TTT GTT CTG GCG CCC AGC TCC AGC CCC CTG CTC ACT 1485
--- --- --- --- --- --- --- --- --- --- --- --- --- --- ---

F   R   P   V   S   *     500   P450 amino acids c18/11ß
TTC CGG CCT GTC AGC TAG   1503  P450c18 nucleotide sequence
--- --- --- --- --- ---         P450118 nucleotide sequence
```

Figure 4.   (Continued)

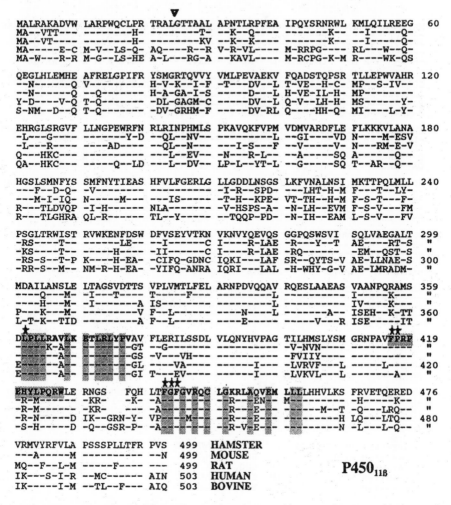

Figure 5. Comparison of the amino acid sequence of hamster P450₁₁β with those of mouse, rat, human, and bovine enzymes. The sequences of mouse, rat, human, and bovine were taken from Domalik et al., (1991), Nonaka et al., (1989), Kawamoto et al., (1990a), and Chua et al., (1987), respectively. The peptides with one, two and three asterisks are the putative heme/steroid binding site, the Ozol's tridecapeptide and the putative heme binding site, respectively.

was calculated between the adrenal P450₁₁β amino acid sequence of the hamster and that of other species with mouse 77% (Domalik et al., 1991), rat 74% (Nonaka et al., 1989), human 63% (Kawamoto et al., 1990a) and bovine 60% (Chua et al., 1987). An even greater homology was even found between the adrenal P450C18 amino acid sequence and that of other species with mouse 86% (Domalik et al.,

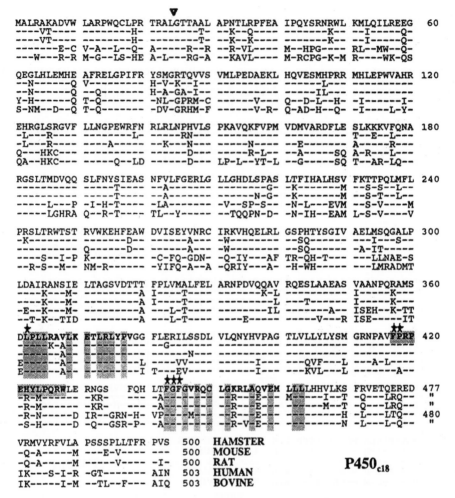

Figure 6. Comparison of the amino acid sequence of hamster P450$_{C18}$ with that of mouse, rat, human, and bovine enzymes. Mouse, rat, human, and bovine sequences were taken from Domalik et al., (1991), Imai et al., (1990), Kawamoto et al., (1990), and Chua et al., (1987), respectively. The peptides with one, two, and three asterisks are the putative heme/steroid binding site, the Ozol's tridecapeptide and the putative heme binding site, respectively.

1991), rat 85% (Imai et al., 1990), human 69% (Kawamoto et al., 1990) and bovine 65% (Chua et al., 1987). Noteworthy is the fact that it is the hamster P450$_{C18}$ and not P450$_{11\beta}$ that shares a greater homology with the bovine P450$_{11\beta}$ which is responsible of the conversion of corticosterone to aldosterone in the bovine adrenal.

The hamster, mouse, and rat P450$_{11\beta}$s are biosynthesized as a precursor of 499 amino-acids and human and bovine P450$_{11\beta}$s as a precursor of 503 amino-acids, all having a 24-amino acid extension peptide. The 4 amino-acids difference is due to the addition of Val290, Gly435, Arg436, Asn437 in human $P450_{11\beta}$ and Val290, Gly435, Ser436, Arg437 in bovine $P450_{11\beta}$. The hamster, mouse and rat $P450_{C18}$ are biosynthesized as precursor of 500 amino-acids and human as a precursor of 503 amino-acids, all having a 24-amino acid extension peptide. In contrast to $P450_{11\beta}$, $P450_{C18}$ possesses a Val290 residue. However, Gly435, Arg436, and Asn437 found in the human $P450_{C18}$ sequence are absent in those of the hamster, mouse and rat. Noteworthy are three highly conserved regions among cytochrome P450 enzymes. These regions include the putative heme/steroid binding site, the Ozol's tridecapeptide and the putative heme binding site, respectively as shown on Figure 5 and Figure 6.

III. PARTICIPATION OF ANGIOTENSIN-II IN THE MECHANISM OF ACTION OF POTASSIUM SUPPLEMENTATION AND SODIUM RESTRICTION

In this section we will discuss the mechanisms involved in the mediation of ZG regulation by potassium and sodium. We will see that AII mediates the ZG response to low sodium intake at the level of $P450_{scc}$, $P450_{C18}$ and AII receptors. AII also mediates the ZG response to high potassium intake at the level of $P450_{C18}$ and AII receptors. AII is the product of two successive catalytic transformations of the precursor angiotensinogen by renin and angiotensin converting enzyme. Two sources of AII are available to the adrenal cortex: AII originating from circulation, and AII formed *in situ* in the adrenal by the local renin-angiotensin system.

It is well known that high potassium or low sodium intake affects plasma renin activity and AII production (Quinn and Williams, 1992). Moreover, the adrenal ZG, which contains all components of the renin-angiotensin system (Campbell and Habener, 1986; Deschepper et al., 1986; Strittmatter et al., 1986; Dzau et al., 1987), is enhanced by potassium supplementation or sodium restriction (Nakamaru et al., 1985; Dzau et al., 1986; Brecher et al., 1989; Shier et al., 1989; Yamaguchi et al., 1990; Tremblay et al., 1992). These findings led to the postulate that the effects on the mineralocorticoid production of alterations in the intake in potassium and sodium were mediated by the generation of AII by circulating renin and by the local renin-angiotensin system. If this hypothesis is correct, inhibitors of the angiotensin converting enzyme, such as captopril which specifically blocks the conversion of angiotensin I into AII, should also block the enhancement of steroidogenesis by low sodium or high potassium intake. Indeed, inhibitors of angiotensin converting enzyme were shown to suppress the response of the adrenal to such changes in the intake of cations.

In the rat, following a high potassium intake, we found that captopril prevented the elevation in the levels of $P450_{C18}$ mRNA (Figure 1) and $P450_{C18}$ protein (analyzed with a polyclonal $P450_{11\beta}$ antibody) thereby indicating a participation of AII in the control of the final step of aldosterone formation by potassium (Tremblay et al., 1992). In contrast, captopril did not prevent the increase in the levels of ZG $P450_{scc}$ mRNA and $P450_{scc}$ protein caused by a high potassium intake. These results imply that the control of the expression of $P450_{scc}$, during high potassium intake is mediated by multiple factors, since the inhibition of AII formation by captopril alone was insufficient to counteract the effect of the high potassium intake. It is noteworthy that plasma renin activity was lowered significantly below control values in potassium supplemented rats (Tremblay et al., 1991). Consequently, the requirement of an intact angiotensin converting enzyme activity to stimulate aldosterone synthesis, despite these low levels of plasma renin activity, indicates that potassium increases $P450_{C18}$ via the local adrenal renin-angiotensin system. This is in agreement with previous reports that potassium increases AII secretion (Shier et al., 1989; Kifor et al., 1991) and renin activity in the isolated ZG (Doi et al., 1984; Shier et al., 1989; Yamaguchi et al., 1990).

In the case of low sodium intake we found that captopril was able to suppress the enhancing effect of this diet on $P450_{C18}$ mRNA (Figure 1) and $P450_{C18}$ protein (analyzed with a polyclonal $P450_{11\beta}$ antibody) levels in rat adrenal ZG (Tremblay and Lehoux, 1992; Tremblay et al., 1992), which is in accord with a report showing an increased level of $P450_{C18}$ protein in the ZG of sodium-depleted rats (Shibata et al., 1991). This indicates that AII participates in the increases in aldosterone levels that accompany changes of sodium intake. In contrast to the high potassium intake, captopril blocked the increase in the levels of $P450_{scc}$ mRNA and $P450_{scc}$ protein caused by a low sodium intake thereby indicating a participation of AII in this process (Tremblay and Lehoux, 1992). These results demonstrate that AII is an important factor in the mediation of the action of low sodium intake at both early and final steps of aldosterone synthesis. In contrast to high potassium intake, plasma renin activity was increased following low sodium intake (Tremblay et al., 1992), suggesting the participation of circulating AII in the increase in ZG steroidogenesis under sodium restriction. However, consistent also with mediation by adrenal angiotensin, enhancement of the local adrenal renin-angiotensin system accompanies the stimulation of aldosterone synthesis during sodium depletion (Doi et al., 1984; Brecher et al., 1989).

The interaction between potassium, sodium, AII and aldosterone synthase is therefore of prime importance in the maintenance of ionic homeostasis. Imbalance or malfunction of any component of the renin-angiotensin system as well as the *in situ* adrenal renin-angiotensin system may result in the development of mild to severe essential hypertension. Consequently, further studies in this field of research should shed light on the development of certain forms of hypertension.

A. Angiotensin-II Receptor in the Control of Aldosterone Synthesis

The first step in the mode of action of AII is extra membrane. AII binds to a cell membrane receptor in the adrenal ZG, thus activating a transduction pathway leading to an increased synthesis of aldosterone (Quinn and Williams, 1992). AII receptors have been found in the adrenal cortex of human (Brown et al., 1980), bovine (Guillemette and Escher, 1983), canine and rat (Capponi and Catt, 1980) adrenals. Two types of AII receptors were found in the adrenal, type 1 (AT_1) and type 2 (AT_2), according to their differential affinity for (Dup753 and PD 123319), two non-peptide antagonists of AII (Timmermans et al., 1991). AT_1 receptors mediate the action of AII on aldosterone synthesis in the adrenal ZG, whereas the role of AT_2 is yet to be elucidated (Chiu et al., 1989; Bumpus et al., 1991). In the rat, a second form of AII receptor type 1 (AT_{1B}) has been reported exhibiting 95% identity in amino acid sequence with AT_1 receptors. The rat adrenal would primarily express AT_{1B} receptors (Kakar et al., 1992).

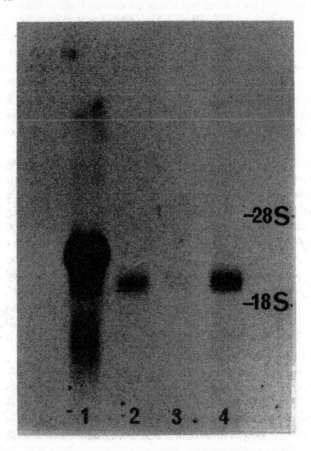

Figure 7. AT_1 receptor mRNA: Comparison between bovine, hamster, and rat adrenals. Total RNA samples isolated from bovine adrenals, lane 1; hamster, lane 2; rat zona fasciculata-reticularis, lane 3; and rat zona glomerulosa, lane 4, were analyzed with a bovine adrenal [^{32}P]-labeled AT_1 receptor cDNA probe.

The sequencing of the cDNAs encoding the rat (Murphy et al., 1991; Iwai et al. 1991; Kakar et al., 1992), bovine (Sasaki et al., 1991) and human (Furuka et al., 1992) AT_1 receptor revealed a high degree of homology. Using northern analysis, we found a single band of AT_1 receptor mRNA at 2.3 kb in the rat adrenal ZG, which is in agreement with the size reported by others (Murphy et al., 1991; Zelezna et al., 1992). When bovine, rat, and hamster adrenal preparations were analyzed on the same blot, AT_1 mRNA migrated at 3.3 kb for the bovine, which is in agreement with a previous report (Sasaki et al., 1991), and at 2.3 kb for the rat and hamster preparations (Figure 7).

By immunoblotting analysis, using an antibody reacting against AT_1 receptors (Zelezna et al., 1992), we identified a specific protein band in the area of 56 kDa (Figure 8) (Lehoux et al., 1994a) in the rat adrenal and of 70 kDa in the hamster adrenal (unpublished observations). These AT_1 receptors are heavier than the 40,889 Da predicted from the cDNA sequence of the rat AT_1 receptor (Murphy et al., 1991). The difference may be attributed to a high degree of post-transcriptional glycosylation. Molecular weights higher than that predicted by the cDNA sequence were also reported by Zelezna et al., (1992) for the rat adrenal, using immunoblotting analysis, and for dog (Guillemette and Escher, 1983) and bovine (Capponi and Catt, 1980) adrenal preparations using photo-affinity labeling followed by electrophoresis.

Potassium influences aldosterone synthesis through changes in AT_1 receptors. Indeed, previous studies on isolated adrenal ZG cells showed that the cells from rats maintained on a high potassium intake bound 3.5 times more $[^{125}I]$angiotensin-II than did controls (Douglas, 1980, 1980a). We also found that a high potassium intake for seven days increased the levels of both ZG AT_1 receptor mRNA and AT_1 receptor protein by about 2-fold (Figure 8). Under such regimen, a concomitant increase in the level of $P450_{C18}$ mRNA was also found (Lehoux et al., 1993). Captopril did not efficiently block aldosterone formation when added to the high potassium diet, in this particular series of experiments. Collectively this indicates that, potassium together with other mechanisms, influences aldosterone synthesis through increased synthesis of AT_1 receptors. More studies will be needed, however, to elucidate the relative importance of this activation pathway by potassium in controlling aldosterone biosynthesis.

There is a participation of AII and AT_1 receptors in the mechanism of action of low sodium intake in increasing aldosterone synthesis in rat adrenal. Similar to the high potassium intake, low sodium intake produced changes in AT_1 receptor synthesis. We found that a low sodium intake led to 3.7-fold and 2.3-fold increases in the level of AT_1 receptor mRNA and AT_1 receptor protein in homogenates of the rat adrenal ZG and that captopril blocked these effects (Lehoux et al., 1993, 1993a). This suggests a transcriptionally mediated process stimulated by AII. Figure 8 also shows increases in the level of AT_1 receptor in adrenal cell membrane preparations of rats fed a low sodium diet. The fact that the increases in the levels of mRNA and protein induced by the low sodium intake were of similar magnitude, indicates that

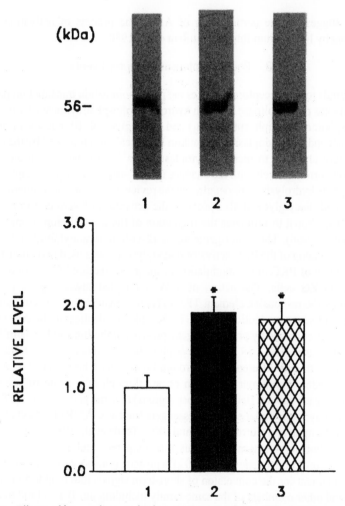

Figure 8. Effects of low sodium or high potassium intake on AT₁ receptor protein in the rat adrenal zona glomerulosa. Rats were maintained for seven days on the following diets: 1, control; 2, low sodium; 3, high potassium. Cell membranes were isolated and analyzed with an anti-AT₁ receptor peptide antiserum. The relative densitometric profiles are shown in the bottom panel where $^{*}p < 0.05$.

AT_1 receptor mRNA was effectively translated into AT_1 receptor protein. Under these conditions increases on the levels of plasma aldosterone and adrenal ZG $P450_{C18}$ mRNA were also observed. Previous works showed an increase of 70% in the number of AII binding sites in rat adrenal ZG after 4 days of sodium restriction (Aguilera et al., 1978; Kifor et al., 1991a). This effect was also prevented by treatment with captopril and was reproduced by infusion of AII (Hauber et al.,

1978), suggesting the participation of AII in the process of activation of AII receptors by low sodium intake (Aguilera et al., 1980).

B. Post Angiotensin Receptor Events

AII binds to AT_1 receptors and uses the phosphoinositide transduction pathway. AII activates phospholipase C which hydrolyzes phosphatidylinositol diphosphate to phosphoinositol triphosphate (IP_3) and diacylglycerol. IP_3 causes a liberation, within seconds, of endoplasmic reticulum bound calcium followed by the opening of calcium channels to increase cytoplasmic Ca^{++}. The increase in aldosterone secretion correlates well with the changes in Ca^{++} associated with phosphoinositol diphosphate hydrolysis. (For details, see the review by Quinn and Williams, 1992).

Ca^{++} and diacylglycerol also provoke the activation of various forms of PKC which was shown to influence the regulation of the transcription of many genes (Nishizuka, 1986). The tumorigenic agent 12-O-tetradecanoylphorbol-13-acetate (TPA), an analog of the PKC activator diacylglycerol, was used to demonstrate the involvement of PKC in the modulation of the genes required for corticosteroido-genesis in ZG cells. The action of TPA is directed towards gene promoters containing the responsive element TRE (TPA responsive element) consisting of a sequence of seven nucleotides TGAGTCA, which is identical to the AP_1 site. The AP_1 site is recognized by proto-oncogene proteins of the Jun and Fos families. The proto-oncogene c-fos, a 39 kDa nuclear phosphoprotein is involved in the control of cell growth and differentiation through binding as a homodimer to AP_1 sites (Muller, 1986). The dimer formation occurs through interaction of leucine-rich domain on each subunit (leucine-zipper regions) (Curran and Franza, 1988). C-fos is induced by intracellular second messengers such as c-AMP (Tsuda et al., 1986), calcium (Morgan and Curran, 1986) and PKC (Tsuda et al., 1986), and in this way converts short-term external stimulus into long term cellular response (Marx, 1986). The proto-oncogene c-jun, a 40 kDa protein, is also a very important cellular control element for the conversion of short-term signals into long term responses. C-jun and other members of the same family including jun-B also bind to the AP_1 consensus site by forming heterodimers with c-fos. C-fos and c-jun interact at their respective leucine-zipper regions to form heterodimers which show increased affinity for AP_1 binding sites (Curran and Franza, 1988). Both c-fos and c-jun are phosphoproteins and it appears that PKC would participate in the dephosphoryla-tion of c-jun to increase the formation of the Jun-Fos complex and its binding to AP_1 (Angel and Karin, 1991).

These findings thus suggest that the action of AII, a biological activator of PKC, would be directed through activation of c-jun towards promoter regions of genes containing AP_1 sites. In fact, putative sites were found in the 5′-non coding regions of bovine *CYP11A* (Ahlgren et al., 1990). The exact mechanism of mediation of AII on the *CYP11B2* gene is still not known, however, since up to now no AP_1 sites were reported in this gene.

IV. ATRIAL NATRIURETIC PEPTIDE (ANP)

Atrial natriuretic peptide has a negative regulatory action on adrenocortical functions. This peptide inhibits the increases in aldosterone secretion *in vivo* which are induced by ACTH, AII, potassium (Chartier et al., 1984), prostaglandins and serotonin (Chartier et al., 1984; De Léan A, et al., 1984; Higuchi et al., 1986; Elliot and Goodfriend, 1987; King and Baertshi, 1989). ANP, a member of a family of homologous peptides, was first isolated from myocardial tissue this being the principal site for its synthesis (De Bold, et al., 1981). cDNAs that encode the precursor of ANP have been cloned from many animal species (reviewed by Baxter et al., 1988). ANP is synthesized as a prohormone which is subsequently converted to its active form. ANP mRNA has been found in various extra-atrial tissues including the kidney, brain, lung, pulmonary veins, pituitary, and the adrenal medulla, suggesting a paracrine role for ANP in these tissues to produce a physiological response (Gardner et al., 1986). This hormone has a very wide range of activities and interacts with many substances which participate in the control of blood volume and arterial pressure. In contrast to vasopressin or AII which are hypertensive, ANP appears to protect the organism against hypertension.

A.   ANP and Corticosteroids

There is a link between the adrenal cortex and ANP, indeed, glucocorticoids increase the synthesis of ANP by inducing the expression of its gene (Gardner et al., 1986a) whereas mineralocorticoids increase ANP synthesis via other mechanisms, most probably through their action on the expansion of plasma volume (Garcia et al., 1985). In return, ANP inhibits aldosterone production and increases sodium and water excretion (Huang et al., 1985), actions that counterbalance those of the renin-angiotensin system. ANP inhibits basal as well as stimulated levels of aldosterone secretion by adrenal ZG cells. In addition, sodium restriction greatly decreases the circulating level of ANP by blocking its release from myocytes (Schwartz et al., 1986; Richards et al., 1987).

In humans many effects of ANP, including aldosterone suppression, seem to be mediated in part by the ability of ANP to suppress the renin-angiotensin system. The secretion of renal renin is suppressed and plasma AII and aldosterone levels are decreased following ANP administration (Richards et al., 1985; Weidmann et al., 1986; Genest et al., 1988). In the other hand, ANP blocks the increases in plasma aldosterone levels provoked by AII administration, suggesting a direct effect (Cuneo et al., 1987; Anderson et al., 1986). Also when the renin-angiotensin system is blocked by enapril, an inhibitor of the conversion enzyme, ANP infusions still suppress aldosterone (Clinkingbeard et al., 1990).

ANP acts directly on the adrenal cortex, by suppressing the capacity for the synthesis of aldosterone (Higuchi et al., 1986; Ohashi et al., 1988). Using bovine adrenal cells in culture for three days, Brochu and co-workers (1991) demonstrated that a 3 h addition of ANP to the incubation media inhibited the formation of pregnenolone and also other enzymatic steps between progesterone and deoxycorticosterone, deoxycorticosterone to corticosterone, and finally between corticosterone and aldosterone. In chicken adrenocortical cells, ANP was shown to inhibit the conversion of cholesterol to pregnenolone and the conversion of corticosterone to aldosterone, actions which are mediated by c-GMP (Rosenberg et al., 1989). Using an animal model, Rebuffat et al., (1988) demonstrated that ANP exerts a long term inhibitory effect on the growth and steroidogenic capacity of the rat adrenal ZG in addition to acutely depressing aldosterone secretion. These authors suggested that ANP may interfere with some of the post-receptor events leading to the AII induced stimulation of the growth of the rat adrenal ZG. At long term, ANP inhibited the effect of ACTH to increase the $P450_{scc}$ mRNA level in bovine adrenal cell cultures, indicating that ANP was able to affect the gene expression of the early enzymatic step controlling corticosteroidogenesis. There are no data on the effects of ANP on the expression of the genes of the intermediary and final steps of aldosterone synthesis, but when available they should contribute to a better understanding of the long term action of ANP on adrenal corticosteroidogenesis at the molecular level.

Although ANP directly stimulates c-GMP accumulation in human and bovine adrenocortical cells, it appears that ANP-induced actions in the adrenal are not mediated solely by c-GMP. Indeed, when c-GMP production was strongly induced by sodium nitroprusside (Inagami, 1989) or when Br-c-GMP (Ganguly et al., 1989) was added to incubation media, aldosterone secretion was not impaired in adrenal cells, whereas it was inhibited in the presence of ANP (Nawaka et al., 1991). Clearly other factors are involved for the complete action of this hormone. In this respect, ANP inhibited c-AMP formation and aldosterone secretion in isolated bovine adrenal ZG cells stimulated either by ACTH or AII. The inhibition of ACTH-stimulated steroid secretion by ANP could be partially reversed by increasing the cellular level of c-AMP while, in contrast, this restoration did not occur in AII stimulated cells (Barret and Isales, 1988). As ACTH and AII control adrenal functions through c-AMP and inositol phosphate, it is not surprising that their transduction pathways are affected differently by ANP. However, ANP did not prevent the formation of inositol triphosphate nor changes in Ca^{++} concentration in response to AII in calf adrenal ZG cells (Ganguly et al., 1989). ANP was reported to inhibit the autophosphorylation of PKC in the plasma membrane preparation of bovine adrenal ZG cells (Pandey et al., 1987) indicating that some effects induced by ANP could be mediated by changes in membrane PKC activity. The effect of ANP on calcium is still controversial although a selective action on different voltage dependent calcium channels cannot be ruled out (McCarthy et al., 1990). Although the ZG is the main adrenal target for ANP, an inhibitory effect on cortisol secretion in human adrenal tissues *in vitro* has also been reported (Naruse et al., 1987). Moreover, ANP stimulated the accumulation of c-GMP and inhibited the secretion of aldosterone, cortisol and dehydroepiandrosterone in bovine and human adrenal cells in culture

(Nawata et al., 1991). Evidently multiple messenger systems underline the action of ANP in the adrenal, and the exact mechanisms of action of ANP are yet to be elucidated.

B. ANP Receptors in the Adrenal Cortex

Three types of ANP receptors have been demonstrated by molecular cloning, ANPRA and ANPRB are implicated in the production of c-GMP (Singh et al., 1988; Chang et al., 1989; Chinkers et al., 1989), while ANPRC is not associated with c-GMP production but is rather involved in its clearance (Takayanagi et al., 1987; Fuller et al., 1988). By hybridization *in situ*, ANP receptors type A have been shown to be expressed only in the ZG, whereas type B mRNA has been found only in the medulla of the rhesus monkey adrenal (Wilcox et al., 1991). ANPRC is present in discrete clusters of cells in the three zones of the adrenal cortex as well as in the medulla and is relatively less abundant than ANPRA. ANPRA mRNA was also detected in the human adrenal by northern blotting analysis (Ohashi et al., 1986). These results suggest that ANPRA is responsible for the inhibitory action of ANP in the adrenal ZG. Nawata et al., (1991) determined that half of the ANP binding sites are occupied in human adrenal under physiological conditions, suggesting that the circulating level of ANP may regulate steroidogenesis under physiological conditions. Moreover, ANP did not suppress aldosterone secretion in patients with Cushing's syndrome (Higuchi et al., 1988) nor those with primary aldosteronism (Higuchi et al., 1986a), this was attributed to a decreased number of ANP receptors and to a lack of capacity to increase c-GMP production.

Data are not yet available on the long term effects on adrenal ANP receptors and the production of c-GMP of ACTH and on other situations that promote increased aldosterone synthesis in normal animals, such as low sodium or high potassium intake and AII administration. When such data become available they will contribute to a better understanding of the mechanism of action of ANP in the normal adrenal.

V. EFFECTS OF ADRENOCORTICOTROPIN ON CORTICOSTEROID FORMATION

Adrenocorticotropin, synthesized in the pituitary, is the main regulator of glucocorticoid formation. This hormone, acts within minutes to increase the output of glucocorticoids, and also has a transient action on the formation of aldosterone in many animal species (Lehoux, 1979). For example, in humans (Rayfield et al., 1973), the infusion of ACTH resulted in a stimulation of aldosterone output, followed 6 hours later by a reduction to baseline. ACTH has also a long-term effect on the integrity of adrenocortical cells. Chronic ACTH administration to rats for many days altered the volume and number of mitochondria (Nussdorfer et al., 1971). Mitochondrial cristae, the site of many enzymatic activities involved in steroid synthesis (Tamaoki, 1973), were modified in ZF-R and also in the ZG, their total surface being increased (Sabatini et al., 1962; Kahri, 1968, 1971) with a concomitant change in the global turnover of mitochondrial proteins (Canick and Villee, 1974; Neri et al., 1978). Under such conditions, the appearance of the mitochondria in ZF-R and ZG was similar. It was thus suggested that the proliferation of adrenocortical cells could originate from the outer to the inner layers (Manuelidis and Mulrow, 1973; Hornsby et al., 1974, 1974a) and that ACTH might increase this transformation. In agreement with this suggestion, ACTH administration to sheep for 5 days provoked a transformation of the aldosterone secreting ZG cells into cells secreting cortisol (McDougall et al., 1980) while ACTH administration to rats for 9 days completely abolished the binding of [^{125}I]angiotensin-II to isolated adrenal ZG cells. The latter phenomenon was completely reversed 19 days after cessation of ACTH treatment (Rao and Lehoux, 1979; Legros and Lehoux, 1983). The basal capacity of isolated ZG cells to produce aldosterone was considerably decreased following chronic ACTH administration whereas their capacity to produce corticosterone was not altered, another indication that these cells had lost their attributes as ZG cells. These results thus demonstrate the action of ACTH to transform ZG cells into functional ZF-R cells.

ACTH at physiological concentrations is essential for the integrity and function of the adrenal cortex. Indeed, in rats following hypophysectomy, the enzymatic activities of $P450_{scc}$, $P450_{C21}$, $P450_{11\beta}$ and adrenodoxin were considerably decreased within 3 days and these activities were partially restored by ACTH administration (Purvis et al., 1973; Brownie et al., 1973).

Furthermore, studies have demonstrated that ACTH increased the levels of the bovine adrenal steroidogenic enzymes $P450_{scc}$, $P450_{C21}$, $P450_{17\alpha}$, and $P450_{11\beta}$ as well as adrenodoxin with 4- to 5-fold increases within 36 hours of stimulation in vitro (Simpson et al., 1988). The mRNA levels of these steroidogenic enzymes were also increased following the addition of ACTH to incubation media (John et al., 1984, 1985, 1986a; Boggaram et al., 1984; Zuber et al., 1986). Similar results were also reported for foetal adrenal cells in culture (DiBlasio et al., 1987; Voutilainen et al., 1988). These changes were attributed to an increase of the initiation of respective gene transcriptions (John et al., 1986), although a stabilization of the $P450_{scc}$ transcripts also occurred under such conditions.

Under physiological conditions in vivo, however, ACTH alters the expression of certain but not all of the adrenocortical steroidogenic enzymes. In the hamster (Lehoux et al., 1987), the response of these genes varies between two situations closely related to physiological conditions: (1) the effect of a single dose of ACTH sufficient to elevate plasma ACTH for less than 5 h, and (2) the effect of multiple ACTH injections that maintained an elevated plasma ACTH level for a period of 24 h. Under the first experimental conditions, with the single ACTH injection, a rapid elevation in plasma cortisol level occurred followed by a decrease to control value 5 h post treatment. At the same time the fluctuations in the adrenal mRNA levels of HMG-CoA reductase, the key regulatory enzyme of cholesterogenesis, were similar to those of plasma cortisol. The various steroidogenic genes responded differently to this challenge. $P450_{scc}$ mRNA levels increased up to 5 h but decreased slowly thereafter to remain still higher than controls at 24 h post treatment. $P450_{17\alpha}$ mRNA levels increased up to 5 h to decrease to control value 15 h post ACTH administration. In contrast 3β-HSD, $P450_{C21}$ and $P450_{11\beta}$ mRNA levels were not affected under such conditions, indicating that the short term effect of ACTH was not directed towards these enzyme systems.

Under the second experimental condition, multiple ACTH injections each 5 h maintained an elevated plasma cortisol level up to 24 hours and provoked sustained elevations in the levels of HMG-CoA reductase mRNA, indicating that cholesterol synthesis also provides a site for regulation of glucocorticoid synthesis. Indeed, parallel fluctuations in HMG-CoA reductase activity, reductase protein, reductase mRNA, and plasma ACTH levels (Lehoux et al., 1990) support a key role for this enzymatic system in the regulation of glucocorticoid formation in the hamster adrenal. Changes in HMG-CoA reductase activity are due to fluctuations in the adrenal cholesterol content rather than to a direct effect of ACTH (Osborne et al., 1988). Cholesterol lowers the transcription of the HMG-CoA reductase gene through a repressor protein which binds to the regulatory region in the presence of sterols (Osborne et al., 1988). Adrenal cholesterogenesis is quantitatively less important in rats, bovines and humans than in hamsters, since the first three species utilize cholesterol mainly originating from plasma for their corticosteroidogenesis under physiological conditions (Lehoux and Lefebvre, 1980). However, ACTH administration to rats was also shown to rapidly and significantly increase adrenal HMG-CoA reductase activity and the level of reductase protein and reductase mRNA, thus indicating the importance of this enzymatic system in the fine control of glucocorticoid formation by ACTH even in the adrenals of those species which predominantly utilize circulating plasma cholesterol for their steroidogenesis.

During the repeated ACTH stimulus, $P450_{scc}$ mRNA levels remained elevated and decreased very slowly after cessation of treatment to reach near control values 30 h after the last injection. In the rat, ACTH treatment also led to an increase in the level of $P450_{scc}$ mRNA (Lehoux and Tremblay, 1992). There was, however, no concomitant increase in hamster adrenal $P450_{scc}$ protein even 30 h after cessation of treatment. Several studies show that ACTH induces changes in $P450_{scc}$ mRNA but does not affect $P450_{scc}$ protein unless the stimulation is prolonged. In cultured bovine adrenal cells a delay between mRNA and protein synthesis was reported when ACTH induced a maximal increase in $P450_{scc}$ mRNA within 6 to 12 h, but despite this the level of $P450_{scc}$ protein remained unchanged and rose only 24 h after exposure to ACTH (Hanukoglu et al., 1990). Also, there was no increase in [35S]methionine incorporation into $P450_{scc}$ protein 12 h after ACTH stimulation of bovine adrenal cells in culture (Dubois et al., 1981), indicating that $P450_{scc}$ mRNA was not immediately translated into $P450_{scc}$ protein. We are not able at present to explain this apparent discrepancy between the rapid increase in $P450_{scc}$ mRNA and the unchanged level of $P450_{scc}$ protein following ACTH stimulation. However, there is a half-life of $P450_{scc}$ protein of several days in rat adrenals *in vivo* (Purvis et al., 1973) and therefore physiological variations occurring during a single day cannot be attributed to changes in the content of $P450_{scc}$ protein. Consequently other factors must be involved in the control of this rate limiting step of glucocorticoid formation by ACTH; the factors that make cholesterol available to the intramito-chondrial $P450_{scc}$ are the most probable candidates for such control. The control of cholesterol metabolism by access of cholesterol to $P450_{scc}$ is discussed in detail in Chapter 5.

Cytochrome $P450_{17\alpha}$ is not expressed in the rat adrenal cortex, and consequently corticosterone is the main glucocorticoid formed in this species. Therefore, circu-lating corticosterone might well be used by rat glomerulosa cells as a substrate for aldosterone synthesis. In contrast, in hamster adrenals, $P450_{17\alpha}$ is deeply involved in the control of glucocorticoid synthesis by ACTH and consequently contributes to divert metabolism away from mineralocorticoids to glucocorticoids. Upon sustained stimulation by ACTH the levels of $P450_{17\alpha}$ mRNA, $P450_{17\alpha}$ protein as well as 17α-hydroxylase activity were increased, in hamster adrenals, and both $P450_{17\alpha}$ mRNA and $P450_{17\alpha}$ protein levels decreased in a similar manner after the last ACTH injection to return to near control values within 15 h. However, the level of $P450_{17\alpha}$ mRNA had already increased maximally at 2.5 h after the first injection of ACTH whereas the $P450_{17\alpha}$ protein level and 17α-hydroxylase activity increased slowly between 5 and 10 h to attain a plateau at 15 h. This indicates that $P450_{17\alpha}$ mRNA was not transcribed immediately into active $P450_{17\alpha}$ enzyme, and a post-transcriptional control mechanism is therefore postulated to explain this delay. Moreover, $P450_{17\alpha}$ was quantitatively increased only under sustained ACTH stimu-lation, in a manner similar to that occurring during diurnal physiological variations of circulating ACTH. Taken collectively, these results indicate a predominant role played by the adrenal 17α-hydroxylase system in the regulation of glucocorticoids in the hamster, a species which like humans, secretes cortisol.

There were no changes in the levels of 3β-HSD, $P450_{C21}$ and $P450_{11\beta}$ following a single injection of ACTH or sustained stimulation by this hormone indicating that these steroidogenic components are not rate-determining enzymes in the hamster adrenal *in vivo* and consequently are not responsible for the fine control of glucocorticoid synthesis mediated by ACTH. Also the transformation of corticos-terone to aldosterone by hamster adrenal mitochondria was significantly decreased by the sustained ACTH stimulus, thereby confirming different control of the formation of mineralocorticoids and glucocorticoids.

A. ACTH Receptors in the Control of Corticosteroid Formation

In addition to its effects on the rate-limiting adrenal enzymes, ACTH partially controls adrenal steroidogenesis through the regulation of its own receptor. The first step in the mechanism of action of ACTH is its binding to specific receptor sites on adrenal cell membranes to increase c-AMP formation, as will be discussed in the next section. ACTH binding sites were found in a variety of adrenocortical cells including mouse tumor cells (Lefkowitz et al., 1971), mouse Y-1 adrenal cells (Osawa and Hall, 1985), rat (Buckley and Ramachandran, 1981; McIlhinney and Schulter, 1975; Gallo-Payet and Escher, 1985; Rani et al., 1983) bovine (Penhoat et al., 1989), rabbit (Durand and Locatelli, 1980) ovine (Durand, 1979) human (Catalano et al., 1986), domestic fowl (Carsia and Weber, 1988) and porcine (Luddens and Havsteen, 1986) adrenal cells. A burst of increased responsiveness to ACTH in the ovine adrenal immediately preceding parturition was shown to be accompanied by an increase in the number of ACTH binding sites (Durand, 1979). Furthermore, the number of ACTH binding sites in membranes prepared from rabbit adrenals was considerably reduced after hypophysectomy (Durand and

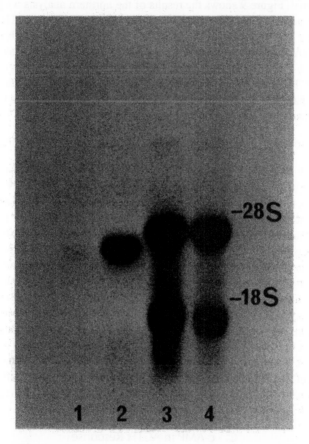

Figure 9. ACTH receptor mRNA: Comparison between bovine, hamster, and rat adrenals. Total RNA samples isolated from bovine adrenals, lane 1; hamster, lane 2; rat zona fasciculata-reticularis, lane 3; and rat zona glomerulosa, lane 4, were analyzed with a mouse adrenal [^{32}P]-labeled ACTH receptor cDNA probe.

Locatelli, 1980), and ACTH administration restored it to presurgical levels. ACTH administration also increased the number of ACTH binding sites in the adrenals of intact rabbits, indicating that ACTH influences its own receptors. ACTH was also shown to increase the number of binding sites in cultured bovine adrenal cells (Penhoat et al., 1989).

Cloning of the ACTH receptor (ACTH-R) cDNA (Montjoy et al., 1992), has permitted an analysis of its transcriptional control. By *in situ* hybridization they found that ACTH-R was expressed primarily across the *zona fasciculata* and in the cortical half of the ZG in the rhesus macaque. A small band was also detected adjacent to the medulla probably in the zona reticularis. By northern analysis, they found a principal ACTH-R mRNA species of approximately 4 kb in the human and rhesus adrenal. Figure 9 shows the results of the northern analysis we performed on adrenal preparations from various species, using a mouse ACTH-R cDNA (provided by Montjoy et al., 1992). Bovine and hamster adrenal hybridized to a single mRNA species of approximately 3.6 kb. In the rat, two species of approximately 4.2 kb and 1.97 kb were detected. As seen in Figure 9, there were comparable levels of ACTH-R mRNA in the rat adrenal ZG and in the rest of the adrenal gland, although it has been reported that the number of ACTH binding sites was greater in the ZG than in the ZF-R (Gallo-Payet and Escher, 1985).

B. Adrenal Responsiveness to ACTH under Low Sodium Intake

When sodium intake is restricted in the rat there is a substantial enhancement of the response to ACTH by the adrenal ZG cells (Underwood et al., 1989). We found no increases, however, in the levels of ACTH-R mRNA in adrenal ZG nor in ZG-R of rats maintained on a low sodium or a high potassium intake for seven days, indicating that the increased responsiveness of the gland was not due to increased ACTH-R expression (Lehoux et al., 1993). Underwood et al., (1989) found that the levels of phosphatidylinositols were increased by ACTH in ZG preparations of rats maintained on a low sodium intake for 5 days whereas there was no change in the adrenal ZG from control rats. This indicates that changes in the levels of phosphoinositides may somehow contribute to the altered adrenal responsiveness to ACTH under sodium restriction. More studies will thus be necessary to understand the mechanisms by which low sodium intake affects the responsiveness of the gland to ACTH. A quantitation of the levels of the ACTH-R mRNA using the above mentioned cDNA, and the levels of the ACTH-R protein, using a specific antibody, should be made before drawing conclusions as to the dynamics of regulation of ACTH-R by variations in sodium intake and by ACTH itself.

C. c-AMP in ACTH Response

After binding to its receptor, ACTH acts primarily through a transduction pathway involving guanidine nucleotide-binding proteins that increase c-AMP formation and protein kinase A activity (Rae et al., 1979; Parissenti et al., 1993).

Within minutes, ACTH or c-AMP activates mechanisms making cholesterol available to P450$_{scc}$, and thus its conversion into pregnenolone (Lambeth and Stevens, 1985). These activation mechanisms require *de novo* protein synthesis, since they are suppressed by the protein synthesis inhibitor cycloheximide (Privalle et al., 1983). In contrast, the transcriptional activation of adrenal steroidogenic P450 genes by ACTH or c-AMP is a long term process that cannot be held responsible for the rapid changes involved in the control of steroidogenesis (Rice et al., 1989; Hanukoglu et al., 1990; Simpson et al., 1990; Lehoux et al., 1992; Waterman et al., 1992).

A tissue specific expression of adrenal steroidogenic enzymes and their activation by c-AMP has been reported. c-AMP responsive elements were described for the steroidogenic P450 genes which are expressed in the adrenals (Simpson et al., 1990) although most of them differ from the classical c-AMP responsive element TGAGCTCA (CRE) in their c-AMP inducible region which was previously described and associated with other genes whose transcription is regulated by c-AMP (Montminy et al., 1987). In addition, Rice et al. (1991) and Lala et al. (1992) reported two regulatory elements in each of the following mouse genes, *CYP11A*, *CYP21A* and *CYP11B2*, which contain variations of a shared motif that interacted with the same or closely related DNA-binding protein(s). These regulatory elements closely resemble the pair of elements, Ad3 and Ad4, which could be involved in the c-AMP induced expression of the bovine *CYP11B* gene (Honda et al., 1990). Lala et al., (1992) identified a 53-kDa protein, termed steroidogenic factor 1 (SF-1), that interacts with the related promoter elements from the three above-mentioned genes. A partially purified protein from a bovine adrenal extract was found to bind to the six regulatory elements, supporting the model that a common steroidogenic cell-selective protein interacts with related promoter elements from three steroidogenic enzymes to regulate their coordinate expression. SF-1 binding protein is most likely to be identical to the 53 kDa protein isolated and purified from bovine adrenal by Morohashi et al., (1992), which strongly bound the cis-acting element Ad4 of the bovine *CYP11B1* gene. Morohashi et al. (1992) and Honda et al. (1993) also suggested that this Ad4-binding protein, expressed only in cells of steroidogenic organs, is a common factor for the regulation of steroidogenic *P450* genes. Lala et al. (1992) isolated a c-DNA expressed only in steroidogenic tissues which encodes a protein interacting with the two related steroidogenic regulatory elements of the mouse *CYP11A*, *CYP21A* and *CYP11B2* genes mentioned above, with a binding specificity indistinguishable from that of SF-1. The sequence of the putative DNA-binding domain of this cDNA perfectly matches the corresponding sequence of the mouse homolog of the *Drosophila* transcriptional factor *fushi tarazu-factor I* involved in early stages of development of *Drosophila* embryos. This factor is a member of the steroid receptor superfamily (Tsukiyama et al., 1992), although its ligand remains undefined. In addition, although this factor is a member of leucine zipper binding protein family, it binds to a single consensus element only and not to two elements as does Jun/Fos. Nuclear receptors have been implicated in a variety of functions in vertebrates, including differentiation of cells, and therefore it is not surprising that SF-1 is a member of the steroid receptor superfamily. In this respect we also found that glucocorticoid receptors are involved in the regulation of the *CYP11A* and *CYP17* genes (see the following section).

The synthesis of adrenal P450s thus appears to be regulated by both common and unique specific cis-acting c-AMP regulatory elements among the genes. The diversity of the mechanisms involved in the control of the transcription of P450 steroidogenic enzymes genes by ACTH or c-AMP provides new avenues for the study of the regulation of the adrenal steroidogenic enzymes.

D. Glucocorticoids and Glucocorticoid Receptors in the Response to ACTH

As mentioned above, glucocorticoids and glucocorticoid receptors appear to be involved in the control of adrenocortical cell function. Indeed, we have shown an inhibitory effect of cortisol, cortisone, corticosterone and dexamethasone on the accumulation of $P450_{scc}$ and $P450_{17\alpha}$ mRNAs in bovine adrenocortical cells in culture when stimulated by ACTH (Trzeciak et al., 1993). Moreover, dexamethasone strongly inhibited the expression of chloramphenicol acetyltransferase in forskolin-treated transfected cells harboring plasmids containing regulatory regions of *CYP11A* and *CYP17* genes. Corticosteroids inhibited the ACTH-induced accumulation of the mRNA of these two key regulatory enzymes at the transcriptional level, an effect mediated by the glucocorticoid receptor (Trzeciak et al., 1993), since pretreatment of cells with the steroid antagonist RU 486 blocked this effect. The precise location of the site of action of glucocorticoid on the two regulatory regions is yet to be determined. It is not known whether in this case glucocorticoid receptors act directly on the gene or indirectly via their interaction with other trans-acting factors. Indeed, it has been reported that transcriptional activation mediated by AP_1 transcription factors can be repressed by glucocorticoid receptors and other nuclear receptors Pearce and Yamamoto (1993).

The magnitude of the inhibitory effect of glucocorticoids in bovine adrenocortical cells implies that it may play a role in maintaining the physiological function of the adrenal cortex. It remains to perform *in vivo* studies in order to validate these *in vitro* findings.

E. Jun/Fos Trans-acting Factors in the Response to ACTH

There is good evidence that members of the Jun and Fos oncogene family play a role in the mechanism of action of *ACTH* in the adrenal. An increased accumulation of c-fos immunoreactive material was observed in ZF-R in rats stressed by immobilization, and this accumulation was attenuated by dexamethasone treatment 2 h prior to immobilization (Yang et al., 1989). In addition, a circadian variation in the expression of the c-fos protein was found by immunocytochemistry in the adrenal ZF-R of adult rats (Koistinaho et al., 1990), whereas there were no immunolabeled cells in the ZG. Furthermore, in hypophysectomized rats, the c-fos mRNA level of the whole adrenal gland increased rapidly after ACTH injection followed by a decline to near control value at 24 h (Imai et al., 1990a). Also, a 4-fold increase in the number of c-fos immunoreactive ZF-R cells was found 45 min after a single injection of ACTH to intact rats (Yang et al., 1990).

By *in situ* hybridization, both c-fos and c-jun mRNAs were found in the rat adrenal ZF-R (Pelto-Huikko et al., 1991), although a very low level of c-fos mRNA was seen compared to that of c-jun. In contrast, the mRNA of the two oncogenes was not detectable in the ZG. In stress induced ACTH release in capsaicin treated rats, a rapid increase in c-fos mRNA levels was found in the three zones of the adrenal cortex and especially in the ZG, whereas small changes in c-jun mRNA levels were seen only in the ZF-R, indicating that these two early genes were differentially expressed under such conditions.

We attempted to further clarify the role of Fos and Jun proto-oncogenes by studying the short-term and long-term effects of ACTH on the expression of their genes *in vivo*. Without stimulation, c-fos mRNA was detectable in both adrenal zones by northern analysis, although it was half only as abundant in the ZF-R as in the ZG (Lehoux and Ducharme, 1995). In the ZG and ZF-R the level of c-fos mRNA increased rapidly by 4.5- and 9.5-fold, 30 min after ACTH administration but it had decreased to below control level at 5 h and remained below detection limits with subsequent twelve-hourly injections of ACTH. Dexamethasone alone provoked a decrease below control values in the level c-fos mRNA in both ZG and ZF-R, when measured 3 h after treatments. Evidently ACTH is needed to maintain a basal mRNA level of this proto-oncogene in rat adrenal and must overcome the feedback inhibition by glucocorticoids synthesized within the gland.

ACTH acts differently on the expression of c-jun in the ZG as compared to the ZF-R. Indeed, we found c-jun mRNA in both the ZG and the ZF-R, although in contrast to c-fos, the level of c-jun was 5-fold higher in the ZF-R than in the ZG (Lehoux and Ducharme, 1995). A slight increase (1.3-fold) in the level of c-jun mRNA occurred in ZF-R within 30 min after the first ACTH injection, followed by a decrease to 50% below control 5 h later. When rechallenged with ACTH at twelve hour intervals, the level of c-jun mRNA in the ZF-R remained at least 50% lower than controls. A different situation was observed in the ZG where a maximal 4-fold increase in c-jun mRNA level occurred within 30 min, which remained elevated at 1 h (3.6-fold) and at even 5 h (2.6-fold) post ACTH administration. Similarly to the ZF-R, however, the long term effect of ACTH provoked a 50% decrease below control value in the level of c-jun mRNA in the ZG 12 h after the first ACTH injection which was again maintained with subsequent treatments. Dexamethasone administration to rats also provoked a 50% reduction in the level of c-Jun mRNA in the ZG compared to control, when measured three hours after injection. Taken together these results indicate that c-fos and c-jun are differently expressed in ZG and ZF-R and that they may participate *in vivo*, under physiological conditions, in the mediation of ACTH effects on adrenal function.

The signaling pathway linking ACTH and Jun/Fos protein is not yet completely elucidated. However, an antagonism between AII and ACTH on the expression of adrenal steroidogenic genes has been demonstrated by *in vitro* studies. In cultured adrenal cells AII was shown to inhibit the induction of $P450_{17\alpha}$, $P450_{11\beta}$ and 3β-HSD by ACTH (Rainey et al., 1991; Naseeruddin and Hornsby, 1990). Moreover, in bovine and ovine ZF-R cells in culture, ACTH produced time- and dose-dependent increase in the levels of c-fos and jun-B but not in c-jun, whereas AII produced increases in the mRNA levels of c-fos and jun-B but also of c-jun (Viard et al., 1992a). AII showed a greater stimulatory effect on c-fos than did ACTH, and ACTH abolished the stimulation of c-jun mRNA by AII. Using inhibitors specific to AT_1 receptors (DUP 753) and to PKC (staurosporine), AII was shown to increase the mRNA of c-jun, jun-B and c-fos through AT_1 receptors by activation of the PKC pathway (Viard et al., 1992). On the other hand, ACTH and $(Bu)_2$c-AMP were shown to increase the c-fos mRNA level and corticosterone secretion in rat adrenal cells in culture (Miyamoto et al., 1992). When the c-AMP-dependent protein kinase was selectively inhibited, however, both basal and ACTH-induced c-fos mRNA were suppressed, indicating that ACTH increased c-fos mRNA by phosphorylation of pre-existing factor(s) via c-AMP-dependent protein kinase.

Collectively these results indicate that ACTH and AII interact with proto-oncogenes of the Jun/Fos family through the A kinase and C kinase pathway respectively, and that their opposing long term action on adrenal cell functions might well be linked to their initial differential effects on the expression of these nuclear proto-oncogene proteins.

F. TBF-β1 in the Response of Adrenal Cells to ACTH

Glucocorticoid synthesis is also regulated by transforming growth factor-β1 (TBF-β1) which is a member of a family of closely related peptides which share similar biological activities. TGF-β1 is synthesized as a large precursor protein which is cleaved to yield a 112 amino acid monomer (Sharples et al., 1987). The active form of the molecule is homodimeric.

In vitro, TGF-β1 exerts a very potent inhibition of both basal and ACTH- and AII-stimulated adrenocortical cortisol production (Feige et al., 1986; Hotta and Baird, 1986; Feige et al., 1987; Rainey et al., 1988, 1989). Furthermore, following the addition of TGF-β1 to bovine adrenocortical cell cultures, ACTH binding sites (Rainey et al., 1989), ACTH-stimulated production of c-AMP (Gupta et al., 1992), AII binding sites (Feige et al., 1987), 17α-hydroxylase (Feige et al., 1986; 1987) and 3β-hydroxysteroid dehydrogenase (Rainey et al., 1991) were decreased and LDL uptake and metabolism (Hotta and Baird, 1987) were inhibited. TGF-β1 also inhibits both basal and ACTH-stimulated aldosterone secretion in bovine adrenocortical cells (Gupta et al., 1992).

TGF-β1 activity is mediated by specific TGF-β membrane receptors (mainly type I and type III in bovine adrenocortical cells) which are positively regulated by ACTH (Cochet et al., 1988; Feige et al., 1991). ACTH therefore sensitizes adrenal cells to the inhibitory action of TGF-β1.

TGF-β1 is synthesized in many tissues but it is also locally synthesized in the adrenal. Indeed, immunohistochemical studies revealed the presence of TGF-β1 in adrenocortical cells (Thompson et al., 1989), and in the bovine adrenal cortex, the concentration is higher in the zona fasciculata than in the ZG (Keramidas et al., 1991). Moreover, TGF-β1 mRNA was detected by northern analysis in mouse adrenal cortex (Thompson et al., 1989), and rat adrenal cortex (unpublished observations), indicating that it can be synthesized in the gland. These results thus suggest that TGF-β1 may act *in vivo* as an autocrine or paracrine regulator of corticosteroidogenesis.

TGF-β1 is also capable of an autoinduction, which is prevented by antisense c-jun or c-fos expression vectors, so demonstrating the participation of AP_1 trans-acting factors in the regulation of this growth factor (Kim et al., 1990). As ACTH has a short term stimulatory effect on c-jun and c-fos mRNA levels in the rat adrenal (see section E, page 5), proto-oncogenes of the Jun/Fos family might well be involved in the antagonistic action between ACTH and TGF-β1 in the control of adrenal steroidogenesis. However, a complete integration of interactions between these various factors remains to be made.

VI. SUMMARY AND CONCLUSIONS

Recent findings clearly demonstrate a different regulation for the biosynthesis of glucocorticoids and mineralocorticoids.

The renin-angiotensin system is the main regulator of aldosterone synthesis. Recent findings have also clearly established that a low sodium or high potassium intake are factors which control the expression of the early and final steps of mineralocorticoid formation. These two intakes provoke increases in the levels of $P450_{scc}$ mRNA and $P450_{scc}$ protein in the rat adrenal ZG. Two genes are expressed in the hamster, mouse, rat and human adrenal which participate in the last steps of aldosterone synthesis; $P450_{11\beta}$ responsible for the 11β-hydroxylation of deoxycorticosterone, and $P450_{C18}$ for the transformation of corticosterone to aldosterone. Up to now, one gene only appears to be involved in both these reactions in the bovine adrenal. In the rat adrenal, $P450_{11\beta}$ is located in ZG and in ZF-R, while 450_{C18} is found in ZG only. Both low sodium or high potassium intake induce the expression of $P450_{C18}$ but not $P450_{11\beta}$ in the rat and also in the hamster adrenal; the effects of these two ionic intakes are mediated through AII, since they are blocked by inhibitors of the angiotensin converting enzyme.

AII exerts its action through its binding to specific AII receptors type 1A in the adrenal, and then uses the phosphoinositides transduction pathway. Recent findings indicate that low sodium as well as high potassium intakes increase the levels of

AT_1 receptor mRNA and AT_1 receptor protein in the rat adrenal, and that this is mediated through AII, at least in the case of the low sodium intake. AII activates phospholipase C which hydrolyzes phosphatidylinositol diphosphate to phospho-inositol triphosphate and diacylglycerol. A rapid mobilization of calcium also occurs followed by an increased secretion of aldosterone. Diacylglycerol provokes the activation of protein kinase C which influences the regulation of the transcription of many genes. The action of protein kinase C is directed towards gene promoters containing the responsive element AP_1, which is recognized by proto-oncogene proteins of the Jun and Fos families, suggesting that the long term effect of AII is mediated by these proto-oncogenes acting at the AP_1 site of steroidogenic genes.

The atrial natriuretic peptide ANP has a negative regulatory action on adrenocortical functions, it has been reported to inhibit the increase in aldosterone secretion induced by ACTH, AII, and potassium administration *in vivo*. For its action, ANP binds to specific receptors in the adrenal ZG and activates the formation of c-GMP.

ACTH is the principal factor involved in the control of glucocorticoid biosynthesis and secretion, whereas many other factors participate in mineralocorticoid metabolism. ACTH binds to specific cell membrane receptors and induces the formation of c-AMP and the activation of protein kinase A. At short term, ACTH activates the transport and the availability of cholesterol to mitochondrial $P450_{scc}$, to facilitate the formation of pregnenolone. The synthesis of new $P450_{scc}$ protein does not appear necessary, at least for the short term effect of ACTH.

A predominant role is played by $P450_{17\alpha}$ in the control of glucocorticoid synthesis. In the hamster, a species, which like humans, secretes cortisol, $P450_{17\alpha}$ is quantitatively increased during a sustained ACTH stimulation, in a manner similar to that occurring during physiological variations in the level of circulating ACTH. The level of other steroidogenic enzymes, 3β-HSD, $P450_{C21}$, and $P450_{11\beta}$ are not changed following a single injection of ACTH or during a sustained stimulation by this hormone for 24 h, indicating that these are not rate-determining enzymes *in vivo*, and therefore are not responsible for the fine control of glucocorticoid secretion mediated by ACTH. ACTH would be necessary, however, to maintain an adequate level of these steroidogenic enzymes in the adrenal.

The transduction pathway involved in ACTH action is not yet completely elucidated. However, specific c-AMP responsive elements have been described for the steroidogenic genes which are expressed in the adrenals. Furthermore, both ACTH and AII are capable of inducing the expression of members of the Jun and Fos families but in different ways. An antagonism has been observed between ACTH and AII on the induction of $P450_{17\alpha}$, $P450_{11\beta}$ and 3β-HSD by ACTH *in vitro*. It is postulated that this opposing action on adrenal cell function might well be linked to the initial differential effects on the expression of the Fos/Jun proteins. Jun and Fos trans-acting factors may also regulate through their effects on the transforming growth factor-β1 which exerts an antagonistic effect on ACTH action. There is therefore a complex interdependency between all these regulatory factors participating in the control of the formation of glucocorticoids. Also, TGF-β1 inhibits basal and both ACTH- and AII- stimulated cortisol production *in vitro*.

Collectively, these findings demonstrate a different regulation for glucocorticoids and mineralocorticoids and illustrate the complexity of the mechanisms participating in the control of corticosteroid formation in the adrenal gland.

ACKNOWLEDGMENTS

Work from our laboratory discussed in this chapter was supported by grants from the Medical Research Council of Canada and the Heart and Stroke Foundation of Canada. The authors thank Madeleine des Chênes for her secretarial help.

REFERENCES

Aguilera, G., Hauger, R.L., & Catt, K.J. (1978). Control of aldosterone secretion during sodium restriction: Adrenal receptor regulation and increased adrenal sensitivity to angiotensin II. Proc. Natl. Acad. Sci. USA 75, 975–979.

Aguilera, G., Schirar, A., Baukal, A., & Catt, K.J. (1980). Angiotensin II receptors: Properties and regulation in adrenal glomerulosa cells. Cir. Res. 46 (Suppl I), 118–127.

Aguilera, G., & Catt, K.J. (1983). Regulation of aldosterone secretion during altered sodium intake. J. Steroid Biochem. 19, 525–530.

Ahlgren, R., Simpson, E.R., Waterman, M.R., & Lund, J. (1990). Characterization of the promoter/regulatory region of the bovine CYP11 (P450$_{scc}$) gene. J. Biol. Chem. 265, 3313–3319.

Anderson, J.V., Struthers, A.D., Payne, N.N., Slater, J.D.H., & Bloom, S.R. (1986). Atrial natriuretic peptide inhibits the aldosterone response to angiotensin II in man. Clin. Sci. 70, 507–512.

Angel, P., & Karin, M. (1991). The role of Jun, Fos and the AP-1 complex in cell-proliferation and transformation. Biochem. Biophys. A. 1072, 129–157.

Barrett, P.Q., & Isales, C.M. (1988). The role of cyclic nucleotides in atrial natriuretic peptide-mediated inhibition of aldosterone secretion. Endocrinology 122, 799–808.

Baxter, J.D., Lewicki, J.A., & Gardner, D.G. (1988). Atrial natriuretic factor. Biotechnology 6, 529–546.

Blair-West, J.R., Coghlan, J.P., Denton, D.A., Goding, J.R., Wintour, M., & Wright, R.D. (1963). The control of aldosterone secretion. Recent Prog. Horm. Res. 19, 311–338.

Boggaram, V., Simpson, E.R., & Waterman, M.R. (1984). Induction of synthesis of bovine adrenocortical P450$_{scc}$, P450$_{11\beta}$, P450$_{C21}$ and adrenodoxin by prostaglandins E$_2$ and F$_{2\alpha}$ and cholera toxin. Arch. Biochem. Biophys. 231, 271–279.

Boyd, J.E., Palmore, W.P., & Mulrow, P.J. (1971). Role of potassium in the control of aldosterone secretion in the rat. Endocrinology 88, 556–565.

Brecher, A.S., Shier, D.N., Dene, H., Wang, S.M., Rapp, J.P., Franco-Saenz, R., & Mulrow, P.J. (1989). Regulation of adrenal renin mRNA by dietary sodium chloride. Endocrinology 124, 2907–2913.

Brochu, M., Ong., H., & De Léan, A. (1991). Sites of action of angiotensin II, atrial natriuretic factor and guanabenz, on aldosterone biosynthesis. J. Steroid Biochem. Molec. Biol. 38, 575–582.

Brown, R.D., Strott, C.A., & Liddle, G.W. (1972). Site of stimulation of aldosterone biosynthesis by angiotensin and potassium. J. Clin. Invest. 51, 1413–1418.

Brown, G., Douglas, J., & Bravo, E. (1980). Angiotensin II receptors and in vitro aldosterone responses of aldosterone-producing adenomas, adjacent nontumorous tissue, and normal human adrenal glomerulosa. J. Clin. Endocrinol. Metab. 51, 718–723.

Brownie, A.C., Alfano, J., & Jefcoate, C.R. (1973). Effect of ACTH on adrenal mitochondrial cytochrome P450 in the rat. Ann. NY Acad. Sci. 212, 344–360.

Buckley, D.I., & Ramachandran, J. (1981). Characterization of corticotropin receptors on adrenocortical cells. Proc. Natl. Acad. Sci. USA 78, 7431–7435.

Bumpus, F.M., Catt, K.J., Chiu, A.T., De Gasparo, M., Goodfriend, T., Husain, A., Peach, M.J., Taylor, Jr. D.G., & Timmermans, P.B.M.W.M. (1991). Special feature nomenclature for angiotensin receptors a report of the nomenclature committee of the council for high blood pressure research. Hypertension 17, 720–721.

Campbell, D.J., & Habener, J.F. (1986). Regional distribution of angiotensinogen messenger RNA in rat adrenal and kidney. J. Hypertension 4, S1–S3.

Canick, J.A., & Villee, D.B. (1974). The effect of adrenocorticotropin on protein degradation in rat adrenal gland and liver. Biochem. J. 144, 397–403.

Capponi, A.M., & Catt, K.J. (1980). Solubilization and characterization of adrenals and uterine angiotensin II receptors after photoaffinity labeling. J. Biol. Chem. 255, 12081–12086.

Carsia, R.V., & Weber, H. (1988). Protein maturation on the domestic fowl induces alterations in adrenocorticotropin receptors. Endocrinology 122, 681–688.

Catalano, R.B., Stuve, L., & Ramachandran, J. (1986). Characterization of corticotropin receptors in human adrenocortical cells. J. Clin. Endocrinol. Metab. 62, 300–304.

Chang, M., Lowe, D.G., Lewis, M., Hellmiss, R., Chen, E., & Goeddel, D.V. (1989). Differential activation by atrial and brain natriuretic peptides of two different receptor guanylate cyclase. Nature 341, 68–71.

Chartier, L., Schiffrin, E., Thibault, G., & Garcia, R. (1984). Atrial natriuretic factor inhibits the effect of angiotensin II, ACTH and potassium on aldosterone secretion in vitro and angiotensin II induced steroidogenesis in vivo. Endocrinology 115, 2026–2028.

Chinkers, M., Garbers, D.L., Chang, M.S., Lowe, D.G., Chin, H., Goeddel, D.V., & Schulz, S. (1989). A membrane form of guanylate cyclase is an atrial natriuretic peptide receptor. Nature 338, 78–83.

Chiu, A.T., Herblin, W.F., McCall, D.E., Ardecky, R.J., Carini, D.J., Duncia, J.V., Pease, L.J., Wong, P.C., Wexler, R.R., Johnson, A.L., & Timmermans, P.B.M.W.M. (1989). Identification of angiotensin II receptor subtypes. Biochem. Biophys. Res. Comm. 165, 196–203.

Chua, S.C., Szabo, P., Vitek, A., Grzeschik, K.H., John, M., & White, P.C. (1987). Cloning of cDNA encoding steroid 11β-hydroxylase P450c11. Proc. Natl. Acad. Sci. USA 84, 7183–7197.

Chung, B.C., Matteson, K.L., & Miller, W.L. (1986). Structure of a bovine gene for P450C21 defines a novel cytochrome P450 gene family. Proc. Natl. Acad. Sci. USA 83, 4243–4247.

Clinkingbeard, C., Sessions, C., & Shenker, Y. (1990). The physiological role of atrial natriuretic hormone in the regulation of aldosterone and salt and water metabolism. J. Clin. Endocrinol. Metab. 70, 582–589.

Cochet, C., Feige, J.J., & Chambaz, E.M. (1988). Bovine adrenocortical cells exhibit high affinity transforming growth factor-β receptors which are regulated by adrenocorticotropin. J. Biol. Chem. 263, 5707–5713.

Cuneo, R.C., Expiner, E., Nicholls, M.G., Yandle, T.G., & Livesey, J.H. (1987). Effect of physiological levels of atrial natriuretic peptide on hormone secretion: Inhibition on angiotensin-induced aldosterone secretion and renin release in normal man. J. Clin. Endocrinol. Metab. 65, 765–772.

Curnow, K.M., Tusie-Luna, M.T., Pascoe, L., Natarajan, R., Gu, J.L., Nadler, J.L., & White, P.C. (1991). The product of the CYP11B2 gene is required for aldosterone biosynthesis in the human adrenal cortex. Molec. Endocr. 5, 1513–1522.

Curnow, K.M., Slutsker, L., Vitex, J., Cole, T., Speiser, P.W., New, M.I., White, P.C., & Pascoe, L. (1993). Mutations in the CYP11B1 gene causing congenital adrenal hyperplasia and hypertension cluster in exons 6, 7, and 8. Proc. Natl. Acad. Sci. USA 90, 4552–4556.

Curran, T., & Franza, B.R. (1988). Fos and Jun: The AP-1 connection. Cell 55, 395–397.

De Bold, A., Borenstein, J.S., Vereas, A.T., & Sonnenberg, H. (1981). A rapid and potent natriuretic response to intravenous injection of atrial myocardial extract in rat. Life Sci. 28, 89–94.

De Léan, A., Racz, K., Gutkowska, J., Nguyen, T.T., Cantin, M., & Genest, J. (1984). Specific receptor-mediated inhibition by synthetic atrial natriuretic factor of hormone-stimulated steroidogenesis in cultured bovine adrenal cells. Endocrinology 115, 1636–1638.

Deane, H.W., Shaw, J.H., & Greep, R.O. (1948). The effect of altered sodium or potassium intake on the width and cytochemistry of the zona glomerulosa of the rat's adrenal cortex. Endocrinology 43, 133.

Debreceni, L., & Csete, B. (1975). *In vitro* production of aldosterone by the rat adrenals after *in vivo* potassium- and sodium- loading and depletion, Endokrinologie 64, 316–322.

Deschepper, C.F., Mellon, S.H., Cumin, F., Baxter, J.D., & Ganong, W.F. (1986). Analysis by immunocytochemistry and *in situ* hybridization of renin and its mRNA in kidney, testis, adrenal and pituitary of the rat. Proc. Natl. Acad. Sci. USA 83, 7552–7556.

DiBlasio, A.M., Voutilainen, R., Jaffe, R.B., & Miller, W.L. (1987). Hormonal regulation of mRNAs for $P450_{scc}$ and $P450_{C17}$ in cultured human fetal adrenal cells. J. Clin. Endocrinol. Met. 65, 170–175.

Dluhy, R.G., Axelrod, L., Underwood, R.H., & Williams, G.H. (1972). Studies on the control of plasma aldosterone concentration in normal man. J. Clin. Invest. 51, 1950–1957.

Doi, Y., Atarashi, K., Franco-Saenz, R., & Mulrow, P.J. (1984). Effects of changes in sodium or potassium balance and nephrectomy on adrenal renin and aldosterone concentrations. Hypertension 6, 1124–1129.

Domalik, M.J., Chaplin, D.D., Kirkman, M.S., Wu, R.C., Liu, W., Howard, T.A., Seldin, M.F., & Parker, K.L. (1991). Different isozymes of mouse 11β-hydroxylase produce mineralocorticoids and glucocorticoids. Mol. Endocrinol. 5, 1853–1861.

Douglas, J.G. (1980). Effects of high potassium diet on angiotensin II receptors and angiotensin-induced aldosterone production in rat adrenal glomerulosa cells. Endocrinology 106, 983–990.

Douglas, J.G. (1980a). Potassium ion as a regulator of adrenal angiotensin II receptors. Am. J. Physiol. 239, E317–E321.

DuBois, R.N., Simpson, E.R., Kramer, R.E., & Waterman, M.R. (1981). Induction of synthesis of cholesterol side chain cleavage cytochrome P-450 by adrenocorticotropin in cultured bovine adrenocortical cells. J. Biol. Chem. 256, 7000–7005.

Durand, P.H. (1979). ACTH receptor levels in lamb adrenals at late gestation and early neonatal stages. Biol. Reprod. 20, 837–845.

Durand, P.H., & Locatelli, A. (1980). Up regulation of corticotropin receptors by $ACTH_{1-24}$ in normal and hypophysectomized rabbits. Biochem. Biophys. Res. Comm. 96, 447–456.

Dzau, V.J., Ingelfinger, J.R., & Pratt, R.E. (1986). Regulation of tissue renin and angiotensin gene expression. J. Cardiovasc. Pharmacol. 8, 511–516.

Dzau, V.J., Ellison, K.E., Brody, T., Ingelfinger, J., & Pratt, R.E. (1987). A comparative study of the distributions of renin and angiotensinogen messenger ribonucleic acid in rat and mouse tissues. Endocrinology 120, 2334–2338.

Elliott, M.E., & Goodfriend, T.L. (1987). Effects of atrial natriuretic peptide, angiotensin, cyclic AMP and potassium on protein phosphorylation in adrenal glomerulosa cells. Life Sci. 41, 2517–2524.

Feige, J.J., Cochet, C., & Chambaz, E.M. (1986). Type β transforming growth factor is a potent modulator of differentiated adrenocortical cell functions. Biochem. Biophys. Res. Commun. 139, 693–700.

Feige, J.J., Cochet, C., Rainey, W.E., Madani, C., & Chambaz, E.M. (1987). Type β transforming growth factor affects adrenocortical cell-differentiated functions. J. Biol. Chem. 262, 13491–13495.

Feige, J.J., Cochet, C., Savona, C., & Shi, D.L. (1991). Transforming growth factor beta 1: An autocrine regulation of adrenocortical steroidogenesis. Endocrine Res. 17, 267–279.

Fuller, F., Porter, J.G., Arfsten, A.E., Miller, J., Sehilling, J.W., Scarborough, R.M., Lewicki, J.A., & Schenk, D.B. (1988). Atrial natriuretic peptide clearance receptor. J. Biol. Chem. 263, 9395, 9401.

Funder, J.W., Blair-West, J.R., Coghlan, J.P., Denton, D.A., Scoggins, B.A., & Wright, R.D. (1969). Effect of plasma (K^+) on the secretion of aldosterone. Endocrinology 85, 381–384.

Furuka, H., Guo, D.F., & Inagami, T. (1992). Molecular cloning and sequencing of the gene encoding human angiotensin II type receptor. Biochem. Biophys. Res. Comm. 183, 8–13.

Gallo-Payet, N., & Escher, E. (1985). Adrenocorticotropin receptors in rat adrenal glomerulosa cells. Endocrinology 117, 38–46.

Ganguly, A., Chiou, S., Leigh, A.W., & Davis, J.S. (1989). Atrial natriuretic factor inhibits angiotensin-induced aldosterone secretion: Not through c-GMP or interference with phospholipase C. Biochem. Biophys. Res. Comm. 159, 148–154.

Gann, D.S., Delea, C.S., Gill, J.R. Jr., Thomas, J.P., & Bartter, F.C. (1964). Control of aldosterone secretion by change of body potassium in normal man. Am. J. Physiol. 207, 104–108.

Garcia, R., Debinski, W., Gutkowska, J., Kuchel, O., Tjibault, G., Genest, J., & Cantin, M. (1985). Gluco- and mineralo-corticoids may regulate the natriuretic effect and the synthesis and release of atrial natriuretic factor by the rat atria *in vitro*. Biochem. Biophys. Res. Commun. 131, 806–814.

Gardner, D.G., Deschepper, C.F., Ganong, W.F., Hane, S., Fiddes, J., Baxter, J.D., & Lewicki, J. (1986). Extra-atrial expression of the gene for atrial natriuretic factor. Proc. Natl. Acad. Sci. USA 83, 6697–6701.

Gardner, D.G., Hane, G.I.S., Trachewsky, D., Schenk, D., & Baxter, J.D. (1986a). Atrial natriuretic peptide mRNA is regulated by glucocorticoids *in vivo*. Biochem. Biophys. Res. Commun. 139, 1047–1054.

Genest, J., Larochelle, P., Cusson, J.R., Gutkowska, J., & Cantin, M. (1988). The atrial natriuretic factor in hypertension. Hypertension, Vol. 11, Supp 1, I3–I7.

Globerman, H., Rosler, A., Theodor, H., New, M.I., & White, P.C. (1988). An inherited defect in aldosterone biosynthesis caused by a mutation in or near the gene for steroid 11-hydroxylase. N. Engl. J. Med. 319, 1193–1197.

Guillemette, G., & Escher, E. (1983). Analysis of the adrenal angiotensin II receptor with the photoaffinity labeling method. Biochemistry 22, 5591–5596.

Griffing, G.T., Dale, S.L., Holbrook, M.M., & Melby, J.C. (1981). 19-nor-deoxycorticosterone excretion in healthy and hypertensive subjects. Trans. Assoc. Am. Physicians 94, 301–309.

Gupta, P., Franco-Saenz, R., Gentry, L.E., & Mulrow, P.J. (1992). Transforming growth factor-$\beta 1$ inhibits aldosterone and stimulates adrenal renin in cultured bovine zona glomerulosa cells. Endocrinology 131, 631–636.

Haning, R., Tait, S.A., & Tait, J.F. (1970). *In vitro* effects of ACTH, angiotensins, serotonin and potassium on steroid output and conversion of corticosterone to aldosterone by isolated adrenal cells. Endocrinology 87, 1147–1167.

Hanukoglu, I., Feuchtwanger, R., & Hanukoglu, A. (1990). Mechanism of corticotropin and c-AMP induction of mitochondrial cytochrome P450 system enzymes in adrenal cortex cells. J. Biol. Chem. 265, 20602–20608.

Hashimoto, T., Morohashi, K.I., & Omura, T. (1989). Cloning and characterization of bovine cytochrome P-450$_{11\beta}$ genes. J. Biochem. 105, 676–679.

Hauber, R.L., Aguilera, G., & Catt, K.J. (1978). Angiotensin II regulates its receptor sites in the adrenal glomerulosa zone. Nature 271, 176–178.

Helmberg, A., Ausserer, B., & Reinhard, K. (1992). Frame shift by insertion of 2 basepairs in codon 394 of *CYP11B1* causes congenital adrenal hyperplasia due to steroid 11β-hydroxylase deficiency. J. Clin. Endocrinol. Metab. 75, 1278–1281.

Higashi, Y., Yoshioka, H., Yamane, M., Gotoh, O., & Fujii-Kuriyama, Y. (1986). Complete nucleotide sequence of two steroid 21-hydroxylase genes tandemly arranged in human chromosome: A pseudogene and a genuine gene. Proc. Natl. Acad. Sci. USA 83, 2841–2845.

Higuchi, K., Nawata, W., Kato, K.I., Ibayashi, W., & Matsuo, H. (1986). Alpha human atrial natriuretic peptide inhibits steroidogenesis in cultured human adrenal cells. J. Clin. Endocrinol. Metab. 62, 941–944.

Higuchi, K., Nawata, H., Kato, K., Ibayashi, H., & Matsuo, H. (1986a). Lack of inhibitory effect of α-human atrial natriuretic polypeptide on aldosteronigenesis in aldosterone-producing adenoma. J. Clin. Endocr. Metab. 63, 192–196.

Higuchi, K., Nawata, J., Kato, K., & Ibayashi, H. (1988). Lack of inhibitory effect of α-human atrial natriuretic polypeptide on cortisol secretion in cultured adrenocortical adenoma cells from the patients with Cushing's syndrome. Horm. Metab. Res. 20, 360–363.

Honda, S.I., Morohashi, K.I., & Omura, T. (1990). Novel c-AMP regulatory elements in the promoter region of bovine P-450$_{11\beta}$ gene. J. Biochem. (Tokyo) 108, 1042–1049.

Honda, S.I., Morohashi, K., Nomura, M., Takeya, H., Kitajima, M., & Omura, T. (1993). Ad4BP regulating steroidogenic P-450 gene is a member of steroid hormone receptor superfamily. J. Biol. Chem. 268, 7494–7502.

Hornsby, P.J., O'Hare, M.J., & Neville, A.M. (1974). Functional and morphological observations on rat adrenal zona glomerulosa cells in monolayer culture. Endocrinology 95, 1240–1251.

Hornsby, P.J., O'Hare, M.J., & Neville, A.M. (1974a). Functional and morphological changes occurring in rat adrenal zona glomerulosa cells in monolayer culture. J. Endocrinol. 61, 97–98.

Hotta, M., & Baird, A. (1986). Differential effects of transforming growth factor type β on the growth and function of adrenocortical cells *in vitro*. Proc. Natl. Acad. Sci. USA 83, 7795–7799.

Hotta, M., & Baird, A. (1987). The inhibition of low density lipoprotein metabolism by transforming growth factor β mediates its effects on steroidogenesis in the bovine adrenocortical cells *in vitro*. Endocrinology 121, 150–159.

Huang, C.L., Lewicki, T., Johnson, L.K., & Cogan, M.G. (1985). Renal mechanism of action of rat atrial natriuretic factor. J. Clin. Invest. 75, 769–773.

Ikushiro, S.I., Kominami, S., & Takemori, S. (1989). Adrenal cytochrome P-45011β-proteoliposomes catalyzing aldosterone synthesis: Preparation and characterization. Biochim. Biophys. Acta 984, 50–56.

Ikushiro, S.I., Kominami, S., & Takemori, S. (1992). Adrenal P-450$_{scc}$ modulates activity of P-450$_{11\beta}$ in liposomal and mitochondrial membranes. J. Biol. Chem. 267, 1464–1469.

Imai, M., Shimada, H., Okada, Y., Yuko, M.H., Ogishima, T., & Ishimura, Y. (1990). Molecular cloning of a cDNA encoding aldosterone synthase cytochrome P450 in rat adrenal cortex. FEBS Letters 263, 299–302.

Imai, T., Seo, H., Murata, Y., Ohno, M., Sotoh, Y., Funanashi, H., Takagi, H., & Matsui, N. (1990a). Adrenocorticotropin increases expression of c-fos and β-actin genes in the rat adrenals. Endocrinology 127, 1742–1747.

Inagami, T. (1989). Atrial Natriuretic Factor. J. Biol. Chem. 264, 3043–3046.

Iwai, N., Yamano, Y., Chaki, S., Konishi, F., Bardhan, S., Tibbetts, C., Sasaki, K., Hasegawa, M., Matsuda, Y., & Inagami, T. (1991). Rat angiotensin II receptor: cDNA sequence and regulation of the gene expression. Biochem. Biophys. Res. Commun. 177, 299–304.

John, M.E., John, M.C., Asley, P., MacDonald, R.J., Simpson, E.R., & Waterman, M.R. (1984). Identification and characterization of cDNA clones specific for cholesterol side-chain cleavage cytochrome P-450. Proc. Natl. Acad. Sci. USA 81, 56528–56532.

John, M.E., John, M.C., Simpson, E.R., & Waterman, M.R. (1985). Regulation of cytochrome P450$_{11\beta}$ gene expression by adrenocorticotropin. J. Biol. Chem. 260, 5760–5767.

John, M.E., John, M.C., Boggaram, V., Simpson, E.R., & Waterman, M.R. (1986). Transcriptional regulation of steroid hydroxylase genes by corticotropin. Proc. Natl. Acad. Sci. USA 83, 4715–4719.

John, M.E., Okamura, T., Dee, A., Adler, B., John, M.C., White, P.C., Simpson, E.R., & Waterman, M.R. (1986a). Bovine steroid-21-hydroxylase: Regulation of biosynthesis. Biochemistry 25, 2846–2853.

Kahri, A.I. (1968). Effect of actinomycin D and puromycin on the ACTH-induced ultrastructural transformation of mitochondria of cortical cells of rat adrenals in tissue culture. J. Cell Biol. 36, 181–195.

Kahri, A.I. (1971). Inhibition by cycloheximide of ACTH-induced internal differentiation of mitochondria in cortical cells in tissue culture of fetal rat adrenals. Anat. Rec. 171, 53–62.

Kakar, S.S., Sellers, J.C., Devor, D.C., Musgrove, L.C., & Neill, J.D. (1992). Angiotensin II type-1 receptor subtype cDNAs: Differential tissue expression and hormonal regulation. Biochem. Biophys. Res. Commun. 183, 1090–1096.

Kaplan, N.M. (1965). The biosynthesis of adrenal steroids: effects of angiotensin II, ACTH and potassium. J. Clin. Invest. 44, 2029–2039.

Kawamoto, T., Mitsuuchi, Y., Onishi, T., Ichikawa, Y., Yokoyama, Y., Sumimoto, Y., Toda, K., Miyahara, K., Kuribayashi, I., Nakao, K., Hosoda, K., Yamamoto, Y., Imura, H., & Shizuta, Y. (1990). Cloning and expression of a cDNA for human cytochrome P-450aldo related to primary aldosteronism. Biochem. Biophys. Res. Commun. 173, 309–316.

Kawamoto, T., Mitsuuchi, Y., Toda, K., Miyahara, K., Yokoyama, Y., Nakao, K., Kosoda, K., Yamamoto, Y., Imura, H., & Shizuta, Y. (1990a). Cloning of cDNA and genomic DNA for human cytochrome P-450$_{11\beta}$. FEBS Letters 269, 345–349.

Keramidas, M., Bourgarit, J.J., Tabone, E., Corticelli, P., Chambaz, E.Z., & Feige, J.J. (1991). Immunolocalization of transforming growth factor $\beta 1$ in the bovine adrenal cortex using antipeptide antibodies. Endocrinology 129, 517–526.

Kifor, I., Moore, T.J., Fallo, F., Sperling, E., Chiou, C.Y., Menachery, A., & Williams, G.H. (1991). Potassium-stimulated angiotensin release from superfused adrenal capsules and enzymatically dispersed cells of the zona glomerulosa. Endocrinology 129, 823–831.

Kifor, I., Moore, T.J., Fallo, F., Sperling, E., Menachery, A., Chiou, C.Y., & Williams, G.H. (1991a). The effect of sodium intake on angiotensin content of the rat adrenal gland. Endocrinology 128, 1277–1284.

Kim, J.J., Angel, P., Lafyatis, R., Hattori, K., Kim, K.Y., Sporn, M.B., Karin, M., & Roberts, A.B. (1990). Autoinduction of transforming growth factor $\beta 1$ is mediated by the AP-1 complex. Mol. Cell. Biol. 10, 1490–1497.

King, M.S., & Baertshi, A.J. (1989). Physiological concentrations of atrial natriuretic factors with intact N-terminal sequences inhibit corticotropin-releasing factor-stimulated adrenocorticotropin secretion from cultured anterior pituitary cells. Endocrinology 124, 286–292.

Kirita, S., Morohashi, K., Hashimoto, T., Yoshioka, H., Fujii-Kuriyama, Y., & Omura, T. (1988). Expression of two kinds of cytochrome P450$_{11\beta}$ mRNA in bovine adrenal cortex. J. Biochem. 104, 633–686.

Koistinaho, J., Roivainen, R., & Yang, G. (1990). Circadian rhythm in c-fos protein expression in the rat adrenal cortex. Mol. Cell. Endocrinol. 71, R1–R6.

Kramer, R.E., Gallant, S., & Brownie, A.C. (1979). The role of cytochrome P-450 in the action of sodium depletion on aldosterone biosynthesis in rats. J. Biol. Chem. 254, 3953–3958.

Lala, D.S., Rice, D.A., & Parker, K.L. (1992). Steroidogenic factor 1, a key regulator of steroidogenic enzyme expression is the mouse homolog of fushi tarazu-factor 1. Mol. Endocrinol. 6, 1249–1258.

Lambeth, J.D., & Stevens, V.L. (1985). Cytochrome P450$_{scc}$: enzymology and the regulation of intra-mitochondrial cholesterol delivery to the enzyme. Endocrine Res. 10, 283–309.

Lauber, M., Sugano, S., Ohnishi, T., Okamoto, M., & Muller, J. (1987). Aldosterone biosynthesis and cytochrome P-45011β: Evidence for two forms of the enzyme in the rat. J. Steroid Biochem. 26, 693–698.

Lauber, M., & Muller, J. (1989). Purification and characterization of two distinct forms of rat adrenal cytochrome P45011β functional and structural aspects. Arch. Biochem. Biophys. 274, 109–119.

Lefkowitz, R.J., Roth, J., & Pastam, I. (1971). ACTH receptor interaction in the adrenal: A model for the initial step in the action of hormones that stimulate adenyl cyclase. Ann. NY Acad. Sci. 185, 195–209.

Legros, F., & Lehoux, J.G. (1983). Changes in characteristics of rat adrenal glomerulosa cells under acute and chronic treatment with ACTH. Can. J. Biochem. Cell. Biol. 61, 538–543.

Lehoux, J.G., Sandor, T., Henderson, I.W., & Chester-Jones, I. (1974). Some aspects of the dietary sodium intake on the regulation of aldosterone biosynthesis in rat adrenals. Can. J. Biochem. 52, 1–6.

Lehoux, J.G. (1979). Factors controlling the biosynthesis of aldosterone. In: Steroid Biochemistry (Hobkirk, R. ed.), pp. 51–80, Vol. II, CRC Press, Boca Raton, FL.

Lehoux, J.G., & Lefebvre, A. (1980). De novo synthesis of corticosteroids in hamster adrenal glands. J. Steroid Biochem. 12, 479–485.

Lehoux, J.G., Lefebvre, A., de Médicis, E., Bastin, M., Bélisle, S., & Bellabarba, D. (1987). Effect of ACTH on cholesterol and steroid synthesis in adrenocortical tissues. J. Steroid Biochem. 27, 1151–1160.

Lehoux, J.G., Lefebvre, A., Bélisle, S., & Bellabarba, D. (1990). Effect of ACTH suppression on adrenal 3-hydroxy-3-methylglutaryl coenzyme A .reductase mRNA in 4-aminopyrazolopyrimidine-treated rats. Molec. Cell. Endocr. 69, 41–49.

Lehoux, J.G., & Ducharme, L. (1992). The differential regulation of aldosterone output in hamster adrenal by angiotensin II and adrenocorticotropin. J. Steroid Biochem. Molec. Biol. 41, 809–814.

Lehoux, J.G., Mason, J.I., & Ducharme, L. (1992). *In vivo* effects of adrenocorticotropin on hamster adrenal steroidogenic enzymes. Endocrinology 131, 1874–1882.

Lehoux, J.G., & Tremblay, A. (1992). *In vivo* regulation of gene expression of enzymes controlling aldosterone synthesis in rat adrenal. J. Biochem. Molec. Biol. 43, 837–846.

Lehoux, J.G., Ducharme, L., & Tremblay, A. (1993). A low sodium intake provokes increases in the levels of angiotensin-II (AII) receptor mRNA and protein in rat adrenal zona glomerulosa and this is mediated by AII. 75th Annual Meeting of The Endocrine Society, June 9–12, Las Vegas, NE, Abst. no 1395.

Lehoux, J.G., Bird, I.M., Rainey, W.E., Tremblay, A., & Ducharme, L. (1993). Both low sodium and high potassium intake increase the level of adrenal angiotensin-II receptor type I but not that of adrenocorticotropin receptor. Endocrinology 134, 776–782.

Lehoux, J.G., Mason, J.I., Bernard, H., Ducharme, L., Véronneau, S., & Lefebvre, A. (1993). On the presence of two cytochrome P450 aldosterone synthase mRNA in the hamster adrenal. J. Steroid Biochem. Mol. Biol. 49, 131–137.

Lehoux, J.G., Lefebvre, A., Ducharme, L., Lehoux, J., Martel, D., & Brière, N. (1996). J. Endocrinol. (in press).

Lehoux, J.G., & Ducharme, L. (1995). *In vivo* effects of adrenocorticotropin on c-jun, jun-B, c-fos, and fos-B in rat adrenal. Endocrine Res. 21, 267–274.

Lifton, R.P., Dluhy, R.G., Powers, M., Rich, G.M., Cook, S., Utick, S., & Lalovel, J.M. (1992). A chimaeric 11β-hydroxylase/aldosterone synthase gene causes glucocorticoid-remediable aldosteronism and human hypertension. Nature 355, 262–265.

Luddens, H., & Havsteen, B. (1986). Characterization of the porcine ACTH receptor with the aid of a monoclonal antibody. Biol. Chem. Hoppe-Zeyler 367, 539–547.

Luetscher, J.A. Jr., & Axelrad, B.J. (1954). Increased aldosterone output during sodium deprivation in normal men. Proc. Soc. Exp. Biol. Med. N.Y. 87, 650–653.

Manuelidis, L., & Mulrow, P. (1973). ACTH effects on aldosterone production and mitochondrial ultrastructure in adrenal gland cultures. Endocrinology 93, 1104–1108.

Marusic, E.T., & Mulrow, P.J. (1967). Stimulation of aldosterone biosynthesis in adrenal mitochondria by sodium depletion. J. Clin. Invest. 46, 2101–2108.

Marx, J.L. (1986). The *fos* gene as "Master Switch." Science 237, 854–856.

Mathew, P.A., Mason, J.I., Trant, J.M., Sanders, D., & Waterman, M.R. (1990). Amino acid substitutions Phe66 → Leu and Ser126 → Pro abolish cortisol and aldosterone synthesis by bovine cytochrome $P450_{11\beta}$. J. Biol. Chem. 265, 20228–20233.

Matsukawa, N., Nonaka, Y., Ying, Z., Higaki, J., Ogihara, T., & Okamoto, M. (1990). Molecular cloning and expression of cDNAs encoding rat aldosterone synthase: Variants of cytochrome $P450_{11\beta}$. Biochem. Biophys. Res. Comm. 169, 245–252.

McCarthy, R.T., Isales, C.M., Bollag, W.B., Rasmussen, J., & Barrett, P.Q. (1990). Atrial natriuretic peptide differentially modulates T- and L-type calcium channels. Am. J. Physiol. 258, F473–F478.

McDougall, J.G., Butkus, A., Coghlan, J.P., Denton, D.A., Muller, J., Oddie, C.J., Robinson, P.M., & Scoggins, B.A. (1980). Biosynthetic and morphological evidence for inhibition of aldosterone production following administration of ACTH to sheep. Acta Endocrinol. 94, 559–570.

McIlhinney, R.A., & Schulster, D. (1975). Studies on the binding of ^{125}I-labelled corticotropin to isolated rat adrenocortical cells. J. Endocrinol. 64, 175–184.

Miyamoto, N., Seo, H., Kanda, K., Hidaka, H., & Matsui, N. (1992). A 3′,5′-cyclic adenosine monophosphate-dependent pathway is responsible for a rapid increase in c-fos messenger ribonucleic acid by adrenocorticotropin. Endocrinology 130, 3231–3236.

Montjoy, K.G., Robbins, L.S., Mortrud, M.T., & Cone, R.D. (1992). The cloning of a family of genes that encode the melanocortin receptors. Science 257, 1248–1251.

Montminy, M.R., & Bilezikjian, L.M. (1987). Binding of a nuclear protein to the c-AMP response element of the somatostatin gene. Nature 328, 175–178.

Morgan, J.I., & Curran, T. (1986). Role of ion flux in the control of c-fos expression. Nature 322, 552–555.

Mornet, E., Dupont, J., Vitek, A., & White, P.C. (1989). Characterization of two genes encoding human steroid 11β-hydroxylase (P450-11β). J. Biol. Chem. 264, 20961–20967.

Morohashi, K., Fujii-Kuriyama, Y., Okada, K., Hirose, T., Inayama, S., & Omura, T. (1984). Molecular cloning and nucleotide sequence of cDNA for mRNA of mitochondrial cytochrome P450scc of bovine adrenal cortex. Proc. Natl. Acad. Sci. USA 81, 4647–4651.

Morohashi, K., Yoshioka, H., Gotoh, O., Okada, Y., Yamamoto, K., Miyata, T., Sogawa, K., Fujii-Kuriyama, Y., & Omura, T. (1987). Molecular cloning and nucleotide sequence of DNA of mitochondrial cytochrome $P450_{11\beta}$ of bovine adrenal cortex. J. Biochem. 102, 559–568.

Morohashi, K., Nonaka, Y., Kirita, S., Hatano, O., Takakusu, A., Okamoto, M., & Omura, T. (1990). Enzymatic activities of $P450_{11\beta}$ expressed by two cDNAs in COS-7 cells. J. Biochem. 107, 635–640.

Morohashi, K., Honda, S., Inomata, Y., Honda, H., & Omura, T. (1992). A common transacting factor, Ad4-binding protein, to the promoters of steroidogenic P-450s. J. Biol. Chem. 267, 17913–17919.

Mukai, K., Imai, M., Shimada, H., Okada, Y., Ogishima, T., & Ishimura, Y. (1991). Structural differences in 5′-flanking regions of rat cytochrome $P450_{aldo}$ and $P450_{11\beta}$ genes. Biochem. Biophys. Res. Comm. 180, 1187–1193.

Muller, J. (1971). Regulation of aldosterone biosynthesis, In: Monographs on Endocrinology, Vol. 5, Springer-Verlag, Berlin.

Muller, R. (1986). Proto-oncogenes and differentiation. Trends Biochem. Sci. 11, 129–132.

Murphy, T.J., Alexander, R.W., Griendling, K.K., Runge, M.S., & Bernstein, K.E. (1991). Isolation of a cDNA encoding the vascular type-1 angiotensin II receptor. Nature 351, 233–236.

Nakajin, S., Shinoda, M., Haniu, M., Shively, J.E., & Hall, P.F. (1984). C21 steroid side chain cleavage enzyme from porcine adrenal microsomes. Purification of the 17α-hydroxylase/C17,20 lyase cytochrome P450. J. Biol. Chem. 259, 3971–3976.

Nakamaru, M., Misono, K.S., Naruse, M., Workman, R.J., & Inagami, T. (1985). A role for the adrenal renin-angiotensin system in the regulation of potassium-stimulated aldosterone production. Endocrinology 117, 1772–1778.

Naruse, M., Obana, K., Naruse, K., Yamaguchi, H., Demura, H., Inagami, I., & Shizume, K. (1987). Atrial natriuretic polypeptide inhibits cortisol as well as aldosterone secretion in vitro from human adrenal tissue. J. Clin. Endocrinol. Metab. 64, 10–16.

Naseeruddin, S.A., & Hornsby, P.J. (1990). Regulation of 11β- and 17α-hydroxylase in cultured bovine adrenocortical cells: 3′,5′-cyclic adenosine monophosphate, insulin-like growth factor-I, and activators of protein kinase C. Endocrinology 127, 1673–1681.

Nawata, H., Ohashi, M., Haji, M., Takayanagi, R., Higuchi, K., Fujio, N., Hashiguchi, T., Atsushi, O., Nakao, R., Ohnaka, K., & Nishi, Y. (1991). Atrial and brain natriuretic peptide in adrenal steroidogenesis. J. Steroid Biochem. Molec. Biol. 40, 367–379.

Neri, G., Gambino, A.M., Mazzochi, G., & Nussdorfer, G.G. (1978). Investigations into the effects of ACTH on the half-life of mitochondrial protein. Experientia 34, 133–134.

Nishizuka, Y. (1986). Studies and perspectives of protein kinase C. Science 233, 305–312.

Nonaka, Y., Matsukawa, N., Morohashi, K.I., Omura, T., Ogihara, T., Tergoka, H., & Okamoto, M. (1989). Molecular cloning and sequence analysis of cDNA encoding rat adrenal cytochrome $P450_{11\beta}$. FEBS Letters 255, 21–26.

Nonaka, Y., & Okamoto, M. (1991). Functional expression of the cDNAs encoding rat 11β-hydroxylase [cytochrome P450(11β)] and aldosterone synthase [cytochrome P450(11β,aldo)]. Eur. J. Biochem. 202, 897–902.

Nussdorfer, G.G., Mazzochi, G., & Rebonato, L. (1971). Further studies on the effect of corticosterone on adrenocortical cells of hypophysectomized ACTH-treated rat. Experientia 27, 1347–1348.

Ogishima, T., Mitani, H., & Ishimura, Y. (1989). Isolation of aldosterone synthase cytochrome P-450 from zona glomerulosa mitochondria of rat adrenal cortex. J. Biol. Chem. 264, 10935–10938.

Ogishima, T., Mitani, F., & Ishimura, Y. (1989a). Isolation of two distinct cytochromes $P450_{11\beta}$ with aldosterone synthase activity from bovine adrenocortical mitochondria. J. Biochem. 105, 497–499.

Ogishima, T., Shibata, H., Shimada, H., Mitani, F., Susuki, H.M., Saruta, T., & Ishimura, Y. (1991). Aldosterone synthase cytochrome P450 expressed in the adrenals of patients with primary aldosteronism. J. Biol. Chem. 266, 10731–10734.

Ogishima, T., Suzuki, H., Hata, J.I., Mitani, F., & Ishimura, Y. (1992). Zone-specific expression of aldosterone synthase cytochrome P-450 and cytochrome P45011β in rat adrenal cortex: Histochemical basis for the functional zonation. Endocrinology 130, 2971–2977.

Ohashi, M., Fujio, N., Kato, K., Nawata, H., Ibayashi, H., & Matsuo, H. (1986). Effect of human α-atrial natriuretic polypeptide on adrenocortical function in man. J. Endocr. 110, 287–292.

Ohashi, M., Fujio, N., Nawata, H., Matsuo, H., & Kato, K. (1988). α-human atrial natriuretic polypeptide binding sites in human adrenal membrane fractions. Reg. Peptide 21, 271–278.

Onishi, T., Wada, A., Lauber, M., Yamano, T., & Okamoto, M. (1988). Aldosterone biosynthesis in mitochondria of isolated zones of adrenal cortex. J. Steroid Biochem. 31, 73–81.

Osawa, S., & Hall, P.F. (1985). Plasma membranes from adrenal cells: Purification and properties. J. Cell. Sci. 77, 57–73.

Osborne, T.F., Gil, G., Goldstein, J.L., & Brown, M.S. (1988). Operator constitutive mutation of 3-hydroxy-3-methylglutaryl coenzyme A reductase promoter abolishes protein binding to sterol regulatory element. J. Biol. Chem. 263, 3380–3387.

Pandey, K.N., Inagami, T., Girard, P.R., Kuo, J.F., & Misono, K.S. (1987). New signal transduction mechanisms of atrial natriuretic factor: inhibition of phosphorylation of protein kinase C and a 240 kDa protein in adrenal cortical plasma membrane by cGTP dependent and independent mechanisms. Biochem. Biophys. Res. Comm. 148, 589–595.

Parissenti, A.M., Parker, K.L., & Schimmer, B.P. (1993). Identification of promoter elements in the mouse 21-hydroxylase (*CYP21*) gene that require a functional cyclic adenosine 3',5'-monophosphate-dependent protein kinase. Mol. Endocrinol. 7, 283–290.

Pascoe, L., Curnow, K.M., Slutsker, L., Connell, J.M.C., Speiser, P.W., New, M.I., & White, P.C. (1992). Glucocorticoid-suppressible hyperaldosteronism results from hybrid genes created by unequal crossovers between CYP11B1 and CYP11B2. Proc. Natl. Acad. Sci. USA 89, 8327–8331.

Pascoe, L., Curnow, K.M., Slutsker, L., Rosler, A., & White, P.C. (1992a). Mutations in the human *CYP11B2* (aldosterone synthase) gene causing corticosterone methyloxidase II deficiency. Proc. Natl. Acad. Sci. USA 89, 4996–5000.

Pearce, D., & Yamamoto, K.R. (1993). Mineralocorticoid and glucocorticoid receptor activities distinguished by non receptor factors at a composite response element. Science 259, 1161–1165.

Pelto-Huikko, M., Dagerlind, A., Ceccatelli, S., & Hokfelt, T. (1991). The immediate-early genes c-fos and c-jun are differentially expressed in the rat adrenal gland after capsaicin treatment. Neuroscience Letters 126, 163–166.

Penhoat, A., Jaillard, C., & Saez, J.M. (1989). Corticotropin positively regulated its own receptors and c-AMP response in cultured bovine adrenal cells. Proc. Natl. Acad. Sci. USA 86, 4978–4981.

Privalle, C.T., Crivello, J.F., & Jefcoate, C.R. (1983). Regulation of intra-mitochondrial cholesterol transfer to side chain-cleavage cytochrome P450 in rat adrenal gland. Proc. Natl. Acad. Sci. USA 80, 702–706.

Purvis, J.L., Canick, J.A., Mason, J.I., Estabrook, R.W., & McCarthy, J.L. (1973). Lifetime of adrenal cytochrome P450 as influenced by ACTH. Ann. NY Acad. Sci. 212, 319–342.

Quinn, S.J., & Williams, G.H. (1992). Regulation of aldosterone secretion. In: The Adrenal Gland (James, V.H.T., ed.), pp. 159–189, Raven Press, New York.

Rae, P.A., Gutmann, N.S., Tsao, J., & Schimmer, B.P. (1979). Mutations in c-AMP-dependent protein kinase and ACTH-sensitive adenylate cyclase affect adrenal steroidogenesis. Proc. Natl. Acad. Sci. USA 76, 1896–1900.

Rainey, W.E., Viard, I., Mason, J.I., Cochet, C., Chambaz, E.M., & Saez, J.M. (1988). Effects of transforming growth factor β on ovine adrenocortical cells. Mol. Cell. Endocrinol. 60, 189–198.

Rainey, W.E., Viard, I., & Saez, J.M. (1989). Transforming growth factor β treatment decreases ACTH receptors on ovine adrenocortical cells. J. Biol. Chem. 264, 21474–21477.

Rainey, W.E., Naville, D., & Mason, J.I. (1991). Regulation of 3β-hydroxysteroid dehydrogenase in adrenocortical cells: Effects of angiotensin-II and transforming growth factor beta. Endocr. Res. 17, 281–296.

Rani, C.S.S., Keri, G., & Ramachandran, J. (1983). Studies on corticotropin-induced desensitization of normal rat adrenocortical cells. Endocrinology 112, 315–320.

Rao, J.A., & Lehoux, J.G. (1979). Effect of ACTH on ^{125}I-labeled angiotensin II binding and response by rat adrenal glomerulosa cells. FEBS Letters 105, 325–328.

Rayfield, E.J., Rose, L.I., Dluhy, R.G., & Williams, G.H. (1973). Aldosterone secretory and glucocorticoid excretory responses to alpha 1-24 (Cortrosyn) in sodium-depleted normal man. J. Clin. Endocrinol. Metab. 36, 30–35.

Rebuffat, P., Mazzocchi, G., Gottardo, G., Meneghelli, V., & Nussdorfer, G.G. (1988). Further investigations on the atrial natriuretic factor (ANF)-induced inhibition of the growth and steroidogenic capacity of rat adrenal zona glomerulosa in vivo. J. Steroid Biochem. 29, 605–609.

Rhéaume, E., Lachance, Y., Zhao, H.E., Breton, N., Dumont, N., de Launot, Y., Trudel, C., Luu-The, V., Simard, J., & Labrie, F. (1991). Structure and expression of a new complementary DNA encoding the almost exclusive 3β-hydroxysteroid dehydrogenase Δ^5-Δ^4-Isomerase in human adrenals and gonads. Molec. Endocr. 5, 1147–1157.

Rice, D.A., Aitken, L.D., Vandenbark, G.R., Mouw, A.R., Franklin, A., Schimmer, B.P., & Parker, K.L. (1989). A c-AMP-responsive element regulates expression of the mouse steroid 11β-hydroxylase gene. J. Biol. Chem. 264, 14011–14015.

Rice, D.A., Kronenberg, M.S., Mouw, A.R., Aitken, L.D., Franklin, A., Schimmer, B.P., & Parker, K.L. (1990). Multiple regulatory elements determine adrenocortical expression of steroid-21-hydroxylase. J. Biol. Chem. 265, 8052–8058.

Rice, D.A., Mouw, A.R., Bogerd, A.M., & Parker, K.L. (1991). A shared promoter element regulates the expression of three steroidogenic enzymes. Mol. Endocrinol. 5, 1552–1561.

Richards, A.M., Nicholls, M.G., Ikram, H., Webster, M.W.I., Yandle, T.C., & Espiner, E.A. (1985). Renal, hemodynamic and hormonal effects of human alpha natriuretic factor in healthy volunteers. Lancet 1, 545–549.

Richards, A.M., Tomalo, G., Cleland, T.G.F., McIntyre, G.D., Leckie, B.J., Dargie, H.T., Ball, S.G., & Robertson, J.I.S. (1987). Plasma atrial natriuretic peptide concentrations during exercise in sodium replete and deplete man. Clin. Sci. 72, 159–164.

Rosenberg, J., Pines, M., & Hurwitz, S. (1989). Inhibition of aldosterone secretion by atrial natriuretic peptide in chicken adrenocortical cells. Biochem. Biophys. Acta 1014, 189–194.

Sabatini, D.D., De Robertis, E.O.P., & Bleichmar, H.B. (1962). Submicroscopic study of the pituitary action on the adrenocortex of the rat. Endocrinology 70, 390–406.

Sandor, T., & Mehdi, A.Z. (1979). Steroids and evolution. In: Hormones and Evolution (Barrington, E.J.W., ed.), pp. 1–61, Academic Press, New York.

Sasaki, K., Yamano, Y., Bardhan, S., Iwai, N., Murray, J.J., Hasegawa, M., Matsuda, Y., & Inagami, T. (1991). Cloning and expression of a complementary DNA encoding a bovine adrenal angiotensin II type-1 receptor. Nature 351, 230–232.

Schneider, E.G., Radke, K.J., Ulderich, D.A., & Taylor, R.E. Jr. (1985). Effect of osmolality on aldosterone secretion. Endocrinology 116, 1621–1626.

Schneider, E.G., & Kramer, R.E. (1986). Effect of osmolality on angiotensin-stimulated aldosterone production by primary cultures of bovine adrenal glomerulosa cells. Biochem. Biophys. Res. Comm. 139, 46–51.

Schwartz, D., Katsube, N.C., & Needleman, P. (1986). Atriopeptin release in conditions of altered salt and water balance in the rat. Biochem. Biophys. Res. Comm. 137, 922–928.

Sharples, K., Plowman, G.D., Rose, T.M., Twardzik, D.R., & Purchio, A.F. (1987). Cloning and sequence analysis of simian transforming growth factor beta cDNA. DNA 6, 239–244.

Shibata, H., Ogishima, T., Mitani, F., Suzuki, H., Murakami, M., Saruta, T., & Ishimura, Y. (1991). Regulation of aldosterone synthase cytochrome P-450 in rat adrenals by angiotensin II and potassium. Endocrinology 128, 2534–2539.

Shier, D.N., Kusano, E., Stoner, G.D., Franco-Saenz, R., & Mulrow, P.J. (1989). Production of renin, angiotensin II and aldosterone by adrenal explant cultures: Response to potassium and converting enzyme inhibition. Endocrinology 125, 486–491.

Simpson, E.R., & Waterman, M.R. (1988). Regulation of the synthesis of steroidogenic enzymes in adrenocortical cells by ACTH. Ann. Rev. Physiol. 50, 427–440.

Simpson, E.R., Lund, J., Ahlgren, R., & Waterman, M.R. (1990). Regulation by cyclic AMP of the genes encoding steroidogenic enzymes: When the light finally shines. . . Mol. Cell. Endocrinol. 70, C25–C28.

Singh, S., Lowe, D.G., Thorpe, D.S., Rodriguez, H., Kuang, W.J., Dangott, L.J., Chinkers, M., Goeddel, D.V., & Jarbers, D.L. (1988). Membrane guanylate cyclase is a cell-surface receptor with homology to protein kinases. Nature 334, 708–712.

Strittmatter, S.M., DeSouza, E.B., Lynch, D.R., & Snyder, S.H. (1986). Angiotensin-converting enzyme localized in the rat pituitary and adrenal glands by [^{3}H] captopril autoradiography. Endocrinology 118, 1690–1699.

Takayanagi, R., Inagami, T., Snajdar, R.M., Imada, T., Tamura, M., & Misono, K.S. (1987). Two distinct forms of receptors for atrial natriuretic factor in bovine adrenocortical cells. J. Biol. Chem. 262, 12104–12113.

Tamaoki, B.I. (1973). Steroidogenesis and cell structure. J. Steroid Biochem. 4, 89–118.

Thompson, N.L., Flanders, K.C., Smith, J.M., Ellingsworth, L.R., Roberts, A.B., & Sporn, M.B. (1989). Expression of transforming growth factor-β1 in specific cells and tissues 669.

Timmermans, P.B.M.W.M., Wong, P.C., Chiu, A.T., & Herblin, W.F. (1991). Nonpeptide angiotensin II receptor antagonists. TIPS Reviews 12, 55–61.

Tremblay, A., & Lehoux, J.G. (1989). Effects of low sodium and high potassium intakes on rat steroid metabolizing enzymes. 71st Annual Meeting of the Endocrine Society, June 21–24, Seattle, WA, Abst. No. 1459.

Tremblay, A., & Lehoux, J.G. (1989a). Regulation of cytochrome P-450 and adrenodoxin mRNA levels by dietary sodium and potassium. 9th International Symposium of the Journal of Steroid Biochemistry May 28–31 Las Palmas, Spain, Abst. No. 23.

Tremblay, A., & Lehoux, J.G. (1989b). Effects of dietary sodium restriction and potassium intake on cholesterol side-chain cleavage cytochrome P-450 and adrenodoxin mRNA levels. J. Steroid Biochem. 34, 385–390.

Tremblay, A., Waterman, M.R., Parker, K.L., & Lehoux, J.G. (1991). Regulation of rat adrenal messenger RNA and protein levels for cytochrome P-450s and adrenodoxin by dietary sodium depletion of potassium intake. J. Biol. Chem. 266, 2245–2251.

Tremblay, A., & Lehoux, J.G. (1992). Influence of captopril on adrenal cytochrome P-450s and adrenodoxin expression in high potassium or low sodium intake. J. Steroid Biochem. 41, 799–802.

Tremblay, A., Parker, K.L., & Lehoux, J.G. (1992). Dietary potassium supplementation and sodium restriction stimulate aldosterone synthase but not 11β-hydroxylase P-450 messenger ribonucleic acid accumulation in rat adrenals and require angiotensin II production. Endocrinology 130, 3152–3158.

Tremblay, A., & Lehoux, J.G. (1993). Transcriptional activation of adrenocortical steroidogenic genes by high potassium or low sodium intake. FEBS 317, 211–215.

Trzeciak, W.H., Lehoux, J.G., Waterman, M.R., & Simpson, E.R. (1993). Dexamethasone inhibits corticotropin-induced accumulation of CYP11A and CYP17 messenger RNAs in bovine adrenocortical cells. Mol. Endocrinol. 7, 206–213.

Tsuda, T., Hamamoni, Y., Yamashita, T., Fukumoto, Y., & Takai, Y. (1986). Involvement of three intracellular messenger systems, protein kinase C, calcium ion and cyclic AMP, in the regulation of c-fos gene expression in Swiss 3T3 cells. FEBS Letters 208, 39–42.

Tsukiyama, T., Heda, H., Hirose, S., & Niwa, O. (1992). Embryonal long terminal repeat-binding protein is a urine homolog of FT2-F1, a member of the steroid receptor superfamily. Mol. Cell. Biol. 12, 1286–1291.

Underwood, R.H., Menachery, A.I., & Williams, G.H. (1989). Effect of sodium intake on phosphoinositides and inositol triphosphate response to angiotensin II, K^+ and ACTH in rat glomerulosa cells. J. Endocrinol. 122, 371–377.

Véronneau, S., Bernard, H., Cloutier, M., Courtemanche, J., Ducharme, L., Lefebvre, A., Mason, J.I., & Lehoux, J.G. (1996). J. Steroid Biochem. Molec. Biol. (in press).

Viard, I., Jaillard, C., Ouali, R., & Saez, J.M. (1992). Angiotensin-II-induced expression of protooncogenes (c-fos, jun-β and c-jun) mRNA in bovine adrenocortical fasciculata cells (BAC) is mediated by AT-1 receptors. FEBS Letters 313, 43–46.

Viard, I., Hall, S.H., Jaillard, C., Berthelon, M.C., & Saez, J.M. (1992a). Regulation of c-fos, c-jun and jun-β messenger ribonucleic acids by angiotensin-II and corticotropin in ovine and bovine adrenocortical cells. Endocrinology 130, 1193–1200.

Voutilainen, R., Picado-Leonard, J., DiBlasio, A.M., & Miller, W.L. (1988). Hormonal and development regulation of human adrenodoxin mRNA in steroidogenic tissues. J. Clin. Endocrinol. Met. 66, 383–388.

Waterman, M.R., Kagawa, N., Zanger, U.M., Momoi, K., Lund, J., & Simpson, E.R. (1992). Comparison of c-AMP-responsive DNA sequences and their binding proteins associated with expression of the bovine CYP17 and CYP11A and human genes. J. Steroid Biochem. Molec. Biol. 43, 931–935.

Weidmann, P., Hasler, L., Gnadinger, M., Lang, R.E., Uehlinger, D.E., Shaw, S., Rascher, W., & Reubi, F.C. (1986). Blood levels and renal effects of atrial natriuretic peptide in normal man. J. Clin. Invest. 77, 734–742.

White, P.C., Dupont, J., New, M.I., Leiberman, E., Hochberg, Z., & Rosler, A. (1991). A mutation in CYP11B1 (Arg- His) associated with steroid 11β-hydroxylase deficiency in jews of moroccan origin. J. Clin. Invest. 87, 1664–1667.

Wilcox, J.N., Augustine, A., Goeddel, D., & Lowe, D.G. (1991). Differential regional expression of three natriuretic peptide receptor genes within primate tissues. Mol. Cell. Biol. 11, 3454–3462.

Yabu, M., Senda, T., Nonaka, Y., Matsukawa, N., Okamoto, M., & Fujita, H. (1991). Localization of the gene transcripts of 11β-hydroxylase and aldosterone synthase in the rat adrenal cortex by in situ hybridization. Histochemistry 96, 391–394.

Yamaguchi, T., Naito, Z., Stoner, G.D., Franco-Saenz, R., & Mulrow, P.J. (1990). Role of the adrenal renin-angiotensin system on adrenocorticotropic hormone- and potassium-stimulated aldosterone production by rat adrenal glomerulosa cells in monolayer culture. Hypertension 16, 635–641.

Yanagibashi, K., Haniu, M., Shively, J.E., Shen, W.H., & Hall, P. (1986). The synthesis of aldosterone by the adrenal cortex. Two zones (fasciculata and glomerulosa) possess one enzyme for 11β-, 18-hydroxylation and aldehyde synthesis. J. Biol. Chem. 261, 3556–3562.

Yang, G., Koistinaho, J., Zhu, S.H., & Hernoven, A. (1989). Induction of c-fos like protein in the rat adrenal cortex by acute stress immunocytochemical evidence. Mol. Cell Endocrinol. 66, 163–170.

Yang, G., Koistinaho, J., Iadarola, M., Zhu, S.H., & Hernoven, A. (1990). Administration of adrenocorticotropin hormone (ACTH) enhances Fos expression in the rat adrenal cortex. Regulatory Peptides 30, 21–31.

Zelezna, B., Richards, E.M., Tang, W., Lu, D., Summers, C., & Raizada, M.K. (1992). Characterization of a polyclonal anti-peptide antibody to the angiotensin II type-1 (AT$_1$) receptor. Biochem. Biophys. Res. Comm. 183, 781–788.

Zuber, M.X., Simpson, E.R., & Waterman, M.R. (1986). Expression of bovine 17α-hydroxylase of cytochrome P450 cDNA in nonsteroids genic (COS) cells. Sciences 234, 1258–1261.

Zuber, M.X., John, M.E., Okamura, T., Simpson, E.R., & Waterman, M.R. (1986a). Bovine adrenocortical cytochrome P450$_{17\alpha}$. Regulation of gene expression by ACTH and elucidation of primary sequence. J. Biol. Chem. 261, 2475–2482.

PHYSIOLOGY AND MOLECULAR BIOLOGY OF P450c21 AND P450c17

Bon-chu Chung

Advances in Molecular and Cell Biology
Volume 14, pages 203–223.
Copyright © 1996 by JAI Press Inc.
All rights of reproduction in any form reserved.
ISBN: 0-7623-0113-9

I. ENZYMOLOGY OF P450c17 AND P450c21

Steroid hormones include mineralocorticoid, glucocorticoid, progestin, androgen, and estrogen which regulate the physiological balances of electrolytes, glucose, sexual development, and maintenance of pregnancy. These hormones are synthesized from a common precursor, cholesterol, through a series of oxidation-reduction reactions. Most of the enzymes catalyzing the conversion of steroids are members of the cytochrome P450 family. Two members of this superfamily, P450c17 and P450c21, will be the focus of discussion in this chapter. P450c17 and P450c21 share many similar properties and are most closely related to each other among all members in the P450 superfamily. Both are tightly bound to the membrane of endoplasmic reticulum in the cell, receiving electrons from NADPH via NADPH-dependent reductase, and use molecular oxygen for enzymatic activities (reviewed in Miller, 1988). It is therefore suitable to discuss their properties in the same chapter.

A. Reactions Catalyzed by P450c17 and P450c21

P450c21 possesses 21-hydroxylase activity, which catalyzes the addition of a hydroxyl group to the C-21 of the steroid backbone of progesterone or 17-hydroxyprogesterone (17-OHP), converting them into deoxycorticosterone and 11-deoxycortisol respectively (Figure 1). These reactions are important for the synthesis of mineralocorticoids and glucocorticoids. Human and bovine P450c21 catalyze the conversion of 17-OHP faster than progesterone. This results in higher amounts of cortisol than aldosterone production in the human and bovine adrenal cortex.

P450c17 possesses two activities, 17α-hydroxylase and 17,20 lyase (Figure 1). 17α-Hydroxylase catalyzes the addition of a hydroxyl group to the C-17 position of the steroid backbone of pregnenolone (a Δ5 substrate) or progesterone (a Δ4 substrate), converting them into 17α-hydroxypregnenolone and 17α-hydroxyprogesterone, respectively. On the other hand, P450c17 can cleave the C-17,20 bond of these two products, further converting them to dehydroepiandrosterone and androstenedione, respectively.

There used to be arguments about whether the 17α-hydroxylase activity and 17,20-lyase activity were catalyzed by one or two different enzymes. Purification of P450c17 to homogeneity and the expression of P450c17 from a single P450c17 cDNA showed that one protein has two intrinsic enzymatic activities (Nakajin and Hall, 1981; Zuber et al., 1986). Purified porcine P450c17 has a V_{max} 4.6 nmol product/min/nmol P450, K_m 1.5 μM for 17α-hydroxylase and a V_{max} of 2.6 nmol product/min/nmol P450, K_m 2.4 μM for 17,20 lyase (Nakajin and Hall, 1981). 17α-Hydroxylase activity is important for the synthesis of glucocorticoid, and 17,20 lyase important for sex steroid synthesis (Figure 1). Therefore the activities of a single P450c17 protein determine the direction of the steroid synthetic pathway towards either glucocorticoids or sex hormones. It acts as an important branch point enzyme.

B. Properties of P450c21 and P450c17

Both P450c21 and P450c17 have been purified from microsomes of various tissues such as bovine and porcine adrenals and testis. Each P450 protein contains one molecule of heme, which acts as the active center for the redox reactions it catalyzes. The reduced form of the hemoprotein upon binding carbon monoxide will exhibit a peak of absorption at 450 nm, hence the name cytochrome P450. Upon binding substrates, these P450s also show a characteristic absorption peak at 380 nm and a trough at 420 nm. These spectrophotometric properties result from

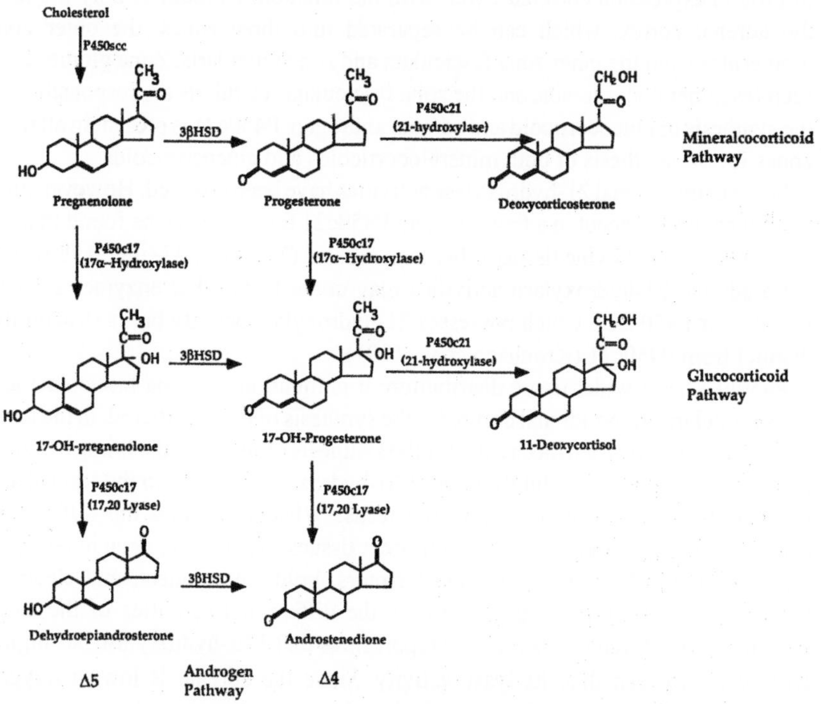

Figure 1. Reactions carried out by P450c17 and P450c21. The common precursor cholesterol is first converted to pregnenolone, which is then converted to other steroids in the mineralocorticoid, glucocorticoid, and androgen synthetic pathways as shown in the figure. The chemical reactions catalyzed by P450c17 and P450c21 are delineated. The Δ4 and Δ5 steroids are also indicated at the bottom of the pathway. P450scc, cholesterol side-chain cleavage enzyme; 3,βHSD, 3β-hydroxysteroid dehydrogenase/$\Delta^{5\rightarrow4}$ isomerase.

interaction of heme and its neighboring amino acids of the P450s and have been useful in the analysis of the structure and property of P450s.

The presence and activities of P450c17 exhibit species-specific differences. P450c17 from rat possesses four enzymatic activities including 17α-hydroxylase activities and 17,20-lyase activities towards both Δ4 and Δ5 substrates. Yet P450c17 from cattle and human has only three activities devoid of the 17,20 lyase activity for the formation of androstenendione. The adrenals of rat, rabbit, and guinea pig do not contain P450c17 and therefore do not synthesize cortisol. These rodents use corticosterone as their major glucocorticoid.

II. REGULATION OF P450c17 AND P450c21

P450c17 and P450c21 are present in steroid-secreting endocrine glands. The sites of protein expression correlates well with the function. P450c21 is only found in the adrenal cortex, which can be separated into three zones, the outer zona glomerulosa and the inner zona fasciculata and zone reticularis. Zona glomerulosa secretes mineralocorticoids, and the zone fasciculata/reticularis are responsible for the synthesis of glucocorticoid and adrenal androgen. P450c21 is present in all three zones for the synthesis of both mineralocorticoids and glucocorticoids.

Some extra-adrenal 21-hydroxylase activities have been reported. However, after careful analysis, except in a few cases, no P450c21 transcript can be found in most of the human and bovine tissues other than adrenal (Mellon and Miller, 1989). The extra-adrenal 21-hydroxylase activities may come from other enzymes such as rabbit liver P450-2C5 which possesses 21-hydroxylase activity but is structurally distinct from P450c21 (Kronbach et al., 1991).

P450c17 has a wider tissue distribution: it is found in the zona fasciculata and zona reticularis of the adrenal cortex for the synthesis of glucocorticoid, in the testis Leydig cells and ovarian theca cells for the synthesis of sex steroids. Being a branch point enzyme, P450c17 exhibits higher 17α-hydroxylase activity in the adrenal and has mainly 17,20 lyase activity in the testis. What determines the differential activities of a single enzyme in two different tissues? There have been hypotheses regarding the differences in hormonal factors, lipid environment, and electron-transfer systems which might determine the differential activities of the same enzyme in two different tissues. One report finds that 17α-hydroxylase has higher affinity for oxygen than its lyase activity. Since the adrenal is low in oxygen pressure (about 7%), the differential affinity of two activities towards oxygen could modulate the amount of cortisol versus androgen secretion in the adrenal (Chabre et al., 1993). Cytochrome b5 increases 17,20-lyase activity much more than 17α-hydroxylase activity. Human testes contain a higher concentration of b5 which can push the pathway all the way to sex steroid production (Katagiri et al., 1995). This could also explain the higher 17,20 lyase activity in the testis than in the adrenal.

A. Regulation in Adrenal Cells

Steroidogenesis of the adrenal cortex is under the control of a pituitary peptide hormone, adrenocorticotropin (ACTH). Binding of ACTH to the surface receptor of the adrenocortical cells illicits a series of reactions including activation of GTP-binding proteins followed by adenylyl cyclase, and an increase of intracellular cAMP concentration. Cyclic AMP as a major intracellular second messenger for ACTH exerts both acute and chronic effects on steroidogenesis (Simpson and Waterman, 1988). The acute effect occurs within minutes after ACTH treatment, involving fast mobilization of cholesterol from the lipid droplet to the inner mitochondrial membrane, therefore increasing substrate concentration for steroidogenesis (Jefcoate et al., 1987). This acute effect could be a mechanism the body evolves in response to stress. Under stressful conditions, high levels of ACTH are released, leading to higher concentration of glucocorticoid in a few minutes. Glucocorticoids then exert various effects to withstand stress.

The chronic action does not occur until hours after ACTH administration. In bovine and human fetal adrenocortical cell cultures, P450scc, P450c11, P450c17, and P450c21 mRNA levels are increased in response to ACTH treatment following a delayed time course in a dose-dependent fashion (Zuber et al., 1985; Simpson and Waterman, 1988). This induction can be mimicked by cAMP analogs or substances which increase intracellular cAMP concentration (Kramer et al., 1984). John et al. (1986) showed that the increased message levels are due to increased transcription of steroidogenic genes. ACTH is secreted by the pituitary in a pulsatile fashion and it is this chronic action on steroidogenic gene activation through cAMP which maintains the specialized function of the adrenal cortex.

In addition to cAMP, protein kinase C plays a modulatory role in the long term maintenance of steroidogenic enzyme and hormone production. In cultured human and bovine fetal adrenocortical cells, stimulation of P450scc, P450c11, P450c17, and P450c21 mRNA by cAMP is inhibited by the addition of a phorbol ester 12-*O*-tetradecanoyl phorbol-13-acetate (TPA), a protein kinase C effector (Ilvesmäki and Voutilainen, 1991). The differential concentrations of protein kinase C in different zones of the adrenal cortex play a key role in adrenal zonation (Hornsby, 1989).

The regulation of mineralocorticoid synthesis in zona glomerulosa of the adrenal cortex is mainly through angiotensin II (Quinn and Williams, 1988). Angiotensin II binds to the receptor at the zona glomerulosa to activate phospholipase C, which in turn generates inositol triphosphates and diacyl glycerol, leading to increases of intracellular calcium concentrations and protein kinase C activation (Barrett et al., 1989). High levels of protein kinase C in zona glomerulosa turns off the expression of P450c17 (Ilvesmäki and Voutilainen, 1991; Bird et al., 1992). Whereas in the zona fasciculata/reticularis, the absence of angiotensin II and the presence of ACTH lead to expression of P450c17. Therefore, the zonation and the expression of steroids in these zones appear to be controlled by angiotensin II receptor, protein kinase C, and P450c17. The zona fasciculata/reticularis has high P450c17 activities, leading to the production of glucocorticoids and adrenal androgen. In zona glomerulosa there is no P450c17, and the major steroid products are mineralocorticoids.

B.   Regulation in Gonads

An important function of P450c17 is to catalyze the synthesis of androgens, mainly in gonads. In the male gonad, testis, androgens are produced in the interstitial Leydig cells. In the female gonad, ovary, androgens are produced in the theca interna cells. P450c17 levels are high in the follicular phase of the ovarian estrus cycle, and decreased drastically in the luteal phase. This correlates well with the level of androgen during estrus cycles (Voss and Fortune, 1993). P450c17 expression in both follicular and luteal cells are subject to leutinizing hormone regulation, via cAMP as a second messenger, similar to the situation in the adrenal (Payne, 1990; Magoffin and Weitsman, 1993).

P450c17 activities in testis Leydig cells are subject to negative regulation by androgen. High levels of androgen can bind to P450c17 and cause its faster degradation through an oxygen-mediated mechanism. Chronic treatment of low concentrations of androgen represses cAMP-induced P450c17 mRNA synthesis by a receptor-mediated mechanism in Leydig cell culture (Hales et al., 1987). These results show that the end products of the P450c17 reaction can go back to the Leydig cells and exert feedback inhibition on P450c17 activities.

Most of the above experiments use cultured cells as an experimental model to study regulation of steroidogenesis. It is worth noting that cultured cells, although easier to manipulate, do not always reflect the situation *in vivo*. Many reports have documented the altered regulation in cultured cells (Hales et al., 1990; LeHoux et al., 1992). It is therefore necessary to validate the *in vitro* data before they can be extended to the physiological situation.

III.   GENE STRUCTURE AND GENE EXPRESSION

A.   Gene Structure

The *P450c21* gene is located on human chromosome 6p21.1, inside the class III region of the major histocompatibility complex, forming a duplicated unit with the neighboring C4 gene which encodes the fourth component of serum complement (Figure 2). The two *P450c21* genes, termed CYP21A1P and CYP21A1, or c21A and *c21B* for short, are more than 98% homologous. Yet the small differences in sequence result in frameshift and nonsense mutations in the *c21A* gene rendering it nonfunctional. Interestingly, in mouse this order is reversed. The mouse *c21A* rather than *c21B* gene is functional. The close homology between these two duplicated genes enables frequent recombination and hence gene conversion events

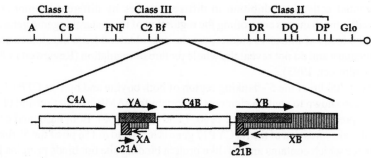

Figure 2. Gene organization of the CYP21 locus at the short arm of human chromosome 6. The upper line shows the organization of the Class I, II, and III regions of the major histocompatibility locus. The second line shows the duplicated C4A/CYP21 locus. Arrows indicate the orientation of transcription of each gene. Complement *C4*, *c21*, *X*, and *Y* genes are shown in open boxes, hatched boxes, boxes in vertical stripes, and dots, respectively.

between them, thus causing apparent mutations of the active gene. Mutations in the *c21B* gene constitute more than 85% of cases of congenital adrenal hyperplasia, a disease characterized by insufficient cortisol synthesis, as will be discussed in the next section.

The *c21* gene locus is packed with many transcription units (Figure 2). A pair of genes termed XA and XB are encoded on the opposite strand of the *c21A* and *c21B* genes, and overlapping the last exon of each. In addition, another pair of genes, termed YA and YB, with the same transcriptional orientation as c21A and c21B, overlap both *c21* and *X* genes. With the exception of XB, which encodes an extracellular matrix protein tenascin-X, the transcripts of XA, YA, and YB are not translated. The reason why so many genes are jammed in a small DNA locus and the significance of the presence of multiple complementary transcripts in the adrenal cells are not clear (Gitelman et al., 1992; Bristow et al., 1993a, 1993b).

The gene encoding P450c17, termed CYP17, is located on human chromosome 10q24.3 (Matteson et al., 1986; Fan et al., 1992). The exon/intron structure of the *CYP17* gene (8 exons) is virtually identical to that of the *CYP21A1* gene (10 exons), except that the first and fourth introns in the *CYP21A1* gene have been deleted in CYP17 (Picado-Leonard and Miller, 1987). Therefore P450c17 and P450c21 are closely related to each other not only in protein properties but also in gene structures.

B. Promoter and Regulatory Elements of P450c17 and P450c21

As discussed in Section II, the expressions of P450c17 and P450c21 are regulated in a tissue-specific fashion by various hormonal and developmental factors. They are generally coordinately stimulated by cAMP, yet in other cases they do exhibit differential activation/inhibition in different tissues by different hormones. The molecular mechanisms controlling the regulation of these steroidogenic genes have been the focus of intensive studies, yet the data obtained from the studies are still fragmentary and do not reveal the whole picture of regulation (Reviewed in Parker and Schimmer, 1993).

About 700 bp of the 5'-flanking region of both bovine and human *CYP17* genes have been shown to contain elements important for cAMP stimulation and phorbol ester repression (Brentano et al., 1990; Lund et al., 1990). Both basal and cAMP-stimulated expression of the rat CYP17 gene is localized to 156 pb of the 5'-flanking sequence which contains an SF1-like protein binding site but binds proteins larger than SF1 (Givens et al., 1994). The cAMP-responsive sequences in the bovine gene have been further located to two regions between –243 to –225 and –80 to –40 relative to the transcriptional start site of the gene, termed CRS1 and CRS2, respectively. These two sites do not have any resemblance to the well defined AMP-responsive element (CRE), nor do they bind any known protein factors. The CRS1 element also mediates phorbol ester repression through the protein kinase C pathway (Bakke and Lund, 1992). These results indicate that the chronic effect of cAMP on gene activation is complex and different from any known characterized mechanism.

For the human *CYP21A1* gene, 4-bp substitution of at the -100 region causes 5-fold decrease in transcriptional activity and the ability of the promoter to bind Sp1 protein (Chang and Chung, 1995). In addition, an ASP protein has been identified to bind to a region about 130 bp upstream from the transcription start site (Kagawa and Waterman, 1991). This binding correlates well with the ability of the reporter gene to respond to cAMP. The mode of action of this protein, however, is not known. In the mouse *Cyp21* gene, multiple regulatory elements located within 300 bp of the 5'- flanking region are bound to proteins *in vitro* and are important for gene expression. Mutation of any of these elements abolishes both basal and cAMP-stimulated gene expression. One of these elements has been characterized in detail. It binds to a protein SF1 which is present mainly in steroidogenic cells and this binding is essential for gene activation (Rice et al., 1990; Lala et al., 1992).

Results from all of the above experiments were obtained mainly by transfections into appropriate cell cultures of the putative regulatory elements linked to a reporter gene. This approach enables the identification of potentially important regulatory elements *in vitro* because of the ease to handle cell culture. The *in vitro* data indicate that important regulatory elements are located within a few hundred base pairs of the transcriptional start site. Transgenic mouse studies, however, showed that efficient transcription of the mouse *Cyp21* gene *in vivo* requires the presence of DNA sequences located between –5.3 and –6.0 kb (Milstone et al., 1992). Transgenic mice bearing the *lacZ* gene under the control of 6.4 kb of the 5'-flanking region of the *Cyp21* gene expressed *lacZ* gene in the adrenal cortex following a correct temporal pattern (Morley et al., 1996). The expression pattern in the adrenal cortex, however, is variegated, indicating that some important control elements are probably missing in this construct. Therefore more transgenic mouse studies are required to get a complete picture of control of steroidogenic gene expression.

IV. DISEASES DUE TO 21-HYDROXYLASE AND 17-HYDROXYLASE DEFICIENCY

P450c17 and P450c21 are important not only because of their roles in steroid hormone biosynthesis, but also because their deficiencies are often the cause of a common genetic disease: congenital adrenal hyperplasia (CAH). Inherited as an autosomal recessive disease, CAH is characterized by hyperplastic growth of the adrenal due to ACTH overstimulation as a result of lack of inhibition of ACTH synthesis by insufficient cortisol. Although deficiency of any of the enzymes in the pathway of cortisol synthesis will lead to CAH, more than 90% of the cases are due to 21-hydroxylase deficiency (Miller and Morel, 1989). 17α-Hydroxylase deficiency occurs at a much lower frequency, with over 120 reported cases (Yanase, 1995).

A. 21-Hydroxylase Deficiency

Clinical Form and Classification

21-Hydroxylase deficiency leads to deficient glucocorticoid and sometimes mineralocorticoid production, and results in accumulation of its precursors, mainly 17-OHP, which can be readily detected in the serum. Lack of glucocorticoid also shunts the reaction pathway to the synthesis of androgen (Figure 3). Overproduction of testosterone causes virilization (masculinization) of patients, even *in utero*. Patients are presented with different degrees of virilization depending on the severity of the disease.

CAH is usually classified as the classical type and the less severe nonclassical type according to the degree of virilization at birth (New and Levine, 1984; Miller and Levine, 1987). Classical CAH has a prevalence rate of 1 in 14,000 in the general population (Pang et al., 1988). Female patients suffer from improper differentiation of the outer genitalia, resulting in clitoris enlargement, and sometimes labile closure, causing problems in sex assignment. In male patients, although genitalia look normal, there is precocious appearance of secondary sexual characterics at the age of 3 or 4. All patients also grow faster than normal at young ages due to high levels of testosterone, but cease growing prematurely, reaching shorter final stature. These symptoms are characteristics of the simple virilizing (SV) type of CAH.

In about half of the classical CAH, patients also have a deficiency of mineralocorticoid, therefore can not conserve salt. In this salt wasting (SW) type of CAH, in addition to virilization, patients usually have to be hospitalized within the first week of life because of electrolyte imbalance.

(A) P450c21 Deficiency

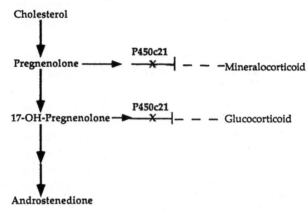

(B) P450c17 Deficiency

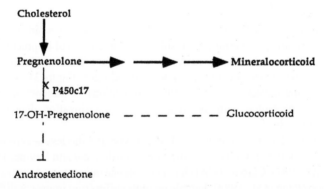

Figure 3. Pathways of steroid synthesis in 21- and 17α-hydroxylase deficiency. Thick arrows and bold letters indicate the major route of the steroid synthesis. Dashed lines indicate the lack of flow of steroids, and the X signs point to the block of synthesis.

The nonclassical (NC) type of CAH has a milder phenotype. Patients usually do not show symptoms until late in childhood or early puberty. The major complaints are menstrual problems, infertility, acne, hirsutism, and so forth. Serum 17-OHP levels are usually higher than normal upon ACTH stimulation. It also has a higher occurrence rate, with 1 in 27 for Ashkenazi Jews, and 1 in 1000 for other Caucasians (Speiser et al., 1985; Sherman et al., 1988).

Table 1. Mutations of the *CYP21B* Gene in Congenital Adrenal Hyperplasia

Site of Mutation	Location	Type
A. Frequent Mutations		
1. Pro-30 to Leu	Pro-30	NC
2. Aberrant Splicing	Intron 2	SV
3. 8 bp deletion	Exon 3	SW
4. Ile-172 to Asn	Ile-172	SV
5. Ile-236/Val-237/Met-239 to Asp/Glu/Lys	Exon 6	SW
6. Val-281 to Leu	Val-281	NC
7. T-Insertion	Exon 7	SW
8. Stop codon 318	Exon 8	SW
9. Arg-356 to Trp	Arg-356	SV
10. Gene deletion	whole *c21B* gene	SW
11. 4 bp substitutions at -94, -101, -104, and -117	Promoter	?
B. Other Rare Mutations		
1. Arg-339 to His		
2. Pro-453 to Ser		
3. Gly-292 to Ser		
4. Arg-484 frameshift (GG to C)		
5. C to T at 5'-untranslated region, Pro-106 to Leu		
6. Stop codon 406		
7. G to C at intron 7 splice donor site		

Source: The data were collected from Helmberg et al. (1992), Speiser et al. (1992), Wedell et al. (1992), Wedell and Luthman (1993), and Chang and Chung (1995).

Molecular Genetics of 21-hydroxylase Deficiency

As mentioned above, frequent recombination between the active *CYP21* gene and its neighboring pseudogene is the major cause of 21-hydroxylase deficiency. The most frequent mutations of the *CYP21A1* gene cover 9 sites plus complete gene deletion, all of which result from gene conversion events changing a small or a large region of the *c21B* gene into the *c21A* gene (Table 1). Among these, the most frequent mutations include aberrant splicing at intron 2 (26%), gene deletion (21%), an Ile-172 to Asn mutation (16%), and a Val-281 to Leu mutation (11%). Altogether, these readily identifiable mutations account for about 95% of the chromosomes being analyzed (Speiser et al., 1992). Rare mutations have also been identified (Table 1B). After sequencing more pseudogenes, it is apparent that the rare mutations are also found in a small percentage of the pseudogenes. Therefore rare mutations can accumulate in the pseudogene and spread to the active gene (Wedell and Luthman, 1993)

Comparing the type of mutations with clinical presentation, there is a general correlation. Gene deletions, frameshift mutations, and nonsense mutations will result in complete loss of enzyme production. These mutations are usually associated with the severe salt-wasting type of CAH (Table 1). Two missense mutations each substituting Arg-356 to Trp and Ile-172 to Asn result in enzymes with about 2% of normal activity. These mutations are associated with the simple virilizing type of CAH. The frequent intron 2 splicing mutation, which causes the production of aberrant transcript, also results in the production of low levels of 21-hydroxylase activity, and is associated with the SV type of CAH. The nonclassical type of CAH is associated with the Pro-30 to Leu and Val-281 to Leu mutations, which result in 10–20% of the wild type activity.

Higher than normal levels of serum 17-OHP and close linkage of the CYP21 locus to HLA have been used for diagnosis and for pedigree analysis. Using newly developed allele-specific polymerase chain amplification of the *c21B* gene, coupled to allele-specific oligonucleotide hybridization or sequencing, the *c21B* gene from patients can now be quickly analyzed and its gene defects identified (Wedell et al., 1992). Using these methods, successful prenatal diagnosis of CAH has been achieved. In a few instances, prenatal treatment has also been performed. In these cases, dexamethasone (a synthetic glucocorticoid) was administered to pregnant mothers starting in the first trimester (New, 1990; Dorr and Sippell, 1993). The external genitalia of the affected females treated this way appeared normal after birth.

In a mouse strain carrying a haplotype *aw18*, about 80 kb of the H-2 class III region spanning both the *C4* and *CYP21* genes were deleted, hence creating a mouse model for the 21-hydroxylase deficiency (Shiroishi et al., 1987; Gotoh et al., 1988). The homozygous *aw18* mice can not synthesize enough steroids and die right after birth. This will be a good mouse model for future gene therapy experiments employing transgenic techniques.

B. 17α-hydroxylase Deficiency

As P450c17 and P450c21 play different roles in the synthesis of steroids, their deficiencies show distinct clinical presentations. Deficiency of 17α-hydroxylase/17,20 lyase activities results in impaired synthesis of glucocorticoid and androgen. All these patients are phenotypically female as the lack of testosterone production causes the formation of female external genitalia in both 46XX and 46XY individuals. Patients are usually diagnosed at puberty when there is a lack of development of secondary sexual characteristics. On top of that, patients suffer from hypertension due to overproduction of mineralocorticoid as a result of a blockage in glucocorticoid synthesis (Figure 3).

The deficient *CYP17* gene has been characterized at the molecular level in more than 20 cases. The gene lesions are usually unique and probably result from spontaneous mutations. Five of these mutations, including nonsense mutations at codons 17, 194 and 239, and a frameshift mutation at codon 120, and a 518 bp deletion with a 469 bp insertion at exon 2, will cause premature termination of translation, and therefore absence of functional protein. Other mutations, including deletion of Phe at codon 53 or 54, six missense mutations (Tyr-64 to Ser, Gly-90 to Asp, Ser-106 to Pro, Pro-342 to Thr, His-373 to Leu, and Arg-496 to Cys), and seven mutations affecting the last exon of the gene would produce P450c17 with greatly reduced enzymatic activities of both 17α-hydroxylase and 17,20 lyase. Interestingly, about half of the mutations characterized so far are clustered in the last exon. These include one base deletion at codon 438, two nonsense mutations at codons 461 and 496, three missense mutation changing Arg-440 to His and Arg-496 into Cys, deletion of codons 487-489, and a GATC duplication at codon 480 forming a mutant P450c17 protein with a different C-terminal sequence. All these mutations result in P450c17 with little or no enzymatic activities (Oshiro et al., 1995; Yanase, 1995).

The treatment of 17α-hydroxylase deficiency is dexamethasone supplementation to suppress ACTH overproduction. Estrogen is given to stimulate the development of secondary sexual characteristics. Sexual hair growth can be stimulated by testosterone therapy.

V. STRUCTURE OF P450c21 AND P450c17

As mentioned above, P450c21 and P450c17 are important enzymes for steroid hormone synthesis, and their deficiencies are the causes of diseases of steroid imbalance. It is therefore desirable to understand the structure of these proteins to gain more understanding of their mode of catalysis. The three-dimensional structures of these two proteins have not been solved due to the difficulty of obtaining large quantities of pure proteins and crystallization. In fact, there are only two known P450 structures from P450cam and P450BM-3, both of which are of bacterial origin and soluble (Poulos et al., 1987; Ravichandran et al., 1993). None of the eukaryotic P450s which are membrane bound has known three dimensional structure. Yet extensive efforts have been made to understand the structures of eukaryotic P450s using existing technology.

A. P450 Folding and Membrane Topology

For many years, P450cam was the only protein with known three-dimensional structure. This protein is composed of 12 α-helices (A to L) and 5 β sheets forming a tight compact structure with a flat triangular shape. The heme is liganded to Cys-357 and sandwiched between the L and I helices. The L helix is located at the proximal side of the heme and the I helix at the distal side. Oxygen is bound to the active center adjacent to an invariable Thr-252 at the I helix during catalysis, while substrate enters the reaction pocket through the opposite direction. This structure has served as a model for the structures of other P450s, yet it is not ideal since P450cam accepts electrons from a donor, putidaredoxin, which is very different from the electron donor of the microsomal P450s, P450 reductase. The interaction of P450cam with putidaredoxin will be distinct from that of microsomal P450s with the P450 reductase. P450BM-3 is a fusion protein between the reductase and P450 moieties, and therefore will serve as a better model for predictions of the tertiary structure of microsomal P450s.

All P450s belong to the same protein superfamily with >15% sequence homology. Nelson and Strobel (1988) aligned 34 vertebrate cytochromes P450 and found 11 conserved hydrophobic regions which would maintain the same helical structure as P450cam. The N-terminal hydrophobic sequence has no homology with P450cam and would be the membrane-binding domain for these membrane-bound P450s. They proposed that the tertiary structure of vertebrate P450s would be similar to that of P450cam with the extension of an N-terminal membrane anchoring domain. Many lines of evidence also suggested that the bulk of the mammalian P450 molecules are located at the cytoplasmic side of the endoplasmic reticulum membrane with the N-terminal hydrophobic domain integrated into the membrane.

The first hydrophobic domain (amino acids 1–35) of P450c21 was shown to be the signal in directing and anchoring the entire protein to the endoplasmic reticulum membrane. This domain also appears to be important in protein folding as truncated P450c17 and P450c21 with the first hydrophobic domain deleted can no longer fold into correct conformation and are targeted to the membrane inefficiently (Clark and Waterman, 1991; Hsu et al., 1993). The same truncated P450l7, however, when expressed in *E. coli*, retained enzymatic activity, indicating that the folding pathways in *E. coli* and in mammalian cells are different (Sagara et al., 1993).

In addition to general overall homology, there are some local clusters of conserved sequences representing important functional domains. The I and L helices are the most conserved regions which together form the heme-binding domain with an invariant Cys in the L helix as the heme-binding ligand. Other conserved clusters are probably important for reductase binding, membrane binding, and substrate binding, as suggested by experiments to be described in the next section.

B.   Heterologous Protein Expression System

One major roadblock in the study of structure-function relationship of proteins is the scarcity of materials. The development of heterologous protein expression systems has alleviated this problem. P450c17 and P450c21 expression plasmids have been transfected into COS-1 cells and directed the expression of active 17α-hydroxylase/17,20 lyase and 21-hydroxylase respectively (Zuber et al., 1986; Hu et al., 1990). This heterologous expression system enables the analysis of various forms of P450s for function. Other mammalian expression systems have also been developed which employ different strategies such as the use of vaccinia virus as a vector (Tusie-Luna et al., 1990) or the establishment of a stable cell line which has the P450c21 cDNA stably integrated and amplified in the genome to ensure high levels of expression (Ricketts et al., 1992). Since both P450c17 and P450c21 are mammalian proteins, their expression in cultured cell lines will most likely produce proteins of identical conformation to that *in vivo*. Yet the expense and slow rate of cell culture make this method difficult for the production of enough proteins for structural studies. Baculovirus produces the most molecules of P450c17 per cell as compared to other expression systems and can be potentially very useful (Barnes et al., 1994).

Yeast cells have the advantage that being a eukaryotic microbe they grow fast and still have the eukaryotic features for protein expression. P450c17 and P450c21 were successfully expressed in yeast (Sakaki et al., 1989, 1990; Wu et al., 1991a). About 10^5 molecules of functional P450s per cell can be produced. This amount of P450 expression is comparable to that found in the adrenal tissue on a per cell basis, and is enough for spectrophotometric measurements for structural studies. This expression system is also useful for the bioconversion of steroid products with industrial applications. Yet the amount of protein produced is still not high enough for crystallization.

Bacteria have been tested for suitability for P450 expression. Earlier attempts resulted in the production of large quantities of proteins which are misfolded and insoluble (Hu and Chung, 1990). A number of modifications have been made including the change of the N-terminal sequence of the P450 to eliminate secondary structure and to simulate bacterial codon usage, lowering of growth temperature, and selection of suitable vectors and hosts. Functional P450c17 and P450c21 were finally expressed in *E. coli* in a reasonable quantity (Barnes et al., 1991; Guzov and Chung, unpublished data). Four histine residues have been added to the C-terminal end of human P450c17 molecule expressed in *E. coli* to facilitate the purification procedure. The resultant pure protein showed the same kinetic characteristics as that of wild type P450c17 (Imai et al., 1993). This pure protein will enable a number of structural and functional studies.

C. Site-Directed Mutagenesis

The availability of a heterologous expression system makes possible the study of protein function when coupled with site-directed mutagenesis. The elucidation of natural mutations associated with P450c17 and P450c21 deficiencies also provide hints to the identification of functionally important residues in the protein. Mutant proteins can exhibit complete loss, partial loss, or no loss of enzymatic activities. P450c21 with Ser-268 substituted by Thr, Cys, and Met still retains full activities. It indicates that Ser-268 is probably located at the surface of the protein and therefore its substitution does not affect its structure and function. It also proves that Ser-268 or Thr-268 sequence is a polymorphic site in the normal population. Mutations of Cys-428 of P450c21 completely eliminates enzymatic activity and P450 absorbance, demonstrating the importance of this heme ligand in heme-binding and catalysis. Ile-172 of P450c21 is a conserved residue which is invariant among all microsomal P450s. Mutation of this residue to Asn is frequent in the simple virilizing form of 21-hydroxylase deficiency. When expressed in yeast, the mutant protein has a reduced P450 content due to its inability to bind heme. It is also more sensitive to proteinase K digestion, indicating the altered conformation. All these lead to greatly reduced catalytic rate (Hsu et al., 1996). Yet most of the information comes from studies of mutant proteins with partial activities and presumably a local disturbance of structure.

Mutations of Val-281, a site associated with nonclassical CAH, lead to formation of mutant enzymes with 15–75% activity depending on the type of residue substituting it. Both P450 absorbance and activity decrease in the order of Val>Ile> Leu >Thr, indicating the importance of the hydroxyl group at the β-carbon of the amino acid side chain. This result, coupled with sequence alignment with other P450s, suggesting that Val-281 is located at the tip of Helix I, and affects the conformation of the heme-binding pocket (Wu and Chung, 1991b).

Many attempts have been made to swap domains of P450c17 and P450c21 with a similar domain from a slightly different protein. This type of experiments have largely been unsuccessful due to the failure to predict the structural changes of the chimeric proteins. Therefore the tertiary structure of a membrane-bound P450 is needed before further insight about the structure of P450s can be gained by mutagenesis.

D. Substrate Binding Site

Both P450c17 and P450c21 have more than one substrate. Computer modeling has not been able to depict the substrate binding sites of P450s in molecular detail, but suggests that the substrate binding pocket is composed of a big hydrophobic pocket, which allows loose binding of different substrates. Mutated P450c21 proteins lose enzymatic activities towards both substrates in parallel, suggesting that there is a single substrate binding pocket in P450c21. The substrate-binding pocket of P450c17, however, seemed to be a two-lobed region where the 17α-hydroxylase and the 17,20 lyase activity can be separated as suggested by computer modeling (Laughton et al., 1990) and the following experimental data.

Amino acids 345–367 of P450c17 (or amino acids 338–360 of P450c21) is in a conserved region which has been implicated to be the steroid-binding site (Picado-Leonard and Miller, 1988). The importance of this region is shown by the identification of the Arg-356 to Trp mutation of P450c21 in the simple virilizing type of CAH, which causes a 50 fold decrease in enzymatic activity (Chiou et al., 1990). Mutation of Arg-346 of P450c17 selectively abolished lyase activity but left the hydroxylase activity relatively intact. Likewise, mutation of Arg-357 affected the hydroxylase but not lyase activity. Therefore, within this putative substrate binding pocket, steroid-protein interaction changes during the course of hydroxylation and side-chain cleavage reactions (Kitamura et al., 1991).

Phe-343 of rat P450c17, which is at the putative steroid-binding site, has been shown to be required for the Δ4-lyase activity (Koh et al., 1993). When changed into Ile as normally found in human and bovine P450c17, the protein no longer uses Δ4-steroids for substrate, although the reaction with Δ5 steroids is still intact. This provides a molecular basis for the species difference in activities of P450c17 and offers a tool to study the enzyme-substrate interactions. Another amino acid which plays a major role in catalysis is Ser-106, which when replaced by Ile as in trout P450c17, had about 5% of the human P450c17 activities (Lin et al., 1993).

VI. SUMMARY

Using tools of molecular biology, we have gained knowledge on the function and structure of P450c17 and P450c21, two similar enzymes engaged in steroid hormone synthesis. Both enzymes share similar properties regarding their structure, intracellular location, substrate specificity, tissue distribution, and mode of regulation of gene expression. Yet small differences do exist. For example, P450c17 possesses two enzymatic activities in one protein molecule, rendering its enzymology more complex. In addition, slightly different functions (synthesis of sex hormones for P450c17 versus mineralocorticoid for P450c21) also require distinct expression pattern and mode of regulation for these two enzymes. The adrenal and gonads use different signaling pathways to regulate the expression of these two enzymes as described in Section II. One protein, SF1, which is a member of the orphan steroid receptor family, is essential for the transcriptional regulation of steroidogenic genes (Parker and Schimmer, 1993). Yet the molecular detail for the regulation of these two genes during development, by hormones, or for tissue-specific distribution is still lacking. More *in vivo* work also needs to be done to compliment the data obtained through cell culture studies.

The determination of the gene structures of P450c17 and P450c21 facilitate greatly the elucidation of the molecular bases of 17- and 21-hydroxylase deficiencies. With the identification of mutations involved in the defect, improved diagnosis method has been devised. When the mechanism of regulation of gene expression is fully understood, future gene therapy experiments perhaps may be designed using available mouse models. The understanding of the tertiary structure is another future goal which will enable the design of new therapeutic agents for the treatment of various diseases related to steroid metabolism.

ACKNOWLEDGMENT

This chapter was written when the author was on sabbatical leave in Dr. Keith Parker's laboratory. Studies carried out by the author and reported in this chapter were supported by NSC-820420B-001 109 M02 from National Science Council and Academia Sincica, Republic of China.

REFERENCES

Bakke, M., & Lund, J. (1992). A novel 3′,5′-cyclic adenosine monophosphate-responsive sequence in the bovine *CYP17* gene is a target of negative regulation by protein kinase C. Mol. Endocrinol. 6, 1323–1331.

Barnes, H.J., Arlotto, M.P., & Waterman, M.R. (1991). Expression and enzymatic activity of recombinant cytochrome P450 17α-hydroxylase in *Escherichia coli*. Proc. Natl. Acad. Sci USA 88, 5597–5601.

Barnes, H.J., Jenkins, C.M., & Waterman, M.R. (1994). Baculovirus expression of bovine cytochrome P450c17 in Sf9 cells and comparison with expression in yeast, mammalian cells, and E. coli. Arch. Biochem. Biophys. 315, 419–494.

Barrett, P.Q., Bollag, W.B., Isales, C.M., McCarthy, R.T., & Rasmussen, H. (1989). Role of calcium in angiotensin II-mediated aldosterone secretion. Endocr. Rev. 10, 496–518.

Bird, I.M., Magness, R.R., Mason, J.I., & Rainey, W.E. (1992). Angiotensin-II acts via the type 1 receptor to inhibit 17α-hydroxylase cytochrome P450 expression in ovine adrenocortical cells. Endocrinology 130, 3113–3121.

Brentano, S.T., Picado-Leonard, J., Mellon, S.H., Moore, C.C.D., & Miller, W.L. (1990). Tissue-specific, cyclic adenosine 3′,5′-monophosphate-induced, and phorbol ester-repressed transcription from the human P450c17 promoter in mouse cells. Mol. Endocrinol. 4, 1972–1979.

Bristow, J., Gitelman, S.E., Tee, M.K., & Miller, W.L. (1993a). Abundant adrenal-specific transcription of the human P450c21A "Pseudogene". J. Biol. Chem. 268, 12919–12924.

Bristow, J., Tee, M.K., Gitelman, S.E., Mellon, S.H., & Miller, W.L. (1993b). Tenascin-X: A novel extracellular matrix protein encoded by the human XB gene overlapping P450c21B. J. Cell Biol. 122, 265–278.

Chabre, O., Defaye, G., & Chambaz, E.M. (1993). Oxygen availability as a regulatory factor of androgen synthesis by adrenocortical cells. Endocrinology 132, 255–260.

Chang, S.-F., & Chung, B.-c. (1995). Difference in transcriptional activity of two homologous CYP12A genes. Mol. Endocrinol. 9, 1330–1336.

Chiou, S.-H., Hu, M.-C., & Chung, B.-c. (1990). A Missense mutation at Ile172→Asn or Arg356→Trp cause steroid 21-hydroxylase deficiency. J. Biol. Chem. 265, 3549–3552.

Clark, B.J., & Waterman, M.R. (1991). The hydrophobic amino-terminal sequence of bovine 17α-hydroxylase is required for the expression of a functional hemoprotein in COS 1 cells. J. Biol. Chem. 266, 5898–5904.

Dorr, H.G., & Sippell, W.G. (1993). Prenatal dexamethasone treatment in pregnancies at risk for congenital adrenal hyperplasia due to 21-hydroxylase deficiency: Effect on midgestational amniotic fluid steroid levels. J. Clin. Endocrinol. Metab. 76, 117–120.

Fan, Y.S., Sasi, R., Lee, C., Winter, J.S., Waterman, M.R., & Lin, C.C. (1992). Localization of the human *CYP17* gene (cytochrome P450(17 alpha)) to 10q24.3 by fluorescence *in situ* hybridization and simultaneous chromosome banding. Genomics 14, 1110–1111.

Gitelman, S.E., Bristow, J., & Miller, W.L. (1992). Mechanism and consequences of the duplication of the human C4/P450c21/gene X locus. Mol. Cell. Biol. 12, 1237–1249.

Givens, C.R., Zhang, P., Bair, S.R., & Mellon, S.H. (1994). Transcriptional regulation of rat cytochrome P450c17 expression in mouse Leydig MA-10 and adrenal Y-1 cells: Identification of a single protein that mediates both basal and cAMP-induced activities. DNA Cell Biol. 13, 1087–1098.

Gotoh, H., Sagai, T., Hata, J.-I., Shiroishi, T., & Moriwaki, K. (1988). Steroid 21-hydroxylase deficiency in mice. Endocrinology, 123, 1923–1927.

Hales, D.B., Sha, L., & Payne, A.H. (1987). Testosterone inhibits cAMP-induced *de novo* synthesis of Leydig cell cytochrome P-45017α by an androgen receptor-mediated mechanism. J. Biol. Chem. 262, 11200–11206.

Hales, D.B., Sha, L., & Payne, A.H. (1990). Glucocorticoid and cyclic adenosine 3′5′-monophosphate-mediated induction of cholesterol side-chain cleavage cytochrome P450 (P450scc) in MA-10 tumor Leydig cells. Increases in mRNA are cycloheximide sensitive. Endocrinology 126, 2800–2808.

Helmberg, A., Tusie-Luna, M.-T., Tabarelli, M., Kofler, R., & White, P.C. (1992). R339H and P453S: CYP21 mutations associated with nonclassic steroid 21-hydroxylase deficiency that are not apparent gene conversions. Mol. Endocrinol. 6, 1318–1322.

Hornsby, P.J. (1989). Regulation of steroid 17α-hydroxylase in adrenocortical cells in culture. Endocr. Res. 12, 159–182.

Hsu, L.-C., Hu, M.-C., Cheng, H.-C., Lu, J.-C., & Chung, B.-c. (1993). The N-terminal hydrophobic domain of P450c21 is required for membrane insertion and enzyme stability. J. Biol. Chem. 268, 14682–14686.

Hsu, L.-C., Hsu, N.-C., Guzova, J.A., Guzov, V.M., Chang, S.-F., & Chung, B.-c. (1996). The common I172N Mutation causes conformational change of cytochrome P450c21 revealed by systematic mutation, kinetic, and structural studies of P450c21. J. Biol. Chem. 271, 3306–3310.

Hu, M.-C. & Chung, B.-c. (1990). Expression of human 21-hydroxylase (P450c21) in bacterial and mammalian cells: A system to characterize normal and mutant enzymes. Mol. Endocrinol. 4, 893–898.

Ilvesmäki, V., & Voutilainen, R. (1991). Interaction of phorbol ester and adrenocorticotropin in the regulation of steroidogenic *P450* genes in human fetal and adult adrenal cell cultures. Endocrinology 128, 1450–1458.

Imai, T., Globerman, H., Gertner, J.M., Kagawa, N., & Waterman, M.R. (1993). Expression and purification of functional human 17α-hydroxylase/17,20-lyase (P450c17) in *Escherichia coli*, use of this system for study of a novel form of combined 17α-hydroxylase/17,20-lyase deficiency. J. Biol. Chem. 268, 19681–19689.

Jefcoate, C.R., DiBartolomeis, M.J., Williams, C.A., & McNamara, B.C. (1987). ACTH regulation of cholesterol movement in isolated adrenal cells. J. Steroid Biochem. 27, 721–729.

John, M.E., John, M.C., Boggaram, V., Simpson, E.R., & Waterman, M.R. (1986). Transcriptional regulation of steroid hydroxylase genes by corticotropin. Proc. Natl. Acad. Sci. USA 83, 4715–4719.

Kagawa, N., & Waterman, M.R. (1991). Evidence that an adrenal-specific nuclear protein regulates the cAMP-responsiveness of the human *CYP21B* (P450c21) gene. J. Biol. Chem. 266, 11199–11204.

Katagiri, M., Kagawa, N., & Waterman, M.R. (1995). The role of cytochrome b5 in the biosynthesis of androgens by human P450c17. Arch Biochem. Biophys. 317, 343–347.

Kitamura, M., Buczko, E., & Dufau, M.L. (1991). Dissociation of hydroxylase and lyase activities by site-directed mutagenesis of the rat $P450_{17\alpha}$. Mol. Endocrinol. 5, 1373–1380.

Koh, Y.C., Buczko, E., & Dufau, M.L. (1993). Requirement of phenylalanine 343 for the preferential Δ^4-lyase *versus* Δ^5-lyase activity of rat CYP17. J. Biol. Chem. 268, 18267–18271.

Kramer, R.E., Rainey, W.E., Funkenstein, B., Dee, A., Simpson, E.R., & Waterman, M.R. (1984). Induction of synthesis of mitochondrial steroidogenic enzymes of bovine adrenocortical cells by analogs of cyclic AMP. J. Biol. Chem. 259, 707–713.

Kronbach, T., Kemper, B., & Johnson, E.F. (1991). A hypervariable region of P450IIC5 confers progesterone 21-hydroxylase activity to P450IIC1. Biochemistry 30, 6097–6102.

Lala, D.S., Rice, D.A., & Parker. K.L. (1992). Steroidogenic factor 1, a key regulator of steroidogenic enzyme expression, is the mouse homolog of *fushi tarazu*-factor I. Mol. Endocrinol. 6, 1249–1258.

Laughton, G.A., Neidle, S., Zvelebil, M.J.M., & Sternberg, M.J.E. (1990). A molecular model for the enzyme cytochrome $P450_{17\alpha}$, a major target for the chemotherapy of prostatic cancer. Biochem. Biophys. Res. Commun. 171, 1160–1167.

LeHoux, J.-G., Mason, J.I., & Ducharme, L. (1992). *In vivo* effects of adrenocorticotropin on hamster adrenal steroidogenic enzymes. Endocrinology 131, 1874–1882.

Lin, D., Black, S.M., Nagahama, Y., & Miller, W.L. (1993). Steroid 17α-hydroxylase and 17,20-lyase activities of P450c17: Contributions of serine[106] and P450 reductase. Endocrinology 132, 2498–2506.

Lund, J., Ahlgren, R., Wu, D., Kagimoto, M., Simpson, E.R., & Waterman, M.R. (1990). Transcriptional regulation of the bovine CYP17 (P-450$_{17\alpha}$) gene: Identification of two cAMP regulatory regions lacking the consensus cAMP-responsive element (CRE). J. Biol. Chem. 265, 3304–3312.

Magoffin, D.A., & Weitsman, S.R. (1993). Differentiation of ovarian theca-interstitial cells *in vitro*: Regulation of 17α-hydroxylase messenger ribonucleic acid expression by leuteinizing hormone and insulin-like growth factor-I. Endocrinology 132, 1945–1951.

Matteson, K.J., Picado-Leonard, J., Chung, B.-c., Mohandas, T.K., & Miller, W.L. (1986). Assignment of the gene for adrenal P450c17 (steroid 17α-hydroxylase/17,20 lyase) to human chromosome 10. J. Clin. Endocrinol. Metab. 63, 789–791.

Mellon, S.H., & Miller, W.L. (1989). Extraadrenal steroid 21-hydroxylation is not mediated by P450c21. J. Clin. Invest. 84, 1497–1502.

Miller, W.L., & Levine, L.S. (1987). Molecular and clinical advances in congenital adrenal hyperplasia. J. Pediatric. 111, 1–17.

Miller, W.L. (1988). Molecular biology of steroid hormones synthesis. Endocr. Rev. 9, 295–318.

Miller, W.L., & Morel, Y. (1989). The molecular genetics of 21-hydroxylase deficiency. Annu. Rev. Genet. 23, 371–393.

Milstone, D.S., Shaw, S.K., Parker, K.L., Szyf, M., & Seidman, J.G. (1992). An element regulating adrenal-specific steroid 21-hydroxylase expression is located within the *slp* gene. J. Biol. Chem. 267, 21924–21927.

Morley, S.C., Viard, I., Chung, B.-c., Ikeda, Y., Parker, K.L., & Mullins, J.J. (1996). Variegated expression of a mouse steroid 21-hydroxylase/β-galactosidase transgene suggests centripetal migration of adrenocorticol cells. Mol. Endocrinol. 10, in press.

Nakajin, S., & Hall, P.F. (1981). Microsomal cytochrome P-450 from neonatal pig testis. Purification and properties of a C_{21} steroid side-chain cleavage system (17α-hydroxylase-$C_{17,20}$ lyase). J. Biol. Chem. 156, 3871–3876.

Nelson, D.R., & Strobel, H.W. (1988). On the membrane topology of vertebrate cytochrome P-450 proteins. J. Biol. Chem. 263, 6038–6050.

New, M.I. (1990). Prenatal diagnosis and treatment of adrenogenital syndrome (steroid 21-hydroxylase deficiency). Dev. Pharm. Therap. 15, 200–210.

Oshiro, C., Takasu, N., Wakugami, T., Komiya, I., Yamada, T., Eguchi, Y., & Takei, H. (1995). Seventeen alpha-hydroxylase deficiency with one base pair deletion of the cytochrome P450c17 (*CYP17*) gene. J. Clin. Endocrinol. Metab. 80, 2526–2529.

Pang, S., Wallace, M.A., Hofman, L., Thuline, H.C., Dorche, C., Lyon, I.C.T., Dobbins, R.H., Kling, S., Fujieda, K., & Suwa, S. (1988). Worldwide experience in newborn screening for classical congenital adrenal hyperplasia due to 21-hydroxylase deficiency. Pediatrics 81, 866–874.

Parker, K.L., & Schimmer, B.P. (1993). Transcriptional regulation of the adrenal steroidogenic enzymes. Trends Endocrinol. Metab. 4, 46–50.

Payne, A.H. (1990). Hormonal regulation of cytochrome P450 enzymes, cholesterol side-chain cleavage and 17α-hydroxylase/C_{17-20} lyase in Leydig cells. Biol. Reproduction 42, 399–404.

Picado-Leonard, J., & Miller, W.L. (1987). Cloning and sequence of the human gene for P450c17 (steroid 17 alpha-hydroxylase/17,20 lyase): Similarity with the gene for P450c21. DNA 6, 439–648.

Picado-Leonard, J., & Miller, W.L. (1988). Homologous sequences in steroidogenic enzymes, steroid receptors and a steroid binding protein suggest a consensus steroid-binding sequence. Mol. Endocrinol. 2, 1145–1150.

Poulos, T.L., Finzel, B.C., & Howar, A.J. (1987). High-resolution crystal structure of cytochrome P450cam. J. Mol. Biol. 195, 687–700.

Quinn, S.J., & Williams, G.H. (1988). Regulation of aldosterone secretion. Ann. Rev. Physiol. 50, 409–426.

Ravichandran, K.G., Boddupalli, S.S., Hasemann, C.A., Peterson, J.A., & Deisenhofer, J. (1993). Crystal structure of hemoprotein domain of P450BM-3, a prototype for microsomal P450's. Science 261, 731–736.

Rice, D.A., Kronenberg, M.S., Mouw, A.R., Aitken, L.D., Franklin, A., Schimmer, B.P., & Parker, K.L. (1990). Multiple regulatory elements determine adrenocortical expression of steroid 21-hydroxylase. J. Biol. Chem. 265, 8052–8058.

Ricketts, M., Chiao, E., Hu, M.-C., & Chung, B.-C. (1992). Amplification of P450c21 expression in cultured mammalian cells. Biochem. Biophys. Res. Commun. 186, 426–431.

Sagara, Y., Narnes, H.J., & Waterman, M.R. (1993). Expression in *Escherichia coli* of functional cytochrome $P450_{c17}$ lacking its hydrophobic amino-terminal signal anchor. Arch. Biochem. Biophys. 304, 272–278.

Sakaki, T., Shibata, M., Yabusaki, Y., Murakami, H., & Ohkawa, H. (1989). Expression of bovine cytochrome P450c17 cDNA in *Saccharomyces cerevisiae*. DNA 8, 409–418.

Sakaki, T., Shibata, M., Yabusaki, Y., Muarkami, H., & Ohkawa, H. (1990). Expression of bovine cytochrome P450c21 and its fused enzymes with yeast NADPH-cytochrome P450 reductase in *Saccharomyces cerevisiae*. DNA Cell Biol. 9, 603–614.

Sherman, S.L., Aston, C.E., Morton, N.E., Speiser, P.W., & New, M.I. (1988). A segregation and linkage study of classical and nonclassical 21-hydroxylase deficiency. Am. J. Hum. Genet. 42, 830–838.

Shiroishi, T., Sagai, T., Natsuume-Sakai, S., & Moriwaki, K. (1987). Lethal deletion of the complement component C4 and steroid 21-hydroxylase genes in the mouse *H-2* class III region, caused by meiotic recombination. Proc. Natl. Acad. Sci. USA 84, 2819–2823.

Simpson, E.R., & Waterman, M.R. (1988). Regulation of the synthesis of steroidogenic enzymes in adrenal cortical cells by ACTH. Ann. Rev. Physiol. 50, 427–440.

Speiser, P.W., Dupont, B., Rubenstein, P., Piazza, A., Kastelan, A., & New, M.I. (1985). High frequency of nonclassical steroid 21-hydroxylase deficiency. Am. J. Hum. Genet. 37, 650–667.

Speiser, P.W., Dupont, J., Zhu, D., Buegeleisen, M., Tusie-Luna, M.-T., Lesser, M., New, M.I., & White, P.C. (1992). Disease expression and molecular genotype in congenital adrenal hyperplasia due to 21-hydroxylase deficiency. J. Clin. Invest. 90, 584–595.

Tusie-Luna, M.-T., Traktman, P., & White, P.C. (1990). Determination of functional effects of mutations in the steroid 21-hydroxylase gene (CYP21) using recombinant vaccinia virus. J. Biol. Chem. 265, 20916–20922.

Voss, A.K., & Fortune, J.E. (1993). Levels of messenger ribonucleic acid for cytochrome P450 17α-hydroxylase and P450 aromatase in preovulatory bovine follicles decrease after the luteinizing hormone surge. Endocrinology 132, 2239–2245.

Wedell, A., Ritzén, E.M., Haglund-Stengler, B., & Luthman, H. (1992). Steroid 21-hydroxylase deficiency: Three additional mutated alleles and establishment of phenotype-genotype relationships of common mutations. Proc. Natl. Acad. Sci. USA 89, 7232–7236.

Wedell, A., & Luthman, H. (1993). Steroid 21-hydroxylase (P450c21): A new allele and spread of mutations through the pseudogene. Human Genetics. 91, 236–240.

Wu, D.-A., Hu, M.-C., & Chung, B.-C. (1991a). Expression and functional study of wild-type and mutant human cytochrome P450c21 in *Saccharomyces cerevisiae*. DNA Cell Biol. 10, 201–209.

Wu, D.-A., & Chung, B.-C. (1991b). Mutations of $P450_{c21}$ (steroid 21-hydroxylase) at Cys 428, Val 281, and Ser 268 result in complete, partial, or no loss of enzymatic activity, respectively. J. Clin. Invest. 88, 519–523.

Zuber, M.X., Simpson, E.R., Hall, P.F., & Waterman, M.R. (1985). Effects of adrenocorticotropin on 17α-hydroxylase activity and cytochrome $P-450_{17\alpha}$ synthesis in bovine adrenocortical cell. J. Biol. Chem. 260, 1842–1848.

Zuber, M., Simpson, E.R., & Waterman, M.R. (1986). Expression of bovine 17α-hydroxylase cytochrome P-450 cDNA in nonsteroidogenic (COS1) cells. Science 234, 1258–1261.

THE AROMATASE REACTION

Evan R. Simpson, Mala S. Mahendroo,
Michael W. Kilgore, Gary D. Means, Serdar E. Bulun,
Margaret M. Hinshelwood, and Carole R. Mendelson

I. INTRODUCTION

The biosynthesis of estrogens occurs throughout the entire vertebrate phylum including mammals, birds, reptiles, amphibians, teleost and elasmobranch fish, and Agnatha (hagfish and lampreys) (Callard, et al., 1990; Callard et al., 1980; Callard,

Advances in Molecular and Cell Biology
Volume 14, pages 225–244.
Copyright © 1996 by JAI Press Inc.
All rights of reproduction in any form reserved.
ISBN: 0-7623-0113-9

1981). It has also been described in the protochordate, Amphioxus (Callard et al., 1984). In most vertebrate species that have been examined, aromatase expression occurs exclusively in the gonads and in the brain. This is true of the fish and avian species that have been examined as well as most mammals such as rodents. In many species estrogen biosynthesis in the brain has been implicated in sex-related behavior such as mating responses, and frequently a marked sexually dimorphic difference has been demonstrated. This is true for example in species of bird in which the song of the male is important in courtship behavior (Hutchinson, 1991). In the case of humans and a number of higher primates, there is a more extensive tissue distribution of estrogen biosynthesis, since this also occurs in the placenta of the developing fetus as well as in the adipose tissue of the adult (Killinger et al., 1987). The ability of the placenta to synthesize estrogen is also the property of a number of ungulate species such as cows, pigs, and horses. However, at least in cattle there is no evidence of estrogen biosynthetic capacity in adipose, whereas in rodent species such as rat and mice, as well as rabbits, neither adipose nor placenta have any ability to synthesize estrogens.

The estrogen produced in each tissue site of biosynthesis is quite specific. For example, the human ovary synthesizes primarily estradiol, whereas the placenta synthesizes estriol and adipose synthesizes estrone. This appears to reflect primarily the nature of the C_{19} steroid presented to the estrogen synthesizing enzyme complex in each tissue site. Thus, in the case of adipose, the principal source of substrate is circulating androstenedione produced by the adrenal cortex. In the case of the placenta, the principle substrate is 16α-hydroxydehydroisoandrosterone sulfate derived by the combined activities of the fetal adrenal and fetal liver. The physiological significance of estrogen biosynthesis in the placenta and adipose of humans is unclear at this time. Although the human placenta produces very large quantities of estrogen, particularly estriol, its role is not understood. This pertains because in pregnancies characterized by placental sulfatase deficiency, the placenta is essentially deprived of C_{19} substrate and hence synthesizes, relatively speaking, minute quantities of estrogen, yet such pregnancies are relatively uncomplicated (France and Liggins, 1969). At most, parturition is delayed by several days. Similarly, at this time no physiological significance has been attributed to estrogen biosynthesis by human adipose; however the latter has been implicated in a number of pathophysiological conditions. Estrogen biosynthesis by adipose not only increases as a function of body weight but as a function of age (Hemsell et al., 1974; Edman et al., 1976), and has been correlated directly with the incidence of endometrial cancer as well as with post-menopausal breast cancer. Furthermore, evidence is accumulating to suggest that the estrogen which is implicated in the development of breast cancer is that which is produced locally within the adipose tissue of the breast itself (Miller and O'Neill, 1987). On the other hand, estrogen biosynthesis in adipose may have beneficial consequences since osteoporosis is more common in small, thin women than in large, obese women. While this may be, in part, the consequence of the bones of the latter being subject to load-bearing exercise, nonetheless it seems likely that the increased production of estrogens by the adipose of obese women is a significant factor.

II. STOICHIOMETRY OF ESTROGEN BIOSYNTHESIS

The biosynthesis of estrogens is catalyzed by an enzyme complex named aromatase (Fig. 1), comprised of aromatase cytochrome P450 (P450arom, the product of the *CYP19* gene) (Thompson and Siiteri, 1974; Mendelson et al., 1985; Nakajin et al., 1986; Kellis and Vickery, 1987, Osawa et al., 1987; Nelson et al., 1993; Simmons et al., 1985) which catalyzes the series of reactions leading to formation of the phenolic A ring characteristic of estrogens. Cytochrome P450arom is presently the sole member of gene family 19, designated CYP19. This designation is based on the fact that the C_{19} angular methyl group is the site of attack by oxygen. The aromatase reaction apparently utilizes three moles of oxygen and three moles of NADPH for every mole of C_{19} steroid metabolized (Thompson and Siiteri, 1974) (Fig. 2). Evidence is accruing that all three oxygen molecules are utilized in the oxidation of the C_{19} angular methyl group to formic acid, which occurs concomitantly with the aromatization of the A ring to give the phenolic structure characteristic of estrogens (Cole and Robinson, 1988; Akhtar et al., 1982). This is the only reaction in vertebrates capable of introducing an aromatic ring into a molecule.

III. STRUCTURAL FEATURES OF CYTOCHROME P450

Understanding the relationship of structure to function in cytochrome P450arom is of great interest because of the novelty of the complex series of reactions involved. cDNA inserts complementary to messenger RNA encoding cytochrome P450 have been isolated and characterized from a broad range of vertebrates including two species of fish, namely rainbow trout (Tanaka et al., 1992) and goldfish (GV Callard, personal communication), chicken (McPhaul et al., 1988); and several

$$\text{NADPH} \longrightarrow \text{Reductase} \longrightarrow \text{CYT P-450}_{\text{arom.}} \left(\begin{array}{c} \text{Androgen} \\ + O_2 \\ \\ \downarrow \\ \text{Estrogen} \\ + H_2O \end{array} \right.$$

Figure 1. Diagram of the aromatase enzyme complex. Cytochrome P450arom is responsible for binding the C19 steroid substrate and modifying its A ring to the phenolic ring characteristic of estrogens. NADPH-cytochrome P450 reductase, a ubiquitous protein of the endoplasmic reticulum, is responsible for transferring reducing equivalents from NADPH to any microsomal form of cytochrome P450 with which it comes into contact.

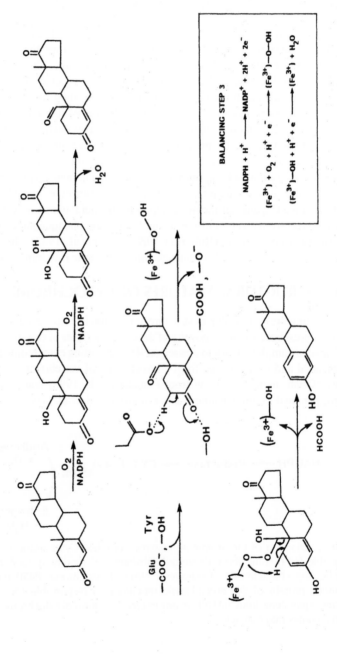

Figure 2. Model of the aromatase reaction. This is based on the proposals of Cole and Robinson (1988), and Akhtar et al. (1982). Reprinted from Graham-Lorence, 1991, with permission.

mammalian species including rat (Hickey et al., 1990), mouse (Terashima et al., 1991), human (Corbin et al., 1988; Harada, 1988; Toda et al., 1989), and most recently bovine, as indicated in Figure 3. The species showing greatest homology to the human is the bovine with an amino acid sequence identity of 86%. The identities of the derived amino acid sequences of P450arom from rat, mouse, chicken, and trout to the human are 77%, 81%, 73%, and 52% respectively. Interestingly, the variation between the two species of teliost fish, namely goldfish and trout (G.V. Callard, personal communication), is almost as great as that between the trout and human, indicating the great range of evolutionary diversity that has occurred in teliost fish since their origins.

Cytochrome P450arom shares a number of structural features common to all cytochrome P450 isoforms. Upstream of the heme-binding domain there is a region of over 20 amino acids which is totally conserved in all P450arom species from chicken to human, except that in the bovine there is the I399L replacement, a conservative change. Upstream of this there is another region of high conservation, namely the portion of the I-helix which is believed to form the substrate-binding pocket proximal to the heme prosthetic group (Graham-Lorence et al., 1991).

In a number of microsomal cytochrome P450 isoforms, the amino-terminus is characterized by a region of hydrophobic amino acids believed to comprise a membrane-anchoring domain. Curiously, in the case of P450arom, this is not apparent, since it has an extension of 20 amino acids with no obvious hydrophobic or amphipathic stretches. Rather, an N-linked glycosylation consensus sequence is present (NIT in the human), and P450arom of human placenta apparently does have attached sugar residues (Sethumadhewan et al., 1991). Instead, the region which can be described as hydrophobic lies between amino acids 20 and 40. In this context it is worth recalling that attempts over 20 years ago to define the subcellular localization of estrogen biosynthetic activity inevitably led to the conclusion that this distribution was bimodal, activity occurring both in membranes sedimenting with the microsomal fraction, as well as those sedimenting with the mitochondrial fraction, as defined by differential centrifugation. The presence of an apparent N-terminal extension containing a glycosylation consensus sequence in most species of P450arom is suggestive that at least some P450arom is present in the plasma membrane. Even more curiously, the cytochrome P450arom clone derived from the rainbow trout has a further 20 amino acid extension at the N-terminus in addition to the extra sequence which is common to all P450arom cDNAS (Tanaka et al., 1992). The importance of this further extension is also unclear at this time.

Comparison of the sequence of the P450arom cDNAS with those of other members of the cytochrome P450 superfamily has led to the conclusion that P450arom is only distantly related to other steroidogenic forms of P450 and indeed is one of the most ancient of the cytochrome P450 lineages, apparently evolving more than 1,000 Myr ago (Nelson et al., 1993). Exactly what the ancestral gene product was doing in those ancient times is unclear; presumably however it was not involved in the synthesis of estrogens. To our knowledge, aromatase has not

Trout   MDLLSPVCGRVMAVVCLDT

Bovine   MLLEVLNPRHYNVTSMVSEVVPIASIAILLLTGFLLLVWNYEDTSSIPGPSYFLGIGPLISHCRFLWMGI 70
Human   *V**M***I***I**I*P*AM*A*TMPV*****LF*******G*********G*CM********G*******
Rat   *F**M***M*****I**P*T**VSAMPL**IM*L***IR*C*SS******G*C*********G*******
Mouse   *F**N***MQ****I**P*T*TVSAMPL**IM*L***I**C*SS******G**********G*******
Chicken   *IP*T***LN*F-**L*PDLM*V*TVP*II*IC**F*I**H*E*******G*CM********G******V
Trout   VIADL*VSESR*A*ATR**GISL*TGSL***LCL**AA*RHT*NN*V***FFC**V***L*YL**I*T**
                                                                            I

Bovine   GSACNYYNKMYGEFMRVVVCGEETLIISKSSSMFHVMKHSHYISRFGSKLGLQFIGMHEKGIIFNNNPAL 140
Human   ********RV********IS***************I***N**S**********C***********E*
Rat   ****************I*S***********V*****N*********R***C******N*******S*
Mouse   ****************I*S**********************R***C******N*******S*
Chicken   *N********T****V***IS****F******V******UN*V**********C***Y*N*********H
Trout   *T*S****SK**DIV***IN****F*L*S**AVH**LRQGR*T******Q**SC***D*R*****S*M**

Bovine   WKAVRPFFTKALSGPGLVRMVTICADSITKHLDRLEEVCNDL-GYVDVLTLMRRIMLDTSNMLFLGIPLD 210
Human   **TT*****M*********V*E*LKT********T*ES-********L*V******T***R****
Rat   *RT*****M***T****I***EV*VE***KQ*****GD*TDNS-*****V*****H********T*******
Mouse   *RTI****M**T*********EV*VE***KQ****G**TDTS-***********H*************
Chicken   **EI****************IA**VE*TIV***K****TTEV-*N*N**N***********K****V***
Trout   **KT*TY*A***T****QKT*DV*VS*TQT***A*QGPDGLMG*Q****S*LRCTVV*I**R*****V**N

Bovine   ESAIVVNIQQYFDAWQALLLKPDIFFKISWLCRKYEKSVKDLKDAMEILIAEKRHRISTAEKLEDSIDFA 280
Human   ******K***********I*********YK************I*V******C****E****ECM***
Rat   **S**KK*****N******I**N*******Y****R*******EI***VEK**QKV*S*******M***
Mouse   *****KK*****N******I*N********Y*****Y*****EIAV*VEK***KV********CM***
Chicken   *****LK**N*****************K***EAA***G*****EQ**QKL**V***DEHM***
Trout   *KELLQK**K***T**TV*I***VY**LD*IHE*HRRAAQE*E**I*S*VDQ**RGLQE*D**DHIN-*T

Bovine   TELIFAEKRGELTRENVNQCILEMLIAAPDTMSVSVFFMLFLIAKHPQVEEAIIREIQTVVGERDIRIDD 350
Human   ****L****D************************L**********N******K****I]****K****
Rat   *******R**D**K****************************TLYV**L***EY*E**T**LK**H****D*****G*
Mouse   *D****R**D**K****************************TLY***L*V*EY*E**A*LK*H****D***K*E*
Chicken   SQ****QN**D**A******V*****H*******L**TL*I**I***DD*T***KMM***E**M*D*EVQS**
Trout   AD****QSH***SA****R**V****V*****L*I*L****L*LKQN*D**LQLLE**D*AI*D*ELHNS*
                                                II

Bovine   MQKLKVVENFINESMRYQPVVDLVMRKALEDDVIDGYPVKKGTNIILNLGRMHRLEFFPKPNEFTLENFA 420
Human   I*****M***Y**********************I*************I******************
Rat   V*N**************L**********R**********I*****Y***********E
Mouse   I*N*********************************I******Y***********E
Chicken   *PN**I****Y**********I****Q*********I****K*********S****E
Trout   L*N*R*L*S*****L*FH****FT**R**S****S**R*P*********H*****S***L*****S*D**E

Bovine   KNVPYRYFQPFGFGPRGCAGKYIAMVMMKVVLVTLLRRFHVQTLQGRCVEKMQKKNDLSLHP-DETRDRL 490
Human   ***************************AI*********K****Q***SI**IH******-***KNM*
Rat   ****************S**********************K****K***K**I*N*P*N*****L-**DSPIV
Mouse   **********************Q*K***K**I*NIP*********-N*D*HLV
Chicken   ****S************V*F******AI*******CR***MK**GLNNI**N*****M**-I*RQPL*
Trout   **I*N*F*****S***S*V**H*******SI******S*YS*CPHE*LTLDCLPQT*N**QQ*VE*EGEPH
                      III

Bovine   EMIFTPRNSDKCLER 505
Human   ***********R***H
Rat   *I**RHIFNTPF*QCLYISL
Mouse   *I**S******Y*QQ
Chicken   **V****SPN*NQSD
Trout   T*K*L**HQARKQS

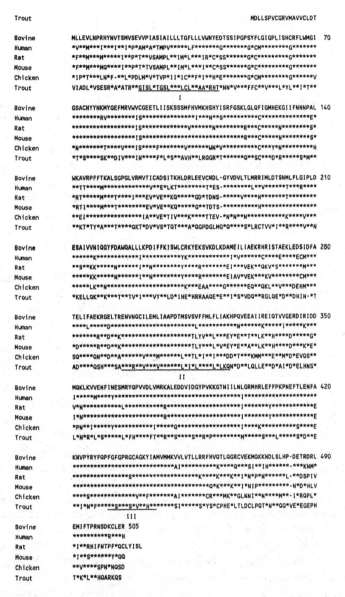

Figure 3. Sequence homologies of P450arom isoforms from a number of species including bovine, human (Corbin et al., 1988), rat (Hickey et al., 1990), mouse (Terashima et al., 1991), chicken (McPhaul et al., 1988), and rainbow trout (Tanaka et al., 1992). Regions of high homology, such as the I-helix (I), an aromatase-specific conserved region (II), and the heme-binding region (III), are indicated.

been described in invertebrates. It would be of considerable interest therefore to know what reactions the ancestral gene product catalyzes in non-vertebrate phyla.

IV. REGULATION OF *CYP19* GENE EXPRESSION IN HUMAN PLACENTA

Screening of clones isolated from human genomic libraries revealed that the open reading frame of the human *CYP19* gene comprises 9 exons, in common with other cytochrome P450 isoforms which generally have between 8 and 10 coding exons (Means et al., 1989; Harada et al., 1990; Toda et al., 1990), (Fig. 4). Also in common with most P450 species, the heme-binding region is located on the last coding exon. Additionally, in the case of human CYP19, there are two polyadenylation sites in the last coding exon downstream from the terminating stop codon. These give rise to the two species of transcript of 3.4 and 2.9 kb which encode human P450arom. Sequences approximately 140 bp upstream of the translation initiation site have homology to TATA and CAAT boxes. However, primer extension analysis using an oligonucleotide primer corresponding to the 5' end of the first coding exon together with polyA+ RNA isolated from human placenta failed to reveal a transcription initiation site that corresponded to the use of these promoter elements. Furthermore, an oligonucleotide corresponding to sequence beginning 39 bp downstream of the putative TATA box failed to hybridize to placental messenger RNA in northern analysis (Means et al., 1989; Toda et al., 1990).

The reason for this became apparent when we sequenced two P450arom cDNA inserts isolated from a primer-extended human placental cDNA library that we had previously constructed. These contained two distinct 5'-termini which were identical downstream of a point of common divergence which was 36 bp 5' of the translation initiation site. Sequences corresponding to these different 5'-termini were then used to screen the human genomic libraries and two other clones were isolated that contained the two alternative exons expressed in the human placental CYP19 mRNA. These we have named untranslated exons I.1 and I.2 (Fig. 4) (Means et al., 1989; Means et al., 1991; Kilgore et al., 1992). Hence we number the coding exons II through X. We have determined that the 5'-termini of P450arom mRNA transcripts in placenta are derived from the use of these two alternative untranslated exons.

When an oligonucleotide corresponding to a portion of exon I.1 was used in primer extension analysis of placental polyA+ RNA, a product was formed that indicated a transcription start site 23 bp downstream of a putative TATA box 5' of exon I.1 (Means et al., 1991). In addition, an oligonucleotide corresponding to exon I.1 hybridized to the same mRNA species in human placental RNA as those hybridizing to probes corresponding to coding exons. In the placenta, almost all the CYP19 transcripts include sequences encoded by exon I.1. Only a minor proportion (<1%) contain sequences encoded by I.2 (Kilgore et al., 1992). Whereas exon I.2 lies 9 kb upstream of exon II, exon I.1 lies at least 35 kb upstream of exon 11, which

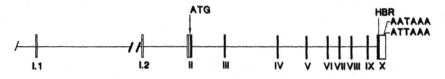

Figure 4. Schematic representation of the human *P450* gene, based on sequences transcribed in ovary and placenta. The closed bars represent translated sequences. The septum in the open bar in exon II represents the splice junction for exons I.1 and I.2, sequences to the left of the septum would be present in mature RNA only when a putative TATA box 149 bp 5′ of the ATG is utilized to promote transcription. The heme-binding region (HBR) is indicated in exon X, as are two alternative polyadenylation sites which give rise to the two species of transcript of 3.4 and 2.9 kb.

contains the translation initiation site (Fig. 4). In addition to placenta, mRNAS containing exon I.1- and I.2-specific sequences also are detected in the human choriocarcinoma cell line, JEG3. The sequence of the entire intron between exons I.1 and I.2 has not been mapped because the genomic clones which contain these exons have not yet been overlapped, so the true distance between them is not established. However, since the coding exons span a distance of some 35 kb, this means the gene is at least 70 kb long, and thus, apparently, the largest cytochrome P450 gene characterized at this time. The human *CYP19* gene has been localized to chromosome 15 (Chen et al., 1988).

V. REGULATION OF *CYP19* GENE EXPRESSION IN HUMAN OVARY

The process of maturation of ovarian follicles occurs in several stages. Small antral follicles are selectively stimulated by small increases in the concentrations of gonadophins FSH and LH such as are found at menses in humans and other primates. Selected follicles progress through a complex process of differentiation to the pre-ovulatory stage. During this stage, increasing stimulation of cAMP by FSH elevates expression of CYP19, CYP11A ($P450_{scc}$) and LH receptor in granulosa cells along with CYP17 in the cal cells (Richards, 1994). The synthesis of estradiol from cholesterol resulting from the coordinate action of these enzymes triggers the surge of LH that effects ovulation and the transition of granulosa cells to luteal cells. During this transition, CYP19 expression drops precipitously while CYP11A progresses to still higher levels independent of cAMP.

We have characterized CYP19 mRNA transcripts in human ovary (Means et al., 1991). In contrast to the placenta, northern analysis of human corpus luteum

```
-955                                                              actga
                              AP-1
-950   aaatgcattt aatgatgact cactctttcc tcactctaca agtttgttca
                                   *                              GRE
-900   acctacacct cttcagctac agactaccta ccatccctga aactctgttc
-850   tgagagtaaa gggattacaa aacctggctg aaaagacaga ttcaatggca
-800   tgttaaaaaa cacagcagaa ccagcacatc agactgtaaa ttgattgtct
-750   tgcacaggat gttagctgct cttcgaatga ggttcctgag tggcacctga
       NF-1                                  GRE
-700   gcctattgct ggtggcatcc tattctgcct gttctctctt tcttcctcct
          *
-650   tccccattcc tttcattctc ttctccctta ttcttcctct gcaattcttt
                         SP-1                               NF-1
-600   ttttccacac taccgttggc cggtccctag ggatactgtt taatctggcc
                         *                          CRE
-550   catggtacaa gagattttag atcttcattg aagtcactag agatggcctg
       *         AP-1
-500   agtgagtcac tttgaattca atagacaaac tgatggaagg ctctgagaag
                                        GRE
-450   acctcaacga tgcccaagaa atgtgttctt actgtagaaa cttactattt
                                                TGF-β
-400   tgatcaaaaa agtcattttg gtcaaaaagg ggagttggga gattgccttt
-350   ttgttttgaa attgatttgg cttcaaggga agaagattgc ctaaacaaaa
                         CRE
-300   cctgctgatg aagtcacaaa atgactccac ctctggaatg agctttattt
-250   tcttataatt tggcaagaaa tttggctttc aattgggaat gcacgtcact
-200   ctacccactc aagggcaaga tgataaggtt ctatcagacc aagcgtctaa
                                       SF-1/Ad4
-150   aggaacctga gactctacca aggtcagaaa tgctgcaatt caagccaaaa
-100   gatctttctt gggcttcctt gttttgactt gtaaccataa attagtcttg
-50    cctaaatgtc tgatcacatt ataaaacagt aagtgaatct gtactgtaca
+1     gcaccctctg aagcaacagg agctatagat gaaccttta ggggattctg
+51    taatttttct gtccctttga tttccacagG ACTCTAAATT GCCCCCTCTG
+101   AGGTCAAGGA ACACAAGATG GTTTTGGAAA TGCTGAACCC GATACATTAT
                      Met ValLeuGluM etLeuAsnPr oIleHisTyr
+151   AACATCACCA GCATCGTGCC TGAAGCCATG CCTGCTGCCA CCATGCCAGT
       AsnIleThrS erIleValPr oGluAlaMet ProAlaAlaT hrMetProVa
```

Figure 5. Sequence of the 5' flanking DNA upstream of exon II corresponding to promoter II. Sequences with similarity to the consensus sequences for binding of known transcriptional activators are underlined, and the putative TATA and CAAT sequences are boxed. Nucleotides that diverge from the consensus sequences are indicated with an asterisk. The transition from lower case to upper case indicates the position of the untranslated exon/exon II boundary. The bases are numbered such the +1 represents the start of transcription using promoter II as defined by the primer extension and S1-nuclease protection experiments described in this manuscript.

polyA+ RNA failed to reveal any transcripts hybridizing to sequences contained in exons I.1 and I.2. To verify these results using a more sensitive approach, the technique of polymerase chain reaction (PCR) amplification was utilized in an attempt to amplify from human corpus luteum RNA exon I.1- and I.2-specific sequences. Oligonucleotides corresponding to sequences within exon II were used as a positive control. Neither of these sequences could be amplified from human ovarian RNA, whereas exon II was readily amplified, indicative that exons I.1 and I.2 are not expressed in human ovary, in contrast to the placenta. Recalling that the DNA sequence immediately upstream of exon II, which contains the translational start site, was previously found to contain sequence elements corresponding to TATA and CAAT boxes (Fig. 5), we utilized primer extension and S1 nuclease protection analyses to determine that the major CYP19 transcript in human corpus luteum was expressed from this promoter (hereafter known as promoter II) (Means et al., 1991). The predicted start site of transcription was estimated to lie some 23 bp downstream of the putative TATA element, which lies 110 bp upstream of the splice junction for exons I.1 and I.2 utilized in placenta (Fig. 5). In addition, northern analysis using corpus luteum mRNA and an oligonucleotide corresponding to sequences upstream of exon II revealed hybridizable CYP19 transcripts, whereas no such hybridizable transcripts were evident in the placental RNA. By contrast, an oligonucleotide corresponding to exon I.1 hybridized to CYP19 mRNA transcripts in placenta, however, no hybridizable transcripts were detectable in corpus luteum mRNA utilizing this probe (Means et al., 1991).

The promoter sequence immediately upstream of exon II (PII) contains several elements that are fully conserved between human and rat; notably, the cAMP responsive element (CRE) and SF1/Ad4 site. The CRE sequence has been shown to bind recombinant CRE binding protein (Richards, 1994) while the SF 1 sequence binds SF1 Zn-finger orphan receptor that is present in most ovarian cell types and is required for gonadal development (Luo et al., 1994). In experiments with bovine luteal cells, we show that cAMP stimulation elevates levels of SF1/Ad4BP together with interactions with this element in the PII promoter (Michael et al., 1995). Recent work (Yujima, unpublished) shows that SF1/Ad4BP is expressed sporadically in the granulosa and surrounding stromal cells prior to CYP19 expression while increasing in each of the major cell types during maturation of the dominant antral follicles. Evidently the selective expression in the granulosa cells of the pre-ovulatory follicle requires some additional regulatory factor.

VI. REGULATION OF *CYP19* GENE EXPRESSION IN HUMAN ADIPOSE

Given this unfolding complexity, it was clearly of great interest to determine the regulatory domains responsible for *CYP19* gene expression in human adipose tissue (Mahendroo et al., 1991). In the first instance, we sought to determine whether or not promoters I.1 or I.2 might be utilized for expression in adipose stromal cells in culture. In order to address this issue, we attempted to amplify transcripts containing these sequences by means of PCR from polyA+ RNA extracted from adipose tissue. Such transcripts were undetectable, indicating that neither exon I.1- nor exon I.2-containing transcripts were present in adipose tissue. In this sense then, CYPI9 transcripts in adipose are similar to those in human corpus luteum and differ from those in human placenta.

In our studies of the regulation of aromatase expression in human adipose, we have utilized adipose stromal cells in culture as a model system (Ackerman et al., 1981; Simpson et al., 1989; Mendelson et al., 1986; Evans et al., 1987). These cells are the stromal elements surrounding the adipocytes and contain most of the aromatase activity present in fat tissue. In culture, they grow and develop as fibroblasts. Treatment with dexamethasone plus serum, or dibutyryl cyclic AMP plus phorbol ester in the absence of serum, give rise to increases in aromatase expression of approximately 20-fold and 150-fold respectively (Mendleson et al., 1986). Primer extension analysis of polyA+ RNA from adipose stromal cells in culture, maximally stimulated with dibutyryl cyclic AMP plus phorbol ester, indicated a transcriptional start site identical to that found in human corpus luteum, namely, 23 bp downstream from the proximal TATA sequence, i.e., promoter II. However, S1 nuclease protection assay led to a somewhat different result (Mahendroo et al., 1991). Although this analysis indicated that a population of the transcripts extended to the same position as the primer-extended product, namely 23 bp downstream from promoter II, at least 50% of the transcripts extended only to the placental intron/exon splice boundary, in other words to the identical position found with transcripts derived from placenta (Mahendroo et al., 1991). These results would indicate that at least 50% of the transcripts in the adipose stromal cells contained a 5' end which was spliced into this junction, indicative of the presence of a 5' untranslated exon. Since PCR analysis had ruled out the presence of exon I.1- or exon I.2-containing 5' termini, we concluded that a third as yet unidentified 5' untranslated exon is spliced into a number of the transcripts present in these adipose stromal cells.

In order to characterize the 5' termini present in CYP19 transcripts in these cells, we decided to prepare primer-extended cDNA libraries by means of the RACE procedure (Mahendroo et al., 1993). Such libraries were prepared from RNA extracted from adipose tissue and adipose stromal cells cultured in the absence or presence of dexamethasone plus serum, or dibutyryl cyclic AMP plus phorbol ester in the absence of serum. Somewhat unexpectedly, sequencing of these primer-extended cDNA clones gave rise to not one, but at least four different 5' termini Table 1. As with the other splicing events previously described, each of these termini is spliced into the common 3'-junction upstream from the translational start site. Thus once again, in each case the nucleotide sequence encoding the open reading frame is unaltered by these splicing events. In adipose tissue, sequences are present which we have called I.4, I.3, and I.3-truncate. Three different libraries were made from adipose tissue: two were from breast adipose of two different patients, the other from thigh/calves. No I.4-containing transcripts were identified in the library made from thigh/calf tissue, whereas no I.3-truncate-containing transcripts were identified in libraries made from breast adipose tissue. The difference in distribution of 5'-ends in the three adipose tissue libraries could be due to patient to patient variation, or else could be a function of tissue localization.

What was even more intriguing was that the distribution of these 5' ends appeared to be a function of the culture conditions under which the cells were maintained, suggesting that promoter usage may be regulated by factors present in the culture medium. In the stromal cells in culture, the choice of 5'-termini appears to be dependent on the hormonal environment of the cells. Most interestingly, in cells expressing promoter II-specific sequences, namely cells treated with cAMP and phorbol esters in the absence of serum, no I.4-specific sequences are detected. Conversely, in cells expressing I.4-specific sequences such as dexamethasone-treated cells in the presence of serum, no promoter II-specific sequences are observed. CYP19 transcripts containing exon I.3 are present in adipose tissue as well as in cells in culture under all conditions. In addition, as can be seen from Table 1, two other 5' termini, which we have called I.4/I.2 and I.3-truncate, are observed less frequently.

Of these various termini, exon I.3 is formed as a result of a splicing event in which a region extending 100 bp upstream from the 3'-splice boundary is removed so that sequence including the promoter II TATA box and the region upstream from it are present as exonic sequence in this cDNA clone (Fig. 6). Exon I.3-truncate is formed as a consequence of a slightly larger splicing event which extends beyond the promoter II TATA box and results in the loss of a fragment 207 bp long. Untranslated exons I.3 and I.3-truncate may be derived from a second TATA box 216 bp upstream from promoter II. Of the other 5' termini, exon I.4 represents a new sequence (Fig.

Table 1. Cellular Distribution of P450arom 5'-Termini Identified by the RACE Procedure

		5'-Sequences Identified by RACE			
Sources of Library	PH	I.3	I.4	I.4/I.2	I.3-truncate
Adipose Tissue:					
Breast - 1	0	0	15	0	0
Breast - 2	0	4	5	0	0
Thigh/Calves	0	16	0	0	10
Adipose Stromal Cells in Culture:					
Control – serum	1	3	2	0	1
Dex-treated + Serum	0	4	7	2	2
Bt$_2$cAMP + PDA treated-Serum	6	9	0	0	1

6). Exon I.4/I.2 contains this I.4 sequence, but spliced downstream of it there is 206 bp of yet another sequence, which turns out to be exon I.2.

Since the RACE procedure utilizes PCR to amplify the appropriate sequences from the transcripts, the observed distribution may not be reflective of the true quantitative distribution of these termini in the CYP19 transcripts. In order to obtain a better estimate of this distribution, we have performed northern analysis on polyA + RNA isolated from adipose stromal cells maintained under the culture conditions described above, using as probes sequences specific to the various 5′ termini which were identified using the RACE procedure. These oligonucleotides were of roughly equal length and labelled as closely as possible to the same specific activity, in order to provide a rough comparison of the band intensities. The results obtained employing exon-specific northern analysis are summarized in Table 2.

On the basis of these results, it appears that three different 5′-termini are present at significant levels in CYP19 transcripts in adipose tissue and adipose stromal cells in culture. Of these, I.4 is present in breast adipose tissue, as well as adipose stromal cells treated with dexamethasone in the presence of serum. Promoter II-specific sequences are present in cells treated with dibutyryl cAMP plus or minus phorbol ester in the absence of serum, but not in cells treated with dexamethasone in the

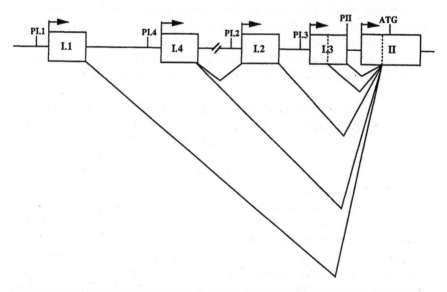

Figure 6. Structure of the human *CYP19* gene upstream of the translational start site showing alternative splicing patterns. The four untranslated exons and first coding exon (exon II) are indicated. Promoters I.1 and II and putative promoters I.4, I.2 and I.3 are also indicated. The size of the genomic region shown spans a distance at least 35 kb, but since the genomic clones containing exons I.1 and I.4 on the one hand, and exon I.2 on the other have not been overlapped, the true distance is still unknown.

Table 2. Summay of Major 5′-Termini in Adipose Cells and Tissues

Tissue/Cells	Major 5′ Terminus
Adipose Tissue	I.4, I.3
Adipose Stromal Cells in Culture:	
Control — Serum	I.3
Control + Serum	I.3
Dex + Serum	I.4, I.3
cAMP — Serum	PII/I.3
cAMP + PDA — Serum	PII/I.3
Ovary	PII
Placenta	I.1

presence of serum. On the other hand, I.3 is present in adipose tissue and cells maintained under all conditions.

We have recently characterized a genomic clone containing exon I.4, and find that it is upstream of exon I.2 and overlaps with the clone containing exon I.1, consistent with the finding of 5′-termini containing exon I.2 sequence fused downstream of I.4. Thus exon I.4 lies between exons I.1 and I.2, some 20 kb downstream of exon I.1. However, we still have not succeeded in overlapping the genomic clones containing I.4 and I.2, so a gap of unknown size remains in the genomic sequence, and the total size of the gene is still not established. Our current understanding of the structure of the 5′-end of the *CYP19* gene is summarized in Fig. 6.

VII. TISSUE-SPECIFIC REGULATION OF HUMAN CYP19 EXPRESSION IS ACHIEVED USING ALTERNATIVE PROMOTERS

The results of this ongoing work can he summarized as follows. Tissue-specific expression of the human *CYP19* gene appears to be regulated by tissue-specific promoters in ovary, placenta, and adipose. This conclusion is based on the presence of specific 5′-termini present in the transcripts encoding P450arom in each of these tissues. Thus, transcripts specific for proximal promoter II are found in the ovary, whereas transcripts specific for the distal promoter I.1 are found uniquely in placenta. Adipose tissue contains two species of transcripts containing I.3 and I.4-specific sequences. The latter appears to be present only in breast adipose. Other elements such as promoter I.2 appear to play at best, minor roles. On the other hand, when adipose stromal cells are placed in culture, other 5′-termini are found. Thus, whereas termini containing I.4-specific sequence are uniquely found in transcripts present in cells treated with dexamethasone in the presence of serum, in cells treated with dibutyryl cAMP (in the presence or absence of phorbol esters) in the absence of serum, the transcripts contain 5'-termini with sequence unique to promoter II. Thus, promoter switching in these cells appears to be regulated by factors present in the culture medium of the cells. It has been reported that transcripts in skin fibroblasts contain a sequence apparently identical to I.4 (Harada, 1992). Aromatase in these cells is also stimulated by glucocorticoids (Berkovitz et al., 1989). On the other hand, aromatase expression in human granulosa cells is not stimulated by glucocorticoids, and ovarian transcripts do not contain I.4-specific sequence.

The expression of P450arom in human ovary appears to utilize a promoter, promoter II, which is proximal to the translation start site, that is to say in the "normal" location. We speculate that this ovarian promoter is the primordial promoter regulating CYP19 expression. This is consistent with results from the laboratory of JoAnne Richards (Hickey et al., 1990) indicating that in rat ovary, CYP19 expression is regulated by a promoter proximal to the translation start site. Similarly, work from the laboratory of Michael McPhaul (Matsumine et al., 1991) indicates that in the chicken ovary, a promoter proximal to the start of translation is also utilized to regulate CYP19 expression.

We speculate that when the human placenta acquired the ability to synthesize estrogens, since this capacity is very great, and since estrogen production by the placenta tends to be a function of placental size, instead of utilizing the ovarian promoter, a powerful, distal promoter was utilized instead, namely, that upstream from exon I.1. Since rat placenta does not synthesize estrogens, and chickens do not have a placenta, one would not expect utilization of tissue-specific alternative promoters in these species. It appears that human P450arom is the first cytochrome P450 to be shown to utilize alternative promoters in the regulation of tissue-specific expression. It is pertinent to ask why this is so. The answer probably relates to the unique tissue-specific distribution of estrogen biosynthesis in the human. In contrast to most other species where estrogen biosynthesis is confined to the gonads and the brain, in the human it is also present in adipose and in placenta. In both these tissues *in situ*, estrogen biosynthesis appears to be relatively unregulated, although dramatic regulation can be achieved utilizing adipose stromal cells in culture. Whereas, the expression of estrogen biosynthesis in the placenta is very great, that in adipose is rather low. It is apparent that the estrogen produced in each of these tissue sites of expression in the human subserves a different function and consequently, the regulation of this biosynthesis must be different in each tissue, and this is apparent from *in vitro* studies. Consequently, it can be hypothesized that a single promoter with a number of upstream regulatory sequences would be inadequate to permit such sophisticated complexity of tissue-specific regulation, and it is for this reason that this particular gene has resorted to the use of alternative promoters to allow for greater versatility in determining tissue-specific regulation of expression. It is worth noting that the 5' untranslated exons which have so far been characterized are spliced into a common intron/exon boundary upstream from the translational start site. This means that the protein which is expressed in each of the various tissue-specific sites of estrogen biosynthesis is identical, in contrast to the situation regarding a number of other genes in which alternative splicing results in differences in the aminoterminus of the protein itself.

Although no other cytochrome P450 has been shown to utilize this form of regulation of expression, a growing number of other proteins do. These include the genes for IGF-II and IGF-I, glucokinase, c-myc, c-fms, aldolase A, PTH-related peptide, GnRH, and prolactin (Battey et al., 1983; DiMattia et al., 1990). Interestingly, at least two other genes have been shown to utilize a different promoter to regulate expression in placenta or uterine tissues, as compared to their adult tissue of expression. Thus, human prolactin expression in decidua appears to be regulated by a distal promoter (DiMattia et al., 1990), as does growth hormone releasing hormone expression in placenta (Gonzalez-Crespo and Baronat, 1991). Thus, at least three genes utilize a distal promoter to regulate their expression in placenta or uterine tissue as compared to other tissues. This, however, does not appear to be a universal truth. For example, α-glycoprotein gene expression in placenta appears to utilize a unique regulatory region upstream from a single promoter rather than an alternative promoter (Steger et al., 1991).

The conclusion that tissue-specific regulation of estrogen biosynthesis is determined, in part, by the use of alternative promoters of the *CYP19* gene, has a number of important consequences. In the first place, we have observed that expression of aromatase in adipose varies with body site of tissue origin and increases dramatically with age, consistent with previous determinations of aromatase activity and *in vivo* studies (Killinger et al., 1987; Hensell et al., 1974; Edman and MacDonald, 1976). As discussed here, regional differences in adipose aromatase expression appear to be dependent on alternative promoter utilization, and so the possibility remains that the age-dependent differences may be regulated in a similar fashion. Secondly, the existence of tissue-specific promoters suggests the possibility that mutations in the 5′-upstream regions of the aromatase gene may exist which inhibit estrogen synthesis in one tissue, but not in another, e.g., in adipose but not in the ovary. Thirdly, the existence of tissue-specific promoters offers the opportunity to devise therapeutic agents which differentially inhibit *CYP19* gene expression in a tissue-specific fashion.

VIII. CYP19 REGULATION AND BREAST CANCER

This work may potentially lead to therapeutic application in the treatment of breast cancer. Estrogen production in breast adipose tissue may be a significant contributor to the etiology of breast cancer. Epidemiological data has suggested such a linkage, although the mechanism is unknown. Certainly the ability of estrogens to stimulate the generation of growth factors that stimulate the proliferation of cancer cells may be a factor. In support of a link to local estrogen production, aromatase activity is elevated in adipose tissue of the breast quadrant containing the tumor (O'Neill et al., 1988). We have used quantitative PCR to show that the highest expression of CYP19 mRNA is also located in the quadrant adjacent to the tumor (Bulun et al.,

1993). Other researchers have shown that breast tumors recruit a distinct population of stromal fibroblasts (Ronnov-Jessen et al., 1995). Disrete populations of these stromal cells probably function as pre-adipocytes. These studies support a model in which crosstalk between breast cancer cells and stromal cells is a key part of tumorigenesis (Simpson et al., 1989). This work increases the importance of the realization that adipose expression of CYP19 is regulated by promoter 1.4. Our recent studies show that this promoter is stimulated by glucocorticoids acting at a GRE element (Zhao et al., 1995a) and additionally by various cytokines (notably IL-6, IL-11, and LIF) acting at a GAS element (γ-interferon activation site) through a Jak/STAT signaling process (Zhao et al., 1995b). Interference with these mechanisms at promoter 1.4 may allow selective suppression of CYP19 in breast adipose and also the potential tumor promoting effect.

ACKNOWLEDGMENTS

We wish to thank past and present members of our laboratory, as well as colleagues from other institutions for their invaluable work summarized here. We also thank Melissa Meister for skilled editorial assistance. This work was supported, in part, by USPHS grant #AG08174.

REFERENCES

Ackerman, G.E., MacDonald, P.C., Gudelsky, G., Mendelson, C.R., & Simpson, E.R. (1981). Potentiation of epinephrine-induced lipolysis by catechol estrogens and their methoxy derivatives. Endocrinology 1O9, 2084–2088.

Akhtar, M., Calder, M.R., Corina, D.L., & Wright, J.N. (1982). Mechanistic studies on C19 demethylation in oestrogen biosynthesis. Biochem. J. 201, 569–580.

Battey, J., Moulding, C., Taub, R., Murphy, W., Stewart, T., Potter, H., Lenoir, G., & Leder, P. (1983). The human c-myc oncogene: structural consequences of translocation into the IgH locus in Burkitt lymphoma. Cell 34, 779–787.

Berkovitz, G.D., Bisat, T., & Carter, K.M. (1989). Aromatase activity in microsomal preparations of human genital skin fibroblasts: Influence of glucocorticoids. J. Steroid Biochem. 33, 341–347.

Bulun, S.E., Price, T.M., Mahendroo, M.S., Aitken, J., & Simpson, E.R. (1993). A link between breast cancer and local estrogen synthesis suggested by quantification of breast adipose tissue aromatase cytochrome P450 transcripts using competitive polymerase chain reaction after reverse transcription. J. Clin. Endocrinol. Metab. 77, 1622–1628.

Callard, G.V. (1981). Aromatization is cyclic AMP-dependent in cultured reptilian brain cells. Brain Res. 204, 451–454.

Callard, G.V., Petro, Z., & Ryan, K.J. (1980). Aromatization and 5α-reduction in brain and non-neural tissues of a cytochrome (Petromyzan marinus). Gen. Comp. Endocrinol. 42, 155–159.

Callard, G.V., Petro, Z., & Ryan, K.J. (1990). Phylogenetic distribution of aromatase and other androgen-converting enzymes in the central nervous system. Endocrinology 103, 2283–2290.

Callard, G.V., Pudney, J.A., Kendell, S.L., & Reinboth, R. (1984). *In vitro* conversion of androgen to estrogen in amphioxus gonadal tissue. Gen. Comp. Endocrinol. 56, 53–58.

Chen, S., Besman, M.J., Sparkes, R.S., Zollman, S., Klisak, I., Mohandes, T., Hall, P.F., & Shively, J.E. (1988). Human aromatase: cDNA cloning, southern blot analysis, and assignment to the gene to chromosome 15. DNA 7, 27–38.

Cole, P.A., & Robinson, C.H. (1988). A peroxide model reaction for placental aromatase. J. Am. Chem. Soc. 110, 1284–1285.

Corbin, C.J., Graham-Lorence, S., McPhaul, M., Mason, J.I., Mendelson, C.R., & Simpson, E.R. (1988). Isolation of a full length cDNA insert encoding human aromatase system cytochrome P-450 and its expression in non-steroidogenic cells. Proc. Natl. Acad. Sci. USA 85, 8948–8952.

DiMattia, G.E., Gellerson, B., Duckworth, M.L., & Friesen, H.G. (1990). Human prolactin gene expression: the use of an alternative noncoding exon in decidua and the IM-9-P3 lymphoblast cell line. J. Biol. Chem. 265, 16412–16421.

Edman, C.D., & MacDonald, P.C. (1976). In: The Endocrine Function of the Human Ovary (James, V.H.T., Serio, M., & Giusti, G., eds.), pp. 135–140. London.

Evans, C.T., Corbin, C.J., Saunders, C.T., Merrill, J.C., Simpson, E.R., & Mendelson, C.R. (1987). Regulation of estrogen biosynthesis in human adipose stromal cells: Effects of dibutyryl cyclic AMP, epidermal growth factor, and phorbol esters on the synthesis of aromatase cytochrome P-450. J. Biol. Chem. 262, 6914–6920.

France, J.T., & Liggins, G.C. (1969). Placental sulfatase deficiency. J. Clin. Endocrinol. Metab. 29, 138–144.

Gonzalez-Crespo, S., & Baronat, A. (1991). Expression of the rat growth hormone-releasing hormone gene in placenta is directed by an alternative promoter. Proc. Natl. Acad. Sci. USA 88, 8749–8753.

Graham-Lorence, S., Khalil, M.W., Lorence, M.C., Mendelson, C.R., & Simpson, E.R. 1991. Structure-function relationships of human aromatase cytochrome P-450 using molecular modeling and site directed mutagenesis. J. Biol. Chem. 266, 11939–11946.

Harada, N. (1988). Cloning of a comparative cDNA encoding human aromatase: immunochemical identification and sequence analysis. Biochem. Biophys. Res. Commun. 156, 725–732.

Harada, N. (1992). A unique aromatase (P450arom) mRNA formed by alternative use of tissue-specific exons I in human skin fibroblasts. Biochem. Biophys. Res. Commun. 189, 1001–1007.

Harada, N., Yamada, K., Saito, K., Kibe, N., Dohmae, S., & Takagi, Y. (1990). Structural characterization of the human estrogen synthetase (aromatase) gene. Biochem. Biophys. Res. Commun. 166, 365–372.

Hemsell, D.L., Grodin, J.M., Brenner, P.F., Siiteri, P.K., & MacDonald, P.C. (1974). Plasma precursors of estrogen. II. Correlation of the extent of conversion of plasma androstenedione to estrone with age. J. Clin. Endocrinol. Metab. 38, 476–479.

Hickey, G.T., Krasnow, J.S., Beattie, W.G., & Richards, J.S. (1990). Aromatase cytochrome P450 in rat ovarian granulosa cells before and after luteinization: Adenosine 3′, 5′-monophosphate-dependent and independent regulation. Cloning and sequencing of rat aromatse cDNA and 5′ genomic DNA. Mol. Endocrinol. 4, 3–12.

Hutchinson, J.B. (1991). Hormonal control of behavior: Steroid action in the brain. Current Opinion in Neurobiology 1, 562–570.

Kellis, J.T., Jr., & Vickery, L.E. (1987). Purification and characterization of human placental aromatase cytochrome P450. J. Biol. Chem. 262, 4413–4420.

Kilgore, M.W., Means, G.D., Mendelson, C.R., & Simpson, E.R. (1992). Alternative promotion of aromatase cytochrome P450 expression in human fetal tissues. Mol. Cell. Endocrinol. 83, R9–R16.

Killinger, D.W., Perel, E., Daniilescu, D., Kherlip, L., & Lindsay, W.R.N. (1987). The relationship between aromatase activity and body fat distribution. Steroids 50, 61–72.

Luo, X., Ikeda, Y., & Parker, K.L. (1994). A cell specific nuclear receptor is essential for adrenal and gonadal development and sexual differentiation. Cell 77, 481–490.

Mahendroo, M.S., Means, G.D., Mendelson, C.R., & Simpson, E.R. (1991). Tissue-specific expression of human P450: The promoter responsible for expression in adipose is different from that utilized in placenta. J. Biol. Chem. 266, 11276–11281.

Mahendroo, M.S., Mendelson, C.R., & Simpson, E.R. (1993). Tissue-specific and hormonally-controlled alternative promoters regulate aromatase cytochrome P450 gene expression in human adipose tissue. J. Biol. Chem. (In press).

Matsumine, H., Herbst, M.A., Ignatius, Ou. S-H., Wilson, J.D., & McPhaul, M.J. (1991). Aromatase mRNA in the extragonadal tissues of chickens with the Henny-feathering trait is derived from a distinct promoter structure that contains a segment of a retroviral long terminal repeat. J. Biol. Chem. 266, 19900–19907.

McPhaul, M.J., Noble, J.F., Simpson, E.R., Mendelson, C.R., & Wilson, J.D. (1988). The expression of a functional cDNA encoding the chicken cytochrome P-450arom (aromatase) that catalyzes the formation of estrogen from androgen. J. Biol. Chem. 263, 16358–16363.

Means, G.D., Kilgore, M.W., Mahendroo, M.S., Mendelson, C.R., & Simpson, E.R. (1991). Tissue-specific promoters regulate aromatase cytochrome P450 gene expression in human ovary and fetal tissues. Mol. Endocrinol. 5, 2005–2013.

Means, G.D., Mahendroo, M., Corbin, C.J., Mathis, J.M., Powell, F.E., Mendelson, C.R., & Simpson, E.R. (1989). Structural analysis of the gene encoding human aromatase cytochrome P-450, the enzyme responsible for estrogen biosynthesis. J. Biol. Chem. 264, 19385–19391.

Mendelson, C.R., Corbin, C.J., Smith, M.E., Smith, J., Simpson, E.R. (1986). Growth factors suppress, and phorbol esters potentiate the action of dibutyryl cyclic AMP to stimulate aromatase activity of human adipose stromal cells. Endocrinology 118, 968–973

Mendelson, C.R., Wright, E.E., Porter, J.C., Evans. C.T., & Simpson, E.R. (1985). Preparation and characterization of polyclonal and monoclonal antibodies against human aromatase cytochrome P-450 (P450arom), and their use in its purification. Arch. Biochem. Biophys. 243, 480–491.

Michael, M.D., Kilgore, M.W., Morohashi, K., & Simpson, E.R. (1995). Ad4BP/SF-1 regulates cyclic AMP-induced transcription from the proximal promotes PID of the human aromatase P450 (CYP19) gene in the ovary. J. Biol. Chem. 270, 13561–13566.

Miller, W.R., & O'Neill, J. (1987). The importance of local synthesis of estrogen within the breast. Steroids 50, 537–548.

Nakajin, S., Shimoda, M., & Hall, P.F. (1986). Purification to homogeneity of aromatase from human placenta. Biochem. Biophys. Res. Commun. 134, 704–710.

Nelson, D.R., Kamataki, T., Waxman, D.J., Guengerich, F.P., Estabrook, R.W., Feyereisen, R., Gonzalez, F.J., Coon, M.J., Gunsalus, I.C., Gotoh, O., Okuda, K., & Nebert, D.W. (1993). The P450 superfamily: Update on new sequences, gene mapping, accession numbers, early trivial names of enzymes, and nomenclature. DNA Cell Biol. 12, 1–51.

O'Neill, J.S., Elton, R.A., & Miller, W.R. (1988). Aromatase activity in adipose tissue from breast quadrants: A link with tumor site. Brit. Med. J. 296, 741–743.

Osawa, Y., Yoshida, N., Franckowiak, M., & Kitawaki, J. (1987). Immunoaffinity purification of aromatase cytochrome P450 from human placental microsomes, metabolic switching from aromatization to 1β and 2β-monohydroxylation, and recognition of aromatase isoenzymes. Steroids 50, 11–28.

Richards, J.S. (1994). Hormonal control of gene expression in the ovary. Endocrine Reviews 15, 725–751.

Rønnov-Jessen, L., Petersen, O.W., Koteliansky, V.E., & Bissell, M.J. (1995). The origin of the myofibroblasts in breast cancer: Recapitulation of tumor environment in culture unravels diversity and implicates converted fibroblasts and recruited smooth muscle cells. J. Clin. Invest. 95, 859–873.

Sethumadhawan, K., Bellino, F.L., & Thotakura, N.R. (1991). Estrogen synthase (aromatase). The cytochrome P450 component of the human placental enzyme is a glycoprotein. Mol. Cell. Endocrinol. 78, 25–32.

Simmons, D.L., Lalley, P.A., & Kasper, C.B. (1985). Chromosomal assignments of gene coding for components of the mixed function oxidase system in mice. J. Biol. Chem. 260, 515–521.

Simpson, E.R., Ackerman, G.E., Smith, M.E., & Mendelson, C.R. (1981). Estrogen formation in stromal cells of adipose tissue of women: induction by glucocorticosteroids. Proc. Natl. Acad. Sci. USA 78, 5690–5694.

Simpson, E.R., Merrill, J.C., Hollub, A.J., Graham-Lorence, S., & Mendelson, C.R. 1989. Regulation of estrogen biosynthesis by human adipose cells. Endocr. Rev. 10, 136–148.

Steger, D.J., Altschmied, J., Buscher, M., & Mellon, P.L. (1991). Evolution of placenta-specific gene expression: Comparison of the equine and human gonadotropin α-subunit genes. Mol. Endocrinol. 5, 243–255.

Tanaka, M., Telecky, T.M., Fukada, S., Adachi, S., Chen, S., & Nagahama, Y. (1992). Cloning a sequence analysis of the cDNA encoding P450aromatase (P450arom) from a rainbow trout (Oncorhynchus mykiss) ovary; relationship between the amount of P450arom mRNA and the production of oestradiol-17β in the ovary. J. Mol. Biol. 8, 53–61.

Terashima, M., Toda, K., Kawamoto, T., Kuribayashi, I., Ogawa, Y., Maeda, T., & Shizuta, Y. (1991). Isolation of a full-length cDNA encoding mouse aromatase P450. Arch. Biochem. Biophys. 285, 231–237.

Thompson, E.A., Jr., & Siiteri, P.K. (1974). The involvement of human placental microsomal cytochrome P450 in aromatization. J. Biol. Chem. 249, 5373–5378.

Toda, K., Terashima, M., Kamamoto, T., Sumimoto, H., Yamamoto, Y., Sagara, Y., Ikeda, H., & Shizuta, Y. (1990). Structural and functional characterization of human aromatase P450 gene. Eur. J. Biochem. 193, 559–565.

Toda, K., Terashima, M., Mitsuuchi, Y., Yamasaki, Y., Yokoyama, Y., Nojiona, S., Ushiro, H., Maeda, T., Yamamoto, Y., Sagara, Y., & Shizuta, Y. (1989). Alternative usage of different poly(A) addition signals for two major species of mRNA encoding human aromatase P450. FEBS Lett. 247, 371–376.

Zhao, Y., Mendelson, C.R., & Simpson, E.R. (1995a). Characterization of the sequences of the human CYP19 (aromatase) gene that mediate regulation by glucocorticoids in adipose stromal cells and fetal hepatocytes. Mol. Endocrinol. 9, 340–349.

Zhao, Y., Nichols, J.E., Bulun, S.E., Mendelson, C.R., & Simpson, E.R. (1995b). Aromatase P450 gene expression in human adipose tissue. Role of a Jak/STAT pathway in regulation of the adipose-specific promoter. Mol. Endocrinol. 9, 340–349.

REGULATION OF CALCIUM METABOLISM BY THE VITAMIN D HYDROXYLASES

H. James Armbrecht, Rama K. Nemani, and
N. Wongsurawat

Advances in Molecular and Cell Biology
Volume 14, pages 245–267.
Copyright © 1996 by JAI Press Inc.
All rights of reproduction in any form reserved.
ISBN: 0-7623-0113-9

ABSTRACT

Serum Ca must be maintained at about 10 mg/100 ml for the proper function of nerve, muscle, and bone. Serum Ca is maintained through the action of vitamin D metabolites and parathyroid hormone (PTH) on intestine, kidney, and bone (DeLuca, 1988; Minghetti and Norman, 1988) (Figure 1). Vitamin D$_3$ (cholecalciferol) itself has very little biological activity. It must first be hydroxylated in the liver to form 25-hydroxy-vitamin D$_3$ (25[OH]D), which has some biological activity. However, the major activation takes place in the kidney, where 25(OH)D is hydroxylated to 1,25-dihy-droxyvitamin D$_3$ (1,25[OH]$_2$D), the metabolite of vitamin D with the most biological activity. 1,25(OH)$_2$D is then inactivated by the 24-OHase in various target tissues. The characteristics of these vitamin D hydroxylases (OHases) and their regulation in health and disease are the subject of this review.

I. REGULATION OF CALCIUM METABOLISM

Serum Ca levels are maintained through the absorption of dietary Ca from the intestine and the resorption of Ca from bone (Fig. 1). A decrease in serum Ca is sensed by the parathyroid glands, which secrete PTH in proportion to the decline in serum Ca (Fig. 1, Step 1). PTH then acts on bone to increase bone resorption (Step 2) and on the kidney to alter vitamin D metabolism (Step 3). PTH increases renal 1-OHase activity (Step 4) which 1-hydroxylates 25(OH)D, itself formed in the liver by the hepatic 25-OHase, to 1,25(OH)$_2$D. 1,25(OH)$_2$D then acts on the intestine to increase Ca absorption (Step 5) and on bone, in concert with PTH, to increase bone resorption (Step 6). These actions of PTH and $1,25(OH)_2D$ tend to normalize serum Ca levels. This then decreases the secretion of PTH, a short-lived peptide hormone, by the parathyroid gland. The action of $1,25(OH)_2D$, a longer-lived steroid hormone, is limited by its metabolism to 1,24,25-trihydroxyvitamin D_3 $(1,24,25[OH]_3D)$ (Step 7). This is accomplished by the 24-OHase found in the intestine, kidney, bone, and other target tissues.

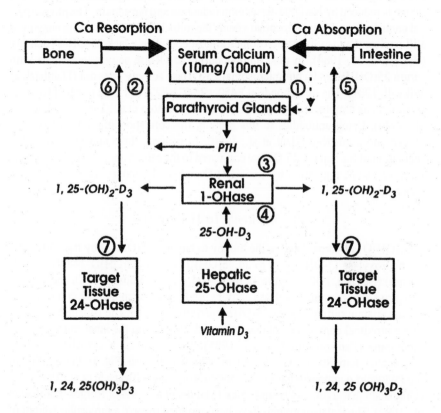

Figure 1. Regulation of serum Ca by vitamin D and PTH. Decreased serum Ca is detected by the parathyroid glands, which secrete PTH (*step 1*). PTH acts on bone to increase Ca resorption (*step 2*) and on the kidney (*step 3*) to increase $1,25(OH)_2D$ production by the renal 1-OHase (*step 4*). $1,25(OH)_2D$ acts on the intestine to increase dietary Ca absorption (*step 5*) and on bone to increase Ca resorption (*step 6*). $1,25(OH)_2D$ is converted in the target tissues by the 24-OHase to $1,24,25(OH)_3D$ (*step 7*), the first product in the catabolic pathway.

II. THE VITAMIN D P450 HYDROXYLASES

A. The Hepatic 25-Hydroxylase

Vitamin D is hydroxylated in the liver to 25(OH)D by the vitamin D-25-hydroxy-lase (25-OHase) (DeLuca, 1988) (Figure 2). Both a microsomal and mitochondrial 25-OHase have been described in the liver. The microsomal form consists of a cytochrome P450 (P450cc25) (Hayashi et al., 1986; Andersson and Jornvall, 1986) and a NADPH-cytochrome P450 reductase, and it is most abundant in the male liver (Saarem and Pedersen, 1987). The mitochondrial 25-OHase consists of a distinct cytochrome P450cc25 (Masumoto et al., 1988; Bjorkhem et al., 1980) which is reduced by NADPH-ferredoxin reductase and ferredoxin. The physiological contribution of the microsomal versus the mitochondrial form of the enzyme is still unclear. Human liver microsomes almost completely lack 25-OHase activity (Saarem et al., 1984). In rats, mitochondrial 25-OHase activity correlates with serum 25(OH)D levels, but microsomal 25-OHase activity does not (Dahlback & Wikvall, 1987). The mitochondrial P450cc25 (CYP27) has been purified from rat liver and has a molecular weight of 52,000 (Masumoto et al., 1988; Bjorkhem et al., 1980). Using antibodies to this protein, the corresponding cDNA has been cloned and sequenced (Usui et al., 1990). In general, there appears to be little regulation of the hepatic 25-OHase compared to the other vitamin D hydroxylases. However, some regulation by 1,25(OH)$_2$D and sex hormones has been reported (see below).

B. The Renal 1α-Hydroxylase

25(OH)D is hydroxylated in the kidney to either 1,25(OH)$_2$D by the 25(OH)D-1α-hydroxylase (1-OHase) or to 24,25(OH)$_2$D by the 25(OH)D-24-OHase (24-OHase) (DeLuca, 1988) (Figure 2). In general, under conditions of Ca and vitamin D deprivation, the renal 1-OHase activity predominates; and, under conditions of Ca and vitamin sufficiency, the 24-OHase activity is greater. The 1-OHase is located in the proximal convoluted tubular cells of the kidney on the inner mitochondrial membrane (Paulson and DeLuca, 1985). It consists of a cytochrome P450 (P450cc1α), ferredoxin, and NADPH-ferredoxin reductase (Fig. 3). Purification of the P450cc1α component has been reported from cow (Hiwatashi et al., 1982), pig (Gray and Ghazarian, 1989), and chick (Mandel et al., 1990b; Burgos-Trinidad et al., 1992). The isolated proteins have been reported to have a molecular weight of 49–52,000 in the bovine (Hiwatashi et al., 1982; Bort and Crivello, 1988) and 57–59,000 in the chick (Mandel et al., 1990b; Burgos-Trinidad et al., 1992). The cloning of the cDNA for the P450cc1α has not yet been reported. The renal 1-OHase is the major site of regulation for vitamin D metabolism. The 1-OHase activity is altered by PTH, 1,25(OH)$_2$D itself, serum Ca and P, and other hormones (Henry and Norman, 1984) (see below).

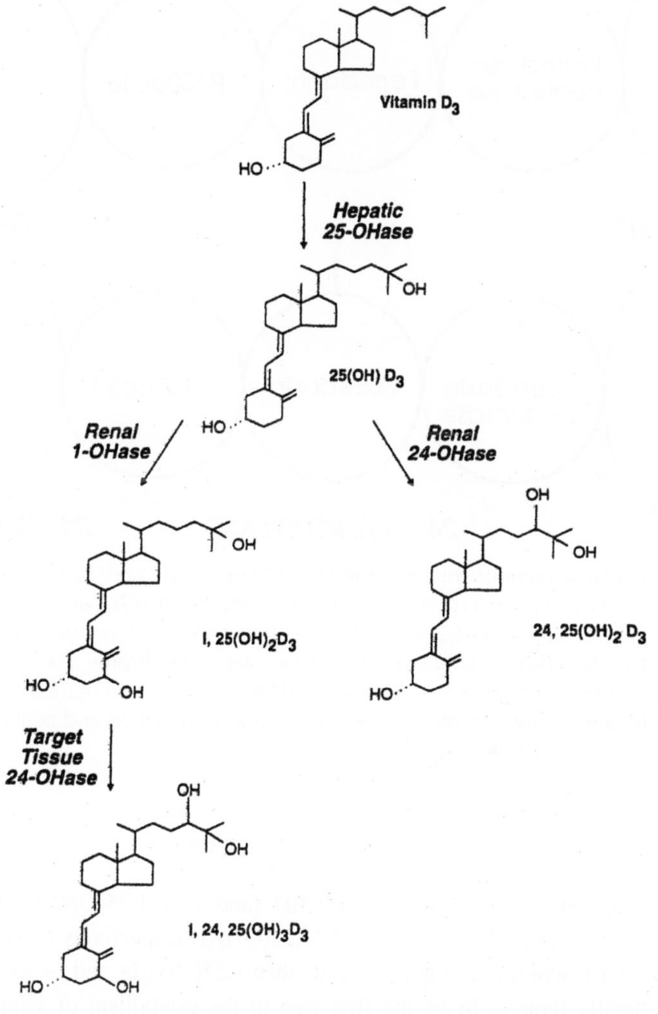

Figure 2. Structures of vitamin D_3 and its hydroxylated metabolites. Vitamin D_3 (cholecalciferol) is converted to 25(OH)D_3 by the hepatic 25-OHase. 25(OH)D_3 is converted to 1,25(OH)$_2D_3$, the most biologically active form of vitamin D, by the renal 1-OHase or to 24,25(OH)$_2D_3$ by the renal 24-OHase. Target tissues for 1,25(OH)$_2D_3$ then 24-hydroxylate the active metabolite to form 1,24,25(OH)$_3$D, which has much less biological activity.

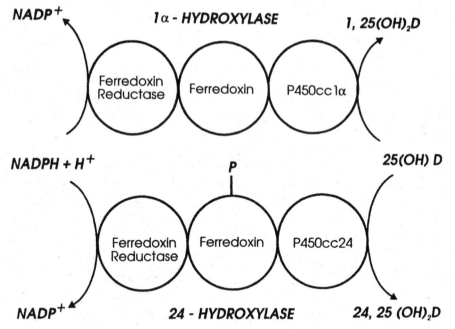

Figure 3. Molecular components of the renal 1α-hydroxylase and 24-hydroxylase. The renal 1α-hydroxylase (*TOP*) and 24-hydroxylase (*BOTTOM*) hydroxylate 25(OH)D to 1,25(OH)₂D and 24,25(OH)₂D respectively. The mitochondrial hydroxylases consist of ferredoxin reductase, ferredoxin, and cytochrome P450cc1α or P450cc24. Hydroxylase activity may be regulated by the expression of these components. In addition, activity may be regulated by the phosphorylation and dephosphorylation of ferredoxin (denoted by P).

C. The Renal 24-Hydroxylase

The renal 24-OHase hydroxylates 25(OH)D (and also 1,25[OH]$_2$D) in the 24-position to produce 24,25(OH)$_2$D and 1,24,25(OH)$_3$D respectively (Figure. 2). These compounds have less biological activity than 1,25(OH)$_2$D, and 24-hydroxylation is generally thought to be the first step in the catabolism of vitamin D compounds. The 24-OHase is located in the proximal tubular cells, but its distribution is different from that of the 1-OHase (Kawashima and Kurokawa, 1983). It is found in proximal straight tubular cells in addition to the proximal convoluted tubular cells. Like the 1-OHase, the 24-OHase is found in the inner mitochondrial membrane and consists of a cytochrome P450 (P450cc24) (CYP24), ferredoxin, and ferredoxin reductase (Fig. 3). The P450cc24 has been purified from the bovine (Bort and Crivello, 1988), pig (Gray et al., 1990), chick (Mandel et al., 1990b; Burgos-Trinidad et al., 1992), and rat (Ohyama and Okuda, 1991). The mammalian

P450cc24 has a molecular weight of 52–53,000 (Bort and Crivello, 1988), and the chick has a weight of 55–59,000 (Mandel et al., 1990b; Burgos-Trinidad et al., 1992). Using antibodies to purified rat P450cc24 (Ohyama and Okuda, 1991), the cDNA for this protein has been cloned and sequenced (Ohyama et al., 1991). Expression of this cDNA in COS cells results in only 24-OHase activity when 25(OH)D is used as a substrate. Recently, the cDNA for the human P450cc24 has also been cloned and expressed (Chen et al., 1993a). Like the 1-OHase, the renal 24-OHase is regulated by PTH, $1,25(OH)_2D$, serum Ca and P, and other hormones, usually in a reciprocal fashion to the 1-OHase (Henry and Norman, 1984).

D. The Extra-Renal 24-Hydroxylase

Like the renal 24-OHase, the extra-renal 24-OHase hydroxylates $1,25(OH)_2D$ to form $1,24,25(OH)_3D$ (Fig. 2). The 24-OHase activity is found in a wide variety of tissues including intestine (Kumar, et al., 1978), lymphocytes (Takeda et al., 1990), fibroblasts (Gamblin et al., 1985), bone (Howard et al., 1981), skin (Bikle et al., 1986), and macrophages (Reichel et al., 1987). In general, it has been assumed that the same cytochrome P450cc24 is found in all tissues with 24-OHase activity. In the rat, DNA probes derived from the renal cytochrome P450cc24 also hybridize with intestinal mRNA of similar size to that of the kidney (Armbrecht and Boltz, 1991). With the cloning of the gene for rat P450cc24 (Ohyama et al., 1993), it has been found that there is only one gene for this P450. This would suggest that all P450cc24 are identical regardless of tissue. The major regulator of the extra-renal 24-OHase is $1,25(OH)_2D$, which markedly increases its activity. Presumedly, this limits the action of $1,25(OH)_2D$ in its target tissue. When 24-OHase activity is blocked by ketoconazole, a P450 inhibitor, the action of $1,25(OH)_2D$ in target tissue is enhanced (Reinhardt and Horst, 1989).

III. REGULATION OF THE RENAL 1-HYDROXYLASE AND 24-HYDROXYLASE

The major regulator of renal hydroxylase activity under physiological conditions is PTH (Fig. 1). Feeding a low Ca diet to vitamin D replete rats results in an increase in renal 1-OHase activity and a decrease in renal 24-OHase activity (Armbrecht et al., 1984a). The effect of the low Ca diet is mediated by the secondary increase in serum PTH. The second important physiological regulator of renal hydroxylase activity is vitamin D status. Feeding a vitamin D-deficient diet to rats also increases renal 1-OHase and decreases 24-OHase activity (Armbrecht et al., 1982a). The effect of the D-deficiency is rapidly reversed by $1,25(OH)_2D$. Feeding a diet low in Ca and vitamin D produces the greatest increase in renal 1-OHase activity (Armbrecht et al., 1982a). This may be due to the decline in serum Ca in these animals. In addition to PTH and $1,25(OH)_2D$, there is evidence that Ca and phosphate may regulate renal hydroxylase activity directly (see below).

A. Regulation by Parathyroid Hormone

Modulation of the renal metabolism of 25(OH)D by the peptide hormone PTH was demonstrated over 20 years ago (Garabedian et al., 1972). The fact that PTH modulates the renal hydroxylase activity directly has been demonstrated in intact rats (Armbrecht et al., 1982b), in renal slices incubated *in vitro* (Armbrecht et al., 1984b), in isolated tubules (Kawashima and Kurokawa, 1983), and in cultured renal cells (Fukase et al., 1982; Henry, 1986). PTH appears to have both short term and long term effects. In some systems, PTH has been demonstrated to act within minutes (Siegel et al., 1986; Kramer and Goltzman, 1982), but in many *in vitro* systems PTH takes 3–4 hours to work (Armbrecht et al., 1984b; Fukase et al., 1982; Henry, 1986). The long term effects of PTH require new protein synthesis (Korkor et al., 1987).

The major pathway of PTH activation is thought to be via cAMP and cAMP-dependent protein kinase. The dose response curves of PTH with regard to cAMP production and cAMP-dependent protein kinase activity are consistent with their involvement in modulating hydroxylase activity (Armbrecht et al., 1984b). The effects of PTH can be mimicked by cAMP itself (Rost et al., 1981) and by forskolin (Armbrecht et al., 1984c), which markedly raises intracellular cAMP levels. The effects of PTH and forskolin are not additive, suggesting that they operate via the same pathway (Armbrecht et al., 1984c). One protein which has been shown to be phosphorylated by a cAMP-dependent kinase is renal ferredoxin (Nemani et al., 1989). Phosphorylation of ferredoxin modulates 1-OHase and 24-OHase activity in reconstituted systems (Nemani et al., 1989; Mandel et al., 1990a). Likewise, phosphorylation of ferredoxin has been reported to alter the activity of adrenal cortical P450s (Monnier et al., 1987). The phosphorylation of ferredoxin has been shown to be modulated by PTH in rat renal slices (Siegel et al., 1986). The effect of PTH is rapid, peaking at 5 min.

In addition to activating the cAMP pathway, PTH also increases protein kinase C (PKC) translocation in the kidney (Nemani et al., 1991). The role of PKC in regulating renal 25(OH)D metabolism has been extensively studied in cultured chick kidney cells using phorbol esters (Henry, 1986; Henry and Luntao, 1989). The phorbol ester TPA alone decreases renal $1,25(OH)_2D$ production and increases $24,25(OH)_2D$ production (Henry and Luntao, 1989). This action of TPA is paralleled by changes in mitochondrial hydroxylase activity and in the phosphorylation state of ferredoxin (Tang et al., 1993). This is further evidence for ferredoxin involvement in the regulation of renal hydroxylase activity. TPA also increases 24-OHase activity in rat renal tubules (Mandla et al., 1990). The interaction between the cAMP and PKC pathways has also been studied in chick kidney cells (Henry and Luntao, 1989). TPA inhibits the effect of forskolin on increasing $1,25(OH)_2D$ and decreasing $24,25(OH)_2D$ production in this system. However, this inhibition can be overcome by the continuous presence of dibutyryl-cAMP. Thus, PTH has the paradoxical effect of activating both the cAMP and the PKC pathways, which have opposite effects on the renal hydroxylases. It may be that the activation of the PKC pathway attenuates the action of PTH via the cAMP pathway. The molecular basis of these interactions, as well as their physiological significance, remain important areas for future investigation.

B. Regulation by 1,25(OH)$_2$D

The effect of the steroid hormone 1,25(OH)$_2$D on renal vitamin D metabolism has also been known for many years (Henry et al., 1974; Tanaka and DeLuca, 1974). The decrease in renal 1-OHase activity and the increase in 24-OHase activity by 1,25(OH)$_2$D has been demonstrated *in vivo* (Armbrecht et al., 1982a) and *in vitro* (Armbrecht et al., 1984b) using slices from rat renal cortex. The effects of 1,25(OH)$_2$D have also been studied in primary cultures of renal cells from chickens (Henry, 1979) and mammals (Fukase et al., 1982).

1,25(OH)$_2$D is thought to act in the kidney via a classical steroid hormone mechanism. Over a time course of hours, it binds to the renal vitamin D receptor, and it modulates renal hydroxylase activity by a process requiring translation (Armbrecht et al., 1983). With the cloning of the cytochrome P450cc24 from the rat kidney (Ohyama et al., 1991), it has been possible to study the mechanism of action of 1,25(OH)$_2$D induction of the renal 24-OHase at the molecular level. Injection of 1,25(OH)$_2$D into intact rats markedly increases the mRNA levels of the cytochrome P450cc24 (Armbrecht and Boltz, 1991; Shinki et al., 1992). In primary cultures of rat renal tubular cells, 1,25(OH)$_2$D rapidly (within 2 hours) increases P450cc24 mRNA levels by a receptor-mediated pathway requiring new mRNA and protein synthesis (Chen et al., 1993b). The rapid increase in mRNA is similar to the rapid rise in 24-OHase activity seen in renal cell lines (Matsumoto et al., 1985). These findings suggest that increased cytochrome P450cc24 mRNA levels account for the induction of renal 24-OHase activity by 1,25(OH)$_2$D.

Recently, it has been shown that phorbol esters modulate the action of 1,25(OH)$_2$D on the renal hydroxylases (Henry and Luntao, 1989). In contrast to their action on the cAMP-dependent pathway, phorbol esters tend to augment the effects 1,25(OH)$_2$D in the kidney. As mentioned previously, phorbol esters alone, like 1,25(OH)$_2$D, tend to decrease renal 1-OHase activity and increase 24-OHase activity. When cultured chick renal cells are incubated with 1,25(OH)$_2$D for 16 hours and then exposed to phorbol esters for 4 hours, the results are additive (Henry and Luntao, 1989). This suggests that phorbol esters work through a pathway independent of 1,25(OH)$_2$D to modulate renal hydroxylase activity.

The interaction of 1,25(OH)$_2$D and phorbol esters has been examined at the molecular level in primary cultures of rat tubular cells (Chen et al., 1993). Preincubation with 1,25(OH)$_2$D followed by a short treatment with TPA markedly increased mRNA levels for cytochrome P450cc24 compared to 1,25(OH)$_2$D alone. TPA alone in the absence of 1,25(OH)$_2$D had no effect. The effect of TPA was blocked by protein kinase inhibitors, suggesting the involvement of PKC. Thus, this effect of 1,25(OH)$_2$D and phorbol esters on renal 24-OHase activity may be explainable in terms of P450cc24 expression. On the other hand, the rapid effect of phorbol esters alone (Mandla et al., 1990) may be mediated by protein phosphorylation.

C. Regulation by Serum Calcium

In vivo studies have long suggested that 1,25(OH)$_2$D production is inversely related to serum Ca concentrations (Boyle et al., 1971; Armbrecht et al., 1982a). It has been proposed that Ca may act directly on the kidney, bypassing the regulatory pathway involving PTH (Figure 1). However, it has been difficult to demonstrate a direct regulatory role of Ca in intact animal studies since changes in serum Ca result in compensatory changes in PTH and serum phosphate (Fig. 1). A more recent study has demonstrated an inverse correlation between serum 1,25(OH)$_2$D and serum Ca in animals with constant serum PTH and phosphate levels (Weisinger et al., 1989). This suggests that serum Ca may directly regulate serum 1,25(OH)$_2$D levels, but these experiments did not differentiate between an effect on renal production or catabolism.

In vitro studies of the effect of Ca on renal 1,25(OH)$_2$D production have been somewhat confusing. In mammals, Ca concentrations above 1 mM inhibit 1,25(OH)$_2$D production by renal slices (Armbrecht et al., 1983) and cultured tubular cells (Fukase et al., 1982). In isolated proximal tubules from rats, maximal production was seen at 0.25 mM Ca, a somewhat lower concentration (Favus and Langman, 1986). In chick tubules, Ca has been reported to inhibit (Bikle and Rasmussen, 1975) or to have no effect (Omdahl and Hunsaker, 1978; Spanos et al., 1981). Part of the discrepancy may be due to the fact that it is difficult to demonstrate an effect of extracellular Ca in preparations from vitamin D-deficient animals (Armbrecht et al., 1983; Spanos et al., 1981). The inhibition of 1,25(OH)$_2$D production by Ca in renal slices is not dependent on new protein synthesis and is additive to the inhibition produced by 1,25(OH)$_2$D itself (Armbrecht et al., 1983).

D. Regulation by Serum Phosphate

A number of *in vivo* studies have shown that renal 1,25(OH)$_2$D production is inversely related to serum phosphate levels (Armbrecht et al., 1982a; Tanaka and DeLuca, 1973). Feeding a low phosphate diet markedly stimulates renal 1,25(OH)$_2$D production (Portale et al., 1986). It has recently been reported that in the isolated perfused kidney low phosphate in the perfusate increases renal production of 1,25(OH)$_2$D (Thomas and Spencer, 1993). However, *in vitro* studies using renal tubules have failed to demonstrate an effect of media phosphate on renal metabolism of 25(OH)D, except at high concentrations. No effect of phosphate was seen in chick tubules (Spanos et al., 1981) or cultured mouse tubules at concentrations lower than 3 mM (Fukase et al., 1982). It may be that some factor other than phosphate itself is mediating the effect of dietary phosphate on renal 1,25(OH)$_2$D production. In the rat, the effect of phosphorus restriction on renal $1,25(OH)_2D$ production is dependent on growth hormone (Gray and Garthwaite, 1985).

E. Regulation by Other Hormones

A number of other hormones are known to alter vitamin D metabolism. These include insulin, growth hormone, calcitonin, prolactin, and the sex hormones (Henry and Norman, 1984). For example, experimental diabetes in rats reduces the renal 1-OHase and increases the 24-OHase, and this is reversed by insulin (Wong-surawat et al., 1983). Hypophesectomy and growth hormone have a similar effect on renal 25(OH)D metabolism (Wongsurawat et al., 1984). In the case of insulin, it has been reported that insulin is necessary for the stimulation of 1-OHase activity by PTH in cultured chick tubule cells (Henry, 1981). However, this observation has not been confirmed and extended. Most of the effects of these and other hormones are probably indirect via their modulation of the calcitropic hormones and/or serum Ca and phosphate.

F. Mechanisms of Renal Hydroxylase Regulation

In discussing the mechanisms by which the renal hydroxylases are regulated, it is important to know whether the hydroxylases have separate cytochrome P450 components. In other words, are P450cclα and P450cc24 separate molecular species, or are they the same molecule which then undergoes post-translational modification according to the Ca needs of the organism? Since only the cytochrome P450cc24 has been isolated and cloned, this question still has not been completely resolved. However, the bulk of the evidence suggests that the two P450s are distinct molecules.

It was originally suggested that the renal hydroxylases were identical on the basis of their reciprocal relationship to each other. Manipulation of dietary Ca and vitamin D tends to increase one hydroxylase activity and decrease the other in a roughly proportional way. Thus, a single cytochrome P450 species, acting on the same 25(OH)D substrate, could produce either $1,25(OH)_2D$ or $24,25(OH)_2D$ depending on some sort of post-translational modification. Modifications which have been suggested include phosphorylation, which has been documented in microsomal P450s (Koch and Waxman, 1989), and cleavage of the C-terminal end of the molecule (Ghazarian, 1990). This concept was strengthened when the isolation of P450cclα and P450cc24 was reported from chick kidney using mono-clonal antibodies (Mandel et al., 1990b; Moorthy et al., 1991). The first ten amino acids of the two proteins were identical, and the amino acid compositions were very similar. However, another study reporting on the isolation of P450cclα and P450cc24 from chick kidney found some differences in the amino terminal se-quences and significantly different amino acid compositions (Burgos-Trinidad et al., 1992). Thus, there is evidence in both directions from the chick kidney.

With the cloning of the rat renal P450cc24 (Ohyama et al., 1991), evidence has mounted that this cytochrome is distinct from the cytochrome responsible for 1-OHase activity. In the intact animal, $1,25(OH)_2D$ increases 24-OHase activity and decreases 1-OHase activity (Armbrecht et al., 1982a), and $1,25(OH)_2D$ also increases mRNA levels for P450cc24 in the kidney (Armbrecht and Boltz, 1991; Shinki et al., 1992). $1,25(OH)_2D$ also increases P450cc24 mRNA levels in primary renal cells (Chen et al., 1993) and in an intestinal cell line (Armbrecht et al., 1993) under conditions which increase 24-OHase activity in isolated cells. In dietary studies in rats, P450cc24 mRNA levels correlate with 24-OHase activity but not 1-OHase activity (Arabian et al., 1993). If the cytochrome P450cc24 actually were capable of both hydroxylase activities, then one would not expect such a close correlation with 24-OHase enzyme activity.

Assuming that the 1-OHase and 24-OHase are separate cytochromes, a model for their regulation in the kidney can be proposed (Fig. 3). This model is based on two mechanisms of regulation-expression of the P450 proteins and phosphorylation of ferredoxin. The two enzyme complexes consist of cytochromes P450cc1α and P450cc24 and shared ferredoxin and ferredoxin reductase proteins. The ferredoxin component may be in the phosphorylated or unphosphorylated state. In this model, $1,25(OH)_2D$ stimulates the 24-OHase by increasing P450cc24 synthesis (Armbrecht and Boltz, 1991; Shinki et al., 1992). The mechanism by which it suppresses the 1-OHase is unknown, but this could be accomplished by inhibiting P450lα expression or phosphorylating ferredoxin, which tends to increase 24-OHase activity *in vitro* (Mandel et al., 1990a). PTH increases 1-OHase activity in this model by either increasing P450clα expression or by dephosphorylating ferredoxin (Siegel et al., 1986). PTH alone does not alter expression of P450cc24 (Shinki et al., 1992). PTH may suppress 24-OHase activity by dephosphorylating ferredoxin (Mandel et al., 1990a). These changes are summarized in Table 1. To test this model more thoroughly will require the isolation of P450cc1α. In addition, the role of phosphorylation of ferredoxin in modulating renal hydroxylase activity must be further characterized. The effect of $1,25(OH)_2D$ and PTH on the expression of ferredoxin and ferredoxin reductase would also be of interest. In the adrenal gland and other steroidogenic tissue, these components are under hormonal regu-

Table 1. Effect of $1,25(OH)_2D$ and PTH on Renal Vitamin D Hydroxylases

	$1,25(OH)_2D$	PTH
1α-Hydroxylase Activity	↓	↑
24-Hydroxylase Activity	↑	↓
P450cc1α Expression	?	?
P450cc24 Expression	↑	—
Ferredoxin Phosphorylation	?	↓

lation. Since they may be shared by two competing hydroxylases in the kidney, it is not clear how they may be regulated in that organ.

IV. REGULATION OF THE EXTRA-RENAL 24-HYDROXYLASE

A. Regulation by 1,25(OH)₂D

The major regulator of the extra-renal 24-OHase is $1,25(OH)_2D$. Injection of $1,25(OH)_2D$ rapidly increases 24-OHase activity in mouse intestine (Tomon et al., 1990a) and rat intestine (Goff et al. 1992). This action of $1,25(OH)_2D$ required ongoing transcription (Tomon et al., 1990a). Feeding a low Ca diet also increases intestinal 24-OHase activity, presumably in response to the elevated serum $1,25(OH)_2D$ levels (Goff et al., 1992). In cultured cells, $1,25(OH)_2D$ has been shown to increase 24-OHase activity in an intestinal cell line (Tomon et al., 1990b), in a bone cell line (Makin et al., 1989), in human fibroblasts (Gamblin et al., 1985), and in human lymphocytes (Takeda et al., 1990).

B. Mechanisms of Extra-Renal Hydroxylase Regulation

The induction of intestinal 24-OHase activity by $1,25(OH)_2D$ is accompanied by an increase in expression of the intestinal cytochrome P450cc24 mRNA. Injection of rats with $1,25(OH)_2D$ results in a very rapid increase in intestinal P450cc24 mRNA levels (Armbrecht and Boltz, 1991; Shinki et al., 1992). High serum PTH has no effect on the induction of intestinal 24-OHase activity (Goff et al., 1992) or P450cc24 mRNA (Shinki et al., 1992) by $1,25(OH)_2D$. Presumedly, this is because there are no demonstrable PTH receptors in the intestine, as in the kidney.

$1,25(OH)_2D$ has also been shown to induce P450cc24 mRNA in the IEC-18 intestinal cell line (Armbrecht et al., 1993). Induction is rapid but requires high concentrations of $1,25(OH)_2D$ (100 nM). Treatment with phorbol ester markedly increases mRNA levels at $1,25(OH)_2D$ concentrations as low as 1 nM. This effect of TPA is similar to that seen in renal tubular cells (Chen et al., 1993), suggesting that TPA may act in multiple cell types to increase sensitivity to $1,25(OH)_2D$.

V. REGULATION OF THE HEPATIC 25-HYDROXYLASE

As mentioned previously, the hepatic 25-OHase activity is the sum of both the microsomal and mitochondrial activities. The relative contribution of each to total 25(OH)D production is not known. There is evidence for some regulation of the hepatic 25-OHase, although this enzyme is under much less control than the renal 1-OHase and 24-OHase. For example, in vitamin D intoxication serum 25(OH)D levels are markedly elevated, but serum $1,25(OH)_2D$ levels are normal or low (Hughes et al., 1976). This suggests that inhibition of the 25-OHase by 25(OH)D

itself is not effective. However, there is some evidence for regulation by
1,25(OH)$_2$D and serum Ca. The physiological significance of this regulation
remains to be established.

A. Regulation by 1,25(OH)$_2$D

In human studies, 1,25(OH)$_2$D has been shown to inhibit the rise in serum
25(OH)D seen after vitamin D administration (Bell et al., 1984). However,
1,25(OH)$_2$D also increases the metabolic clearance of 25(OH)D, suggesting that
production of 25(OH)D may not be affected (Halloran et al., 1986). In *in vitro*
studies, 1,25(OH)$_2$D inhibits 25(OH)D production in the perfused rat liver (Baron
and Milne, 1983) and in isolated hepatocytes (Baron and Milne, 1986).

B. Regulation by Calcium

The studies in hepatocytes demonstrate that, in addition to inhibiting 25(OH)D
production, 1,25(OH)$_2$D also increases cytosolic Ca levels in hepatocytes (Baran
and Milne, 1986). Calcium ionophores also inhibit hepatocyte production of
25(OH)D (Baran and Milne, 1986). This suggests that the effect of 1,25(OH)$_2$D in
hepatocytes may be mediated by Ca. On the other hand, a later study found that Ca
ionophore stimulated 25-OHase activity in hepatocytes from normocalcaemic
vitamin D-depleted rats (Benbrahim et al., 1988). It may be that the Ca and vitamin
D status of the donor are important in determining the response of hepatocytes to
stimuli.

C. Regulation by Sex Hormones

The distribution of the hepatic 25-OHase is sex-dependent, and mitochondrial
25-OHase activity may be modulated by sex hormones. In the female, the mitochondrial form of the enzyme predominates, and in the male the microsomal form
is most abundant (Saarem and Pedersen, 1987). In the adult female, microsomal
activity is very low (Andersson & Jornvall, 1986) and may be a separate species
(Hayashi et al., 1988). Testosterone decreases the mitochondrial form of the enzyme
in female rats, and estradiol increases the mitochondrial form in male rats (Saarem
and Pedersen, 1987). Microsomal enzyme activity is relatively unaffected by sex
hormone treatment.

D. Mechanisms of Hepatic Hydroxylase Regulation

There have been no published studies on the mechanism of 25-OHase regulation
at the molecular level. Such studies are possible with the availability of antibodies
to P450cc25 (Hayashi et al., 1986; Bergman and Postlind, 1990; Usui et al., 1990)
and a cDNA probe (Usui et al., 1990). In addition, regulation of the 25-OHase is
seen in isolated hepatocytes (Baran and Milne, 1986), making possible a study of
mechanisms *in vitro*. It would be of interest to compare the regulation of the

25-OHase in liver by 1,25(OH)$_2$D with the effect of 1,25(OH)$_2$D on the 1-OHase and 24-OHase in kidney and intestine.

VI.   SEQUENCES OF THE VITAMIN D HYDROXYLASES

The cDNAs for cytochromes P450cc25 (CYP27) and P450cc24 (CYP24) have been cloned and sequenced (Usui et al., 1990; Ohyama et al., 1991). The CYP27 also participates in bile acid synthesis. Each cDNA consists of a 5' promoter region, a translated presequence, the mature protein, a 3' noncoding region, and a poly A tail. The 3' noncoding region of P450cc25 is short (58 bp), but the 3' noncoding region of P450cc24 is unusually long (343 bp) and GC rich, which may contribute to mRNA stability. Both proteins have a 32–35 amino acid presequence which is amphophilic in nature and characteristic of presequences which direct proteins into mitochondria. The mature P450cc25 protein consists of 501 amino acids with a molecular weight of 57,182, and the P450cc24 protein consists of 479 amino acids corresponding to a molecular weight of 55,535. Both proteins have the characteristic features of cytochrome P450 proteins such as the heme binding domain. These two proteins share the greatest homology with each other compared to other known P450s. However, the low degree of homology (30%) indicates that they are distinct P450 families.

VII.   GENE STRUCTURE OF CYTOCHROME P450cc24

The first gene structure for a vitamin D hydroxylase, cytochrome P450cc24, has recently been elucidated (Ohyama et al., 1993a). The *P450cc24* gene (CYP24) is a single gene in the rat and contains 12 exons spanning about 15 kb. The gene structure of CYP24 is most similar to that of the *CYP11* gene family, which contains the mitochondrial P450scc and P45011β. Seven of the 11 introns found in CYP24 are also found in identical positions in CYP11 with the same phases. CYP24, therefore, has four additional introns, and it lacks only intron 1 from CYP11.

Analysis of the 5' flanking region of the *CYP24* gene is of special interest because of the multifactorial regulation of the expression of the cytochrome P450cc24. In the region -516 to -1, two sequences similar to the vitamin D responsive element were found. A typical binding site for transcriptional factor Sp1 was also found in this region, but consensus regulatory elements for cAMP and protein kinase C were not found.

Preliminary transfection experiments have been performed to characterize the function of the CYP24 promoter (Ohyama et al., 1993b). Various promoter constructs were fused to the CAT reporter gene and transfected into COS cells. The response to 1,25(OH)$_2$D (about 30-fold) was not affected by removal of the region greater than -167, but all activity was lost below -102. In gel retardation assays, a fragment (−167 to −129) produced a specific retarded band in the presence of 1,25(OH)$_2$D and vitamin D receptor. These findings suggest that the region from

−167 to −129 confers responsiveness. This region contains several vitamin D response elements similar to those seen in osteocalcin.

VIII. CLINICAL RELEVANCE

Many diseases involve altered vitamin D metabolism with a subsequent effect on Ca homeostasis (Aurbach et al., 1992). Some of these involve altered serum levels of $1,25(OH)_2D$. Specifically, there may be either too little or too much production of $1,25(OH)_2D$. Too little production of $1,25(OH)_2D$ does not usually result in decreased serum Ca, but it does result in increased bone resorption. The increased bone resorption is necessary to maintain serum Ca in the face of decreased intestinal Ca absorption. Examples of this include aging, diabetes, and hypophosphatemia. Animal models have been developed for these diseases, and changes in renal $1,25(OH)_2D$ production have been studied in these models. Overproduction of $1,25(OH)_2D$ usually results in hypercalcemia and hypercalciuria. An example of this includes absorptive hypercalciuria, which may involve excess renal production of $1,25(OH)_2D$. Another example is the hypercalcemia of granulomatous diseases such as sarcoidosis, which is due to production of $1,25(OH)_2D$ by affected tissue. Although there are many other disorders involving Ca and vitamin D metabolism, the ones discussed here directly involve the 1-OHase enzyme and its regulation.

A. Aging

The rat has been used as an animal model of the aging process. Renal $1,25(OH)_2D$ production decreases with age in the rat as measured using renal slices (Armbrecht et al., 1980) and isolated renal mitochondria (Ishida et al., 1987). This decrease in renal $1,25(OH)_2D$ production results in decreased serum $1,25(OH)_2D$ levels (Armbrecht et al., 1984a), decreased intestinal Ca absorption, and negative Ca balance (Armbrecht et al., 1984a). This age-related decrease in renal $1,25(OH)_2D$ production is seen despite an increase in serum PTH and a decrease in serum phosphorus with age (Armbrecht et al., 1984a). Since increased PTH and decreased serum phosphorus are normally potent stimulators of the renal 1-OHase, this indicates altered regulation of the enzyme with age. PTH administration to the intact animal *in vivo* or to isolated renal slices *in vitro* does not stimulate renal $1,25(OH)_2D$ production in adult rats as it does in young rats (Armbrecht et al., 1982b). Clinical studies have shown that the capacity of PTH to raise serum $1,25(OH)_2D$ levels also decreases with age in humans (Tsai et al., 1984). This lack of response to PTH is not due to decreased stimulation of cAMP production or cAMP-dependent protein kinase activity by PTH with age (Armbrecht et al., 1986). Rather, the capacity of PTH to alter the phosphorylation state of ferredoxin declines with age (Wong-surawat et al., 1989). Since the phosphorylation state of ferredoxin may regulate 1-OHase activity, this could account for the decrease in PTH stimulation with age.

Decreased renal $1,25(OH)_2D$ production has been proposed as an important factor in age-related (Type II) osteoporosis.

B. Diabetes

The streptozotocin rat has been used as a model for juvenile-onset (Type I) diabetes. This diabetic model has low serum $1,25(OH)_2D$ levels and decreased intestinal Ca absorption. Renal $1,25(OH)_2D$ production is decreased in the diabetic rat, and production is elevated toward normal by insulin (Wongsurawat et al., 1983). Since serum PTH levels do not change with diabetes and insulin treatment, these results suggest that the diabetic kidney is refractory to PTH in terms of 1-OHase activity. This has been confirmed in intact animal studies (Wongsurawat and Armbrecht, 1985). Diabetic rats do not respond to PTH by increasing renal $1,25(OH)_2D$ production, but this response is restored by insulin treatment (Wongsurawat and Armbrecht, 1985). As in aging animals, decreased PTH responsiveness is not due to decreased cAMP production or cAMP-dependent protein kinase activity (Wongsurawat et al., 1991). However, the capacity of PTH to alter the phosphorylation state of ferredoxin is blunted (Wongsurawat et al., 1991). These alterations in $1,25(OH)_2D$ production could contribute to the osteopenia seen in the diabetic population.

C. Hypophosphatemia

Hypophosphatemic rickets is characterized by hypophosphatemia, hyperphosphaturia, and low serum $1,25(OH)_2D$ levels. Patients with this disease do not respond to PTH with normal increases in serum $1,25(OH)_2D$ levels (Lyles and Drezner, 1982). The biochemical basis of this disease has been studied using the mutant hypophosphatemic mouse. In this mouse model, renal 1-OHase activity is decreased compared to control animals with the same degree of hypophosphatemia (Lobaugh and Drezner, 1983). In addition, the capacity of PTH to stimulate renal 1-OHase activity in the mutant mouse is diminished (Nesbitt et al., 1986).

D. Absorptive Hypercalciuria

Absorptive hypercalciuria is a common cause of kidney stones. In some cases, this is accompanied by increased serum $1,25(OH)_2D$ levels. This has been shown to be due to increased production of $1,25(OH)_2D$ (Insogna et al., 1985). Further studies have shown that these patients do not respond appropriately to a long term increase in dietary Ca by suppressing serum $1,25(OH)_2D$ levels (Broadus et al., 1984).

E. Sarcoidosis and other Granulomatous Diseases

Sarcoidosis is characterized by hypercalcemia which is due to increased serum $1,25(OH)_2D$ levels due to lack of regulation (Bell et al., 1979). In this disease, the production of $1,25(OH)_2D$ is extrarenal. Production of $1,25(OH)_2D$ has been demonstrated in cultures of alveolar macrophages from patients with sarcoidosis (Adams et al., 1983).

ACKNOWLEDGMENTS

This work was supported by the Medical Research Service and the Geriatric Research, Education, and Clinical Center of the Department of Veterans Affairs, NIH grant AG-12587, and by NSF grant 91-18003. The editorial assistance of Carol McCleary is gratefully acknowledged.

REFERENCES

Adams, J.S., Sharma, O.P., Gacad, M.A., & Singer, F.R. (1983). Metabolism of 25-hydroxyvitamin D_3 by cultured pulmonary alveolar macrophages in sarcoidosis. J. Clin. Invest. 72, 1856–1860.

Andersson, S., Holmberg, I., & Wikvall, K. (1983). 25-Hydroxylation of C_{27}-steroids and vitamin D_3 by a constitutive cytochrome P-450 from rat liver microsomes. J. Biol. Chem. 258, 6777–6781.

Andersson, S., & Jornvall, H. (1986). Sex differences in cytochrome P-450-dependent 25-hydroxylation of C_{27}-steroids and vitamin D_3 25-hydroxylase of rat liver microsomes. J. Biol. Chem. 261, 16932–16936.

Arabian, A., Grover, J., Barre, M.G., & Delvin, E.E. (1993). Rat kidney 25-hydroxyvitamin D_3 1 alpha- and 24-hydroxylases: Evidence for two distinct gene products. J. Steroid Biochem. Mol. Biol. 45, 513–516.

Armbrecht, H.J., Zenser, T.V., & Davis, B.B. (1980). Effect of age on the conversion of 25-hydroxyvitamin D to 1,25-dihydroxyvitamin D by the kidney of the rat. J. Clin. Invest. 66, 1118–1123.

Armbrecht, H.J., Gross, C.J., & Zenser, T.V. (1981). Effect of dietary calcium and phosphorus restriction on calcium and phosphorus balance in young and old rats. Arch. Biochem Biophys. 210, 179–185.

Armbrecht, H.J., Zenser, T.V., & Davis, B.B. (1982a). Modulation of renal production of 24,25- and 1,25-dihydroxyvitamin D_3 in young and adult rats by dietary calcium, phosphorus, and 1,25-dihydroxyvitamin D_3. Endocrinology 110, 1983–1988.

Armbrecht, H.J., Wongsurawat, N., Zenser, T.V., & Davis, B.B. (1982b). Differential effects of parathyroid hormone on the renal 1,25-dihydroxyvitamin D_3 and 24,25-dihydroxyvitamin D_3 production of young and adult rats. Endocrinology 111, 1339–1344.

Armbrecht, H.J., Wongsurawat, N., Zenser, T.V., & Davis, B.B. (1983). In vitro modulation of renal 25-hydroxyvitamin D_3 metabolism by vitamin D_3 metabolites and calcium. Arch. Biochem. Biophys. 220, 52–59.

Armbrecht, H.J., Forte, L.R., & Halloran, B.P. (1984a). Effect of age and dietary calcium on renal 25(OH)D metabolism, serum $1,25(OH)_2D$, and PTH. Am. J. Physiol. 246, E266–270.

Armbrecht, H.J., Wongsurawat, N., Zenser, T.V., & Davis, B.B. (1984b). Effect of PTH and $1,25(OH)_2D_3$ on renal 25(OH)D_3 metabolism, adenylate cyclase, and protein kinase. Am. J. Physiol. 246, E102–E107.

Armbrecht, H.J., Forte, L.R., Wongsurawat, N., Zenser, T.V., & Davis, B.B. (1984c). Forskolin increases 1,25-dihydroxyvitamin D_3 production by rat renal slices in vitro. Endocrinology 114, 644–649.

Armbrecht, H.J., Boltz. M.A., & Foric, L.R. (1986). Effect of age on parathyroid hormone and forskolin stimulated adenylate cyclase and protein kinase activity in the renal cortex. Exptl. Gerontol. 21, 515–522.

Armbrecht, H.J., & Boltz, M.A. (1991). Expression of 25 hydroxyvitamin D 24-hydroxylase cytochrome P450 in kidney and intestine. Effect of 1,25-dihydroxyvitamin D and age. FEBS Lett. 292, 17–20.

Armbrecht, H.J., Hodam, T.L., Boltz, M.A., & Chen, M.I. (1993). Phorbol ester markedly increases the sensitivity of intestinal epithelial cells to 1,25-dihydroxyvitamin D_3. FEBS Letters 327, 13–16.

Aurbach, G.D., Marx, S.J., & Spiegel, A.M. (1992). Parathyroid hormone, calcitonin, and the calciferols. In: Williams Textbook of Endocrinology (Wilson, J.D. and Foster, D.W., eds.), pp. 1397–1476. Saunders, Philadelphia. 1992.

Baran, D.T., & Milne, M.L. (1983). 1,25-dihydroxyvitamin D-induced inhibition of [3]H-25-hydroxyvitamin D production by the rachitic rat liver *in vitro*. Calcif. Tissue Int. 35, 461–464.

Baran, D.T., & Milne, M.L. (1986). 1,25-dihydroxyvitamin D increases hepatocyte cytosolic calcium levels: A potential regulator of vitamin D-25 hydroxylase. J. Clin. Invest. 77, 1622–1626.

Bell, N.H., Stern, P.H., Pantzer, E., Sinha, T.K., & DeLuca, H.F. (1979). Evidence that increased circulating 1α,25-dihydroxyvitamin D is the probable cause for abnormal calcium metabolism in sarcoidosis. J. Clin. Invest. 64, 218–225.

Bell, N.H., Shaw, S., & Turner, R.T. (1984). Evidence that 1,25-dihydroxyvitamin D_3 inhibits the hepatic production of 25-hydroxyvitamin D in man. J. Clin. Invest. 74, 1540–1544.

Benbrahim, N., Dube, C., Vallieres, S., & Gasron-Barre, M. (1988). The calcium ionophore A23187 is a potent stimulator of the vitamin D_3-25-hydroxylase in hepatocytes isolated from normocalcemic vitamin D-depleted rats. Biochem. J. 255, 91–97.

Bergman, T., & Postlind, H. (1990). Characterization of pig kidney microsomal cytochrome P-450 catalyzing 25-hydroxylation of vitamin D_3 and C_{27} steroids. Biochem. J. 270, 345–350.

Bikle, D.D., & Rasmussen, H. (1975). The ionic control of 1,25-dihydroxyvitamin D_3 production in isolated chick renal tubules. J. Clin. Invest. 55,292–298.

Bikle, D.D., Nemanic, M.K., Gee, E., & Elias, P. (1986). 1,25-Dihydroxyvitamin D_3 production by human keratinocytes, kinetics, and regulation. J. Clin. Invest. 78, 557–566.

Bjorkhem, I., Holmberg, I., Oftebro, H., & Pedersen, J.I. (1980). Properties of a reconstituted vitamin D_3 25-hydroxylase from rat liver mitochondria. J. Biol. Chem. 255, 5244–5249.

Bort, R.E., & Crivello, J.F. (1988). Characterization of monoclonal antibodies specific to bovine renal vitamin D hydroxylases. Endocrinology 123, 2491–2498.

Boyle, I.T., Gray, R.W., & DeLuca, H.F. (1971). Regulation by calcium of *in vivo* synthesis of 1,25-dihydroxycholecalciferol and 21,25-dihydroxycholecalciferol. Proc. Nat. Acad. Sci. USA 68, 2131–2134.

Broadus, A.E., Insogna, K.L, Lang, R., Ellison, A.F., & Dreyer, B.E. (1984). Evidence for disordered control of 1,25-$(OH)_2$D production in absorptive hypercalciuria. N. Engl. J. Med. 311, 73–80.

Burgos-Trinidad, M., Ismail, R., Ettinger, R.A., Prahl, J.M., & DeLuca, H.F. (1992). Immunopurified 25-hydroxyvitamin D 1α-hydroxylase and 1,25-dihydroxyvitamin D 24-hydroxylase are closely related but distinct enzymes. J. Biol. Chem. 267, 3498–3505.

Chen, K.S., Prahl, J.M., & DeLuca, H.F. (1993a). Isolation and expression of human 1,25-dihydroxyvitamin D_3 24-hydroxylase cDNA. Proc. Natl. Acad. Sci. 90, 4543–4547.

Chen, M.L., Boltz, M.A., & Armbrechr, H.J. (1993b). Effects of 1,25-dihydroxyvitamin D_3 and phorbol ester on 25-hydroxyvitamin D_3 24-hydroxylase cytochrome P450 messenger ribonucleic acid levels in primary cultures of rat renal cells. Endocrinology 132, 1782–1788.

Dahlback, H., & Wikvall, K. (1987). 25-Hydroxylation of vitamin D_3 in rat liver: Roles of mitochondrial and microsomal cytochrome P-450. Biochem. Biophys. Res. Commun. 142, 999–1005.

DeLuca, H.F. (1988). The vitamin D story: A collaborative effort of basic science and clinical medicine. FASEB J. 2, 224–236.

Favus, M.J., & Langman, C.B. (1986). Evidence for calcium-dependent control of 1,25-dihydroxyvitamin D_3 production by rat kidney proximal tubules. J. Biol. Chem. 261, 11224–11229.

Fukase, M., Birge, S.J., Rifas, L., Avioli, L.V., & Chase, L.R. (1982). Regulation of 25 Hydroxyvitamin D_3 1-hydroxylase in serum-free monolayer culture of mouse kidney. Endocrinology 110, 1073–1075.

Gamblin, G.T., Liberman, U.A., Eil, C., Downs, R.W. Jr., DeGrange, D.A., & Marx, S.J. (1985). Vitamin D-dependent rickets type II: Defective induction of 25-hydroxyvitamin D_3-24-hydroxylase by 1,25-dihydroxyvitamin D_3 in cultured skin fibroblasts. J. Clin. Invest. 75. 954–960.

Garabedian, M., Holick, M.F., DeLuca, H.F., & Boyle, I.T. (1972). Control of 25-hydroxycholecalciferol metabolism by parathyroid glands. Proc. Natl. Acad. Sci. USA 69, 1673–1676.

Ghazarian, J.G. (1990). The renal mitochondrial hydroxylases of the vitamin D_3 endocrine complex: How are they regulated at the molecular level? J. Bone Mineral Res. 5, 897–903.

Goff, J.P., Reinhardt, T.A., Engstrom, G.W., & Horst, R.L. (1992). Effect of dietary calcium or phosphorus restriction and 1,25-dihydroxyvitamin D administration on rat intestinal 24-hydroxylase. Endocrinology 131, 101–104.

Gray, R.W., & Garthwaite, T.L. (1985). Activation of renal $1,25(OH)_2D$ synthesis by phosphate deprivation: Evidence for a role for growth hormone. Endocrinology 116, 189–193.

Gray, R.W., & Ghazarian, J.G. (1989). Solubilization and reconstitution of kidney 25-hydroxyvitamin D_3 1α- and 24-hydroxylases from vitamin D-replete pigs. Biochem. J. 259, 561–568.

Gray, R.W., Omdahl, J.L., Ghazarian. J.G., & Horst, R.L. (1990). Induction of 25-OH-vitamin D_3 24- and 23-hydroxylase activities in partially purified renal extracts from pigs given exogenous $1,25(OH)_2D_3$. Steroids 55, 395–398.

Halloran, B.P., Bikle, D.D., Levens, M.J., Castro, M.E., Globus, R.K., & Holton, E. (1986). Chronic 1,25-dihydroxyvitamin D_3 administration in the rat reduces the serum concentration of 25-hydroxyvitamin D by increasing metabolic clearance rate. J. Clin. Invest. 78, 622–628.

Hayashi, S., Noshiro, M., & Okuda, K. (1986). Isolation of a cytochrome P-450 that catalyzes the 25-hydroxylation of vitamin D_3 from rat liver microsomes. J. Biochem. 99, 1753–1763.

Hayashi, S., Usui, E., & Okuda, K. (1988). Sex-related difference in vitamin D_3 25-hydroxylase of rat liver microsomes. J. Biochem. 103, 863–866.

Henry, H.L., Midgett, R.J., & Norman, A.W. (1974). Studies on calciferol metabolism X. Regulation of 25-hydroxyvitamin D_3-1-hydroxylase, in vivo. J. Biol. Chem. 249, 7584–7592.

Henry, H. (1979). Regulation of the hydroxylation of 25-hydroxyvitamin D_3 in vivo and in primary cultures of chick kidney cells. J. Biol. Chem. 254, 2722–2729.

Henry, H.L. (1981). Insulin permits parathyroid hormone stimulation of 1,25-dihydroxyvitamin D_3 production in cultured kidney cells. Endocrinology 108, 733–735.

Henry, H.L., & Norman, A.W. (1984). Vitamin D: Metabolism and Biological Actions. Ann. Rev. Nutr. 4, 493–520.

Henry, H.L. (1986). Influence of a tumor promoting phorbol ester on the metabolism of 25-hydroxyvitamin D_3. Biochem. Biophys. Res. Commun. 139, 495–500.

Henry, H.L., & Luntao, E.M. (1989). Interactions between intracellular signals involved in the regulation of 25-hydroxyvitamin D_3 metabolism. Endocrinology 124, 2228–2234.

Hiwatashi, A., Nishii, Y., & Ichikawa, Y. (1982). Purification of cytochrome P-4501α (25-hydroxyvitamin D_3-1α-hydroxylase) of bovine kidney mitochondria. Biochem. Biophys. Res. Commun. 105, 320–327.

Howard, G.A., Turner, R.T., Sherrard, D.J., & Baylink, D.J. (1981). Human bone cells in culture metabolize 25-hydroxyvitamin D_3. J. Biol. Chem. 256, 7738–7740.

Hughes, M.R., Baylink, D.J., Jones, P.G., & Haussler, M.R. (1976). Radioligand receptor assay for 25-hydroxyvitamin D_2/D_3 and 1,25-dihydroxyvitamin D_2/D_3: application to hypervitaminosis D. J. Clin. Invest. 58, 61–70.

Insogna, K.L., Broadus. A.E., Dreyer, B.E., Ellison, A.F., & Gertner, J.M. (1985). Elevated production rate of 1,25-dihydroxyvitamin D in patients with absorptive hypercalciuria. J. Clin. Endocrinol. Metab. 61, 490–495.

Ishida, M., Bulos, B., Takamoto, S., & Sacktor, B. (1987). Hydroxylation of 25-hydroxyvitamin D by renal mitochondria from rats of different ages. Endocrinology 121, 443–448.

Kawashima, H., & Kurokawa, K. (1983). Unique hormonal regulation of vitamin D metabolism in the mammalian kidney. Miner. Electrolyte Metab. 9, 227–235.

Koch, J.A., & Waxman, D.J. (1989). Posttranslational modification of hepatic cytochrome P-450. Phosphorylation of phenobarbitol-inducible P-450 forms PB-4(IIB1) and PB-5 (IIB2) in isolated rat hepatocytes and *in vivo*. Biochemistry 28, 3145–3152.

Korkor, A.B., Gray, R.W., Henry, H.L., Kleinman, J.G., Blumenthal, S.S., & Garancis, J.C. (1987). Evidence that stimulation of 1,25(OH)$_2$D$_3$ production in primary cultures of mouse kidney cells by cyclic AMP requires new protein synthesis. J. Bone Miner. Res. 2, 517–524.

Kramer, R., & Goltzman, D. (1982). Parathyroid hormone stimulates mammalian 25-hydroxyvitamin D$_3$-1-α-hydroxylase *in vitro*. Endocrinology 110, 294–296.

Kumar, R., Schoer, H.K., & DeLuca, H.F. (1978). Rat intestinal 25-hydroxyvitamin D$_3$ and 1α, 25-dihydroxyvitamin D$_3$-24-hydroxylase. J. Biol. Chem. 253, 3804–3809.

Lobaugh, B., & Drezner. M.K. (1983). Abnormal regulation of renal 25-hydroxyvitamin D-1α-hydroxylase activity in the X-linked hypophosphatemic mouse. J. Clin. Invest. 71, 400–403.

Lyles, K.W., & Drezner, M.K. (1982). Parathyroid hormone effects on serum 1,25-(OH)$_2$D levels in patients with X-linked hypophosphatemic rickets: evidence for abnormal 25-(OH)D 1-hydroxylase activity. J. Clin. Endocrinol. Metab. 54, 638–644.

Makin, G., Lohnes, D., Byford, V., Ray, R., & Jones, G. (1989). Target cell metabolism of 1,25-dihydroxyvitamin D$_3$ to calcitroic acid. Evidence for a pathway in kidney and bone involving 24-oxidation. Biochem. J. 262, 173–180.

Mandel, M.L., Moorthy, B., & Ghazarian, J.G. (1990a). Reciprocal post-translational regulation of renal 1α- and 24-hydroxylases of 25-hydroxyvitamin D$_3$ by phosphorylation of ferredoxin. Biochem. J. 266, 385–392.

Mandel, M.L., Swartz, S.J., & Ghazarian, J.G. (1990b). Avian kidney mitochondrial hemeprotein P-4501α: Isolation, characterization and NADPH-ferredoxin reductase-dependent activity. Biochim. Biophys. Acta 1034, 239–246.

Mandla, S., Boneh, A., & Tenenhouse, H.S. (1990). Evidence for protein kinase C involvement in the regulation of renal 25-hydroxyvitamin D$_3$-24 hydroxylase. Endocrinology 127, 2639–2647.

Masumoto, O., Ohyama, Y., & Okuda, K. (1988). Purification and characterization of vitamin D 25-hydroxylase from rat liver mitochondria. J. Biol. Chem. 263, 14256–14260.

Matsumoto T., Kawanobe, Y., & Ogata, E. (1985). Regulation of 24,25-dihydroxyvitamin D-3 production by 1,25-dihydroxyvitamin D-3 and synthetic human parathyroid hormone fragment 1-34 in a cloned monkey kidney cell line (JTC-12). Biochim. Biophys. Acta 845, 358–365.

Minghetti, P.P., & Norman, A.W. (1988). 1,25(OH)$_2$-Vitamin D$_3$ receptors: Gene regulation and genetic circuitry. FASEB J. 2, 3043–3053.

Monnier, N., Defaye, G., & Chambaz, E.M. (1987). Phosphorylation of bovine adrenodoxin. Structural study and enzymatic activity Eur. J. Biochem. 169, 147–153.

Moorthy, B., Mandel, M.L., & Ghazarian, J.G. (1991). Amino-terminal sequence homology of two chick kidney mitochondrial proteins immunoisolated with monoclonal antibodies to the cytochrome P450 of 25-hydroxyvitamin D$_3$-1α-hydroxylase. J. Bone Miner. Res. 6, 199–204.

Nemani, R., Ghazarian, J.G., Moorthy, B., Wongaurawat, N., Strong, R., & Armbrecht, H.J. (1989). Phosphorylation of ferredoxin and regulation of renal mitochondrial 25-hydroxyvitamin D-1-hydroxylase activity *in vitro*. J. Biol. Chem. 264, 15361–15366.

Nemani, R., Wongsurawat, N., & Armbrecht, H.J. (1991). Effect of parathyroid hormone on rat renal cAMP-dependent protein kinase and protein kinase C activity measured using synthetic peptide substrates. Arch. Biochem. Biophys. 285, 153–157.

Nesbitt, T., Drezner, M.K., & Lobaugh, B. (1986). Abnormal parathyroid hormone stimulation of 25-hydroxyvitamin D-1α-hydroxylase activity in the hypophosphatemic mouse. Evidence for a generalized defect of vitamin D metabolism. J. Clin. Invest. 77, 181–187.

Ohyama, Y., & Okuda, K. (1991). Isolation and characterization of a cytochrome P-450 from rat kidney mitochondria that catalyzes the 24-hydroxylation of 25-hydroxyvitamin D$_3$. J. Biol. Chem. 266, 8690–8695.

Ohyama, Y., Noshiro, M., & Okuda, K. (1991). Cloning and expression of cDNA encoding 25-hydroxy-vitamin D_3 24-hydroxylase. FEBS Lett. 278, 195–198.

Ohyama, Y., Noshiro, M., Eggertsen, G., Gotoh, O., Kato, Y., Bjorkhem, I., & Okuda, K. (1993a). Structural characterization of the gene encoding rat 25-hydroxyvitamin D_3 24-hydroxylase. Biochemistry 32, 76–82.

Ohyama, Y., Ozono, K., Uchida, M., Noshiro, M., & Kato, Y. (1993b). Sequence elements in the rat 25-hydroxyvitamin D_3 24-hydroxylase which confer responsiveness to vitamin D. J. Bone Mineral Res. 8, S137.

Omdahl, J.L., & Hunsaker, L.A. (1978). Direct modulation of 25-hydroxyvitamin D_3 hydroxylation in kidney tubules by 1,25-dihydroxyvitamin D_3. Biochem. Biophys. Res. Comm. 81, 1073–1079.

Paulson, S.K., & DeLuca, H.F. (1985). Subcellular location and properties of rat renal 25-hydroxyvi-tamin D_3-1α-hydroxylase. J. Biol. Chem. 260, 11488–11492.

Portale, A.A., Halloran, B.P., Murphy, M.M., & Morris, R.C. (1986). Oral intake of phosphorus can determine the serum concentration of 1,25-dihydroxyvitamin D by determining its production rate in humans. J. Clin. Invest. 77, 7–12.

Reichel, H., Koeffler, H.P., & Norman, A.W. (1987). Synthesis *in vitro* of 1,25-dihydroxyvitamin D_3 and 24,25-dihydroxyvitamin D_3 by interferon-c-stimulated normal human bone marrow and alveolar macrophages. J. Biol. Chem. 262, 10931–10937.

Reinhardt, T.A., & Horst, R.L. (1989). Ketoconazole inhibits self-induced metabolism of 1,25-dihy-droxyvitamin D_3 and amplifies 1,25-dihydroxyvitamin D_3 receptor up-regulation in rat osteosar-coma cells. Arch. Biochem. Biophys. 272, 459–465.

Rost, C.R., Bikle, D.D., & Kaplan, R.A. (1981). *In vitro* stimulation of 25-hydroxycholecalciferol 1-α-hydroxylation by parathyroid hormone in chick kidney slices: Evidence for a role for adenosine-3', 5'-monophosphate. Endocrinology 108, 1002–1006.

Saarem, K., Bergseth, S., Oftebro, H., & Pederson, J.I. (1984). Subcellular localization of vitamin D_3 25-hydroxylase in human liver. J. Biol. Chem. 259, 10936–10940.

Saarem, K., & Pedersen, J.I. (1987). Sex differences in the hydroxylation of cholecalciferol and of 5 beta-cholestane-3 alpha-7 alpha-12 alpha-triol in rat liver. Biochem. J. 247, 73–78.

Shinki, T., Jin, C.H., Nishimura, A., Nagai, Y., Ohyama, Y., Noshiro, M., Okuda, K., & Suda, T. (1992). Parathyroid hormone inhibits 25-hydroxyvitamin D_3-24-hydroxylase mRNA expression stimu-lated by 1,25-dihydroxyvitamin D_3 in rat kidney but not in intestine. J. Biol. Chem. 267, 13757–13762.

Siegel, N., Wongsurawat, N., & Armbrecht, H.J. (1986). Parathyroid hormone stimulates dephospho-rylation of the renoredoxin component of the 25-hydroxyvitamin D_3-1-hydroxylase from rat renal cortex. J. Biol. Chem. 261, 16998–17003.

Spanos, E., Freake, H., MacAuley, S.J., & MacIntyre, I. (1981). Regulation of vitamin D metabolism by calcium and phosphate ions in isolated renal tubules. Biochem. J. 196, 187–193.

Takeda, E., Yokota, I., Ito, M., Kobashi, H., Saijo, T., & Kuroda, Y. (1990). 25-Hydroxyvitamin D-24-hydroxylase in phytohemagglutinin-stimulated lymphocytes: Intermediate bioresponse to 1,25-dihydroxyvitamin D_3 of cells from parents of patients with vitamin D-dependent rickets type II. J. Clin. Endocrin. Metab. 70, 1068–1074.

Tanaka, Y., & DeLuca, H.F. (1973). The control of 25-hydroxyvitamin D metabolism by inorganic phosphorus. Arch. Biochem. Biophys. 154, 566–574.

Tanaka, Y., & DeLuca, H.F. (1974). Stimulation of 24,25-dihydroxyvitamin D_3 production by 1,25-di-hydroxyvitamin D_3. Science 183, 1198–1200.

Tang, C., Kain, S.R., & Henry, H.L. (1993). The phorbol ester 12-O-tetrodecanoyl-phorbol-13-acetate stimulates the dephosphorylation of mitochondrial ferredoxin in cultured chick kidney cells. Endocrinology 133: 1823–1829.

Thomas, B.R., & Spencer, E.M. (1993). 1,25-Dihydroxyvitamin D_3 production in the isolated perfused rat kidney in response to changing perfusate phosphorous concentrations and insulin-like growth factor. J. Nutr. Biochem. 4, 158–161.

Tomon, M., Tenenhouse, H.S., & Jones, G. (1990a). 1,25-Dihydroxyvitamin D_3-inducible catabolism of vitamin D metabolites in mouse intestine. Am. J. Physiol. 258, G557–G563.

Tomon, M., Tenenhouse, H.S., & Jones, G. (1990b). Expression of 25-hydroxyvitamin D_3-24-hydroxylase activity in Caco-2 cells. An *in vitro* model of intestinal vitamin D catabolism. Endocrinology 126, 2868–2875.

Tsai, K.S., Heath, H., Kumar, R., & Riggs, B.L. (1984). Impaired vitamin D metabolism with aging in women. J. Clin. Invest. 73, 1668–1672.

Usui, E., Noshiro, M., & Okuda, K. (1990). Molecular cloning of cDNA for vitamin D_3 25-hydroxylase from rat liver mitochondria. FEBS Lett. 262, 135–138.

Weisinger, J.R., Favus, M.J., Langman, C.B., & Bushinsky, D.A. (1989). Regulation of 1,25-dihydroxyvitamin D_3 by calcium in the parathyroidectomized, parathyroid hormone-replete rat. J. Bone and Mineral Res. 4, 929–935.

Wongsurawat, N., Armbrecht, H.J., Zenser, T.V., Davis, B.B., Thomas, M.L., & Forte, L.R. (1983). 1,25-Dihydroxyvitamin D_3 and 24,25-dihydroxyvitamin D_3 production by isolated renal slices is modulated by diabetes and insulin in the rat. Diabetes 32, 302–306.

Wongsurawat, N., Armbrecht, H.J., Zenser, T.V., Forte, L.R., & Davis, B.B. (1984). Effects of hypophysectomy and growth hormone treatment on renal hydroxylation of 25-hydroxycholecalciferol in rats. J. Endocrinol. 101, 333–338.

Wongsurawat, N., & Armbrecht, H.J. (1985). Insulin modulates the stimulation of renal 1,25-dihydroxyvitamin D_3 production by parathyroid hormone. Acta. Endocr. 109, 243–248.

Wongsurawat, N., Armbrecht, H.J., & Siegel, N.A. (1989). Effects of diabetes and age on PTH-stimulated phosphorylation state of renoredoxin. J. Bone Mineral Res. 4, S145.

Wongsurawat, N., Armbrecht, H.J. & Siegel. N.A. (1991). Effects of diabetes mellitus on parathyroid hormone-stimulated protein kinase activity, ferredoxin phosphorylation, and renal 1,25-dihydroxyvitamin D production. J. Lab. Clin. Med. 117, 319–324.

THE REGULATION OF CHOLESTEROL
CONVERSION TO BILE ACIDS

John Y. L. Chiang and Z. Reno Vlahcevic

Advances in Molecular and Cell Biology
Volume 14, pages 269–316.
Copyright © 1996 by JAI Press Inc.
All rights of reproduction in any form reserved.
ISBN: 0-7623-0113-9

I. INTRODUCTION

Cytochrome P450 isozymes are known to have broad and overlapping substrate specificity toward endogenous compounds, such as cholesterol, fatty acids, and xenobiotics such as drugs and carcinogens. Three unique cytochrome P450 isozymes are involved in the conversion of cholesterol to bile acids which are physiological agents needed for the transport and disposal of sterols, lipid-soluble vitamins, drugs and other xenobiotics in mammals. The major pathway for the catabolism of cholesterol to bile acids in the liver requires a cytochrome P450 monooxygenase, cholesterol 7α-hydroxylase, as the first and rate-limiting enzyme (Myant and Mitropoulos, 1977; Björkhem, 1985; Russell and Setchell, 1992). Microsomal cholesterol 7α-hydroxylase (P450c7) is a product of the *CYP7* gene (Nelson et al., 1993). The rate of bile acid synthesis and cholesterol 7α-hydroxylase activity are under negative feedback regulation by bile acids returning to the liver via enterohepatic circulation of bile (Carey and Cahalane, 1988; Vlahcevic et al., 1989). Because cholesterol 7α-hydroxylase is the rate-limiting enzyme in bile acid synthesis, a major pathway for elimination of cholesterol from the body, the regulation of this enzyme has been a subject of intense study within last two decades. Recently the highly purified enzyme, specific antibody and cDNA clone have become available to study its regulation at the molecular level (Ogishima et al., 1987; Noshiro et al., 1989; Chiang et al., 1990; Li et al., 1990; Jelinek et al., 1990). The second microsomal cytochrome P450 involved in bile acid synthesis is sterol-12α-hydroxylase, which is required for the synthesis of cholic acid, and may play a role in the regulation of the ratio of cholic acid to chenodeoxycholic acid in the bile. A purified P450 isozyme capable of catalyzing 12α-hydroxylase activity was obtained recently (Ishida et al., 1992). However, cDNA encoding 12α-hydroxylase has not been obtained. Without sequence information it is not possible to confirm a *P450* gene for 12α-hydroxylase. The third P450 enzyme in bile acid synthetic pathway is a mitochondrial sterol 27-hydroxylase which catalyzes the side-chain oxidation of cholesterol or bile acid intermediates and is known to be a product of the *CYP27* gene (Cali and Russell, 1991; Andersson et al., 1989). This gene is expressed in various tissues at a high level and has multiple functions other than bile acid synthesis, including 25-hydroxylation of vitamin D_3 and 24- and 25-hydroxylation of cholesterol in extrahepatic tissues (Usui et al., 1990; 1990a; Su, 1990). The 27-hydroxylase may not play a major role in the regulation of bile acid synthesis, however, it may regulate cholesterol homeostasis via the synthesis of oxysterols which are known to regulate the genes in cholesterol synthesis pathway and the low-density lipoprotein (LDL) receptor gene.

Cholesterol homeostasis is maintained by two input and two output pathways in the liver: the LDL receptor-mediated uptake mechanism and *de novo* synthesis of cholesterol provide input of cholesterol, whereas conversion of cholesterol to bile acids and storage of cholesterol as cholesteryl esters are the output of cholesterol. The conversion of cholesterol to bile acids is the predominant pathway for the elimination of cholesterol from the body. Biliary secretion of cholesterol requires bile acids for intrahepatic transport and is also a significant pathway for elimination of cholesterol. Purification of cholesterol 7α-hydroxylase P450 and cloning of the gene have contributed to the understanding of molecular mechanism of regulation of bile acid synthesis and cholesterol homeostasis. In this chapter, we will review the major pathways of bile acid synthesis, its physiological regulation, possible molecular mechanisms of gene regulation and disorders of bile acid metabolism.

II. BILE ACIDS, TRANSPORT AND ENTEROHEPATIC CIRCULATION

A. Physiological Function

Bile salts have a number of important physiologic functions: (a) their formation from cholesterol represents a major pathway responsible for elimination of cholesterol; about 50% of cholesterol eliminated via its biotransformation to bile acids, (b) bile acids are responsible for maintenance of bile flow, an important pathway for solubilization and excretion of organic compounds, endogenous metabolites, drugs and a variety of xenobiotics, (c) the rates of cholesterol and lecithin secretion are dependent on bile acid secretion, (d) bile salts play an important role in the maintenance of hepatic cholesterol homeostasis by directly or indirectly affecting a number of enzymes of hepatic cholesterol metabolism, such as 3-hydroxy-3-methylglutaryl-CoA-reductase (HMG-CoA reductase), the rate-determining enzyme in cholesterol synthesis, cholesterol 7α-hydroxylase, the rate-determining enzyme in bile acid synthesis, acyl CoA:cholesterol acyltransferase (ACAT), an enzyme which esterifies free cholesterol and cholesterol ester hydrolase (CEH), an enzyme which hydrolyzes cholesteryl esters, and (e) bile salts are responsible for the intraluminal solubilization, transport and absorption of cholesterol, fat soluble vitamins and other lipids from the intestine. A decreased rate of bile salt secretion into the intestine such as observed in cholestatic liver diseases, may result in poor absorption of cholesterol and other lipids (Watt and Simmonds, 1976).

B. Chemistry

Bile acids (5β-colonic acids) are formed in the liver as a final degradation product of cholesterol. Their distinguishing characteristics include a saturated 19-carbon sterol nucleus; a β-oriented hydrogen at position 5; a branched, saturated 5-carbon side chain terminating in carboxylic acid and an α-oriented hydroxyl group in several positions of the sterol ring or side chain (Danielsson and Sjövall, 1975; Danielsson and Gustafsson, 1981; Danielsson and Wikvall, 1986; Björkhem, 1985). The term "primary" bile acids refers to those bile acids synthesized *de novo* by the liver; in human they include choleic acid (3α,7α,12α-trihydroxy-5β-colonic acid) and chenodeoxycholic acid (3α,7α-dihydroxy-5β-colonic acid). Choleic and chenodeoxycholic acids are dehydroxylated in the intestines by bacteria at the C-7 position which results in the formation of "secondary" bile acids, deoxycholic ($3\alpha,12\alpha$-dehydroxy-5β-colonic acid) and lithocholic acids (3α-monohydroxy-5β-colonic acid), respectively (Danielsson and Sjövall, 1975; Danielsson and Gustafsson, 1981; Danielsson and Wikvall, 1986; Björkhem, 1985). These bile acids are conjugated with taurine or glycine and represent major bile acids in humans (Figure 1). The term bile acids refers to the form in which the carboxylic acid side chain is protonated, while bile salts refers to the ionized form. In this chapter, the terms are used interchangeably.

In the aqueous solutions (bile or intestines), bile salts aggregate in the form of micelles, with their hydrophilic parts exposed to the surrounding aqueous media while their hydrophobic surfaces are sequestered facing the interior of the micelle. Aggregation of bile salts occurs at certain threshold concentrations of bile salts ("critical micellar concentration" or CMC). CMC is different for each bile salt and in many instances, it determines the physiologic role of individual bile salt in the liver and in the intestine. Micelles composed of bile salts alone are called "simple" micelles while "mixed" micelles consists of bile salts, phosphatidylcholine (lecithin) and cholesterol, the two additional lipids found in bile (Carey, 1985). The ability of mixed micelles of bile salts and lecithin to solubilize biliary cholesterol in bile is defined as "cholesterol saturation index" or CSI (Carey, 1985). When the ability of mixed micelles to solubilize cholesterol is exceeded, CSI of bile is high. The supersaturation of bile with cholesterol could occur either due to decrease in bile salt and phospholipid secretion or due to excess of cholesterol secretion.

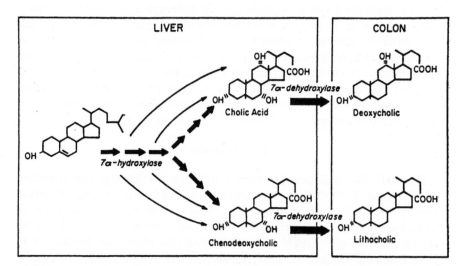

Figure 1. "Primary" (cholic and chenodeoxycholic) and "secondary" (deoxycholic and lithocholic) bile acids in man. In man, these bile acids are conjugated with glycine or taurine in the ratio of 3:1.

Increase in cholesterol saturation of bile leads to precipitation of cholesterol from the mixed micelles into the aqueous medium in form of cholesterol crystals, which in turn aggregate and form cholesterol gallstones (Admirand and Small, 1968).

The biological properties of bile salts are determined by their relative affinity for aqueous vs lipid environment ("hydrophobic-hydrophilic balance"). The hydrophobic-hydrophilic balance is evaluated by determining the partition coefficient between polar and non-polar solvents, or by measuring the retention of bile salts on reverse phase high performance liquid chromatography. The relative hydrophobicity of individual bile salts or a mixture of bile salts is defined as the "hydrophobicity index." The hydrophobicity index of individual bile salts is generally determined by number and orientation of hydroxyl groups on the bile salt molecule. Hydrophobic bile salts have a higher tendency to self associate in aqueous medium, are better solubilizers of cholesterol and are more powerful detergents than hydrophilic bile salts. As a result of these physicochemical properties, hydrophobic bile salts also exhibit higher cell and membrane toxicity (Heuman, 1988). In addition, hydrophobic bile salts have also been shown to be potent regulators of enzymes participating in the maintenance of cholesterol homeostasis in the liver whereas hydrophilic bile salts have no effect on the regulation of these enzymes (Vlahcevic et al., 1993a).

C. Bile Acid Transport

Bile is an aqueous solution containing approximately 90% of water and 10% of solutes. In man, about 85% of solute is represented by bile salts, cholesterol and phospholipid with the ratio of 65%:10%:25%. Phospholipids consist mostly of phosphatidylcholine (lecithin) while cholesterol is present in free rather than esterified form (Carey, 1985). Bile also contains proteins, vitamins, steroid hormones, drugs, other xenobiotics and conjugated bilirubin. Primary bile acids synthesized from the cholesterol are conjugated with two amino acids (glycine and taurine) in the liver and secreted into the biliary canaliculi via active transport mechanisms.

The elimination of bile salts from the portal blood takes place in the liver and involves uptake at the sinusoidal (basolateral) membrane by transporters ("hepatic uptake"); intracellular transport from sinusoids to the canalicular membrane, and the transcanalicular transport ("canalicular" secretion). These various phases of hepatic transport of bile salts are integral part of the enterohepatic circulation (Figure 2).

Hepatic Uptake

After absorption from the intestines, bile salts are transported via portal circulation into the liver. The first-pass hepatic extraction of bile salts ("hepatic uptake") is highly efficient for trihydroxy bile salts (90%) and somewhat less efficient for dihydroxy bile salts (65–75%) (Iga and Klaassen, 1982). Basolateral (sinusoidal)

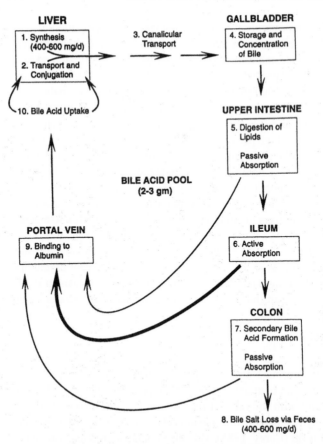

Figure 2. Schematic presentation of enterohepatic circulation of bile acids. The following processes are involved in the enterohepatic circulation: Liver: Synthesis of bile acids (1); sinusoidal uptake of free and conjugated bile acids returning to the liver via the portal vein (10); intrahepatic transport of newly synthesized and recirculated bile acids to the canaliculi (2); canalicular transport of bile acids from the hepatocytes into biliary canaliculi and on into bile ductules and ducts (3). Gallbladder: Storage and the concentration of bile (4). Upper Intestines: Micellar solubilization of ingested lipids by bile salts and passive absorption of bile salts (5). Ileum: Active absorption of bile salts (6). Colon: Secondary bile acid formation and passive absorption of bile salts (7). Fecal Excretion: Bile salt loss (8). Portal Recirculation to the Liver: Binding to the plasma albumin (9).

uptake extracts bile salts together with organic and inorganic solutes from the plasma. The uptake of bile salts from the hepatocytes may involve both sodium-dependent and sodium-independent transport processes (Nathanson and Boyer, 1991). The kinetics of bile salt uptake has been well characterized in a number of study models (Reichen and Paumgartner, 1976; Schwartz et al., 1975; Van Dyke et al., 1982; Inoue et al., 1982). Several putative bile acid transporting polypeptides have been identified and isolated. Their molecular weights have been estimated to range between 48 and 54 kDa (Kramer et al., 1982; Wieland, 1984; Von Dippe and Levy, 1983). A 49 kDa polypeptide has been reconstituted and identified as an essential component of sodium-dependent bile acid transport system (Anantha-narayanan et al., 1988). Polyclonal antibodies raised against 48 kDa protein inhibited the initial rate of Na^+-dependent uptake in cultured rat hepatocytes. This suggests that this protein is a component of the basolateral Na^+-dependent transport system (Ananthanarayanan, 1991). Von Dippe et al. (1993) showed that a 49 kDa Na^+-dependent bile acid transporter is indistinguishable from microsomal epoxide hydrolase. An even greater uncertainty exists about sodium-independent basolateral bile acid uptake. Photoaffinity and chemical labeling studies indicated that a 54 kDa protein may be an important part of sodium-independent organic anion uptake (Frimmer and Ziegler, 1988). In order to further elucidate bile acid transport, expression cloning strategy using xenopus laevis oocytes were used (Hagenbuch et al., 1991). The same authors reported successful expression cloning and characterization of a cDNA encoding the rat liver basolateral Na^+/taurocholate cotransporter polypeptide (Hagenbuch et al., 1990; Hagenbuch et al., 1991).

Intracellular Transport

Following hepatic uptake, bile salts are translocated from the basolateral to the canalicular (apical) membranes. Bile salts could traverse the hepatocytes coupled to cytosolic proteins, by partitioning between the organelles and the cytosol, or via the vesicular transport. The estimated hepatic transcellular transport is very rapid (2 min). Recently, several proteins with high affinity for bile salts have been isolated and characterized from rat liver cytosol and their role in the intrahepatic bile acid transport is being studied (Takikawa et al., 1986; Takikawa et al., 1986a; Stolz et al., 1987). A 37 kDa cytosolic bile acid-binding protein has been identified in rat liver as 3α-hydroxysteroid dehydrogenase (Stolz et al., 1989).

Transcanalicular Transport

Hepatic secretion of bile acids from the liver and into the biliary canaliculi is the major determinant of water and solute excretion in bile. This step of bile acid transport in the liver is rate-limiting for the entire process of hepatobiliary elimination of bile salts. A fraction of water and solute secreted into the bile which is associated with the secretion of bile acids is called bile acid-dependent bile flow. The steepest uphill transport of bile acids against concentration gradient is from the hepatocyte to bile (Boyer et al., 1992). Taurocholate, the major bile acid constituent in human bile, was thought previously to be transported into the bile through the canalicular membrane by a carrier, which is driven by a membrane potential (Boyer et al., 1992; Inoue et al., 1984). The canalicular specific glycoprotein with molecular weight of about 100,000 was isolated and antibodies against it was raised. The results using specific antibodies inhibited taurocholate transport from canalicular but not basolateral vesicles (Ruetz et al., 1988). However, the membrane potential is too small a force to be solely responsible for overcoming the actual gradient of bile salt across the canalicular membrane (Meier et al., 1984). More recently, several investigators have demonstrated ATP-dependent transport of taurocholate and other bile salts across the canalicular membrane (Stieger et al., 1992; Adachi et al., 1991; Muller et al., 1991; Nishida et al., 1991, 1991a; Akerboom et al., 1991). The ATP-dependent transport of bile salts is necessary to overcome the uphill bile salt concentration gradient from blood (100 μMol/l) to bile (10 mMol/l). Photoaffinity labeling of canalicular plasma membrane subfraction identified a 110 kDa glycoprotein as the best candidate for the bile salt carrier (Muller et al., 1991). The carrier protein responsible for this transport has been recently purified and characterized (Fricker et al., 1987; Ruetz et al., 1987; Sippel et al., 1990). ATP-dependent taurocholate transport may act synergistically with electrochemical potential-dependent taurocholate transport (Nishida et al., 1991). High affinity ATP-dependent taurocholate transport into both Canalicular and basolateral human liver plasma membrane (Wolters et al., 1992). Cyclosporin A is a potent inhibitor of ATP-dependent bile salt transport across the canalicular membrane which may explain the cholestatic effects of this drug (Kadmon et al., 1993). In addition to bile salt-dependent canalicular transport, a bile salt-independent canalicular transport was described (Nathanson and Boyer,1991). The contribution of bile salt-independent transport to total bile flow appears to be species specific. Recently, the canalicular multispecific organic anion transporter (cMOAT) have been cloned (Paulusma et al., 1996). It is a homolog of the multidrug resistance gene, the defective of which causes congenital jaundice in Dubin-Johnson syndrom.

Biliary Secretion and Intestinal Absorption

Canalicular secretion of bile acids is coupled with secretion of phosphatidylcholine and cholesterol; in fact, bile acid secretion is a driving force for biliary secretion of cholesterol and lecithin (Wagner et al., 1976). After secretion of bile into the biliary canaliculi and larger bile ducts, hepatic bile is either stored in the gallbladder or directly secreted into the duodenum. In gallbladder, bile is concentrated and approximately 80% of water is absorbed. Following digestion of food, the hormone cholecystokinin is released from the intestines. Its release is followed by a simultaneous contraction of the gallbladder and relaxation of the sphincter of Oddi with subsequent emptying of bile into the duodenum. In the proximal intestine, bile salts participate in the micellar solubilization and transport of digested lipids from the lumen to the intestinal brush border where they are subsequently absorbed. In the upper intestine (jejunum), a relatively small amount of bile salt is absorbed (Carey et al., 1983). The most important absorption site for the bile salts is in the terminal ileum. Ileal absorption of bile salts satisfies a number of criteria for active transport such as movement against electrochemical gradient, saturation kinetics, inhibition by competitive inhibitors, and sodium dependency

(Weinberg et al., 1986; Kramer et al., 1983). Na$^+$-dependent bile acid transporter has recently been identified to be an integral 93-kDa peptide and a peripheral 14 kDa bile acid-binding protein in the terminal ileum of rabbit (Kramer et al., 1993). Sequence analysis of 14 kDa bile acid binding protein showed a significant homology with gastrotropin, a cytosolic protein possibly involved in its intracellular transport of bile salts into ileum. A hamster ileal sodium-dependent bile acid transporter has recently been characterized by expression cloning of a cDNA encoding a 348 amino acid protein (Wong et al., 1994).

D. Enterohepatic Circulation

The process of bile salt synthesis, transport and cycling combines physiologic utilization of bile salts with their preservation and re-utilization. This process is referred to as enterohepatic circulation of bile salts. The total amount of bile salts circulating in the enterohepatic circulation is defined as a bile acid pool. In man, bile acid pool consists of glycine and taurine conjugates of bile acids (choleic: chenodeoxycholic) in the ratio of 3:1 and of conjugates of "secondary" bile acids (deoxycholic and lithocholic acids). Bile acid composition (cholic:chenodeoxycholic:deoxycholic:lithocholic acids) in man is approximately 40:40:20:<1. Bile acid pool size and bile acid kinetics were estimated in man by the isotope dilution technique. Using this technique, the total bile acid pool size was shown to be between 2.5–3.0 gm (Vlahcevic et al., 1971). Intestinal conservation of bile acids circulating in the enterohepatic circulation is highly efficient; approximately 95% of circulating bile salts are absorbed in each pass through the intestine and returned to the liver. Approximately 400–600 mg of bile salts are lost into the stool each day. This daily loss of bile salts is compensated by identical amounts of newly synthesized bile salts from cholesterol in the liver. The rate of bile acid synthesis is inversely proportional to the bile salts concentration in the portal blood, which in turn is determined by the rate of bile salt loss in the intestine. The size of the bile acid pool is influenced by the rate of bile acid synthesis, the cycling frequency, the extent of bile salt absorption, the rate of gallbladder emptying and the intestinal transit time. Under most physiologic circumstances, the size of the circulating bile acid pool remains constant as the bile salt loss is compensated by bile acid synthesis. Different phases of enterohepatic circulation are discussed in detail in two recent reviews (Vlahcevic et al., 1989; Carey and Cahalane, 1988) and are pictorially depicted in Figure 2.

E. Cholesterol Homeostasis

The maintenance of hepatic cholesterol is achieved by several integrated feedback mechanisms in the liver which involve regulation of LDL receptor biosynthesis, HMG-CoA reductase activity and synthesis, cholesterol 7α-hydroxylase activity and biosynthesis, ACAT and CEH activity and biliary cholesterol secretion. The bile salts circulating in the enterohepatic circulation play an important role in

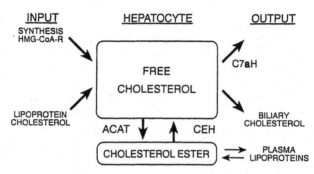

Figure 3. Factors playing a role in the maintenance of cholesterol homeostasis in the liver; HMG-CoA-reductase (HMG-CoA-R); cholesterol 7α-hydroxylase (C7αH); Acyl-CoA-cholesteryl acyltransferase (ACAT); cholesteryl ester hydrolase (CEH); low density lipoprotein (LDL) receptors and biliary cholesterol secretion.

the regulation of some of these individual enzymes and in the integration of these processes. A recent review of this topic is available (Vlahcevic et al., 1993). The major factors playing a role in the maintenance of hepatic cholesterol homeostasis are depicted in Figure 3.

III. CHOLESTEROL DEGRADATION TO BILE ACIDS

A. Bile Acid Biosynthetic Pathways

Cholesterol degradation to primary bile acids can be initiated by either a microsomal cholesterol 7α-hydroxylase (neutral pathway) or by mitochondrial sterol 27-hydroxylase (acidic pathway) (Figure 4). In the major pathway (neutral pathway), modifications of cholesterol nucleus precede oxidation and cleavage of the side chain. The existence of an alternative ("acidic") pathway was known for some time, but its contribution to total bile acid synthesis was thought to be small. The acidic pathway is characterized by the initial 27-hydroxylation of cholesterol side chain, which is followed by changes in the cholesterol nucleus. The end products of cholesterol degradation via the neutral pathway are cholic and chenodeoxycholic acids, which are synthesized in roughly similar amounts. By contrast, the end product of the acidic pathway is predominantly chenodeoxycholic acid (Vlahcevic et al., 1991, 1992; Björkhem, 1992; Stravitz et al., 1993; Russell and Setchell, 1992). Recently, Axelson and Sjövall (1990) and Axelson et al. (1988) reported that putative intermediates of acidic pathway can be found in plasma, which raised the possibility that contribution of this pathway to the total bile acid synthesis may be more significant than previously thought (Axelson et al., 1988). These authors have suggested that neutral pathway may be the main pathway under most physiologic conditions, while acidic pathway might be a more significant contributor to bile acid synthesis in the conditions where bile acid synthesis is low, such as in certain diseases of the liver (Axelson and Sjövall, 1990). This interesting hypothesis has yet to be experimentally proven.

"Neutral" Pathway

The biosynthesis of bile acids from cholesterol via the neutral pathway requires at least fourteen different steps catalyzed by a variety of enzymes located in cytosol, microsomes, mitochondria and peroxisomes. However, in different species, bile acid biosynthesis pathways and types of bile acids synthesized varies somewhat. The first and initial step in cholesterol degradation to bile acids is characterized by hydroxylation of cholesterol at the C-7 position leading to formation of 7α-hydroxycholesterol. This reaction is catalyzed by microsomal cholesterol 7α-hydroxylase (P-450c7). Cholesterol 7α-hydroxylase is thought to be the rate determining enzyme in the bile acid biosynthetic pathway. This enzyme is localized in the smooth endoplasmic reticulum and it is a member of the cytochrome P-450 enzyme family, based on its sensitivity to carbon monoxide inhibition, and the requirement for NADPH, oxygen, lecithin and NADPH-cytochrome P-450 reductase (Myant and Mitropoulos, 1977). This enzyme is believed to be feedback repressed by the flux of hydrophobic bile acids returning to the liver via portal circulation (Vlahcevic et al., 1991, 1992; Myant and Mitropoulos, 1977).

The conversion of 7α-hydroxycholesterol, a product of cholesterol 7α-hydroxylase, to 7α-hydroxy-4-cholesten-3-one, the second step in the neutral pathway, is catalyzed by a microsomal 3β-hydroxy-Δ^5-C_{27}-steroid oxidoreductase\isomerase (Björkhem, 1985). The 3β-hydroxy-Δ^5-C_{27}-steroid oxidoreductase shows a preference for C_{27} sterols (Wikvall, 1981). 7α-Hydroxy-4-cholesten-3-one is a branch point intermediate in the bile acid biosynthesis pathway which can be converted into the precursors of either choleic or chenodeoxycholic acids. This intermediate may be hydroxylated at the C-12 position by a microsomal cytochrome P-450 monooxygenase (sterol 12α-hydroxylase) and subsequently reduced by two cytosolic enzymes Δ^4-3-ketosteroid-5β-reductase and 3α-hydroxysteroid dehydrogenase to 5β-cholestane-$3\alpha,7\alpha,12\alpha$-triol (Furuebisu et al., 1987; Onishi et al., 1991). Intermediates hydroxylated at the C-12 position are always converted to cholic acid. Alternatively, 7α-hydroxy-4-cholesten-3-one may be reduced by Δ^4-3-ketosteroid-5β-reductase and 3α-hydroxysteroid dehydrogenase to 5β-cholestane-$3\alpha,7\alpha$-diol which is ultimately converted to chenodeoxycholic acid. Δ^4-3-ketosteroid 5β-reductase has been recently cloned by Onishi et al. (1991). The next step in this pathway is the reduction of the 3-oxo group of 7α-hydroxy-5β-cholestan-3-one and $7\alpha,12\alpha$-dihydroxy-5β-cholestan-3-one to the corresponding 3α-hydroxy group. This biotransformation is catalyzed by a soluble NADPH-dependent 3α-hydroxysteroid oxidoreductase (Pawlowski et al., 1991; Stolz et al., 1991; Deyashiki et al., 1992). This enzyme also serves as a bile acid-binding protein and has been purified and cloned (Stolz et al., 1987, 1991, 1993; Pawlowski et al., 1991; Cheng et al., 1991). 5β-cholestane-$3\alpha,7\alpha,12\alpha$ triol and 5β-cholestane-$3\alpha,7\alpha$

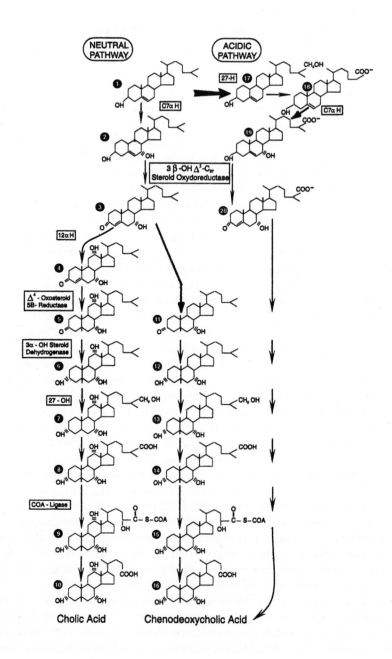

Cholic Acid   Chenodeoxycholic Acid

Figure 4. Cholesterol degradation to bile acids via "neutral" and "acidic" bile acid pathways. Neutral Pathway: Cholesterol (1); 7α-hydroxycholesterol (2); 7α-hydroxy-4-cholesten-3-one (3); 7α,12α-dihydroxy-4-cholesten-3-one (4); 5β-cholestane-7α,12α-diol-3-one (5); 5β-cholestane-3α,7α,12α-triol (6); 3α,7α,12α-trihydroxy-5β-cholestane-26-ol (7); 3α,7α,12α-trihydroxy-cholestanoic acid (8); 3α,7α,12α-tetrahydroxy-5β-cholestanoyl-CoA (9); Cholic Acid (10); 5β-cholestane-7α-hydroxy-3-one (11); 5β-cholestane-3α,7α-diol (12); 5β-cholestane-3α,7α,26-triol (13); 3α,7α-dihydroxy-5β-cholestanoic acid (14); 3α,7α,24-trihydroxy-5β-cholestanoyl-CoA (15); and Chenodeoxycholic Acid (16). Acidic Pathway: 27-hydroxycholesterol (17); 3β-hydroxy-5-cholestenoic acid (18); 3β,7α-dihydroxy-5-cholestenoic acid (19); 7α-hydroxy-3-oxo-cholestenoic acid (20).

---

diol are the most important substrates for the side-chain cleavage enzymes. This sequence of reactions completes the alteration of the sterol nucleus of cholesterol.

The initial step in the side-chain oxidation involves hydroxylation of C_{27} position which is catalyzed by mitochondrial cytochrome P-450 monooxygenase (Wikvall, 1984). Until recently, this enzyme was designated as 26-hydroxylase; however, studies of the stereochemistry of this biotransformation show that mitochondrial ω-hydroxylase exclusively attacks the C_{27} of the 25-pro-S methyl group (Björkhem, 1985). There are two possible mechanisms for the oxidation of the CH_2OH group at C_{27} to a carboxylic acid. First, the 27-hydroxylase has the unique ability to both hydroxylase and to oxidize the CH_2OH group to COOH. However, the rate of oxidation of this CH_2OH group to COOH in the hepatocytes is not known. Alternatively, cytosolic NAD^+-dependent alcohol and aldehyde dehydrogenase may sequentially oxidize the alcohol to a carboxylic acid (Russell and Setchell, 1992). The next step in side-chain removal involves the ligation of 3α,7α,12α-trihydroxy-5β-cholestanoic acid or 3α,7α-dihydroxy-5β-cholestanoic acid to co-enzyme A. This reaction is catalyzed by an ATP-dependent microsomal coenzyme A ligase (Björkhem, 1985). The final steps in side-chain oxidation occur mainly in the liver peroxisomes and are believed to be similar to the β-oxidation of fatty acids (Pedersen and Gustafsson, 1980). These reactions involve the introduction of a double bond at C_{24}-C_{25} by an oxidase (Schepers et al., 1990) followed by hydration by a bifunctional hydrates/dehydrogenase (Prydz et al., 1986) to yield either 3α,7α,12α,24-tetrahydroxy-5β-cholestanoyl-CoA or 3α,7α,-24-trihydroxy-5β-cholestanoyl-CoA for the cholic acid and chenodeoxycholic acid pathway, respectively. The C_{24}-hydroxy group then is oxidized to an oxo-derivative by the action of the dehydrogenase activity of the bifunctional enzyme (Prydz et al., 1986). This reaction is followed by thiolysis yielding propionyl-CoA and cholyl-CoA or chenodeoxycholyl-CoA (Schram et al., 1987). Prior to secretion from the hepatocyte, free bile acids are conjugated at C_{24} to either glycine or taurine. This reaction involves the initial activation of cholic acid and chenodeoxycholic acid to CoA by a microsomal CoA synthetase (Lim and Jordan, 1981). The final conjugation to either glycine or taurine is catalyzed by bile acid CoA: amino acid N-acyltransferase (Johnson et al., 1991).

"Acidic" Pathway

Bile acid biosynthesis via acidic pathway is initiated by mitochondrial sterol 27-hydroxylase enzyme to form 27-hydroxycholesterol. Little is known about the individual enzymes in this pathway and relative contribution of this pathway to bile acid synthesis is controversial. In circumstances where 27-hydroxylation of cholesterol occurs first, 7α-hydroxylation of C_{27} sterol intermediates must occur at a later stage in the pathway. Recently, a model was proposed for the biosynthesis of bile acids in humans (Axelson and Sjovall, 1990). This model was derived from studies of bile acid precursors in healthy subjects (Axelson and Sjovall, 1990; Axelson et al., 1988), and patients with an altered bile acid production (Axelson et al., 1989; 1989a). Axelson et al. (1988) demonstrated the accumulation of $3\alpha,7\alpha$-dihydroxy-5-cholestenoic acid in the plasma of patients with low cholesterol 7α-hydroxylase activity, suggesting the possibility of a different 7α-hydroxylase specific for 27-hydroxylated intermediates. Shoda et al. (1993) reported a formation of 7α and 7β-hydroxylated bile acid precursors from 27-hydroxycholesterol in human liver microsomes and mitochondria, while Toll et al. (1992), reported 7α-hydroxylation of 27-hydroxycholesterol, 3β-hydroxy-5-cholestenoic acid and 3β-hydroxy-5-cholenoic acid by two cytochrome P-450 fractions in pig liver microsomes. This same group of investigators has identified a third pig liver 7α-hydroxylase in isolated mitochondria (Axelson et al., 1992). Surprisingly, mitochondria metabolized ^{13}C-labeled cholesterol not only to 27-hydroxy-cholesterol, but to 7α-hydroxy-3-oxo-4-cholestenoic acid; these incubations did not, however, yield detectable 7α-hydroxycholesterol. Using various synthesized intermediates, these investigators proposed a pathway beginning with 27-hydroxylation of cholesterol, 7α-hydroxylation by a specific mitochondrial C_{27} steroid 7α-hydroxylase, further oxidation of the side chain and oxidation/isomerization of the A-ring, thereby bypass microsomal cholesterol 7α-hydroxylase as the rate-limiting enzyme.

These data strongly suggest the existence of mitochondrial 27-hydroxylase activity capable of side-chain oxidation, as well as a 7α-hydroxylase and 3β-hydroxysteroid dehydrogenase/isomerase active on 27-hydroxylated substrates (Axelson et al., 1992). A 7α-hydroxylase activity specific for 27-hydroxycholesterol has been found in microsomes (Björkhem et al., 1992), and mitochondria, (Shoda et al., 1993) of human liver and in HepG2 cell (Martin et al., 1993). Microsomal 7α-hydroxylase activity toward 27-hydroxycholesterol was similar in cholestyramine- and untreated patients, while cholesterol 7α-hydroxylase activity was 4-fold higher in those receiving cholestyramine indicating that 7α-hydroxylase of 27-hydroxycholesterol is different from cholesterol 7α-hydroxylase (Björkhem et al., 1992). Princen et al. (1991), have suggested that the "acidic pathway" may account for nearly 50% of total bile acid biosynthesis in primary cultures of rat and human hepatocytes treated with cyclosporin A, a selective inhibitor of mitochondrial 27-hydroxylase. Recent evidences suggested that cyclosporin A may interfere with binding of cholesterol to the active site of the enzyme (Dahlbäck-Sjöberg et al., 1993). The regulation of sterol 27-hydroxycholesterol and other enzymes in the acidic pathway is not known.

Other Pathways

It was demonstrated previously that 7α-hydroxycholesterol was rapidly converted to chenodeoxycholic acid and to a lesser extent to cholic acid in cultured human hepatoblastoma (HepG2) cells (Javitt and Budat, 1989; Javitt, 1990). A cholic acid precursor, 3α, 7α, 12α-trihydroxy-5β-cholestanoic acid (THCA) has been found secreted into the medium. It was reported that THCA-CoA ligase and THCA-CoA oxidase were absent in HepG2 cells (Ostlund Farrants et al., 1993). Axelson et al. (1991) identified 29 bile acid and steroid intermediates in HepG2 cells. Most intermediates in cholic acid and chenodeoxycholic acid synthesis pathways were found and cholic acid and chenodeoxycholic acid were synthesized in equal amount. They suggested that in HepG2 cells the major bile acid biosynthetic pathways start with 7α-hydroxylation of cholesterol and oxidation to 7α-hydroxy-4-cholesten-3-one followed by hydroxylation at either the 27 or 12α position. Chenodeoxycholic acid is formed by the sequence of 27-hydroxylation, oxidation and degradation of the side-chain and A-ring reduction. Cholic acid is formed by the sequence of 12α-hydroxylation, 27-hydroxylation, oxidation, and degradation of the side-chain and reduction of the A-ring. In these pathways, oxidation and side-chain cleavage occur prior to 5β-reduction. An alternative pathway to cholic acid included A-ring reduction of 7α, 12α-dihydroxy-3-oxo-4-cholestenoic acid to form 3α, 7α, 12α-trihydroxy-5β-cholestenoic acid prior to side-chain cleavage. They believe that these pathways also exist in humans. However, a pathway to chenodeoxycholic acid starting with 27-hydroxylation of cholesterol was not observed in HepG2 cells. This is in contrast to previous observations that 27-hydroxylase pathway is the major pathway for synthesis of chenodeoxycholate in HepG2 cells (Javitt, 1990). It is interesting to note that dexamethasone increases 12α-hydroxylation and/or decreases 27-hydroxylation and thus shift synthesis toward predominantly cholic acid.

An alternative pathway for cholic acid synthesis involves a microsomal 25-hydroxylation (Shefer et al., 1976). It was shown that 5β-cholestane-3α, 7α, 12α-triol is efficiently 25-hydroxylated in the microsomes. This pathway for the side-chain cleavage in cholic acid biosynthesis does not involve 27-hydroxylation. In cerebrotendinous xanthomatosis patients (CTX, see section IV), 5β-cholestane—3α, 7α, 12α, 25-tetrol accumulated in bile and feces. Salen et al. (1979) suggested that the 25-hydroxylase pathway may be a major pathway for cholic acid synthesis in humans. However, recent *in vivo* studies measuring [^{14}C]-acetone from [26-^{14}C]-cholesterol have shown that 25-hydroxylation of cholic acid probably account for <5% of choleic acid synthesis in rats and in humans (Duane et al., 1988). It seems unlikely that 25-hydroxylation pathway is of importance in bile acid synthesis.

B. Cytochrome P450 Hydroxylases in Bile Acid Biosynthetic Pathway

Cholesterol 7α-Hydroxylase

Biochemistry.   Figure 5 illustrates the structural features of the rat choles-terol 7α-hydroxylase gene, mRNA and polypeptide.

CYP7 Gene: Cholesterol 7α-hydroxylase is a product of the unique *CYP7* gene which belongs to the P450 superfamily (Nelson et al., 1993). Only one gene has been found in this family. The gene has been cloned from five species, rat, human, rabbit, mouse, (Tzung, et al., 1994) and hamster. The human *CYP7* gene was mapped to q11-q12 of the chromosome 8 (Cohen et al., 1992). The *CYP7* gene spans about 11 kb of the genome and consists of six exons and five introns, which are fewer than other P450 genes typically having 8 to 12 exons. The sizes and distribution of exons and introns are similar in these homologous genes except that the intron 3 of the human gene is about 1 kb longer than that of the rat and hamster genes (Figure 6). The 5'-flanking sequences of these genes have been determined (Jelinek and Russell, 1990; Nishimoto et al., 1991; Chiang et al., 1992; Molowa et al., 1992; Cohen et al., 1992; Nishimoto et al., 1993; Thompson et al., 1993; Crestani et al., 1993). Numerous sequence discrepancies were noted in the over-

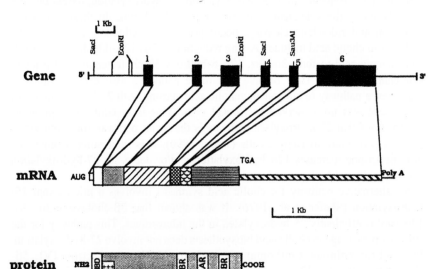

Figure 5. Structural features of the rat cholesterol 7α-hydroxylase gene, mRNA and protein. MBD: membrane-binding domain; SBR: substrate-binding region; AAR: aromatic amino acid region; HBR: heme-binding region.

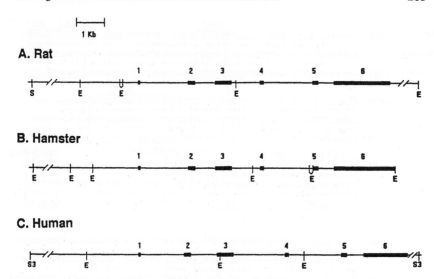

Figure 6. Gene organization of rat, human, and hamster *CYP7* gene. Heavy lines are exons. E. EcoRI: S3, Sau3A1; S, Sac I.

lapping regions of reported human gene sequences (Thompson et al., 1993; Wang and Chiang, 1994). The proximal promoter sequences from the start codon to about 250 bp upstream are highly homologous. Alignment of sequences in this region revealed a sequence identity of 83% between hamster and rat, 77% between human and hamster and 72% between rat and human (Crestani et al., 1993). This conserved region contains highly homologous sequence motifs, many of them are consensus sequences for liver-enriched transcription factor binding sites (Figure 7). These include HNF1/LFB1 (Tronche and Yaniv, 1992), HNF3/TGT3 (Paulweber et al., 1991), HNF4/LFA1 (Sladek, 1990) and C/EBP (Grange et al., 1991). This proximal promoter must be important not only for liver-specific gene transcription but also for the regulation of *CYP7* gene expression by physiological agents such as bile acids. Upstream sequence of flanking region of the genes become diverged considerably. There is an overall sequence identity of 60% between human and rat as well as human and hamster. The rat and hamster share 70% sequence identity. Many repetitive sequences, ubiquitous transcription factor recognition motifs and steroid/thyroid hormone response elements are present in these genes (Chiang et al., 1992). Three transcription start sites in the rat gene have been determined by Jelinek and Russell (1990), however, only one start site (G) located 60 bases upstream of the ATG start codon was determined by Nishimoto et al. (1991). The same start site was also determined in the human gene (Molowa et al., 1992). The typical TATA box and a reversed CCAAT box sequences are located 24 bp and 48, respectively, upstream of the start site G in all three genes.

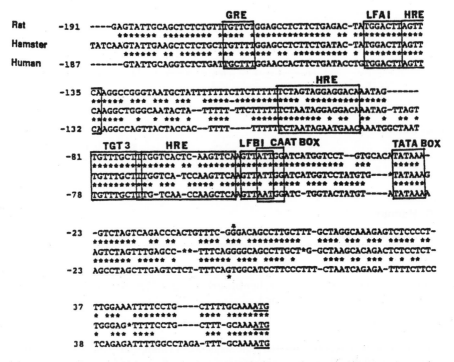

Figure 7. Alignment and analysis of proximal promoter regions of human, rat and hamster genes. GRE, glucocorticoid response element; HRE, steroid/thyroid hormone response element; LFAI, liver factor AI; LFBI, liver factor BI, TGT₃, liver-enriched element, same as HNF₃. Transcription start sites are indicated by*. Translation start codons ATG are underlined. (Reproduced from Crestani et al., 1993, with permission.)

mRNAs: The *CYP7* gene transcribes multiple mRNA species in rat liver (Li et al., 1990). Two major species of 4 kb and 2.5 kb mRNAs are present together with several minor species of mRNAs in rat liver (Li et al., 1990; Jelinek et al., 1990; Sundseth and Waxman, 1990). Only a single mRNA of 2.8 kb was found in the human liver (Noshiro and Okuda, 1990), two mRNA species (2.8 and 1.8 Kb) were found in HepG2 cells (Crestani et al., 1994) and one major mRNA of about 4.0 kb was found in hamster livers (Crestani et al., 1993). These multiple mRNA species apparently are derived from using different polyadenylation signals located in the 3'-untranslated region of the rat mRNA, since two types of cDNAs with poly (A+) tails of different lengths in the 3'-untranslated region were isolated from a rat cDNA library (Li et al., 1990). Many AUUUA motifs are found in the 3'-untranslated region of mRNAs. These motifs are present in many short-lived mRNAs and may contribute to the multiple species of mRNAs and short half-life of mRNAs reported (Li et al., 1990; Noshiro et al., 1990).

Protein: The rat cholesterol 7α-hydroxylase mRNA encodes 503 amino acid residues whereas human and hamster mRNAs encode 504 amino acid residues. Alignment of amino acid sequences revealed a high sequence identity of 92% between hamster and rat, and 82% between hamster and human (Figure 8). The putative steroid-binding site (residues 344 to 356), the aromatic amino acid region (residues 401 to 412) and the heme-binding region (residues 437 to 450) are completely identical in these homologous proteins (Figure 8). The N-terminal hydrophobic signal sequence is typical of P450 enzymes. This sequence is believe to be the membrane-anchor domain which functions as a stop-transfer signal (Kemper et al., 1989) and is not required for the catalytic activity (Li and Chiang, 1991). This sequence is followed by a stretch of charged amino acid residues. This N-terminal domain is encoded by the exon I. Amino acid residues encoded by exon II are completely identical in the rat and hamster sequences and are also highly conserved in the human sequence. Exon III is the largest coding exon which encodes amino acid residues with a sequence identity of about 80% among these species. The exon IV has a rather low sequence identity of about 60%. The conserved, putative sterol-binding site is partially located in this region. The putative sterol-

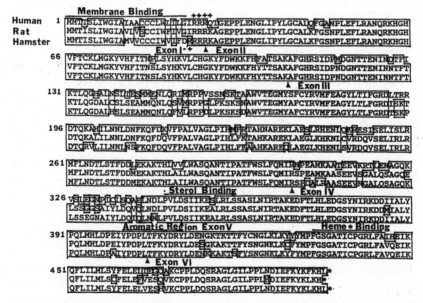

Figure 8. Alignment of amino acid sequences of human, rat and hamster cholesterol 7α-hydroxylases. Identical amino acid residues are enclosed in boxes. N-terminal membrane-binding domain, putative sterol binding site, aromatic amino acid region and home-binding site are overlined and labeled. + signs are positive amino acid residues. Amino acid splitted in the intron/exon boundaries are indicated by arrowheads. (Reproduced from Crestani et al., 1993, with permission.)

binding sequence is conserved in many P450 isozymes which metabolize cholesterol substrate, such as P450scc (Morohaski et al., 1987), P45011β (Chua et al.,
1987), P45017 (Chung et al., 1987), P45021 (White et al., 1984), P450laω (Hardwick, 1987), P45014DM (Kalb et al., 1987) and steroid 27-hydroxylase (Andersson
et al., 1989). It is interesting to note that P450c7 has the highest sequence identity
(about 20 to 25%) with cholesterol-metabolizing P450 isozymes. Amino acid
residues encoded by exon V are the most conserved with a sequence identity of
about 97%. This region contains part of sterol-binding site and the conserved
aromatic amino acid region which may involve in electron transport reaction. The
C-terminal heme-binding region is encoded by exon VI. The close localization of
the aromatic amino acid region to both the sterol-binding site and heme-binding
site indicate the importance of this residues in substrate binding and catalysis.

Cholesterol 7α-hydroxylase activity has been transiently expressed in COS cells
using mammalian expression vector carrying rat cDNA (Noshiro et al., 1989;
Jelinek et al., 1990). This expression system expressed a very low level of activity
and enzyme which could be measured only by activity assay and immunoblot. To
express a level high enough for enzyme purification and structural determination,
permanent expression in a heterologous system such as *E. coli*, yeast and baculovirus is most desirable. P450 enzymes are highly hydrophobic and are associated
with microsomal membranes. The lack of an internal membrane system in *E. coli*
prevented the expression of active P450 enzymes in high yield. One strategy to
overcome this problem is to express a truncated form which lacks the N-terminal
membrane-binding domain in *E. coli*. Rat cholesterol 7α-hydroxylase which lacks
the N-terminal membrane-binding 24 amino acid residues were expressed in *E. coli*
(Li and Chiang, 1991). The expressed enzymes were purified and were active in
the reconstitution of cholesterol 7α-hydroxylase activity. This bacterially-expressed
enzyme had a similar Km but a lower Vmax than those of intact enzyme purified
from rat liver microsomes (Li and Chiang, 1991). These results provide evidences
that the hydrophobic membrane-binding domain is not required for catalysis.
Cholesterol 7α-hydroxylase has a very short half-life which was estimated to be
about 2 to 3 hours (Myant and Mitropoulos, 1977). Cholestyramine treatment
immediately stimulated 7α-hydroxylase activity and removal of cholestyramine
from the diet immediately reduced activity, mRNA and protein levels (Li et al.,
1990). However, 7α-hydroxylase contains no apparent PEST sequence which is
present in many short-lived enzymes (Rogers et al., 1986).

The lack of suitable human liver tissues and the expression of extremely low level
of cholesterol 7α-hydroxylase in human liver prevented the purification of human
7α-hydroxylase. Two laboratories have reported partial purification of the human
enzyme (Nguyen et al., 1990; Erikson and Bosterling, 1981). The same strategy
used for the expression of rat 7α-hydroxylase was applied to express the human
enzyme, however, the level of expression was extremely low. Different expression
vector (pJL) and bacteria strain (TOPP3) were used to express a truncated human
7α-hydroxylase to a level similar to the rat enzyme (Karam and Chiang, 1994). The

Km for cholesterol (5.9 μM) and Vmax (0.13 n mol/min) of the expressed human enzymes were similar to those of the expressed rat enzyme indicating that extremely low activity and enzyme level in human liver is due to the low level of enzyme expressed, not the result of different kinetic properties of 7α-hydroxylase isozymes in the rat and human livers. This low Km for cholesterol also suggests that cholesterol 7α-hydroxylase should be saturated *in vivo*.

Polymorphisms of *CYP7* gene: Three possible polymorphisms in amino acid codons were identified by Karam and Chiang (1992). Two different codons 100, TTT (Phe) and TCT (Ser), have been found in cDNA clones isolated from a human liver cDNA library. However, this codon was a TTT in a genomic clone (Wang and Chiang, 1994) and in ten human liver genomic DNAs (Karam and Chiang, unpublished result). It is possible that TCT is encoded by a rare allele in a human liver. Codon 347 was a GAT (Asp) in the cDNA sequence published by Karam and Chiang (1992), in a genomic clone by Cohen et al. (1992) and by Nishimoto et al. (1993), but codon 347 was a AAT (Asn) in a cDNA clone reported by Noshiro and Okuda (1990). It was suggested that a polymorphism might exist in the codon 347 (Nishimoto et al.,1993). The third polymorphism was in codon 385 which was a GAC (Asp) in all cDNA clones isolated by Karam and Chiang (1992), but a AGC (Ser) in a cDNA isolated by Noshiro and Okuda (1990). Nishimoto et al. (1993) reported that this discrepancy in codon 385 was due to cDNA sequencing error reported previously (Noshiro and Okuda, 1990). The observed polymorphisms in codon 100 and codon 347 are drastic amino acid conversions. The codon 100 is located in a region which has been shown to be important in substrate-binding in other *P450* genes (Kronbach et al., 1989). The codon 347 is located in the putative sterol-binding site. The effect of these amino acid substitutions on the enzyme activity and substrate binding is not known at present.

Four single strand conformation polymorphisms (SSCP) in the 5'-flanking region, intron 2 and intron 4, and one length polymorphism in the *Alu* sequence of the 3'-untranslated region of the human *CYP7* gene were identified by Cohen et al. (1992). They found that each polymorphism was biallelic and segregated in a Mendelian fashion in several families and 80% of unrelated subjects were heterozygous for at least one of the polymorphisms (Cohen et al., 1992). A *Mae* II polymorphism was identified in the 5'-flanking region of the *CYP7* gene (Thompson et al., 1993). No genetic diseases have been linked to the mutation in *CYP7* gene yet. It is quite remarkable that the *CYP7* gene is so well conserved in humans and during evolution. Nevertheless, these polymorphisms could be used as genetic markers for tracing segregation of the *CYP7* gene in family and are candidates for diseases related to bile acid and cholesterol metabolism in humans (Cohen et al., 1992).

Mechanisms of Regulation

Cholesterol 7α-hydroxylase is a highly regulatable enzyme which appears to be rate-determining for the entire bile acid biosynthetic pathway (Vlahcevic et al.,

1992). The multitude of data indicate that bile acid synthesis is under a complicated, multifactorial regulation by physiological regulators including bile acids, cholesterol and thyroxine and dexamethasone. Possible mechanisms of regulations are (1) availability of substrate (Straka et al., 1990), (2) direct inhibition of activity by bile salts, (3) transcriptional regulation, (4) post-transcriptional regulation by phosphorylation/dephosphorylation, and (5) stability of mRNA and enzyme.

Cholesterol. Under most physiologic and pathophysiologic circumstances, HMG-CoA reductase and cholesterol 7α-hydroxylase specific activities change in tandem, as if they respond to the same regulatory signal (Bjökhem and Akerlund, 1988). The sole exception is cholesterol which is a powerful repressor of HMG-CoA reductase, but appears to up-regulate cholesterol 7α-hydroxylase activity in rats and humans (Li et al., 1990; Pandak et al., 1991; Mitchell et al., 1991; Dueland, 1993). The interrelationship between cholesterol and bile acid biosynthetic pathway has been recently explored in the rats with chronic bile fistula administered lovastatin and AY9944, two known inhibitors of cholesterol synthesis (Pandak et al., 1990; 1990a). Administration of a single dose of lovastatin, an competitive inhibitor of HMG-CoA reductase, resulted in a rapid (3 hours) down-regulation of cholesterol 7α-hydroxylase activity (approximately 50% decrease) and inhibition of bile acid synthesis. This effect was abolished by continuous infusion of mevalonate. These latter data provide suggestive evidence that newly synthesized cholesterol plays an important role in the regulation of cholesterol 7α-hydroxylase (Pandak et al., 1990). Administration of AY9944, an inhibitor of the last step in the cholesterol biosynthetic pathway, resulted in a similar down-regulation of cholesterol 7α-hydroxylase. However, the addition of mevalonate did not reverse the down-regulation of cholesterol 7α-hydroxylase activity or inhibition of bile acid synthesis caused by AY9944 (Pandak et al., 1990a). The role of newly synthesized cholesterol in the regulation of cholesterol 7α-hydroxylase was further explored in rats with acute bile fistula. In these rats, continuous infusion of lovastatin prevented the up-regulation of cholesterol 7α-hydroxylase activity or bile acid synthesis (Vlahcevic et al., 1993a). These data provided additional evidence for the importance of newly synthesized cholesterol in the maintenance of basal levels of cholesterol 7α-hydroxylase gene transcription.

Recent studies in the rat have shown that cholesterol feeding up-regulates cholesterol 7α-hydroxylase at the level of gene transcription (Pandak et al., 1991). Because of the possibility that cholesterol feeding may affect cholesterol 7α-hydroxylase indirectly i.e., by altering bile acid absorption in the intestines (Björkhem et al., 1991). Jones et al. (1993) carried out studies in the chronic bile fistula rats which received a single injection of lovastatin, while rats with intact enterohepatic circulation received mevalonate. In the first set of experiments, the availability of newly synthesized cholesterol was reduced, while in the second experiment, excess of newly synthesized cholesterol was presented to the enzyme. In chronic bile fistula rats, specific activity, enzyme mass, mRNA levels and gene transcriptional activity markedly decreased three hours after injection of a single dose of lovastatin. In contrast, continuous infusion of mevalonate over a period of 24 hours to rats with intact enterohepatic circulation resulted in a marked increase in cholesterol 7α-hydroxylase specific activity, mRNA levels and gene transcriptional activity. These data provide additional evidence that cholesterol (or an metabolic product of cholesterol) has an important stimulatory function on cholesterol 7α-hydroxylase. Taken together, these data indicate the existence of homeostatic mechanism whereby cholesterol supply may modulate cholesterol elimination by affecting bile acid synthesis. Under circumstances in which cholesterol is present in excess, cholesterol 7α-hydroxylase is up-regulated and degradation of cholesterol to bile acids is facilitated. Conversely, when cholesterol availability is decreased, cholesterol 7α-hydroxylase is suppressed, leading to a decrease in elimination of cholesterol. In both instances, cholesterol homeostasis is effectively maintained in the rat. The role of cholesterol substrate availability in the regulation of cholesterol 7α-hydroxylase was widely debated and controversial (Davis et al., 1983a; Einarsson et al., 1989). In contrast, high cholesterol diet supposed cholesterol 7-hydroxylases activity and mR/vA level in the liver of African green monkey (Rudel et al., 1994) and had no effect in the hamster (Horton et al., 1995). In Watanabe rabbit—a model of familial hypercholesterolemia with defective LDL receptor, cholesterol 7α-hydroxylase activity was found to be much lower than that in New Zealand white rabbits and high cholesterol diets did not further reduce enzyme activity (Xu et al., 1995). In hypercholesterolemia-resistant rabbits, cholesterol 7α-hydroxylase activity and mRNA levels were substantially higher (Poorman et al., 1993). It seems that the response to dietary cholesterol input varies greatly with different species and individuals.

Bile salts. The negative feedback control of cholesterol 7α-hydroxylase by bile salts has been proposed for over three decades by a number of investigators. Erickson (1957) and Thompson and Vars (1953) reported increase in bile acid synthesis after complete biliary diversion. Interruption of enterohepatic circulation also resulted in an increase in cholesterol 7α-hydroxylase activity (Danielsson et al., 1967). Conversely, Shefer et al. (1969, 1970, 1973) reported that intraduodenal infusion of bile salts down-regulated cholesterol 7α-hydroxylase activity, while Mosbach et al. (1971) provided additional evidence that cholesterol 7α-hydroxylase is a rate-determining enzyme in bile acid biosynthesis. Failure to exhibit negative feedback control in cultured rat hepatocytes (Davis et al., 1983; Kubaska, et al., 1985) led to several new studies of bile acid biofeedback in the rat, which demonstrated that hydrophobic but not hydrophilic bile salts down-regulated HMG-CoA reductase and cholesterol 7α-hydroxylase in roughly similar proportions (Heuman et al., 1988a, 1988b, 1989). Similar results were obtained in the rat by Shefer et al. (1990). The relationship between the hydrophobicity and 7α-hydroxylation of cholesterol was recently confirmed in humans (Bertolotti et al., 1991).

These original studies have not provided information on the molecular basis of regulation of cholesterol 7α-hydroxylase or whether bile acids affect cholesterol 7α-hydroxylase directly or indirectly, i.e., via primary effect on HMG-CoA reductase (Heuman et al., 1988a, 1989). Pandak et al. (1991) was first to report that in the rat, taurocholate decreases cholesterol 7α-hydroxylase specific activity, enzyme mass, mRNA levels and gene transcriptional activity. Similar results were obtained with other bile salts leading the authors to conclude that bile salts affect cholesterol 7α-hydroxylase predominantly at the level of gene transcription (Pandak et al., 1993). Stravitz et al. (1993a) and Twisk et al. (1993) have shown the relationship between bile salt hydrophobicity and gene transcriptional activity. These latter studies provided strong support that bile salts regulate cholesterol 7α-hydroxylase predominantly at the level of gene transcription.

Twisk et al. (1995) reported that the position and orientation of hydroxyl groups on the steroid nucleus of bile acids might determine the potency of bile acid down-regulation of cholesterol 7α-hydroxylase mRNA expression.

Additional studies by Pandak et al. (1992) have shown that taurocholate represses cholesterol 7α-hydroxylase transcription even in the presence of excess mevalonate, a precursor of cholesterol which is eventually converted to cholesterol. Similar data were obtained by Stravitz et al. (1993) in cultured rat hepatocytes suggesting that taurocholate represses cholesterol 7α-hydroxylase directly rather than indirectly by suppressing the HMG-CoA reductase, which in turn would modulate a supply of newly synthesized cholesterol. Dueland et al. (1993) reported that taurocholate suppressed both the activity and mRNA of 7α-hydroxylase in two inbred strains of mice. They also found that adding taurocholate to the cholesterol-rich diet suppressed both the activity and mRNA of 7α-hydroxylase. These authors concluded that the repressive effect of taurocholate overcomes the inductive effect of cholesterol on 7α-hydroxylase activity and mRNA. However, Shefer et al. (1992) reported down-regulation of cholesterol 7α-hydroxylase activity in the face of unchanged levels of steady-state cholesterol 7α-hydroxylase mRNA levels in rats fed taurocholate and cholesterol. These authors concluded that cholesterol 7α-hydroxylase mRNA level is dependent upon the supply of cholesterol, whereas bile acids repress the enzyme at the post-transcriptional level. Spady and Cuthbert (1992) have reported that sodium cholate repressed cholesterol 7α-hydroxylase activity and mRNA levels in the rat and cholate did not suppress 7α-hydroxylase activity when fed to rats on a high cholesterol diet. They concluded that the effect of cholesterol dominates the effect of cholate on 7α-hydroxylase in rat liver. The reason behind these discrepant conclusions in whole animals is not known. Dietary or *de novo* synthesized cholesterol may be responsible for a basal level of gene expression while the hydrophobic bile acids play a major role in modulating the level of *CYP7* gene expression *in vivo*. The level of *CYP7* gene expression in the liver is ultimately determined by the balance of these two mechanisms. The high rate of cholesterol synthesis and less efficient bile acid feedback in the rat may explain the dominate effect of cholesterol, while the converse is true in the mouse and cholate have a dominant effect on the *CYP7* gene expression.

Hormones. Glucocorticoids have been implicated in the regulation of cholesterol 7α-hydroxylase because diurnal variation of the enzyme can be abolished by adrenalectomy (Van Cantford, 1973). Hylemon et al. (1992) have recently demonstrated that in cultured rat hepatocytes cholesterol 7α-hydroxylase is not expressed and that addition of dexamethasone and thyroxine increased cholesterol 7α-hydroxylase mRNA to the levels similar to rats treated with cholestyramine. Dexamethasone increased activity and mRNA level in a rat hepatoma cell L35 which is resistant to 25-hydroxycholesterol and over-expressed cholesterol 7α-hydroxylase activity (Leighten et al., 1991). Princen et al. (1989) also reported the stimulation of cholesterol 7α-hydroxylase activity and bile acid synthesis in primary rat hepatocytes. In contrast, the addition of glucagon decreased cholesterol 7α-hydroxylase mRNA level and gene transcriptional activity probably via decrease of cyclic AMP (Hylemon et al., 1992). The physiologic role of glucagon in regulation of cholesterol 7α-hydroxylase is yet to be established. Ness et al. (1990) reported a rapid increase in cholesterol 7α-hydroxylase mRNA levels in hypophysectomized rats following injection of thyroid hormone. Pandak et al. (1993a) have shown that hypophysectomy and thyroidectomy plus adrenalectomy, decrease cholesterol 7α-hydroxylase specific activity, steady state mRNA levels and gene transcriptional activity in the rat. In contrast, thyroidectomy and adrenalectomy did not affect any of these parameters. These *in vitro* and *in vivo* data suggest an important physiologic role of thyroxine and dexamethasone in regulation of cholesterol 7α-hydroxylase. Table 1 summarizes the regulation of cholesterol 7α-hydroxylase activity, protein, mRNA and transcription rate by various physiological factors.

Recently, Vistani et al. (1995) studied transcriptional activity of the rat CYP7/lucerifase reporter constructs by hormone in transient transcription assays in HepG2 cells. There investigators found that glucocorticoid and retinoic acid stimulated the promoter activity and identified several response elements mediating their effects. Thyroid hormone did not have any effect. They also identified response elements mediating the inhibition by insulin and phorbol 12-myristate 13-acetate (PMA). cAMP/protein kinase A stimulated promoter activity. Both positive and negative response elements were formed. They concluded that the proximal promoter contained a pleitropic domain that regulate the effects of multiple signals.

Molecular Basis of Regulation

It is evident now that bile acid synthesis and cholesterol 7α-hydroxylase activity are physiologically regulated by bile acids, cholesterol (or its metabolites) and thyroid/steroid hormones at the gene transcriptional level. The *trans*-activating transcription factors and *cis*-acting regulatory elements involved in the regulation of the *CYP7* gene are not known at present. The –200 bp 5'-flanking promoter

Table 1. Regulation of Cholesterol 7α-hydroxylase Activity, Protein, mRNA and
Transcription Rate in the Rat

	Cholesterol 7α-hydroxylase			
Treatment	*Activity*	*Protein*	*mRNA*	*Transcription*
Bile Salts	↓	↓	↓	↓
Cholestyramine	↑	↑	↑	↑
Bile Fistula	↑	↑	↑	↑
Cholesterol	↑	↑	↑	↑
Mevinolin	↓	↓	↓	↓
Mevalonate	↑	?	↑	↑
Diurnal Rhythm				
Dark	↑	↑	↑	↑
Light	↓	↓	↓	↓
Starvation	↓	↓	↓	?
Thyroid Hormone	↑	↑	↑	?
Glucocorticoid				
High dose	↓	↓	↓	?
Low dose	↑	?	↑	↑
cAMP	?	?	↓	↓
Glucagon	?	?	↓	↓

region of the *CYP7* gene is highly conserved and likely to play an important role in the regulation of the *CYP7* gene.

Molowa et al. (1992) construct nested deletion mutants of the human *CYP7* promoter/CAT reporter chimeric genes and transiently transfected in Hep G2 cells. Three HNF3 binding sites have been located in the region from nucleotide –432 to –220 in the human gene and were shown to confer *CYP7* gene transcription activity when cotransfected with HNF3 (Molowa et al., 1992). However, deletion of the promoter sequence to –200 completely abolished promoter activity in Hep G2 cells. This result is surprising, since basal transcriptional activity and tissue-specific expression of most genes are usually regulated by *cis*- elements located in the proximal promoter region. On the other hand, Hoekman et al. (1993) reported that major promoter activity of the rat *CYP7* gene is located in –145 region which could confer taurocholate inhibition of the *CYP7* promoter activity when *CYP7*/CAT gene constructs were transiently transfected in rat primary hepatocytes. A bile acid responsive element was located between –49 and –79. Crestani et al. (1994, 1995) studied effects of bile acids and steroid/thyroid hormones on the promotor activity of *CYP7* gene fused to the leuciferase reporter gene which was transiently transfected into confluent culture of HepG2 cells. These investigators found that taurocholate and tauroursodeoxycholate did not repress promoter activity, whereas taurodeoxycholate and taurochenodeoxycholate suppressed the promoter activity.

Bile acid responsive elements were located in the −160 fragment and also in the upstream region. Ramirez et al. (1994) localized a liner-specific enhancer located 7 Kb upstream of the transcriptional initiation site of the rat *CYP7* gene. This upstream 1 Kb enhancer was necessary for expression of the rat *CYP7* gene and for regulation of the proximal promoter by bile and cholesterol as demonstrated in a conditionally transformed hepatocyte cell line H2.35 and in transgenic mice.

Lavery and Schibler (1993) identified a high affinity binding site for DBP, GTTATGTCAG which is centered around −225 of the rat *CYP7* gene. DBP is a liver-enriched transcription factor which was originally identified in the albumin gene promoter D site. The expression of DBP follows a stringent circadian rhythm (Wuarin and Schibler, 1990). Cotransfection of *CYP7*/CAT constructs with DBP plasmid in HepG2 cells stimulated promoter activity by more than ten-fold. Their data suggest that DBP may play an important role in the regulation of *CYP7* gene expression.

Steroid responsive elements (SRE), CAC(C/G)(C/T)CAC, which are present in the promoters of steroid-regulated genes such as HMG-Co A reductase, LDL receptor and HMG-Co A synthase genes are not found in the promoters of these *CYP7* genes (Goldstein and Brown, 1990). However, several modified SRE repeats (one or two mismatches) were identified in the 5'-upstream region of the rat gene or in the first intron of the hamster genes (Chiang et al., 1992; Crestani et al., 1993). It should be noted that oxysterols suppress the expression of genes in cholesterol synthesis pathway, whereas cholesterol or oxysterols activates the *CYP7* gene in bile acid synthesis pathway. Therefore, different steroid responsive motifs and specific transcription factors might be regulating the *CYP7* gene. The stimulation of cholesterol 7α- hydroxylase gene transcription by high cholesterol diet or by the infusion of 7-ketocholesterol, a competitive inhibitor of cholesterol 7α-hydroxylase, has been interpreted as a compensatory mechanism in response to oxysterol repression of the HMG-CoA reductase and LDL receptor genes (Breuer et al., 1993).

Two mechanisms could be proposed to explain the negative regulation of the *CYP7* gene transcription by bile acids (Figure 9). The general transcription factor (TFIID) of the *CYP7* gene is *trans*-activated by an activator A which binds to its binding site, i.e., SRE, and support a basal level of gene expression. Bile acid responsive element (BARE) may be located adjacent to SRE and TFIID complexes in the promoter. Hydrophobic bile acids transported to the hepatocytes may stimulate a nuclear bile acid receptor (BAR) which binds to the BARE and represses the activation of gene transcription by the activator A. In this mechanism, the BAR is a negative regulator. The second model is that an activator B binds to the BARE, which interacts with the normal activator A and subsequently activates the *CYP7* gene transcription. Bile acids prevent the binding of activator B to the BARE directly or via a receptor BAR, thus suppressing the *CYP7* gene expression (squelching effect). In this mechanism, the BARE is a conditional positive-regulatory element.

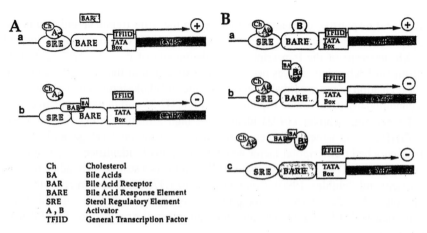

Ch	Cholesterol
BA	Bile Acids
BAR	Bile Acid Receptor
BARE	Bile Acid Response Element
SRE	Sterol Regulatory Element
A , B	Activator
TFIID	General Transcription Factor

Figure 9. Mechanisms of regulation of the *CYP7* gene. TFIID represent the general transcription complex which binds to the TATA box of the *CYP7* gene. SRE represents sterol or other normal transcription factor A binding element. In this diagram, cholesterol (Ch) binds to factor A and trans-activates the SRE of the *CYP7* gene (a). In mechanism A, BARE, bile acid response element, is a negative element which recognizes bile acid receptors (BAR) and bile acid (BA) complex and suppresses gene transcription (b). In mechanism B, the BARE is a positive element which recognizes an activator B (a). BA binds to activator B and prevents its binding to the BARE (b). Alternatively, the BAR-BA complex inactivates activator B and prevents its interaction with normal activator A (c).

Dnase I footprinting experiment revealed several protected footprints in the proximal promoter region (Chiang and Stroup, 1994). One of these transcription factor binding sites covers nucleotides from −81 to −37 which consist of TGT3, TRE-like motif, LFB1, and CAAT box (Figure 7). TGT3, LFB1 and CAAT motifs are the binding sites for HNF3, HNF1, and C/EBP, respectively. These motifs partially overlapped with an imperfect direct repeat, TGGTCANNNNAGTTCA, which is similar to the thyroid hormone responsive element (TRE), AGGTCA (Chiang et al., 1992). Gel mobility shift assays were applied to study DNA protein interaction and revealed several specific band shifts when oligonucleotides corresponding to this footprint were incubated with rat liver extracts (Chiang and Stroup, 1994). It is interesting that nuclear extracts isolated from livers of rats treated with 0.25% deoxycholate abolished gel shifts of the TRE-like sequence. These preliminary results suggest that bile acid response is located in this TRE-like repeat which may bind a specific activating protein factor. Bile acid feedback prevents the binding of this activator and reduces the gene transcription (Mechanism B in Figure 9). We propose that bile acid responsive protein factor(s) may be related to the steroid/thyroid hormone super gene family. Other transcription factors, repressors or adaptors binds to distal region of the promoter must also participate in the interaction and regulation of the general transcription machinery. This complicated gene regulation mechanism remains to be elucidated.

Based on the observation that protein kinase (CPKC) inhibitors prevented the down-regulation of cholesterol 7α-hydroxylase mRNA by taurocholate and PKC activator PMA decreased mRNA and transcriptional activity, Stravitz et al. (1995) suggested that bile acids down-regulate *CYP7* gene transcription indirectly, through activation of PKC. The PKC isoforms and down stream transcription factors mediating this signal transduction pathway remain to be elucidated.

Integration of Various Regulatory Mechanisms

Cholesterol 7α-hydroxylase is expressed and maintained at the functional level by cholesterol, thyroxine plus dexamethasone and possibly some other hormones. Hydrophobic bile salts circulating in the enterohepatic circle down-regulate cholesterol 7α-hydroxylase. The stimulatory effect of cholesterol provides an effective mechanism for removal of excess cholesterol from the body via its conversion to bile acids. In contrast, under the circumstances of decreased cholesterol supply, cholesterol 7α-hydroxylase gene is turned off with the reduction of cholesterol degradation to bile acids. In either case, cholesterol homeostasis is maintained. Unexpected bile salts losses such as occur in diseases of terminal ileum, where bile salt absorption takes place, are compensated by stimulation of cholesterol 7α-hydroxylase gene transcription and subsequent increase in bile acid synthesis. In case of excess bile salt intake, such as with bile acid feeding, cholesterol 7α-hydroxylase is turned off, albeit never completely. It is conceivable that cholesterol, thyroxine and dexamethasone are responsible for the basal level of cholesterol 7α-hydroxylase gene expression, while hydrophobic bile acids mediated repression may fine-tune cholesterol 7α-hydroxylase gene transcription.

Liver appears to maintain cholesterol homeostasis at all cost. This finely tuned mechanism has always fascinated the investigators. Scallen and Sanghvi (1983) have postulated that HMG-CoA reductase, cholesterol 7α-hydroxylase, and acyl CoA:cholesterol acyltransferase (ACAT) are all regulated by phosphorylation-dephosphorylation. Tang and Chiang (1986) have reported the activation of cholesterol 7α-hydroxylase activity by protein kinase and inhibition of activity by alkaline phosphatase in a reconstituted cholesterol 7α-hydroxylase. Cholesterol 7α-hydroxylase in rat liver microsomes and in the purified form could be phosphorylated *in vitro* by cAMP-dependent protein kinase (Chiang and Li, 1991). This mechanism may not play a major role in the regulation, however, it may provide a short-term modulation of catalytic activity. Nevertheless, it appears that at least the rates of cholesterol and bile acid synthesis are integrated, since under most physiologic and pharmacologic manipulations, HMG-CoA reductase and cholesterol 7α-hydroxylase change in the same direction. The understanding of integrative mechanisms responsible for the maintenance of cholesterol homeostasis in the liver will provide the basis for the future therapeutic approaches to the treatment of diseases associated with hypercholesterolemia.

There is apparently a cross-talk among three rate-limiting enzymes or protein in cholesterol metabolism, cholesterol 7α-hydroxylase, HMG-CoA reductase and LDL-receptor. Cholestyramine treatment stimulated and bile acid feedback inhibited all three activities. Coordinated regulation between cholesterol 7α-hydroxylase and HMG-CoA reductase has been discussed before. Recent reports also suggest that cholesterol 7α-hydroxylase and LDL-receptor may be coordinately regulated (Dueland et al., 1992). Transfection of non-hepatic Chinese hamster ovary cells with 7α-hydroxylase plasmid over-expressed 7α-hydroxylase mRNA and activity in these cells. These investigations found, in the presence of 5% serum, expression of the LDL receptor by transfected cells was 20 times that of non-transfected cells. They suggest that either 7α-hydroxylase is responsible for the change in LDL receptor expression or that both genes are regulated by a common factor. This is in line with their previous report that screening of 7α-hydroxylase deficient rat hepatoma cells (H35) for resistance to 25-hydroxycholesterol yielded L35 cells which expressed a high level of 7α-hydroxylase mRNA and activity by activation of the silent endogenous *CYP7* gene (Leighton et al., 1992). Over-expression of cholesterol 7α-hydroxylase may metabolize and inactivate oxysterols in these cells. This may explain the insensitivity of the liver toward down-regulation of LDL receptor and the parallel induction of the LDL receptor and the HMG-CoA reductase when 7α-hydroxylase is over-expressed (Dueland et al., 1992).

Sterol 12α-Hydroxylase

A cytochrome P-450 dependent microsomal 12α-hydroxylase is another important enzyme in the bile acid biosynthetic pathway. This enzyme is generally considered to determine the ratio of cholic to chenodeoxycholic acid synthesis and therefore, the ratio of these two bile acids in bile. The enzyme is insensitive to carbon monoxide and is not inducible by phenobarbital (Einarsson, 1968; Suzuki et al., 1968). This microsomal cytochrome P-450 enzyme 12α-hydroxylates 7α-hydroxy-4-cholesten-3-one ultimately resulting in the synthesis of cholic acid. Bernhardson et al. (1973) reported that partially purified microsomal cytochrome P-450 catalyzed 12α-hydroxylation of substrate in the presence of NADPH, cytochrome P-450 reductase and phosphatidylcholine. The activity of 12α-hydroxylase increases with the addition of sterol carrier proteins probably as a result of facilitated delivery of 7α-hydroxy-4-cholesten-3-one to the enzyme site (Grabowski et al., 1976). Gustafsson (1985) has shown that 12α-hydroxylation of 5β-cholestane 3α,7α-diol can be accomplished by both microsomal and mitochondrial fractions in human fetal tissue. However, the mitochondrial 12α-hydroxylase may be lost during ontogeny.

Sterol 12α-hydroxylase activity is regulated independently of the 7α-hydroxylase activity. The only common mechanism shared is bile acid feedback inhibition of both 7α- and 12α-hydroxylase activities (Björkhem, 1985). Biliary drainage of bile acids or cholestyramine stimulated the rate of bile acid synthesis and also 12α-hydroxylase activity. This may explain that cholic acid synthesis is stimulated more than chenodeoxycholic acid synthesis (Björkhem, 1985; Fears and Sabine, 1986; Kuroki et al., 1983). In contrast to 7α-hydroxylase, 12α-hydroxylase activity is markedly stimulated in starvation (Ishida, 1992). Streptozotocin treatment also increased 12α-hydroxylase activity. Cholesterol feeding or thyroid hormone treatment decreases steroid 12α-hydroxylase activity (Fears and Sabine, 1986). The rate of bile acid synthesis, the bile acid pool size, steroid 12α-hydroxylase activity, and the ratio of cholic acid to chenodeoxycholic acid are increased in diabetes mellitus and in diabetic rats (Ogura et al., 1986). The increase in cholic acid synthesis was shown to involve 12α-hydroxylation of 5β-cholestane-3α, 7α-diol. Insulin treatment normalized bile acid synthesis in diabetic rats (Kimura et al., 1988).

There are marked differences in 12α-hydroxylase activity in different species. The activity is the highest in rabbit which has cholic acid as a predominant bile acid and is lower in guinea pig which has chenodeoxycholic acid as the major bile acids. Kuroki and Hoshita (1983) found that cholic acid, chenodeoxycholic acid or ursodeoxycholic acid feeding down-regulated cholesterol 7α-hydroxylase activity in hamsters. The same authors also provided evidence that 12α-hydroxylase activity in microsomal fractions was higher in female than in male hamsters and reported a strong correlation between 12α-hydroxylase activity and bile acid composition (Kuroki et al., 1983). It was suggested that sterol 12α-hydroxylase may regulate the ratio of cholic acid to chenodeoxycholic acid, hence the hydrophobicity of bile acids in the bile. Therefore, 12α-hydroxylase may also play an important role in the regulation of bile acid synthesis by regulating bile acid hydrophobicity of the bile. However, *in vivo* study that used a labeled substrate for the 12α-hydroxylase did not support the role 12α-hydroxylase play in the regulation of cholic acid to chenodeoxycholic acid ratio (Björkhem et al., 1983). This may be explained by the existence of separate cholesterol pools in the liver for the synthesis of cholic acid and chenodeoxycholic acid (Ahlberg et al., 1979). Recent study by Einarsson et al. (1992) revealed that sterol 12α-hydroxylase activity was increased two-fold in liver microsomes of patients treated with cholestyramine or undergone ileal resection. The increase of the ratio of cholic acid to chenodeoxycholic acid in these patients is likely due to a compensatory increase of the sterol 12α-hydroxylase activity.

The sterol 12α-hydroxylase has been partially purified from rabbit liver by Murakami et al. (1982), and found to be free from sterol 25-, 26-, or 27-hydroxylase activity. The estimated molecular weight of this protein was 56,000. Recently, purification and characterization of sterol 12α-hydroxylase P450 has been accomplished by Ishida et al. (1992) from rabbit liver microsomes. The NH_2-terminal amino acid sequence appears to be unique among other cytochrome P-450s. These authors found that 12α-hydroxylase activity in rat liver microsomes was not inhibited by thyroid hormone, but was stimulated in streptozotocin induced diabetes, by starvation and by cholestyramine. However, immunoblot using an antibody against the purified 12α-hydroxylase revealed no correlation between the amount of enzyme and enzyme activity in microsomes (Ishida et al., 1992). These authors concluded that 12α-hydroxylase might be regulated at the post-translational level. The cDNA encoding 12α-hydroxylase has not been cloned, therefore, it is not possible to measure the steady-state level of 12α-hydroxylase mRNA under these experimental conditions. It cannot be ruled out completely, however, that the polyclonal antibody obtained might not be specific or enzyme preparation used for antibody preparation might not be homogeneous.

Mitochondrial 27-Hydroxylase

Bile acid biosynthesis can be initiated by the hydroxylation of cholesterol at either C_7 or C_{27} by cytochrome P-450 hydroxylases. Mitochondria isolated from liver and other tissues contain a cytochrome P-450 mixed function monooxygenase which hydroxylates several C_{27} steroids. Steroid substrates recognized by this enzyme include: 5β-cholestane-3α,7α,12α-triol, 5β-cholestane-3α,7α-diol, 7α-hydroxy-4-cholesten-3-one, 5-cholesten-3α,7α-diol and cholesterol. The mitochondrial 27-hydroxylase can initiate 27-hydroxylation of cholesterol (the initial reaction of the "acidic" pathway) or participate in the side-chain cleavage of bile acid intermediates formed via the "neutral" pathway. Though cholesterol is a substrate for this enzyme, kinetic studies indicate that bile acid intermediates are probably the preferred substrates *in vivo* (Björkhem and Gustafsson, 1974). Substitution at either of the two terminal methyl groups of the C_{27} steroid side chain creates an asymmetric carbon at C_{25}. The mitochondrial 27-hydroxylation of cholesterol appears to be stereospecific for the C_{27} methyl group (25 pro-S). There is also evidence that the intermediates in the bile acid biosynthetic pathway are also hydroxylated exclusively at C_{27} by the mitochondrial enzyme (Batta et al., 1983). Therefore, it has been recommended that the name should no longer be used and that this mitochondrial enzyme should be referred to as a 27-hydroxylase.

The physiological function of 27-hydroxylase in non-liver cells is not clear. Javitt and Budat (1989) suggested that 27-hydroxycholesterol may function as a repressor of HMG-CoA-reductase, the rate-limiting enzyme in cholesterol biosynthesis. Hence, 27-hydroxycholesterol could serve as an endogenous regulator of rates of cholesterol biosynthesis. 27-Hydroxycholesterol can be found as a normal component of LDL and other serum lipoproteins (Javitt, 1990a). 27-hydroxylase has been shown to catalyze 25-hydroxylation of vitamin D_3 (Usui, 1990; Su, 1990). In pig liver, a mitochondrial 27-hydroxylase was able to catalyze 27-, 25-, or 24-hydroxylation of cholesterol at a relative rate of about 1:0, 2:0.04 (Lund, 1993).

Wikvall (1984) was the first to isolate and partially purify 27-hydroxylase from the rabbit liver. More recently, the rabbit mitochondrial cytochrome P-450 steroid 27-hydroxylase has been purified, cloned and nucleotide sequence of the cDNA determined by Andersson et al. (1989). The enzyme is a product of a unique P450 gene family, *CYP27*, which has been mapped to q33-qter interval of human chromosome 2 (Cali et al., 1991). The enzyme has 499 amino acids plus a 36 amino acid mitochondrial signal sequence. The enzyme is located in the inner mitochondrial membrane and requires ferredoxin, ferredoxin reductase and NADPH for activity (Pedersen et al., 1977). In the presence of ferredoxin and ferredoxin reductase, sterol 27-hydroxylase will oxidize several sterol substrates. Unlike cholesterol 7α-hydroxylase and 12α-hydroxylase, sterol 27-hydroxylase are present in many different tissues, suggesting that this enzyme activity is not limited to bile acid synthesis (Skrede et al., 1986; Su et al., 1990). The relative amounts of 27-hydroxylase mRNA (1.9 kb) were found to be the highest in liver, duodenum, adrenal gland, lung, kidney and spleen, respectively (Andersson et al., 1989). The level of 27-hydroxylase mRNA does not appear to be regulated by bile acid, cholesterol or hormones. The mitochondrial 27-hydroxylase has been expressed in COS-M6 cells and shown to hydroxylase 5β-cholestane-3α,7α,12α-triol. Surprisingly, the transfected COS-M6 cells metabolized the tetrol to the C_{27} cholestanoic acid (Cali and Russell, 1991). Recent evidence suggest that the purified 27-hydroxylase is also capable of oxidizing the tetrol to an acid (Andersson et al., 1989; Dahlback and Holmberg, 1990).

IV. BILE ACID SYNTHESIS IN PATHOLOGIC CONDITIONS

A decrease in bile acid synthesis has been described in a number of pathologic conditions, some as a primary defect in the enzymes involved in bile acid biosynthesis pathways, the others occurring as a secondary phenomenon, i.e., as a result of a disease process. Abnormality in bile acid biosynthesis have been detected in several inherited diseases including defects in modification of the cholesterol nucleus (3β-hydroxy-Δ^5-C_{27}-sterol oxidoreductase and Δ^4-3-oxosteroid-5β-reductase), defects in side-chain oxidation (27-hydroxylase) and defects in peroxisomal β-oxidation (Zellweger Syndrome and related diseases) (Clayton, 1991; Balistreri, 1991). Deficiency in bile acid metabolism could cause malabsorption of vitamins and fats, cholesterol gallstone disease, cirrhosis, hepatitis, cholestasis and cholesterol storage disease in humans (Björkhem, 1992; Björkhem and Skrede, 1989; Hanson et al., 1979).

A. Defects in Modification of Cholesterol Nucleus

An inherited deficiency of cholesterol 7α-hydroxylase has not as yet been described. A decrease in bile acid synthesis (Vlahcevic et al., 1970, 1972; Bell et al., 1972) and the reduction of cholesterol 7α-hydroxylase activity have been reported in patients with cholesterol gallstones (Einarsson et al., 1986; Bertolotti et al., 1988). It was postulated that inappropriately low bile acid synthesis in patients with cholesterol gallstones were due to "insensitive" bile acid biofeedback. As a result of a decreased bile acid synthesis and increase in fractional daily turnover of bile acids, patients with cholesterol gallstones have a decrease in the circulating bile acid pool which is thought to be a contributing factor in the pathogenesis of cholesterol gallstones (Vlahcevic et al., 1970, 1972; Bell et al., 1972; Berr et al., 1992).

The pharmacological therapy of cholesterol gallstones with bile salts administered orally was based on the rationale that increase in bile acid pool size will increase cholesterol solubility in bile. Indeed, the oral administration of chenodeoxycholate and ursodeoxycholate over a period of up to 1–2 years resulted in an increase in bile acid pool size and dissolution of cholesterol gallstones in approximately 40% of patients. The studies done subsequently have shown however, that cholesterol gallstone dissolution occurred with these two bile acids mostly because of a decrease in biliary cholesterol secretion (Heuman et al., 1990).

Patients with chronic liver disease have also been reported to have significantly diminished bile acid pool size and decreased cholic acid synthesis with the marked shift to predominance of chenodeoxycholic acid in bile (Vlahcevic et al., 1971a, 1972a; McCormick et al., 1973). Because of marked predominance of chenodeoxycholic acid in bile in patients with cirrhosis, it has been postulated that those patients may have a deficiency in 12α-hydroxylase alone or in combination with other enzymes (Goldman et al., 1982; Vlahcevic et al., 1981).

Clayton et al. (1987) described a defect of 3β-Δ^5-C_{27}-hydroxysteroid oxidoreductase in a child with giant cell hepatitis. Additional patients with this autosomal recessive disease have been recently described by Setchell et al. (1990). These patients excrete in urine multiple C_{24} bile acids with 3β-hydroxy-Δ^5 structure. The specific activity of the enzyme was negligible in the patient and low in parents which was consistent with a heterozygous carrier phenotype (Buchmann et al., 1990).

Δ^4-3-oxosteroid-5β-reductase deficiency is another autosomal recessive disorder of bile acid synthesis described by Setchell et al. (1988, 1990) in a small number of infants with neonatal hepatitis and cholestasis. The abnormal bile acids retain the Δ^4-3-oxo configuration of sterol ring. The normal bile acids are present in negligible amounts, attesting to the defect of this enzyme. The immunoblot analysis shows lack of enzyme presence in the liver of affected patients (Russell and Setchell, 1992). These patients also have significant amounts of 5α-cholenoic acids, which are of hepatic origin and are generally not found in humans. The pathogenesis of liver disease in this patient is not certain, but it has been speculated that it could occur as a result of accumulation of toxic bile salts and a decrease in bile acid generated bile flow. A significant improvement in the symptoms, laboratory data and normalization of hepatic morphology occurred following feeding of primary bile acids to these patients (Ichimiya et al., 1990). The administered bile acid was effective most likely because of suppression of bile acid synthesis via negative feedback control and subsequent reduction of production of abnormal bile acids.

B. Defect in Sterol 27-Hydroxylase

Cerebrotendinous xanthomatosis (CTX) is a rare autosomal recessive defect of cholesterol metabolism manifested by xanthoma, progressive neurologic dysfunction, accumulation of cholesterol in the tissues, premature coronary heart disease and high frequency of cholesterol gallstones (Björkhem and Skrede, 1989). Setoguchi et al. (1974) were first to report that patients with CTX excreted abnormal C_{27} bile alcohols coupled with decreased levels of bile acids, and in particular, chenodeoxycholic acid. They interpreted these findings as consistent with a defect in 25-hydroxylase enzyme (Salen et al., 1979). The assumption was that 25-hydroxylase is a major route of bile acid biosynthesis in man. Other investigators raised questions about this interpretation and demonstrated that only about 5% of bile acid biosynthesis takes place via 25-hydroxylation pathway in man (Duane et al., 1988). Oftebro et al. (1980) reported marked decrease in the mitochondrial 27-hydroxylase activity in CTX patients, while Javitt et al. (1982) found reduced or absent levels of 27-hydroxycholesterol in humans. These results were confirmed by others (Miki et al., 1986).

Mitochondrial 27-hydroxylase has been cloned and sequenced from the rat, rabbit and most recently human liver (Cali and Russell, 1991). These investigators also identified the molecular defects in mitochondrial 27-hydroxylase in two patients with CTX (Cali et al., 1991). The two observed mutations appear to be either in ferredoxin or heme ligand binding site of 27-hydroxylase. Transfection of COS cells with sterol 27-hydroxylase cDNA from CTX patients could not express active enzyme. These latter data provide evidence that abnormality of bile acid biosynthesis in CTX patients results from a point mutation of the mitochondrial 27-hydroxylase gene which encoded an inactive 27-hydroxylase. Leitersdorf et al. (1993) reported a CTX patient who has no detectable sterol 27-hydroxylase mRNA. Analysis of the 27-hydroxylase gene revealed a deletion of thymidine in exon 4 and a G to A substitution at the 3' splice acceptor site of intron 4 of the gene, which cause a frame shift and splice-junction mutation. The defect in 27-hydroxylase leads to excessive accumulation of a 7α-hydroxycholesterol, 7α-hydroxy-4-cholesten-3-one, 5β-cholestane-3α, 7α, 12α-triol, cholesterol and cholestanol in CTX. The reduced bile acid synthesis in CTX patient may up-regulate cholesterol 7α-hydroxylase activity and lead to the accumulation of both 7α-hydroxy-cholesterol and 7α-hydroxy-4-cholesten-3-one. Since bile acid feedback also inhibits HMG-CoA reductase activity, *de novo* cholesterol synthesis is stimulated in CTX. 7α-hydroxy-4-cholesten-3-one cannot be metabolized and is converted to cholestanol (Björkhem and Skrede, 1989; Björkhem, 1992).

C. Defect in Peroxisomal β-Oxidation

Zellweger syndrome and related diseases, infantile Refsum disease, neonatal adrenoleukodystrophy, are characterized by accumulation of intermediates of bile acid synthesis and very long chain fatty acids in serum and urine. Altered bile acid metabolism leads to the overproduction of 3α, 7α, 12α-trihydroxy-5β-cholestanoic acid and 3α, 7α-dihydroxy-5-β-cholestanoic acid (Hanson et al., 1979). Peroxisomes are absent in the liver and kidney of Zellweger patients (Goldfischer et al.,

1973, 1986). These indicate that the oxidative cleavage of the side-chain of corprostanoic acids are occurred in the peroxisomes and perosixomal 3-oxo-acyl-CoA thiolase is defective in Zellweger Syndrome (Schrame et al., 1987). A generalized impairment of peroxisome biogenesis is believed to cause Zellweger and related syndromes (Lazarow and Moser, 1989). Recently, X-linked adrenoleuko-dystrophy gene was identified to encode a peroxisomal membrane protein which shares sequence homology to the ATP-binding cassette transporter (Mosser et al., 1993).

ACKNOWLEDGMENTS

This work was supported by the National Institutes of Health Grants GM31584, DK44442, PO1DK38030 and Biomedical Research Support Grant; and grants from the American Heart Associations, Ohio Affiliates, Hoechst, AG, Veterans Administration, and Research Challenge grant of the Ohio Board of Regents.

REFERENCES

Adachi, Y., Kobayashi, H., Kurumi, Y., Shouji, M., Kitano, M., & Yamamoto, T. (1991). ATP-dependent taurocholate transport by rat liver canalicular membrane vesicles. Hepatology 14, 655–659.

Admirand, W.H., & Small, D.M. (1968). The physical chemical basis of cholesterol solubility in bile. Relationship to gallstone formation and dissolution in man. J. Clin. Invest. 47, 1043–1052.

Ahlberg, J.R., Angelin, B., Björkhem, I., Einarsson, K., & Leijd, B. (1979). Hepatic cholesterol metabolism in normo- and hyperlipidemic patients with cholesterol gallstones. J. Lipid Res. 20, 107–115.

Akerboom, T.P.M., Narayanaswami, V., Kunst, M., & Sies, H. (1991). ATP-dependent S-(2,4-dini-trophenyl)glutathione transport in canalicular plasma membrane vesicles from rat liver. J. Biol. Chem. 266, 13147–13152.

Ananthanarayanan, M., Von Dippe, P., & Levy, D. (1988). Identification of the hepatocyte Na+-dependent bile acid transport protein using monoclonal antibodies. J. Biol. Chem. 263, 8338–8343.

Ananthanarayanan, M., Bucuvalas, J.C., Shneider, B.L., Sippel, C.J., & Suchy, F.J. (1991). An ontogenically regulated 48 kDa protein is a component of the Na+-bile acid cotransporter of rat liver. Am. J. Physiol. 261, G810–G817.

Andersson, S., Davis, D. L., Dahlback, H., Jornvall, H., & Russell, D.W. (1989). Cloning, structure, and expression of the mitochondrial cytochrome P-450 sterol 26-hydroxylase, a bile acid biosynthetic enzyme. J. Biol. Chem. 264, 8222–8229.

Axelson, M., Mork, B., & Sjövall, J. (1988). Occurrence of 3β-hydroxy-5-cholestenoic acid, 3β,7α-di-hydroxy-5-cholestenoic acid and 7α-hydroxy-3-oxo-4-cholestenoic acid as normal constituents in human blood. J. Lipid. Res. 29, 629–641.

Axelson, M., Aly, A., & Sjövall, J. (1988). Levels of 7α-hydroxy-4-cholesten-3-one in plasma reflect rates of bile acid biosynthesis in man. FEBS Lett. 239, 324–328.

Axelson, M., Mork, B., Aly, A., Wisen, O., & Sjövall, J. (1989). Concentration of cholestenoic in plasma from patients with liver disease. J. Lipid Res. 30, 1877–1882.

Axelson, M., Mork, B., Aly, A., Walldivs, G., & Sjövall, J. (1989a). Concentration of cholestenoic acid in plasma from patients with reduced intestinal absorption of bile acids. J. Lipid Res. 30, 1883–1887.

Axelson, M., & Sjövall, J. (1990). Potential bile acid precursors in plasma-possible indicators of biosynthetic pathways to cholic and chenodeoxycholic acids in man. J. Steroid. Biochem. 36, 631–640.

Axelson, M., Shoda, J., Sjövall, J., Toll, A., & Wikvall, K. (1992). Cholesterol is converted to 7α-hydroxy-3-oxo-4-cholestenoic acids in liver mitochondria. J. Biol. Chem. 267, 1701–1704.

Axelson, M., Mork, B., & Everson, G.T. (1991). Bile acid synthesis in cultured human hepatoblastoma cells. J. Biol Chem. 266, 17770–17777.

Balistreri, W.F. (1991). Fetal and neonatal bile acid synthesis and metabolism-clinical implications. J. Inher. Metab. Dis. 14, 459–477.

Batta, A.K., Salen, G., & Shefer, S. (1983). Configuration of C_{25} in 3α,7α,12α-trihydroxy-5β-cholestan-26-oic acid isolated from human bile. J. Lipid. Res. 24, 94–96.

Bell, C.C., Jr., McCormick, W.C., Gregory, D.H., Law, O., Vlahcevic, Z.R., & Swell, L. (1972). Relationship of bile acid pool size to the formation of lithogenic bile in the male Indians of the Southwest. Surg. Gynecol. Obstet. 134, 473.

Bernhardson, C., Björkhem, I., Danielsson, H., & Wikvall, K. (1973). 12α-hydroxylation of 7α-hydroxy-4-cholesten-3-one by a reconstituted system of rat liver microsomes. Biochem. Biophys. Res. Commun. 54, 1030–1038.

Berr, F., Pratschke, E., Fischer, S., & Paumgartner, G. (1992). Disorders of bile acid metabolism in cholesterol gallstone disease. J. Clin. Invest. 90, 859–868.

Bertolotti, M., Iori, R., Zironi, F., Montanari, M., Tripodi, A., & Carulli, N. (1988). Bile acids and cholesterol metabolism: Effect of deoxycholic acid on plasma lipoprotein and biliary cholesterol concentration. Ital. J. Gastroenterol. 20, 240–245.

Bertolotti, M., Abate, N., Loria, P., Dilengite, M., Carubbi, F., Pinetti, A., Digresolo, A., & Carulli, N. (1991). Regulation of bile acid synthesis in human: Effect of treatment with bile acids, cholestyramine or simvastatin on cholesterol 7α-hydroxylation rates *in vivo*. Hepatology 14, 830–837.

Björkhem, I., & Gustafsson, J. (1974). Mitochondrial ω-hydroxylation of the cholesterol side chain. J. Biol. Chem. 249, 2528–2535.

Björkhem, I., Eriksson, M., & Einarsson, K. (1983). Evidence for a lack of regulatory importance of the 12α-hydroxylase in formation of bile acids in man: An *in vivo* study. J. Lipid. Res. 24, 1451–1456.

Björkhem, I. (1985). Mechanism of bile acid biosynthesis in mammalian liver. In: Steroid and Bile Acid (Danielsson H., & Sjovall J., eds.), pp. 231–278. Elsevier Science Publishers, The Netherlands.

Björkhem, I., & Akerlund, J.E. (1988). Studies on the link between HMG-Co A reductase and cholesterol 7α-hydroxylase in rat liver. J. Lipid Res. 29, 136–143.

Björkhem, I., & Skrede, S. (1989). Cerebrotendinous xanthomatosis and phytosterolemia. In: The Metabolic Basis of Inherited Diseases (Scriver, C.R., Beaudet, A.L., Sly, W.S., & Valle, D., eds.), pp. 1283–1302. McGraw-Hill, New York.

Björkhem, I., Eggertsen, G., & Andersson, U. (1991). On the mechanism of stimulation of cholesterol 7α-hydroxylase by dietary cholesterol. Biochim. Biophys. Acta 1085, 329–335.

Björkhem, I., Nyberg, B., & Einarsson, K. (1992). 7α-Hydroxylation of 27-hydroxycholesterol in human liver microsomes. Biochim. Biophys. Acta 1128, 73–76.

Björkhem, I. (1992). Mechanism of degradation of the steroid side chain in the formation of bile acids. J. Lipid Res. 33, 455–471.

Boyer, J.L., Graf, J., & Meier, P.J. (1992). Hepatic transport system regulating pH, cell volume and bile secretion. Ann. Rev. Physiol. 54, 415–438.

Breuer, O., Sudjana-Sugiaman, E., Eggertsen, G., Chiang, J.Y.L., & Björkhem, I. (1993). Cholesterol 7α-hydroxylase is upregulated by the competitive inhibitor 7-oxocholesterol in rat liver. Eur. J. Biochem. 215, 705–710.

Buchmann, M.S., Kvittingen, E.A., Nazer, H., Gunasekaran, T., Clayton, P.T., Sjövall, J., & Björkhem, I. (1990). Lack of 3β-hydroxy-Δ^5-C_{27}-steroid dehydrogenase/isomerase in fibroblasts from a child with urinary excretion of 3β-hydroxy-Δ^5-bile acids. J. Clin. Invest. 86, 2034–2037.

Cali, J.J., Hsich, C.L., Francke, U., & Russell, D.W. (1991). Mutation in the bile acid biosynthetic enzyme sterol 27-hydroxylase underlie cerebrotendinous xanthomatosis. J. Biol. Chem. 266, 7779–7783.

Cali, J.J., & Russell, D.W. (1991). Characterization of human steroid 27-hydroxylase. J. Biol. Chem. 266, 7774–7778.

Carey, M.C., Small, D.M., & Bliss, C.M. (1983). Lipid digestion and absorption. Ann. Rev. Physiol. 45, 651–677.

Carey, M.C. (1985). Physical-chemical properties of bile acids and their salts. In: Sterols and Bile Acids (Danielsson, H., & Sjövall, J., eds.)., pp. 345–403. Elsevier, New York.

Carey, M.C., & Cahalane, M.J. (1988). The enterohepatic circulation. In: The Liver: Biology and Pathobiology (Arias, I., Popper, H., & Schachter, D., eds.), pp. 573–616. Raven Press, New York.

Cheng, K., White, P., & Qin, K. (1991). Molecular cloning and expression of rat liver 3α-hydroxysteroid dehydrogenase. Mol. Endocrinol. 5, 823–828.

Chiang, J.Y.L., Miller, W.F., & Lin, G.M. (1990). Regulation of cholesterol 7α-hydroxylase in the liver:purification of cholesterol 7α-hydroxylase and the immunochemical evidence for the induction of cholesterol 7α-hydroxylase by cholestyramine and circadian rhythm. J. Biol. Chem. 265, 3889–3897.

Chiang, J.Y.L., & Li, Y.C. (1991). Regulation of bile acid synthesis: Purification, cloning, and regulation of cholesterol 7α-hydroxylase. In: Bile acids as therapeutic agents (Paumgartner, G., Stiehl, A., & Gerok, W. eds.), pp. 29–44. MTP Press.

Chiang, J.Y.L., & Stroup, D. (1994). Identification of a liver nuclear protein factor responsive to bile acid inhibition of cholesterol 7α-hydroxylase gene. J. Biol. Chem. 269, 17502–17507.

Chiang, J.Y.L., Yang, T.P., & Wang, D.P. (1992). Cloning and 5'-flanking sequence of a rat cholesterol 7α-hydroxylase gene. Biochim. Biophys. Acta. 1132, 337–339.

Chua, S.C., Szabo, P., Vitek, A., Grzeschik, K.H., John, M., & White, P.C. (1987). Cloning of cDNA encoding steroid 11β-hydroxylase (P450c11). Proc. Natl. Acad. Sci. USA 84, 7193–7197.

Chung, B.C., Picado-Leonard, J., Hanin, M., Bienkowski, M., Hall, P.F., & Miller, W.L. (1987). Cytochrome P45017 (steroid 17α-hydroxylase/17,20 lyase): Cloning of human adrenal and testis cDNAs indicates the same gene is expressed in both tissues. Proc. Natl. Acad. Sci. USA. 84, 407–411.

Clayton, P.T., Leonard, J.V., Lawson, A.M., Setchell, K.D.R., Andersson, S., Egestad, B., & Sjovall, J. (1987). Familial giant cell hepatitis associated with synthesis of 3β,7α-dihydroxy- and 3β,7α,12α-trihydroxy-5-cholenoic acids. J. Clin. Invest. 79, 1031–1038.

Clayton, P.T. (1991). Inborn errors of bile acid metabolism. J. Inher. Metab. Dis. 14, 478–496.

Cohen, J.C., Cali, J.J., Jelinek, D.F., Mehrabian, M., Sparkes, R.S., Lusis, A.J., Russell, D.W., & Hobbs, H.H. (1992). Cloning of the human cholesterol 7α-hydroxylase gene (CYP7) and localization to chromosome 8q11–q12. Genomics 14, 153–161.

Crestani, M., Galli, G., & Chiang, J.Y.L. (1993). Genomic cloning, sequencing and analysis of the hamster cholesterol 7α-hydroxylase gene (CYP7). Arch. Biochem. Biophys. 306, 451–460.

Crestani, M., Karam, W.G., & Chiang, J.Y.L. (1994). Effects of bile acids and steroid/thyroid hormones on the expression of cholesterol 7α-hydroxylase mRNA and the CYP7 gene in HepG2 cells. Biochem. Biophys. Res. Commun. 198, 546–553.

Crestani, M., Stroup, D., & Chiang, J.Y.L. (1995). Hormonal regulation of the rat cholesterol 7α-hydroxylase gene (CYP7). J. Lipid Res. 36, 2419–2432.

Dahlbäck, H., & Holmberg, I. (1990). Oxidation of 5β-cholestane-3α,7α,12α-triol into 3α,7α,12α-trihydroxy-5β-cholestanoic acid by cytochrome $P450_{26}$ from rabbit liver mitochondria. Biochem. Biophys. Res. Commun. 167, 391–395.

Danielsson, H., Einarsson, K., & Johansson, G. (1967). Effect of biliary drainage on individual reactions in the conversion of cholesterol to taurocholic acid. Bile acids and steroids 180. Eur. J. Biochem. 2, 44–49.

Dahlbäck-Sjöberg, H., Björkhem, I., & Prinen, H.M.G. (1993). Selective inhibition of mitochondrial 27-hydroxylation of bile acid intermediates and 25-hydroxylation of vitamin D_3 by cyclosporin A. Biochem. J. 293, 203–206.

Danielsson, H., & Sjövall, J. (1975). Bile acid metabolism. Annu. Rev. of Biochem. 233–253.

Danielsson, H., & Gustafsson, S. (1981). Biochemistry of bile acids in health and disease. In: Pathobiology Annual Vol. 11 (Ioachim, H.L., ed.), pp. 259–298. Raven Press, New York.

Danielsson, H., & Wikvall, K. (1986). Biosynthesis of bile acids. In: Cholesterol 7α-Hydroxylase (7α-monoxygenase) (Fears, R., & Sabine, J.R., eds.), pp. 9–21. CRC Press, Boca Raton, FL.

Davis, R.A., Highsmith, S.E., Malone-McNeal, M., Archambault-Schexnayder, J., & Kuan, J.-C.W. (1983). Bile acid synthesis by cultured hepatocytes: Inhibition by mevinolin but not by bile acids. J. Biol. Chem. 258, 4079–4082.

Davis, R.A., Hyde, P.M., Kuan, J.W., Malone-McNeal, M., & Archambault-Schexnayder, J. (1983a). Bile acid secretion by cultured rat hepatocytes:regulation by cholesterol availability. J. Biol. Chem. 258, 3661–3667.

Deyashiki, Y., Taniguchi, H., Amano, T., Nakayama, T., Hara, A., & Sawada, H. (1992). Structural and functional comparison of two human liver dihydrodiol dehydrogenases associated with 3α-hydroxysteroid dehydrogenase activity. Biochem. J. 282, 741–746.

Duane, W.C., Pooler, P.A., & Hamilton, J.N. (1988). Bile acid synthesis in man. *In vivo* activity of the 25-hydroxylation pathway. J. Clin. Invest. 82, 82–85.

Dueland, S., Trawick, J.D., Nenseter, M.S., MacPhee, A.A., & Davis, R.A. (1992). Expression of 7α-hydroxylase in non-hepatic cells results in liver phenotypic resistance of the low density lipoprotein receptor to cholesterol repression. J. Biol. Chem. 267, 22695–22698.

Dueland, S., Drisko, J., Grat, L., Machleder, D., Lusis, A.J., & Davis, R.A. (1993). Effect of dietary cholesterol and taurocholate on cholesterol 7α-hydroxylase and hepatic LDL receptors in inbred mice. J. Lipid Res. 34, 923–931.

Erickson S. (1957). Biliary excretion of bile acids and cholesterol in bile fistula rats. Proc. Soc. Exp. Biol. Med. 94, 578–582.

Einarsson K. (1968). On the properties of the 12α-hydroxylase in cholic acid biosynthesis. Bile acids and steroid 198. Eur. J. Biochem. 5, 101–108.

Einarsson, K., Angelin, B., Ewerth, S., Nilsell, K., & Bjorkhem, I. (1986). Bile acid synthesis in man: Assay of hepatic microsomal cholesterol 7α-hydroxylase activity by isotope dilution-mass spectrometry. J. Lipid Res. 27, 82–88.

Einarsson, K., Akerlund, J.E., & Bjorkhem, I. (1989). The pool of free cholesterol is not of major importance for the regulation of cholesterol 7α-hydroxylase activity in rat liver microsomes. J. Lipid Res. 29, 136–143.

Einarsson, K., Akerlund, J.E., Reihner, E., & Bjorkhem, I. (1992). 12α-hydroxylase activity in human liver and its relation to cholesterol 7α-hydroxylase activity. J. Lipid Res. 33, 1591–1595.

Erikson, S.K., & Bosterling, B. (1981). Cholesterol 7α-hydroxylase from human liver: Partial purification and reconstitution into defined phospholipid-cholesterol vesicles. J. Lipid. Res. 22, 872–876.

Everson, G.T., & Polokoff, M.A. (1986). HepG2. A human hepatoblastoma cell line exhibiting defects in bile acid synthesis and conjugation. J. Biol. Chem. 261, 2197–2201.

Fears, R., & Sabine, J.R. (1986). Cholesterol 7α-Hydroxylase (7α-Monooxygenase). CRC Press, Boca Raton, FL.

Fricker, G., Schneider, S., Gerok, W., & Kurz, G. (1987). Identification of different transport systems for bile salts in sinusoidal and canalicular membranes of hepatocytes. Biol. Chem. Hoppe-Seyler 368, 1143–1150.

Frimmer, M., & Ziegler, K. (1988). The transport of bile acids in liver cells. Biochim. Biophys. Acta 947, 75–99.

Furuebisu, M., Deguchi, S., & Okuda, K. (1987). Identification of cortisone 5β-reductase as Δ^4-3-ketosteroid 5β-reductase. Biochim. Biophys. Acta 912, 110–114.

Goldfischer, S., Moore, C.L., Johnson, A.B., Spiro, A.I., Valsamis, M.P., Wisniewski, H.K., Ritch, R.H., Norton, W.T., Rapin, I., & Gartner, L.M. (1973). Peroxisomal and mitochondrial defects in the cerebro-hepato-renal syndrome. Science 182, 62–64.

Goldfischer, S., Collins, J., Rapin, I., Neumann, P., Neglia, W., Spiro, A.J., Ishii, I., Roels, F., Vamecq, I., & van Hoof, F. (1986). Pseudo-Zellweger Syndrome: Deficiencies in several peroxisomal oxidation activities. J. Pediatr. 108, 25–32.

Goldstein, J.L., & Brown, S.M. (1990). Regulation of the mevalonate pathway. Nature 343, 425–430.

Goldman, M., Vlahcevic, Z.R., Schwartz, C.C., Gustafsson, J., & Swell, L. (1982). Bile acid metabolism in cirrhosis. VIII. Quantitative evaluation of bile acid synthesis from [7β-^{3}H] 7α hydroxy-cholesterol and [6-^{3}H] 26 hydroxycholesterol. Hepatology 2, 59–66.

Grabowski, G.A., McCoy, K.E., Williams, G.C., Dempsey, M.E., & Hanson, R.F. (1976). Evidence for carrier proteins in bile acid synthesis. The effects of squalene and sterol carrier protein and albumin on the activity of 12α-hydroxylase. Biochim. Biophys. Acta 441, 380–390.

Grange, T., Rohx, J., Rigand, G., & Pictet, R. (1991). Cell-type specific activity of two glucocorticoid responsive units of rat tyrosine amino transferase gene is associated with multiple binding sites for C/EBP and a novel liver specific nuclear factor. Nucleic Acid Res. 19, 131–139.

Gustafsson, J. (1985). Bile acid synthesis during development: Mitochondrial 12α-hydroxylation in human liver. J. Clin. Invest. 75, 604–607.

Hagenbuch, B., Stieger, B., Foguet, M., Lubbert, H., & Meier, P.J. (1990). Expression of the hepatocyte Na+/bile acid cotransporter in xenopus laevix oocytes. J. Biol. Chem. 265, 5357–5360.

Hagenbuch, B., Stieger, B., Foguet, M., Lubbert, H., & Meier, P.J. (1991). Functional expression cloning and characterization of the hepatocyte Na+/bile acid cotransport system. Proc. Natl. Acad. Sci. USA 88, 10629–10633.

Hanson, R.F., Szczepanik-Van Leeuwen, P., Williams, G.C., Grabowski, G., & Sharp, H.L. (1979). Defects of bile acid synthesis in Zellweger's syndrome. Science 203, 1107–1108.

Hardwick, J.P., Song, B.J., Huberman, E., & Gonzalez, F.J. (1987). Isolation, complementary DNA sequence and regulation of rat hepatic lauric acid ω-hydroxylase (cytochrome P450law): Identi-fication of a new cytochrome P450 gene family. J. Biol. Chem. 262, 801–810.

Heuman, D.M. (1988). Quantitative estimation of the hydrophilic-hydrophobic balance of mixed bile salt solutions. J. Lipid Res. 30, 719–730.

Heuman, D.M., Vlahcevic, Z.R., Bailey, M.L., & Hylemon, P.B. (1988a). Regulation of bile acid synthesis: II. Effect of bile acid feeding on enzyme regulating hepatic cholesterol and bile acid synthesis in the rat. Hepatology 8, 892–897.

Heuman, D.M., Hernandez, C.R., Hylemon, P.B,. Kubaska, W., Hartman, C., & Vlahcevic, Z.R. (1988b). Regulation of bile acid synthesis: I. Effects of conjugated ursodeoxycholate and cholate on bile acid synthesis in chronic bile fistula rat. Hepatology 8, 358–365.

Heuman, D.M., Hylemon, P.B., & Vlahcevic, Z.R. (1989). Regulation of bile acid synthesis. III. Correlation between biliary bile salt hydrophobicity index and the activities of enzymes regulating cholesterol and bile acid synthesis in the rat. J. Lipid Res. 30, 1161–1171.

Heuman, D.M., Moore, E.W., & Vlahcevic, Z.R. (1990). Pathogenesis and dissolution of gallstones. In: Hepatology: A Textbook of Liver Disease (Zakim, D., & Boyer, T., eds), pp. 1480–1515. W.B. Saunders, Philadelphia.

Hoekman, M.F.M., Rientjes, J.M.J., Twisk, J., Planta, R.J., Princen, H.M.G., & Mager, W.H. (1993). Translational regulation of the gene encoding cholesterol 7α-hydroxylase in the rat. Gene 130, 217–223.

Horton, J.D., Cuthbert, J.A., & Spady, D.K. (1995). Regulation of hepatic 7α-hydroxylase expression and response to dietary cholesterol in the rat and hamster. J. Biol. Chem. 270, 5381–5387.

Hylemon, P.B., Gurley, E.C., Stravitz, R.T., Litz, J.S., Pandak, W.M., Chiang, J.Y.L., & Vlahcevic, Z.R. (1992). Hormonal regulation of cholesterol 7α-hydroxylase mRNA levels in primary rat hepatocyte cultures. J. Biol. Chem. 267, 16866–16871.

Ichimiya, H., Nazer, H., Gunasekaran, T., Clayton, P., & Sjovall, J. (1990). Treatment of chronic liver disease caused by 3β-hydroxy-Δ^5-C_{27}-steroid dehydrogenase deficiency with chenodeoxycholic acid. Arch. Disease in Childhood 65, 1121–1124.

Iga, T., & Klaassen, C.D. (1982). Hepatic extraction of bile acids in rat. Biochem. Pharmacol. 31, 205–209.

Inoue, M., Kinne, R., Tran, T., & Arias, I.M. (1982). Taurocholate transport by rat liver sinusoidal membrane vesicles. Evidence for sodium cotransport. Hepatology 2, 572–579.

Inoue, N., Kinner, R., & Arias, I.M. (1984). Taurocholate transport by rat liver canalicular membrane vesicles. J. Clin. Invest. 73, 659–663.

Ishida, H., Noshiro, M., Okuda, K., & Coon, M.J. (1992). Purification and characterization of 7α-hydroxy-4-cholesten-3-one 12α-hydroxylase. J. Biol. Chem. 267, 21319–21323.

Javitt, N.B., Kok, E., Cohen, B., & Burstein, S. (1982). Cerebrotendinous xanthomatosis: reduced serum 26-hydroxycholesterol. J. Lipid Res. 23, 627–630.

Javitt, N.B., & Budat, K. (1989). Cholesterol and bile acid synthesis in Hep G2 cells. Metabolic effects of 26- and 7α-hydroxylation. Biochem. J. 262, 989–992.

Javitt, N.B. (1990). HepG2 cells as a resource for metabolic studies: Lipoprotein, cholesterol and bile acids. FASEB J. 4, 161–168.

Javitt, N.B. (1990a). 26-hydroxycholesterol: Synthesis, metabolism and biologic activities. J. Lipid Res. 31, 1527–1533.

Jelinek, D.F., Andersson, S., Slaughter, C.A., & Russell, D.W. (1990). Cloning and regulation of cholesterol 7α-hydroxylase, the rate-limiting enzyme in bile acid synthesis. J. Biol. Chem. 265, 8190–8197.

Jelinek, D.F., & Russell, D.W. (1990). Structure of the rat gene encoding cholesterol 7α-hydroxylase. Biochem. 29, 7782–7785.

Johnson, M.R., Barnes, S., Kwockye, J.B., & Diasio, R.B. (1991). Purification and characterization of bile acid-CoA:Amino acid N-acyltransferase from human liver. J. Biol. Chem. 266, 10227–10233.

Jones, M.P., Pandak, W.M., Hylemon, P.B., Chiang, J.Y.L., Heuman, D.M., & Vlahcevic, Z.R. (1993). Cholesterol 7α-hydroxylase: Evidence for transcriptional regulation by cholesterol and/or metabolic products of cholesterol in the rat. J. Lipid Res. 34, 885–892.

Kadmon, M., Klunemann, C., Bohme, M., Ishikawa, T., Gorgas, K., Otto, G., Herfarth, C., & Keppler, D. (1993). Inhibition by cyclosporin A of ATP-dependent transport from the hepatocyte into bile. Gastroenterology 104, 1507–1514.

Kalb, V.F., Woods, C.W., Turi, T.G., Dey, C., Sutter, T.R., & Loper, J.C. (1987). Primary structure of the lanosterol demethylase gene from *Saccharomyces cerevisiae*. DNA 6, 529–537.

Karam, W.G., & Chiang, J.Y.L. (1992). Polymorphisms of human cholesterol 7α-hydroxylase. Biochem. Biophys. Res. Commun. 185, 588–595.

Karam, W.G., & Chiang, J.Y.L. Expression and purification of human cholesterol 7α-hydroxylase in *E. coli*. J. Lipid Res. 35, 1222–1231.

Kemper, B., & Szczesna-Skorupa, E. (1989). Drug Metab. Rev. 20, 811–820.

Kimura, K., Ogura, Y., & Ogura, M. (1988). Increased rate of cholic acid formation for 3α, 7α-dihydroxy-5β-cholestane in perfused livers from diabetic rats. Biochim. Biophys. Acta 963, 329–332.

Kramer, W., Bicke, U., Buscher, H.P., Gerok, W., & Kurz, G. (1982). Bile salt binding polypeptides in plasma membranes of hepatocytes revealed by photoaffinity labeling. Eur. J. Biochem. 129, 13–24.

Kramer, R.W., Burckhardt, G., Wilson, F.A., & Kurz, G. (1983). Bile salt-binding polypeptides in brush-border membrane vesicles from rat small intestine revealed by photoaffinity labeling. J. Biol. Chem. 258, 3623–3627.

Kramer, W., Girbig, F., Gutjahr, U., Kowalewski, S., Jouvenal, K., Muller, G., Tripier, D., & Wess, G. (1993). Intestinal bile acid absorption. Na^+-dependent bile acid transport activity in rabbit small intestine correlates with the co-expression of an integral 93 kDa and a peripheral 14 kDa bile acid-binding membrane protein along the duodenum-ilium axis. J. Biol. Chem. 268, 18035–18046.

Kronbach, T., Larabee, T.M., & Johnson, E.F. (1989). Hybrid cytochrome P450 identify a substrate binding domain in P450 IIC5 and P450 IIC4. Proc. Natl. Acad. Sci. USA 86, 8262–8265.

Kubaska, W.M., Gurley, E.C., Hylemon, P.B., Guzelian, P., & Vlahcevic, Z.R. (1985). Absence of negative feedback control of bile acid biosynthesis in cultured rat hepatocytes. J. Biol. Chem. 260, 13459–13463.

Kuroki, S., Muramoto, S., Kuramoto, T., & Hoshita, T. (1983). Sex differences in gallbladder bile composition and hepatic steroid 12α-hydroxylase activity in hamsters. J. Lipid Res. 24, 1543–1549.

Kuroki, S., & Hoshita, T. (1983). Effect of bile acid feeding on hepatic steroid 12α-hydroxylase activity in hamsters. Lipid 18, 789–794.

Lavery, D.J., & Schibler, U. (1993). Circadian transcription of the cholesterol 7α-hydroxylase gene may involve the liver-enriched bZIP protein DBP. Genes & Dev. 7, 1871–1884.

Lazarow, P.B., & Moser, H.W. (1989). Disorder of peroxisomes biogenesis. In: The Metabolic Basis of Inherited Disease 6th Edition (Scriver, C.R., Beaudet, A.L., Sly, W.S., & Valle, D., eds.), pp. 1479–1502. McGraw Hill, New York.

Leighton, J.K., Dueland, S., Straka, M.S., Trawick, J., & Davis, R.A. (1991). Activation of the silent endogenous cholesterol 7α-hydroxylase gene in rat hepatoma cells: A new complementation group having resistance to 25-hydroxycholesterol. Mol. Cell. Biol. 11, 2049–2056.

Leitersdorf, E., Reshef, A., Meiner, V., Levitzki, R., Schwartz, S.P., Dann, E.J., Berkman, N., Cali, J.J., Klapholz, L., & Berginer, V.M. (1993). Frameshift and splice—Junction mutations in the sterol 27-hydroxylase gene cause cerebrotendinous xanthomatosis in Jew of Moroccan origin. J. Clin. Invest. 91, 2488–2496.

Li, Y.C., Wang, D.P., & Chiang, J.Y.L. (1990). Regulation of cholesterol 7α-hydroxylase in the liver: cloning, sequencing and regulation of cholesterol 7α-hydroxylase cDNA. J. Biol. Chem. 265, 12012–12019.

Li, Y.C., & Chiang, J.Y.L. (1991). The expression of a catalytically active cholesterol 7α-hydroxylase cytochrome P450 in E. coli. J. Biol. Chem. 266, 19186–19191.

Lim, W.C., & Jordan, T.W. (1981). Subcellular distribution of hepatic bile acid-conjugating enzymes. Biochem. J. 197, 611–618.

Lund, E., Björkhem, I., Furster, C., & Wikvall, K. (1993). 24-, 25-, and 27-hydroxylation of cholesterol by a purified preparation of 27-hydroxylase from pig liver. Biochim. Biophys. Acta. 1166, 177–182.

Martin, K.O., Budai, K., & Javitt, N.B. (1993). Cholesterol and 27-hydroxycholesterol 7α-hydroxylation:evidence for two different enzymes. J. Lipid. Res. 34, 581–588.

McCormick, W.C., Bell, C.C., Jr., Swell, L., & Vlahcevic, Z.R. (1973). Cholic acid synthesis as an index of severity of liver disease in man. Gut 14, 895–902.

Meier, P.J., Meier-Abt, A.S., Barret, C., & Boyer, J.L. (1984). Mechanism of taurocholate transport in canalicular and basolateral rat liver plasma membrane vesicles. Evidence for an electrogenic canalicular organic anion carrier. J. Biol. Chem. 259, 10614–10622.

Miki, H., Takeuchi, H., Yamada, A., Nishioka, M., Matsuzawa,Y., Hamamoto, I., Hiwatashi, A., & Ichikawa, Y. (1986). Quantitative analysis of the mitochondrial cytochrome P-450 linked monooxygenase system: NADPH-hepatoredoxin reductase, hepatoredoxin, and cytochrome P-450$_s$27 in livers of patients with cerebrotendinous xanthomatosis. Clin. Chim. Acta 160, 255–263.

Mitchell, J.C., Stone, B.G., Logan, G.M., & Duane, W.C. (1991). Role of cholesterol synthesis in regulation of bile acid synthesis and biliary cholesterol secretion in humans. J. Lipid Res. 32, 1143–1149.

Molowa, D.T., Chen, W.S., Cimis, G.M., & Tan, C.P. (1992). Transcriptional regulation of the human cholesterol 7α-hydroxylase gene. Biochem. 31, 2539–2544.

Morohaski, K. Sogawa, K., Omura, T., & Fujii-Kuriyama, T. (1987). Gene structure of human cytochrome P450scc, cholesterol desmolase. J. Biochem. (Tokyo). 101, 879–887.

Mosbach, E.H., Rothschild, M.A., Bekersky, I., & Hauser, S. (1971). Bile acid synthesis in the isolated, perfused rabbit liver. J. Clin. Invest. 50, 1720–1730.

Mosser, J., Donar, A.M., Sarde, C.-O., Kloschis, P., Fell, R., Moser, H., Poustka, A.-M., Mandel, J.-L., & Aubourg, D. (1993). Putative X-linked adrenoleukodystrophy gene shares unexpected homology with ABC transporters. Nature 361, 726–730.

Muller, M., Ishikawa, T., Berger, U., Klunemann, C., Lucka, L., Schreyer, A., Kannicht, C., Reutter, W., Kurz, G., & Keppler, D. (1991). ATP-dependent transport of taurocholate across the hepatocyte canalicular membrane mediated by a 110 kDa glycoprotein binding ATP and bile salt. J. Biol. Chem. 266, 18920–18926.

Murakami, K., Okada, Y., & Okuda, K. (1982). Purification and characterization of 7α-hydroxy-4-cholesten-3-one-12α-monooxygenase. J. Biol. Chem. 257, 8030–8035.

Myant, N.B., & Mitropoulos, K.A. (1977). Cholesterol 7α-hydroxylase. J. Lipid Res. 18, 135–153.

Nathanson, M.H., & Boyer, J.L. (1991). Mechanism and regulation of bile secretion. Hepatology 14, 551–566.

Ness, G.C., Pendleton, L.C., Li, Y.C., & Chiang, J.Y.L. (1990). Effect of thyroid hormone on hepatic cholesterol 7α-hydroxylase, LDL receptor, HMG-Co A reductase, farnesyl pyrophosphate synthetase and apolipoprotein A-1 mRNA levels in hypophysectomized rats. Biochim. Biophys. Res. Commun. 172, 1150–1156.

Nelson, D.R., Kamataki, T., Waxman, D.J., Guengerich, F.P., Estabrook, R.W., Feyerersen, R., Gonzalez, F.J., Coon, M.J., Gunsalus, I.C., Gotoh, O., Okuda, K., & Nebert, D.W. (1993). The P450 super family: Update on new sequences, gene mapping, accession number, early trivial names of enzymes and nomenclature. DNA and Cell Biol. 12, 1–51.

Nguyen, L.B., Shefer, S., Salen, G., Ness, G., Tanaka, R.D., Packin, V., Thomas, P. Shore, V., & Batta, A. (1990). Purification of cholesterol 7α-hydroxylase from human and rat liver and production of inhibiting polyclonal antibodies. J. Biol. Chem. 265, 4541–4546.

Nishida, T., Gatmaitan, Z., Che, C., & Arias, I.M. (1991). ATP and membrane potential act synergistically in canalicular bile acid transport. Hepatology 14, 145A.

Nishida, T., Gatmaitan, Z., Che, M., & Arias, I.M. (1991a). Rat liver canalicular membrane vesicles contain an ATP-dependent bile acid transport system. Proc. Natl. Acad. Sci. USA 88, 6590–6594.

Nishimoto, M., Gotoh, O., Okuda, K., & Noshiro, M. (1991). Structural analysis of the gene encoding rat cholesterol 7α-hydroxylase, the key enzyme for bile acid biosynthesis. J. Biol. Chem. 266, 6467–6471.

Nishimoto, M., Noshiro, M., & Okuda, K. (1993). Structure of the gene encoding human liver cholesterol 7α-hydroxylase. Biochim. Biophys. Acta. 1172, 147–150.

Noshiro, M., Nishimoto, M., Morohashi, K., & Okuda, K. (1989). Molecular cloning of cDNA for cholesterol 7α-hydroxylase from rat liver microsomes: Nucleotide sequence and expression. FEBS Lett. 257, 97–100.

Noshiro, M., & Okuda, K. (1990). Molecular cloning and sequence analysis of cDNA encoding human cholesterol 7α-hydroxylase. FEBS Lett. 268, 137–140.

Noshiro, M., Nishimoto, M., & Okuda, K. (1990). Rat liver cholesterol 7α-hydroxylase: Pretranslational regulation for circadian rhythm. J. Biol. Chem. 265, 10036–10041.

Oftebro, H., Björkhem, I., Skrede, S., Schreiner, A., & Pedersen, J. (1980). Cerebrotendinous xanthomatosis. A defect in mitochondrial 26-hydroxylation required for normal biosynthesis of cholic acid. J. Clin. Invest. 65, 1418–1430.

Ogishima, T., Deguchi, S., & Okuda, K. (1987). Purification and characterization of cholesterol 7α-hydroxylase from rat liver microsomes. J. Biol. Chem. 262, 7646–7650.

Ogura, Y., Ito, T., & Ogura, M. (1986). Effect of diabetes and of 7α-hydroxycholesterol infusion on the profile of bile acids secreted by the isolated rat livers. Biol. Chem. Hoppe-Seyler 367, 1095–1099.

Onishi, Y., Noshiro, M., Shimasato, T., & Okuda, K. (1991). Molecular cloning and sequence analysis of cDNA encoding Δ^4-3-ketosteroid-5β-reductase of rat liver. FEBS Lett. 283, 215–218.

Östlund Farrants, A-K., Nilsson, A., & Pedersen, J.I. (1993). Human hepatoblastoma cells (HepG2) and rat hepatoma cells are defective in important enzyme activities in the oxidation of the C_{27} steroid side chain in bile acid formation. J. Lipid Res. 34, 2041–2050.

Pandak, W.M., Heuman, D.M., Hylemon, P.B., & Vlahcevic, Z.R. (1990). Regulation of bile acid synthesis. IV. Interrelationship between cholesterol and bile acid biosynthesis pathways. J. Lipid Res. 31, 79–90.

Pandak, W.M., Heuman, D.M., Hylemon, P.B., & Vlahcevic, Z.R. (1990a). Regulation of bile acid synthesis:V. inhibition of conversion of 7-dehydrocholesterol to cholesterol is associated with down-regulation of cholesterol 7α-hydroxylase activity and inhibition of bile acid synthesis. J. Lipid Res. 31, 2149–2158.

Pandak, W.M., Li, Y.C., Chiang, J.Y.L., Studer, E.J., Gurley, E.C., Heuman, D.M., Vlahcevic, R.Z., & Hylemon, P.B. (1991). Regulation of cholesterol 7α-hydroxylase mRNA and transcriptional activity by bile salts in the chronic bile fistula rat. J. Biol. Chem. 266, 3416–3421.

Pandak, W.M., Vlahcevic, Z.R., Chiang, J.Y.L., Heuman, D.M., & Hylemon, P.B. (1992). Bile acid synthesis VI: Regulation of cholesterol 7α-hydroxylase by taurocholate and mevalonate. J. Lipid Res. 33, 659–668.

Pandak, W.M., Vlahcevic, Z.R., Heuman, D.M., Stravitz, R.T., Chiang, J.Y.L., & Hylemon, P.B. (1993a). Hormonal regulation of cholesterol 7α-hydroxylase specific activity, mRNA levels and transcriptional activity. Hepatology 18, 179A.

Pandak, W.M., Vlahcevic, Z.R., Heuman, D.M., Reford, K.S., Chiang, J.Y.L., & Hylemon, P.B. (1994). Effect of different bile salts on the steady-state mRNA levels and transcriptional activity of cholesterol 7α-hydroxylase. Hepatology 19, 941–947.

Paulusma, C.C., Bosma, P.J., Zaman, G.J.R., Bakker, C.T.M., Otter, M., Scheffer, G.L., Scheper, R.J., Borst, P., & Elferink, R.J.P. (1996). Congenital jaundice in rats with a mutation in a multidrug resistance-associated protein gene. Science 271, 1126–1128.

Paulweber, B., Onasch, M.A., Nagy, B.P., & Levy-Wilson, B. (1991). Similarity and differences in the function of regulatory elements at the 5'-end of the human apolipoprotein B gene in cultured hepatocytes (HepG2) and colon carcinoma (CaCO-2) cells. J. Biol. Chem. 266, 24149–24160.

Pawlowski, J.E., Huizinga, M., & Penning, T.M. (1991). Cloning and sequencing of the cDNA for rat liver 3α-hydroxysteroid/dihydrodiol dehydrogenase. J. Biol. Chem. 266, 8820–8825.

Pedersen, J., Oftebro, H., & Vanngard, T. (1977). Isolation from bovine liver mitochondria of a soluble ferredoxin active in a reconstituted steroid hydroxylating system. Biochem. Biophys. Res. Commun. 26, 666–673.

Pedersen, J.L., & Gustafsson, J. (1980). Conversion of 3α,7α-12α-trihydroxy-5β-cholestanoic acid into cholic acid by rat liver proxisomes. FEBS Lett. 121, 345–348.

Poorman, J.A., Buck, R.A., Smith, S.A., Overturf, M.L., & Loose-Mitchell, D.S. (1993). Bile acid excretion and cholesterol 7α-hydroxylase expression in hypercholesterolemia-resistant rabbits. J. Lipid Res. 34, 1675–1685.

Princen, H.M.G., Meijer, P., & Hoffee, B. (1989). Dexamethasone-regulates bile acid synthesis in monolayer cultures of rat hepatocytes by induction of cholesterol 7α-hydroxylase. Biochem. J. 262, 341–348.

Princen, H.M.G., Meijer, P., Wolthers, B.G., Vonk, R.J., & Kuipers. (1991). Cyclosporin A blocks bile acid synthesis in cultured hepatocytes by specific inhibition of chenodeoxycholic acid synthesis. Biochem. J. 275, 501–505.

Prydz, K., Kase, B.F., Bjorkhem, I., & Pedersen, J.I. (1986). Formation of chenodeoxycholic acid from 3α,7α-dihydroxy-5β-cholestanoic acid by rat liver peroxisomes. J. Lipid Res. 27, 622–628.

Ramirez, M.I., Karaoglu, D., Haro, D., Barillas, C., Bashirzadeh, R., & Gil, G. (1994). Cholesterol and bile acid regulate cholesterol 7α-hydroxylase expression at the transcriptional level in culture and in transgenic mice. Mol. Cell Biol. 14, 2809–2821.

Reichen, J., & Paumgartner, G. (1976). Uptake of bile acids by the perfused rat liver. Am. J. Physiol. 231, 734–742.

Rogers, S., Wells, R., & Rechsteiner, M. (1986). Amino acid sequence common to rapidly degraded proteins: The PEST hypothesis. Science 234, 364–368.

Rudel, L., Deckelman, C., Wilson, M., Scobey, M., & Anderson, R. (1994). Dietary cholesterol and down regulation of cholesterol 7α-hydroxylase and cholesterol absorption in African green monkeys. J. Clin. Invest. 93, 2463–2472.

Ruetz, S., Fricker, G., Hugentobler, G., Winterhalter, K., Kurz, G., & Meier, P.J. (1987). Isolation and characterization of the putative canalicular bile salt transport system of rat liver. J. Biol. Chem. 262, 11324–11330.

Ruetz, S., Hugentobler, G., & Meier, P.J. (1988). Functional reconstitution of the canalicular bile acid transport system of the rat liver. Proc. Natl. Acad. Sci. USA 85, 6147–6151.

Russell, D.W., & Setchell, D.R. (1992). Bile acid biosynthesis. Biochem. 31, 4737–4749.

Salen, G., Shefer, S., Cheng , F.W., Dayal, B., Batta, A., & Tint, G. (1979). Cholic acid biosynthesis. The enzymatic defect in cerebrotendinous xanthomatosis. J. Clin. Invest. 63, 38–44.

Scallen, T.J., & Sanghvi, A. (1983). Regulation of three key enzymes in cholesterol metabolism by phosphorylation-dephosphorylation. Proc. Natl. Acad. Sci. USA 80, 2477–2480.

Schepers, L., van Veldhoven, P.P., Casteels, M., Eyssen, H.J., & Mannaerts, G.P. (1990). Presence of three acyl-CoA oxidases in rat liver peroxisomes. An inducible fatty acid-CoA oxidase, a noninducible fatty acid acyl-CoA oxidase and a noninducible trihydroxy coprostanonyl-CoA oxidase. J. Biol. Chem. 265, 5242–5246.

Schrame, A.W., Goldfischer, S., van Roermund, C.W.T., Brouwer-Kelder, E.M., Collins, J., Hashimoto, T., Heymans, H.S.A., van Den Bosch, H., Shutgeus, R.B.H., Tager, J.M., & Wanders, R.J.A. (1987). Human peroxisomal 3-oxoacyl-CoA thiolase deficiency. Proc. Natl. Acad. Sci. USA 84, 2494–2496.

Schwartz, L.R., Burr, R., Schwenk, M., Pfaff, E., & Greim, H. (1975). Uptake of taurocholic acid into isolated rat liver cells. Eur. J. Biochem. 55, 627–623.

Setchell, K.D.R., Suchy, F.J., Welsh, M.B., Zimmer-Nechemiss, L., Heubi, J., & Balistreri, W.F. (1988). Δ^4-3-Oxosteroid 5β-reductase deficiency described in identical twins with neonatal hepatitis. J. Clin. Invest. 82, 2148–2157.

Setchell, K.D.R., Flick, R., Watkins, J.B., & Piccoli, D. (1990). Chronic hepatitis in a 10-year old due to an inborn error in bile acid synthesis-diagnosis and treatment with oral bile acids. Gastroenterology 98, A631.

Setoguchi, T., Salen, G., Tint, G., & Mosbach, E. (1974). A biochemical abnormality in Cerebrotendinous Xanthomatosis. J. Clin. Invest. 53, 1393–1401.

Shefer, S., Hauser, S., Bekersky, I., & Mosbach, E.H. (1969). Feedback regulation of bile acid biosynthesis in the rat. J. Lipid Res. 10, 646–655.

Shefer, S., Hauser, S., Bekersky, I., & Mosbach, E.H. (1970). Biochemical site of regulation of bile acid synthesis in the rat. J. Lipid Res. 11, 404–411.

Shefer, S., Hauser, S., Lapar, V., & Mosbach, E.H. (1973). Regulatory effects of sterols and bile acids on hepatic 3-hydroxy-3-methylglutaryl CoA reductase and cholesterol 7α-hydroxylase in the rat. J. Lipid Res. 14, 573–580.

Shefer, S., Cheng, F.W., Dayal, B., Hanser, S., Tint, G.S., Salen, G., & Mosbach, E.H. (1976). A 25-hydroxylation pathway of cholic acid biosynthesis in man and rat. J. Clin. Invest. 57, 897–903.

Shefer, S., Nguyen, L., Salen, G., Batta, A.K., Brooker, D., Zaki, F.G., Rani, I., & Tint, G.S. (1990). Feedback regulation of bile acid synthesis in the rat: Differing effects of taurocholate and tauroursocholate. J. Clin. Invest. 85, 1191–1198.

Shefer, S., Nguyen, L.B., Salen, G., Ness, G.C., Tint, S., Batta, A.K., Hauser, S., & Rani, I. (1991). Regulation of cholesterol 7α-hydroxylase by hepatic 7α-hydroxylated bile acid flux and newly synthesized cholesterol supply. J. Biol. Chem. 266, 2693–2696.

Shefer, S., Nguyen, L.B., Salen, G., Ness, G.C., Chowdhary, I.R, Lerner, S., Batta, A.K., & Tint, G.S. (1992). Differing effects of cholesterol and taurocholate on steady-state hepatic HMGCoA reductase and cholesterol 7α-hydroxylase activities and mRNA levels in the rat. J. Lipid Res. 33, 1193–1200.

Shoda, J., Toll, A., Axelson, M., Pieper, F., Wikvall, K., & Sjovall, J. (1993). Formation of 7α-and 7β-hydroxylated bile acid precursors from 27-hydroxycholesterol in human liver microsomes and mitochondria. Hepatology. 17, 395–403.

Sippel, C.J., Ananthanarayanan, M., & Suchy, F.J. (1990). Isolation and characterization of the canalicular membrane bile acid transport protein of rat liver. Am. J. Physiol. 258, G728–G737.

Skrede, S., Björkhem, I., Kvittingen, E.A., Buchmann, M.S., Lie, S.O., East, C., & Grundy, S. (1986). Demonstration of 26-hydroxylation of C_{27}-steroids in human skin fibroblasts, and a deficiency of this activity in cerebrotendinous xanthomatosis. J. Clin. Invest. 78, 729–735.

Sladek, F.M., Zhong, W., Lai, E., & Darnell, E. (1990). Liver-enriched transcriptional factor HNF-4 is a novel member of the steroid hormone receptor superfamily. Gene and Dev 4, 2353–2365.

Spady, D.K., & Cuthbert, J.A. (1992). Regulation of hepatic sterol metabolism in the rat: Parallel regulation of activity and mRNA for 7α-hydroxylase but not 3-hydroxy-3-methylglutaryl-Co A reductase for low density lipoprotein receptor. J. Biol. Chem. 267, 5584–5591.

Stieger, B., O'Neill, B., & Meier, P.J. (1992). ATP-dependent bile salt transport in canalicular rat liver plasma membrane vesicles. Biochem. J. 284, 67–74.

Stolz, A., Takikawa, H., Sugiyama, Y., Kuhlenkamp, J., & Kaplowitz, N. (1987). 3α-hydroxysteroid dehydrogenase activity of the Y' bile acid binders in rat liver cytosol. Identification, kinetics and physiological significance. J. Clin. Invest. 79, 427–434.

Stolz, A., Takikawa, H., Ookhtens, M., & Kaplowitz, N. (1989). The role of cytoplasmic proteins in hepatic bile acid transport. Ann. Rev. Physiol. 51, 161–176.

Stolz, A., Rahimi-Kiani, M., Ameis, D., Chan, E., Ronk, M., & Shively, J. (1991). Molecular structure of rat hepatic 3α-hydroxy steroid dehydrogenase. J. Biol. Chem. 266, 15253–15257.

Stolz, A., Hammond, L., Lou, H., Takikawa, H., Rouk, M., & Shively, J.E. (1993). cDNA cloning and expression of the human hepatic bile acid-binding protein. J. Biol. Chem. 268, 10448–10457.

Straka, M.S., Junker, L.H., Zacarro, L., Zogg, D.L., Dueland, S., Everson, G.T., & Davis, R.A. (1990). Substrate stimulation of 7α-hydroxylase, an enzyme located in the cholesterol-poor endoplasmic reticulum. J. Biol. Chem. 265, 7145–7149.

Stravitz, R.T., Hylemon, P.B., & Vlahcevic, Z.R. (1993). The catabolism of cholesterol. Curr. Opin. in Lipidology 4, 223–229.

Stravitz, R.T., Hylemon, P.B, Heuman, D.M., Hagey, L.R., Schteingart, C.D., Ton-Nu, H.-T., Hofmann, A.F., & Vlahcevic, Z.R. (1993a). Transcriptional regulation of cholesterol 7α-hydroxylase mRNA by conjugated bile acids in primary cultures of rat hepatocytes. J. Biol. Chem. 268, 13987–13993.

Stravitz, R.T., Vlahcevic, Z.R., Gurley, E.C., & Hylemon. (1995). Repression of cholesterol 7α-hydroxylase transcription by bile acids is mediated through protein kinase C in primary cultures of rat hepatocytes. J. Lipid Res. 36, 1359–1368.

Su, P., Rennert, H., Shayiq, R.M., Yamamoto, R., Zheng, Y.-M., Addya, S., Strauss, J.F., & Avadhani, N.G. (1990). cDNA encoding a rat mitochondrial cytochrome P-450 catalyzing both the 26-hydroxylation of cholesterol and 25-hydroxylation of Vitamin D_3: Gonadotropic regulation of the cognate mRNA in ovaries. DNA & Cell Biol. 9, 657–665.

Sundseth, S.S., & Waxman, D.J. (1990). Hepatic P450 cholesterol 7α-hydroxylase: Regulation in vivo at the protein and mRNA level in response to mevalonate, diurnal rhythm and bile acid feedback. J. Biol. Chem. 265, 15090–15095.

Suzuki, M., Mitropoulos, K.A., & Myant, N.B. (1968). The electron transport mechanism associated with 12α-hydroxylation of C_{27} steroids. Biochem. Biophys. Res. Commun. 30, 516–521.

Takikawa, H., Sugiyama, Y., & Kaplowitz, N. (1986). Binding of bile acids by glutathione-S-transferase. J. Lipid Res. 27, 955–966.

Takikawa, H., Stolz, A., Sugimoto, M., Sugiyama, Y., & Kaplowitz, N. (1986a). Comparison of the affinities of newly identified human bile acid binder and cationic glutathione-S-transferase for bile acids. J. Lipid Res. 27, 652–657.

Tang, P.M., & Chiang, J.Y.L. (1986). Modulation of reconstituted cholesterol 7α-hydroxylase by phosphatase and protein kinase. Biochem. Biophys. Res. Commun. 134, 797–802.

Thompson, J.C., & Vars, H.M. (1953). Biliary excretion of cholic acid and cholesterol in hyper-, hypo-, and euthyroid rats. Proc. Soc. Exp. Biol. Med. 83, 246–248.

Toll, A., Shoda, J., Axelson, M., Sjovall, J., & Wikvall, K. (1992). 7α-Hydroxylation of 26-hydroxy-cholesterol, 3β-hydroxy-5-cholestenoic acid and 3β-hydroxy-5-cholenoic acid by cytochrome P450 in pig liver microsomes. FEBS Lett. 296, 73–76.

Tompson, J.F., Lira, M.E., Lloyd, D.B., Hayes, L.S., Williams, S., & Elsenboss, L. (1993). Cholesterol 7α-hydroxylase promoter separated from cyclophilin pseudogene by Alu sequence. Biochim. Biophys. Acta. 1168, 239–242.

Tronche, F., & Yaniv, M. (1992). HNF1, a homeoprotein member of the hepatic transcription regulatory network. BioEssays 14, 579–587.

Twisk, Z.J., Lehmann, E.M., & Princen, H.M.G. (1993). Differential feedback regulation of cholesterol 7α-hydroxylase mRNA and transcriptional activity by rat bile acids in primary monolayer cultures of rat hepatocytes. Biochem. J. 290, 685–691.

Twisk, J., Hoekman, M.F., Muller, L.M., Iida, T., Tamaru, T., Ijzerman, A., Mager, W.H., & Princen, H.M. (1995). Structural aspects of bile acids involved in the regulation of cholesterol 7α-hydroxylase and sterol 27-hydroxylase. Eur. J. of Biochem. 228, 596–604.

Tzung, K.W., Ishimura-Oka, K., Kihara, S., Oka, K., & Chan, L. (1994). Structure of the mouse cholesterol 7α-hydroxylase gene. Genomics 21, 244–247.

Usui, E., Noshiro, M., & Okuda, K. (1990). Molecular cloning of cDNA for vitamin D3 25-hydroxylase from rat liver mitochondria. FEBS Lett. 262, 135–138.

Usui, E., Noshiro, M., Ohyama, Y., & Okuda, K. (1990a). Unique property of liver mitochondrial P450 to catalyze the two physiologically important reaction involved in both cholesterol catabolism and vitamin D activation. FEBS Lett. 274, 175–177.

Van Cantford, J. (1973). Controle par les glucocortico-steroides de l'activite circadienne da la choles-terol 7α-hydroxylase. Biochemie 55, 1171–1173.

Van Dyke, R.W., Stephens, J.E., & Scharschmidt, B.F. (1982). Bile acid transport in cultured rat hepatocytes. Am. J. Physiol. 243, G484–G492.

Vlahcevic, Z.R., Bell, C.C., Jr., Buhac, I., Farrar, J.T., & Swell, L. (1970). Diminished bile acid pool size in patients with gallstones. Gastroenterology 59, 165–173.

Vlahcevic, Z.R., Buhac, I., Farrar, J.T., Bell, C.C., & Swell, L. (1971a). Bile acid metabolism in patients with cirrhosis. I. Kinetic aspects of cholic acid metabolism. Gastroenterology 60, 491–498.

Vlahcevic, Z.R., Miller, J.R., Farrar, J.T., & Swell, L. (1971). Kinetics and pool size of primary bile acids in man. Gastroenterology 61, 85–90.

Vlahcevic, Z.R., Judijudata, P., Bell, C.C., Jr., & Swell, L. (1972). Bile acid metabolism in patients with cirrhosis. II. Cholic and chenodeoxycholic acid metabolism. Gastroenterology 62, 1174–1181.

Vlahcevic, Z.R., Bell, C.C., Jr., Gregory, D.H., Buker, G., Juttijudata, P., & Swell, L. (1972a). Relationship of bile acid pool size to the formation of lithogenic bile in female Indians of the Southwest. Gastroenterology 62, 73–83.

Vlahcevic, Z.R., Goldman, M., Schwartz, C.C., Gustafsson, J., & Swell, L. (1981). Bile acid metabolism in cirrhosis. VII. Evidence for defective feedback control of bile acid synthesis. Hepatology 1, 146–150.

Vlahcevic, Z.R., Heuman, M.D., & Hylemon, P.B. (1989). Physiological and pathophysiology of enterohepatic circulation of bile acids. In: Hepatology: A Textbook of Liver Diseases (Zakim, D., & Boyer, T.P., eds.), pp. 341–377. W.B. Saunders, Philadelphia.

Vlahcevic, Z.R., Heuman, D.M., & Hylemon, P.H. (1991). Regulation of bile acid synthesis. Hepatology 13, 590–600.

Vlahcevic, Z.R., Pandak, W.M., Heuman, D.M., & Hylemon, P.B. (1992). Function and regulation of hydroxylases involved in the bile acid biosynthesis pathways. Sem Liver Dis. 12, 403–419.

Vlahcevic, Z.R., Hylemon, P.B., & Chiang, J.Y.L. (1993). Hepatic cholesterol metabolism. In: The Liver: Biology and Pathobiology Third Edition (Arias, I., Boyer, N., Fausto, N., Jakoby, W., Schacter, D., & Schafritz, D., eds). pp. xx–xx. Raven Press, New York.

Vlahcevic, Z.R., Pandak, W.M., Hylemon, P.B., & Heuman, D.M. (1993a). Role of newly synthesized cholesterol or its metabolites on the regulation of bile acid biosynthesis following after short-term biliary diversion in the rat. Hepatology 18, 660–668.

Von Dippe, P., & Levy, D. (1983). Characterization of the bile acid transport system in normal and transformed hepatocytes. J. Biol. Chem. 258, 8896–8901.

Von Dippe, P., Amoui, M., Alves, C., & Levy, D. (1993). Na^+-dependent bile acid transport by hepatocytes is mediated by a protein similar to microsomal epoxide hydrolase. Am. J. Physiol. 264, G528–G534.

Wagner, C.L., Trotman, B.W., & Soloway, R.D. (1976). Kinetic analysis of biliary lipid excretion in man and dog. J. Clin. Invest. 57, 473–477.

Wang, D.P., & Chiang, J.Y.L. (1994). Structure and nucleotide sequences of the human cholesterol 7α-hydroxylase gene (CYP7). Genomics 20, 320–323.

Watt, S.M., & Simmonds, W.J. (1976). The specificity of bile salts in the intestinal absorption of micellar cholesterol in the rat. Clin. Exp. Pharmacol. Physiol. 3, 305.

Weinberg, S.L., Burckhard, G., & Wilson, F.A. (1986). Taurocholate transport by rat intestinal basolateral membrane vesicles. Evidence for the presence of an anion exchange transport system. J. Clin. Invest. 78, 44.

White, P.C., New, M.I., & Dupont, B. (1984). Cloning and expression of cDNA encoding a bovine adrenal cytochrome P450 specific for sterol 21-hydroxylase. Proc. Natl. Acad. Sci. USA 81, 1986–1990.

Wieland, T., Nassal, M., Kramer, W., Fricker, G., Bickel, U., & Kurz, G. (1984). Identity of hepatic membrane transport systems for bile salts, phalloidin, and antamanide by photoaffinity labeling. Proc. Natl. Acad. Sci. USA 81, 5232–5236.

Wikvall, K. (1981). Purification and properties of a 3β-hydroxy-Δ^5-C_{27}-steroid oxidoreductase from rabbit liver microsomes. J. Biol. Chem. 256, 3376–3380.

Wikvall, K. (1984). Hydroxylations in biosynthesis of bile acids: Isolation of a cytochrome P-450 from rabbit liver mitochondria catalyzing 26-hydroxylation of C_{27} steroids. J. Biol. Chem. 259, 3800–3804.

Wolters, H., Kuipers, F., Sloof-Maarten, J.H., & Vonk Roel, J. (1992). ATP-dependent taurocholate transport in human liver plasma membrane. J. Clin. Invest. 90, 2321–2326.

Wong, M.H., Oelkers, P., Craddock, A.L., & Dawson, P.A. (1994). Expression cloning and characterization of the hamster ileal sodium-dependent bile acid transporter. J. Biol. Chem. 269, 1340–1347.

Wuarin, J., & Schibler, U. (1990). Expression of the liver-enriched transcriptional activator protein DBP follows a stringent circadian rhythm. Cell 63, 1257–1266.

Xu, G., Salen, G., Shefer, S., Ness, G.C., Nguyen, L.B., Parker, T.S., Zhao, Z., Donnelly, T.M., & Tint, G.S. (1995). Unexpected inhibition of cholesterol 7α-hydroxylase by cholesterol in New Zealand white and Watanabe heritable hyperlipidemia rabbits. J. Clin. Invest. 95, 1497–1504.

EICOSANOID METABOLISM AND BIOACTIVATION BY MICROSOMAL CYTOCHROME P450

Jorge H. Capdevila, Darryl Zeldin, Armando Karara, and John R. Falck

I. INTRODUCTION

Mammalian cells contain substantial amounts of Arachidonic acid (8,11,14-eicosa-tetraenoic acid) esterified to cellular glycerophospholipids. Under basal, unstimu-lated conditions, the cell levels of free, non-esterified, arachidonic acid are low and

Advances in Molecular and Cell Biology
Volume 14, pages 317–339.
Copyright © 1996 by JAI Press Inc.
All rights of reproduction in any form reserved.
ISBN: 0-7623-0113-9

nearly undetectable (1,2). Most cells have developed an elaborate enzymatic machinery that, in response to a variety of stimuli, catalyzes: (a) the hydrolytic cleavage of the arachidonic acid molecule from selected, hormonally sensitive, phospholipid pools, (b) the transduction of chemical information into the fatty acid molecular template by means of regio and stereoespecific oxygenation reactions and, (c) the decoding of that chemical information either by receptor mediated processes or, alternatively, by the direct effects of these metabolites on metabolic pathways (1–4). Metabolism by prostaglandin synthetase generates an unstable cyclic endoperoxide (prostaglandin H^2, PGH_2) that rearranges enzymatically or chemically to prostaglandins (PGs), prostacyclin (PGI_2) or thromboxane A_2 (TXA_2) (1,2) (Figure 1). Metabolism by the lipoxidases generates several regio-isomeric allylic hydroperoxides containing a *cis,trans* conjugated diene function-ality. One of these, the 5-hydroperoxyeicosatetraenoic acid (5-HETE), serves as the precursor for the biosynthesis of leukotrienes (LT) (2,3) (Figure 1). The physiologi-cal and biomedical significance of prostanoids and leukotrienes has been exten-sively documented (1–3). The reactions catalyzed by prostaglandin synthetase and lipoxidases are mechanistically similar to those of the free radical mediated autoxi-dation (1). Reactions are initiated by substrate activation with the generation of a

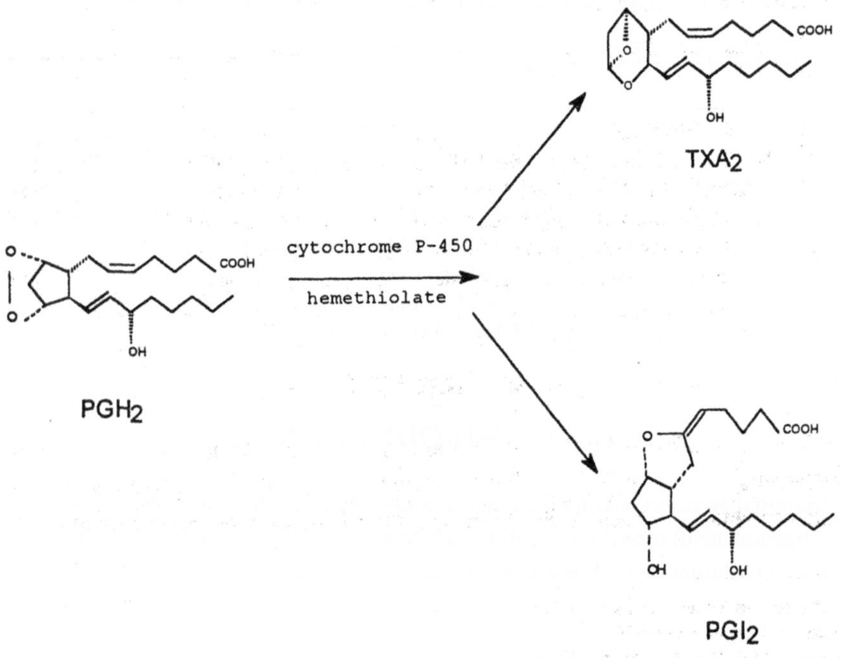

Figure 1. Metabolic transformations catalyzed by the cyclooxygenase and 5-lipoxy-genase members of the arachidonic acid cascade.

carbon centered radical, products are then formed by coupling of the carbon radical to ground state molecular oxygen. In addition to kinetic effects, the enzymes control the regiochemistry of carbon activation and oxygen insertion and provide the chiral environment needed for asymmetric catalysis and formation of optically active metabolites (1).

The participation of microsomal cytochrome P450 in the hydroxylation of the ultimate and penultimate carbon atom of short and mid chain saturated fatty acids (ω/ω1-oxidations) has been known for quite some time (5,6). However, it was not until 1981 that its role in the oxygenated metabolism of arachidonic acid was unequivocally demonstrated (7–10). Interest in these novel reactions of P450 was stimulated by: a) the initial demonstration of the potent biological activities of some of its products (11,12) and, b) the documentation of its role in the *in vivo* metabolism of endogenous arachidonic acid pools (13,14). These earlier observations served to move these reactions from the realm of an *in vitro* biochemical curiosity to that of a metabolic pathway. Additionally, they established microsomal P-450 as a member of the "arachidonic acid metabolic cascade" and, perhaps more importantly, suggested functional roles for the enzyme system in the bioactivation of the fatty acid and thus, in cell and organ physiology (15–17). The last few years have witnessed a renewed interest in defining the contribution that cytochrome P450 derived metabolites of arachidonic acid play in organ and body physiology. The study of the biological significance of these reactions has developed into an area of intense and, some times controversial research. This chapter we will focus mainly on recent advances in the biochemistry of cytochrome P450 catalyzed eicosanoid metabolism. The potential physiological importance of these reactions and metabolites has been reviewed (15–17).

Microsomal cytochrome P450 participates in the metabolic transformation of several eicosanoids by catalyzing both, NADPH-independent and NADPH-dependent reactions. This distinction reflects the marked differences in types of oxygen chemistry involved in these reactions, i.e., the NADPH dependent activation of atmospheric oxygen or the NADPH-independent isomerization of arachidonic acid hydroperoxides.

II. NADPH-INDEPENDENT METABOLISM OF EICOSANOIDS

Cytochrome P450 is an active peroxidase that catalyzes the metabolism of a wide variety of organic hydroperoxides, including fatty acid hydroperoxides (18–20). This peroxidase activity, initially described by O'Brien and collaborators (19), is associated with the ferric, Fe^{+3}, state of microsomal P450, does not involve electron transfer from NADPH and exhibits high catalytic rates (18–20). The mechanism by which the hemoprotein cleaves the peroxide O-O bond i.e., homolytic or heterolytic scission, plays a decisive role in determining the catalytic outcome of these reactions and is highly dependent on the nature of P450 isoform, the chemical properties of the organic hydroperoxide and the nature of the oxygen acceptor (18–20). A homolytic pathway was proposed to account for the formation, from 15-HPETE, of 11- and, 13-hydroxy-14,15-epoxyeicosatrienoic acids by rat liver microsomes (21). Spectral studies (22) and sequence homology analysis showed the presence in human lung (23) and platelet (24) thromboxane synthetase of several structural features, including a heme-thiolate prostetic group, typical of P450 type hemoproteins. The heterolytic cleavage of the PGH_2 endoperoxide and an oxygen atom transfer or oxenoid mechanism (25) has been proposed by Ullrich and

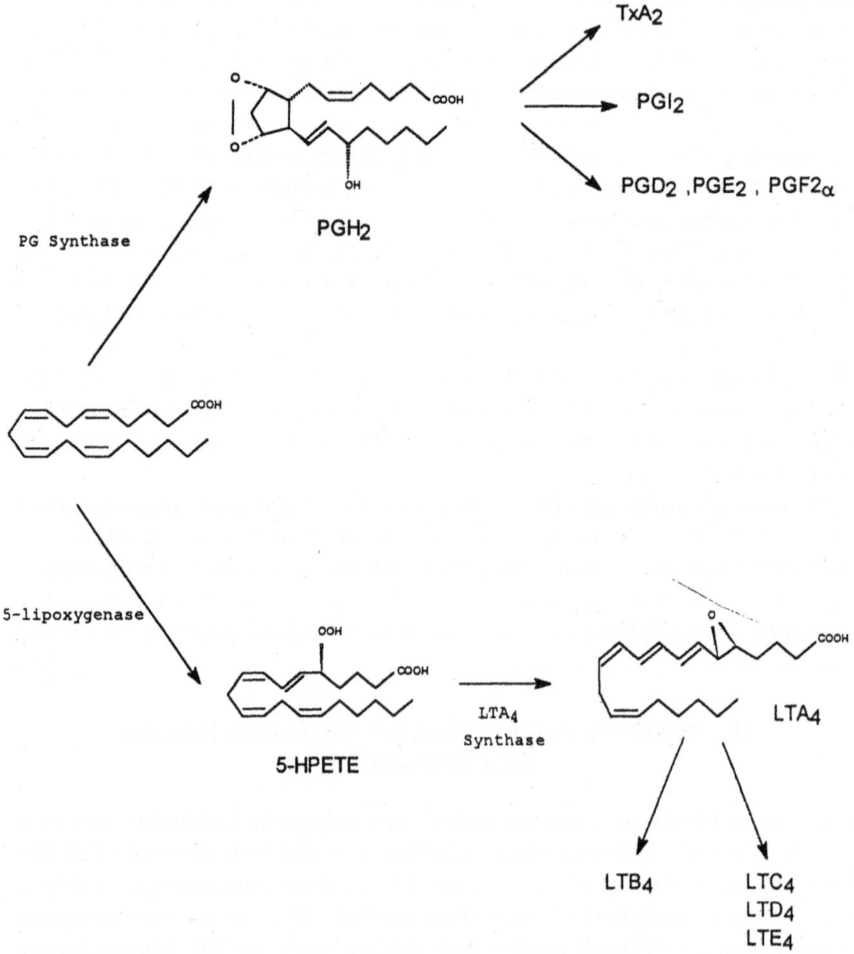

Figure 2. Cytochrome P450 catalyzed isomerization of PGH_2 to thromboxane A_2 (TXA_2) or to prostaglandin I_2 (PGI_2. Prostacyclin).

collaborators to account for the enzymatic formation of PGI_2 and TXA_2 from PGH_2 (22)(Figure 2). The biomedical importance of PGI_2 and TXA_2 for vascular physiology has been extensively documented (1,2). Recently, a plant P450-like peroxidase was isolated and characterized (26). This enzyme catalyzes the heterolytic cleavage of 13-hydroperoxy linoleic acid to the corresponding allene oxide, a precursor in jasmonic acid biosynthesis (26). The deduced primary structure of the allene oxide synthase shows $\leq 25\%$ identity to other P450's, with several segments of the protein matching conserved regions of the P450 gene superfamily (27).

As mentioned, a distinctive feature of all the P450-like fatty acid peroxide isomerases is their inability to catalyze the activation of molecular oxygen. More over, while all these enzymes share with P450s a heme-thiolate prostetic group, their overall homology to other members of the P450 gene superfamily is limited and suggests an early evolutionary functional specialization.

III. NADPH-DEPENDENT METABOLISM OF EICOSANOIDS

Microsomal P450 oxidizes a variety of eicosanoids that, in addition to arachidonic acid, includes prostanoids, leukotrienes and EETs. For the most part, these reactions result in the hydroxylation of the eicosanoid ultimate or penultimate carbon atoms (C-19 and C-20; ω and ω-1 oxidation). While it has been generally accepted that these reactions result in an attenuation of the biological activities of their substrates and that they may be important in eicosanoid catabolism, recent studies have indicated that some ω/ω-1 oxidized prostanoids shown unique and potent biological properties (28).

A. Metabolism of Oxygenated Eicosanoids by Cytochrome P450

The presence of prostanoids hydroxylated at the C-19 position in human semen was initially reported by Samuelsson and Hamberg in 1966 (29). Since then, C-19 and C20 hydroxylations are recognized routes for the metabolism of several prostanoids (30). Interest in the potential significance of these reactions for prostanoid activation and/or disposition lead to the studies of their enzymology and to the demonstration of the role of NADPH, of the microsomal fractions and of microsomal P450 (30) in these reactions. More recently, reconstitution studies utilizing purified components of the microsomal electron transport chain, as well as the enzymatic characterization of recombinant P450s utilizing heterologous expression have shown that these reactions are catalyzed, predominantly, by members of the P450 4A gene subfamily and that substrate specificity and regioselectivity are more or less isoform specific (31–37).

The ω-oxidation of leukotriene B_4 (LTB_4) as well as peptidoleukotrienes has been documented in whole animals, isolated cells and subcellular fractions (38,39). As with prostanoids, these reactions are catalyzed by microsomal P450 and may play important roles in the catabolism and disposition of leukotrienes (2,3). Inasmuch as the ω and/or ω-1 oxidation of eicosanoids may significantly influence the tissue concentrations and/or the functional properties of bioactive eicosanoids, the study of the physiological significance of these reactions and of the P450 isoforms involved will continue to be an active and important area of research.

B. Cytochrome P450 and the Metabolism of Arachidonic Acid: The Cytochrome P450 Arachidonic Acid Monooxygenase

Introduction

The investigation of the role of P450 in the metabolism and bioactivation of arachidonic acid were preceded by the early spectral documentation of the interactions between the fatty acid and liver microsomal P450 (40), as well as by studies, apparently motivated by the then growing pharmacological and toxicological importance of P450, of the effects that dietary or exogenously added polyunsaturated fatty acids had in the microsomal metabolism of drugs (41). In 1976, Cinti and Feinstein (42) demonstrated the presence of P450 in microsomal fractions isolated from human platelets and, more importantly, that the arachidonic acid-induced platelet aggregation could be blocked by known P450 inhibitors. This study suggested, for the first time, a functional role for this enzyme system as a participant in cellular eicosanoid metabolism.

The demonstration that microsomal fractions or purified and reconstituted components of the microsomal electron transport chain actively catalyzed the NADPH-dependent oxygenation of arachidonic acid was first reported in 1981 (7). Almost immediately thereafter, the structural characterization of most of the reaction products generated by incubates containing liver or kidney microsomal fractions was completed (8–10). The physiological significance of many prostanoids and leukotrienes was well known prior to the characterization of their biosynthetic pathways. The potent activities of many of these molecules served as powerful incentive for the extensive studies of the enzymes of the arachidonate cascade carried out during the last 20–25 years. In contrast, while the P450 system had been by then extensively characterized at the molecular level, the products formed by the enzyme system during arachidonic acid metabolism were of unknown function. However, the potential functional significance of this P450 activity was suggested early on by the demonstration that the epoxyeicosatrienoic acids (EETs) could a) serve as potent and selective *in vitro* stimuli for the release of several peptide hormones from either organ or cell preparations (11,12) and, b) modify the permeability of renal epithelia to ions such as Na^+ and K^+ (43). Since these early observations, and thanks to the contribution of several[1] research groups, the list of

---

[1]The term arachidonic acid monooxygenase identifies all NADPH-dependent oxygenation reaction catalyzed by cytochrome P-450 during arachidonic acid metabolism.

biological activities associated with the products of the P450 arachidonic acid monooxygenase reaction has grown continuously (15–17).

During arachidonic acid metabolism, the P450 enzyme system catalyzes the NADPH-dependent, redox coupled activation of molecular oxygen an its delivery to the substrate ground state carbon skeleton. This feature, i.e., oxygen activation as opposed to carbon activation, distinguishes the P450 enzyme system from the other enzymes of the arachidonate cascade. Catalytic turnover entails a sequence of steps that include: (a) the binding of the eicosanoid substrate to ferric P450, (b) the NADPH-Cytochrome P-450 oxido-reductase catalyzed one electron reduction of the P450-substrate complex (c) the binding of O_2 by the substrate bound ferrous P450, (d) the NADPH-cytochrome P450 oxido-reductase mediated transfer of a second electron from NADPH to the substrate bound ferrous P450, (e) the redox coupled cleavage of the O_2 molecule, (f) the regio and stereoselective transfer of an active form of atomic oxygen to the eicosanoid substrate and, (g) the release of the other oxygen atom as water and the regeneration of ferric P450. Thus, while P450 functions as an active arachidonic acid monooxygenase, prostaglandin synthetase and lipoxygenases are arachidonic acid dioxygenases.

Reactions Catalyzed

Arachidonic acid is metabolized by liver microsomal P450s at rates that varied from 1 to 6 nmols of product/min/mg of microsomal protein and with apparent K_m values for liver microsomes of 20 to 40 µM (8,44,45). As with most of the enzymes of the arachidonic acid cascade, P450 will not oxidize, at significant rates, the fatty acid when esterified to phospholipids (44). An analysis of the reaction products generated by rat liver microsomal P450 indicated that during arachidonic acid oxygenation the enzyme system catalyzed three type of reactions: (a) allylic oxidation (**lipoxygenase like reaction**), to generate six regioisomeric hydroxyeicosatetraenoic acids containing a *cis,trans*-conjugated dienol functionality (HETEs) (5-, 8-, 9-, 11-, 12-, and 15-HETE), (b) hydroxylations at the sp^3 carbons adjacent to the fatty acid methyl end. (ω/ω-1 oxygenase reaction) to generate 16-, 17-, 18-, 19-, and 20- hydroxyeicosatetraenoic acids (ω-4, ω-3, ω-2, ω-1 and ω-alcohols)(16-, 17-, 18-, 19-, and 20-OH-AA) and, (c) olefin epoxidation (**epoxygenase reaction**), to generate for regioisomeric *cis*-epoxyeicosatrienoic acids (EETs)(5,6-, 8,9–, 11,12–, and 14,15-EET) (Figure 3).

As with most P450 activities, early studies demonstrated that the type of reaction catalyzed, as well as, the profile of metabolites generated was highly dependent on the animal and tissue source of microsomal enzymes, the nutritional and hormonal state of the animal, its sex, age and prior exposure to drugs, pesticides or to known inducers of microsomal P450 (17,44–46). For example, while rat and rabbit liver microsomal fractions catalyze preferentially arachidonic acid epoxidation, rat and rabbit kidney microsomes metabolize the fatty acid mainly to generate mainly ω and ω-1 alcohols (Figure 4) (16,17). Based on studies done utilizing either microsomal fractions or reconstituted system containing several solubilized and purified

P450 isoforms it was concluded that P450 controls, in a protein specific fashion, the regioselectivity of arachidonic acid oxidation at two different levels: a) the type of reaction catalyzed, i.e., olefin epoxidation (EETs), allylic oxidation fflETEs) or hydroxylation at the C_{16}-C_{20} sp^3 carbons (16-, 17-, 18-, 19-, and 20-OH-AA) ( ) and, b) to a lesser extent, in selecting the regio-selectivity of oxygen insertion (44).

Allylic oxidation (Lipoxygenase-like reaction). This microsomal P450 activity leads to the formation of six regioisomeric allylic alcohols, the HETEs (47) (Figure 3). The HETEs are structurally similar to those generated by lipoxygenases however, for the P450 catalyzed reaction no evidence has been found for the formation of intermediate hydroperoxides (20). The stereochemical characterization of the HETEs generated by rat liver microsomal P450 showed that while the formation of 12(R)-HETE (Figure 3) was highly enantioselective, 5-, 8-, 9-, 11-, and 15-HETE were generated as nearly racemic mixtures (47). HETE formation by microsomal P450 may involve hydroxylation at the *bis*-allylic C_6, C_{10} and C_{13} carbon atoms followed by enzymatic or chemical isomerization of the resulting *bis*-allylic hydroxyeicosatetraenoic acids to either chiral or racemic HETEs (48). The NADPH-dependent formation of 13-hydroxyeicosatetraenoic acid and of 15-, 11- and 12-HETE (with 90, 59 and 90% optical purity for 15(R)-, 11(R)-, and 12(R)-HETE, respectively) by human liver microsomes has been demonstrated

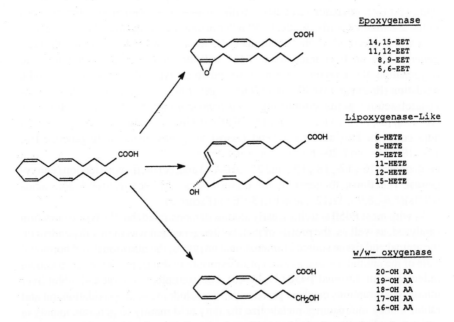

Figure 3. Reactions catalyzed by rat liver microsomal cytochrome P450 during the NADPH-dependent metabolism of arachidonic acid.

(48). The catalysis of 12(R)-HETE formation appears to be unique to the microsomal P450 system; its enantiomer, 12(S)-HETE, is the characteristic product of mammalian lipoxygenases (49). The presence of 12(R)-HETE in human skin and its increased formation during psoriatic inflammation and the potent inhibition of Na^+/K^+ ATPase activity by 12(R)-HETE have been reported (50,51). While a role for P450 in the catalysis of 12(R)-HETE formation by microsomal fractions from bovine cornea epithelium has been proposed (51), the role of microsomal P450 and the chirality of the 12-HETE product remains controversial (52). Although the contribution of P450 to the *in vivo* formation of HETEs remains to be demonstrated, the unique chirality of these products and their associated biological activities continues to stimulate interest in the study of these reactions. Issues remaining to be defined include: (a) the role of P450 in the *in vivo* catalysis of HETE formation, (b) the characterization of the individual P450 isoforms responsible for this reactions, (c) the uniqueness of 12(R)-HETE formation by P450 enzymes and, (d) the clarification of the reaction mechanism(s).

Hydroxylations at C_{19} and C_{20} (ω/ω-oxygenase reaction). As mentioned, the omega and omega-1 oxidation of fatty acids (ω/ω-1 oxidation) has been considered to participate in the catabolism of mid-chain fatty acids prior to urinary excretion or degradation by β-oxidation. The fatty acid ω/ω-1 oxidation is regulated *in vivo* by a variety of factors including, animal age, diet, starvation, the administration of fatty acids, hypolipidemic drugs, aspirin, steroids and diabetes (17,46,53–56). The role of microsomal P450 in these reactions was demonstrated after the reconstitution of the lauric acid ω-hydroxylase utilizing the first solubilized and purified form of liver microsomal P450 (6). The catalysis of arachidonic acid ω and ω-1 oxidation by P450 was first documented, unequivocally, in 1981 when rabbit kidney cortex microsomes were shown to catalyze the NADPH-dependent formation of 19- and 20- hydroxyeicosatetraenoic acids (9)(Figure 3). Since then, the arachidonic acid ω/ω-1 oxygenase reaction has been demonstrated in several tissues, including human liver and kidney, as well as rat and rabbit liver and kidney (16,17). More recently, the 16-, 17-, and 18-hydroxyeicosatetraenoic acids have been added to the list of products generated by the P450 arachidonic acid ω/ω-1 oxygenase activity (44,57). While the oxygen chemistry and reaction mechanisms responsible for the ω/ω-1 oxygenation of arachidonic acid and that of saturated fatty acids are probably similar, for arachidonic acid these reactions impose additional steric requirements on the P450 protein catalyst. Hydroxylation at the thermodynamically less reactive C_{16} through C_{20} and not at the chemically comparable C_2 through C_4 suggest a highly rigid and structured binding site for the arachidonic acid template. Thus, the binding site must be capable of positioning the acceptor carbon atoms not only in optimal proximity to the heme-bound active oxygen but also with complete segregation of the reactive 5,6-, 8,9-, 11,12-, and 14,15- olefins and of the bis-allylic C_7, C_{10} and C_{13} methylene carbons. A demonstration of the degree of regiochemical control exerted by these enzymes is

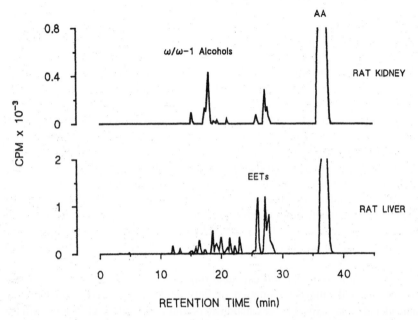

Figure 4. Reversed phase HPLC chromatograms of arachidonic acid metabolites generated by incubates containing rat liver or kidney microsomal fractions. Microsomal fractions isolated from male rat liver or whole kidneys (0.5 and 1.0 mg of microsomal protein/ml, respectively) were incubated for 5 (liver) or 15 min (kidney) at 30°C with $[1-^{14}C]$arachidonic acid (100 μM, 1–5 μ Ci/μmole) in the presence of NADPH (1 mM). The ethyl ether soluble products were dissolved in ethanol and injected onto a 5 μm Dynamax Microsorb C_{18} column (4.6 × 250 mm, Rainin Instruments Co. Inc., MA). Products were resolved by a linear solvent gradient from $H_2O/CH_3CN/HOaC$ (49.95:49.95:0.1, v/v) to $CH_3CN/HOAc$ (99.9:0.1, v/v) in 40 min at 1 ml/min. The eluent radioactivity was monitored with an on-line Radiomatic Flo-One β-detector (Radiomatic Instruments, FL).

illustrated by the fact that, at present, these reactions cannot be reproduced chemically, i.e., no methods means are currently available for the selective oxidation of the arachidonic acid $C_{17}-C_{20}$ carbon atoms.

Several rat, rabbit or human members of the P450 4A gene family have been purified and/or cloned and expressed (31–37). While individual P450 4A isoforms show regioselectivity for either the ω or the ω-1 hydroxylation of lauric acid or prostanoids (31–37), to date, all the P450 4A isoforms so far characterized, metabolize arachidonic acid to either 20-OH-AA or to a mixture of 20-OH-AA containing small quantities of 19-OH-AA. None of these P450 4A isoforms show exclusive regioselectivity for the fatty acid C_{19} position (31–37). Interestingly, studies with inducers of microsomal P450, as well as, reconstitution of the hydroxylase with purified enzymes showed that P450s 1A1, 1A2 and 2E1 may be responsible for the hydroxylations occurring at the C_{16} through C_{19} carbon atoms of arachidonic acid (44). Thus, while P450 1A1 and 1A2 are more or less selective for hydroxylations at C_{19} and C_{16}, respectively (44), P450 2E1 catalyzes stereoselective hydroxylations at C_{18} and C_{19} (58).

The notion that the arachidonic acid ω/ω-1 oxygenase may play a role in the control of vascular tone has stimulated interest in the study of the biological properties of its products. Changes in the activities of the renal ω/ω-1 oxygenase have been linked to the onset of hypertension in the spontaneously hypertensive rat model (SHR/WKY) (16). The cyclooxygenase dependent vasoconstrictor activity of 20-OH-AA has been reported (16,28), as well as the modulation of Na^+/K^+ ATPases by both 19- and 20-OH-AA (16). The urinary excretion of 20-OH-AA has been reported (59), however the majority of the 20-OH-AA present in rat urine is conjugated to glucuronic acid, a well established route for the excretion of hydroxylated compounds (60).

Olefin epoxidation (epoxygenase reaction). Of the three reactions catalyzed by P450 during arachidonic acid metabolism the one that has attracted the most attention is that of the arachidonic acid epoxygenase. This is based, in part on the potent biological activities of its metabolites and the early demonstration of the EETs as endogenous constituents of several organ tissues, including human kidney and plasma (17). In mammals, the epoxidation of polyunsaturated fatty acids to *bis*-allylic, *cis*-epoxides is unique to the P450 enzyme system and, at difference with the fatty acid ω/ω-1 oxygenase, more or less selective for arachidonic acid (53). Thus, while the enzymatic or non-enzymatic isomerization of polyunsaturated fatty acid hydroperoxides can yield epoxides or epoxy-alcohol derivatives (3,62,63), these products are structurally different to those generated by the P450 enzymes system. Arachidonic acid epoxidation by P450 entails the reductive cleavage of molecular oxygen and the regio- and stereoselective delivery of an active species of atomic oxygen to a fatty acid ground state olefin. The fact that the P450 enzymes generate only *cis*-epoxides indicates that either the enzymatic reaction proceeds by a concerted process or that, alternatively, the enzyme's active site restricts the freedom of C-C rotation for the transition state.

Enzymology of EET Formation

The catalysis of EET formation by either isolated cell preparations, microsomal fractions or by purified and reconstituted P450 isoforms has been demonstrated in numerous tissues, including liver, kidney, brain, lung, pituitary, adrenals and endothelium (15–17). The P450 isoform heterogeneity of the microsomal epoxygenase was initially suggested by studies indicating organ specific regioselectivity of epoxidation, thus, for example, while rat brain microsomes form 5,6-EET as their major epoxygenase product ($\geq 90\%$ of the total EETs) (11), the 11,12-EET is the predominant regioisomer formed by rat liver and kidney microsomal fractions (42

Table 1. Effect of Animal Treatment with Phenobarbital (PB) or β-Naphthoflavone (β-NF) on the Stereochemical Selectivity of the Rat Liver Microsomal AA Epoxygenase. The condition for animal treatment and metabolism are as indicated in reference 25. Values shown (in nmols of product/mg of microsomal protein/min, at 25°C) are means calculated from at least three different experiments with S.E. ≤ 10% of the mean for all cases.

Enantiomer	Control	PB	β-NF
8(R),9(S)-EET	0.26	0.13	0.10
8(S),9(R)-EET	0.12	0.45	0.05
11(R),12(S)-EET	0.48	0.13	0.17
11(S),12(R)-EET	0.11	0.62	0.06
14(R),15(S)-EET	0.14	0.74	0.07
14(S),15(R)-EET	0.28	0.24	0.08
Total	1.40	2.33	0.53

and 53% of the total EETs, respectively) (45,64). The molecular heterogeneity, as well as, the P450 isoform specific control of the regio- and stereoselectivity of the arachidonic acid epoxygenase was demonstrated by experiments in which specific inducers of microsomal P450 were utilized in order to alter the inventory of microsomal P450 isoforms or, alternatively, by studies of the regio- and stereo-selectivity of arachidonic acid epoxidation by purified P450 isoforms (45). As shown in Table 1, animal treatment with phenobarbital resulted in a nearly doubling of the epoxygenase activity, an increase in the overall enantioselectivity of the microsomal epoxygenase, and a concomitant inversion in the absolute configuration of the EETs (45). Thus, phenobarbital induces the biosynthesis of those P450 isoform(s) responsible for the formation of a single enantiomer of 8,9-, 11,12-, and 14,15-EET (Table 1). On the other hand, microsomal fractions from β-naphtofla-vone treated animals showed a reduced epoxygenase activity (45), an overall attenuation of the enantioselectivity of the microsomal epoxygenase, and a prefer-ential decrease in the contents of those P450 isoform(s) responsible for the genera-tion of 8(R),9(S)-, 11(R),12(S)- and 14(S),15(R)-EET (Table 1) (45). The presence in the endoplasmic reticulum of multiple P450 epoxygenases with unique enan-tioselective properties was, unequivocally, demonstrated by reconstituting the reaction using solubilized and purified P450 isoforms (45). For example, while P450 1A2 catalyzed the highly asymmetric epoxidation of the *re,si* face of the 11,12-olefin to 11(R),12(S)-EET (93% optical purity), P450 2B2 catalyzed the asymmetric epoxidation of the opposite enantiotopic face, the *si,re* face, and generated the 11(S),12(R)-EET enantiomer with an optical purity of 90% (45).

The stereochemical properties of the EETs formed by purified or microsomal P450s revealed an unprecedented high stereochemical selectivity for the oxidation

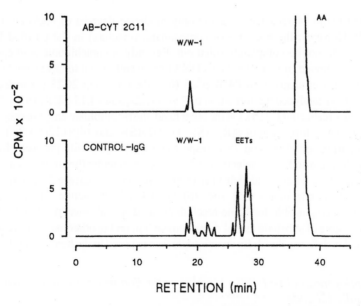

Figure 5. Selective inhibition of the rat liver microsomal epoxygenase by anti-cyto-chrome P450 2C11 IgG. Microsomal fractions isolated from rat liver (0.5 μM cyto-chrome P450) were incubated with [1-^{14}C]arachidonic acid (100 μM, 1 μCi/μmole) and NADPH (1 mM) in the presence of nonimmune or anti-cytochrome P450 2C11 IgG (20 mg of protein/ml, each). After 5 min at 30ºC, the reaction products were extracted, resolved and detected as in Figure 4. Shown are the radioactivity profiles obtained from control (*bottom*) and antibody containing incubates (*top*).

of an unbiased, acyclic molecule such as arachidonic acid. It therefore follows that, the active site molecular coordinates responsible for the heme-arachidonic acid spatial orientation are remarkably rigid and highly structured and point to this fatty acid as one of the natural substrates for these proteins (44,45). An important corollary of these studies was the demonstration that, at difference with the cyclooxygenase and lipoxygenase members of the arachidonate cascade, the regio- and enantioselectivity of the epoxygenase was variable and under *in vivo* regulatory control (45,14). Furthermore, the selective inhibition of the rat kidney (64) and liver microsomal epoxygenase reaction (Figure 5) by polyclonal antibodies raised against liver P450 2C11 and not by anti-P450s 1A1, 1A2, 2A3, 4A1, 2B1, 2B2 and 2E1 indicates that the liver and kidney predominant microsomal epoxygenase(s) are members of the P450 2C gene family (64,65).

An important, unresolved issue, is that of the existence of P450 epoxygenase isoforms with distinct regioselectivity for the individual olefins of the arachidonate molecule, i.e., epoxygenase isoforms selective for the metabolism of either the 5,6-, 8,9-, 11,12- or 14,15 olefins. Most of the purified isoforms so far characterized, although highly stereoselective show only moderate degrees of regioselectivity (45,58). However, the presence of contaminant, protein(s) in the purified P450 isoforms can not be completely ruled out. Recently, an arachidonic acid epoxygenase was cloned from a rat kidney cDNA library and shown to have a nucleotide sequence nearly identical to P450 2C23 (65). Recombinant 2C23 (expressed in COS-1 cells) catalyzed the formation of 8,9-, 11,12-, and 14,15-EET in a 14:28:10 molar ratio, respectively (65). On the other hand, the recombinant enzyme catalyzed the asymmetric formation of 8(R),9(S)-, 11(R)12(S)-, and 14(S),12(R)-EET with optical purities of 95, 85 and 75%, respectively (65). Since with the recombinant protein the regio and stereoselectivity of the reaction is under the control of a single protein catalyst, we concluded that (a) a single protein can catalyze the epoxidation of more than one olefin (65) and, (b) the P450 2C23 active site molecular coordinates responsible for heme-arachidonic acid spatial orientation allow for a limited degree of substrate lateral displacement but, on the other hand, are remarkably rigid in defining the olefin enantiotopic face exposed to the heme-bound active oxygen.

The Endogenous EETs: P450 a Member of the Endogenous Arachidonate Metabolic Cascade

Whereas *in vitro* biochemical studies are important for the characterization of metabolic pathways, they provide only limited information concerning the *in vivo* significance of the enzymes and products involved. Rat liver P450, which accounts for approximately 4 to 5% of the total liver microsomal protein, has been extensively characterized during the last 20 years. However, information concerning a clearly defined endogenous role for these hemoproteins is as yet lacking. It was therefore clear to us from the outset that the importance and uniqueness of the catalysis of arachidonic acid metabolism by P450 was going to be defined, primarily, by its potential significance to cell and tissue physiology. To address these important questions and, since asymmetric synthesis is an accepted requirement for the biosynthetic origin of most eicosanoids, a method was developed for the quantification and chiral characterization of the EET pools present in biological samples (66). The epoxidation of endogenous pools of arachidonic acid was first demonstrated in 1984 when rat liver was shown to contain approximately 1 μg of total EETs/g of wet tissue (13). Chiral analysis of the liver endogenous EETs showed the biosynthesis of 8,9-, 11,12-, and 14,15-EET in a 4:1, 2:1 and 3:1 ratio of antipodes, respectively (14)(Table 2) (Figure 6). Animal treatment with Phenobarbital resulted in 3.7-fold increase in the concentrations of microsomal P450 and a concomitant, regioselective 6.8- and 3.4-fold increase in the liver concentrations of 8,9- and 14,15-EET (14). As with the microsomal enzymes, phenobarbital treatment induced the endogenous biosynthesis of a single EET enantiomer, with 8(S),9(R)- and 14(R),15(S)-EET biosynthesized as nearly optically pure enantiomers (14)(Table 2). While qualitatively similar, the enantiofacial selectivity of the microsomal rat kidney epoxygenase is markedly higher than that of the liver

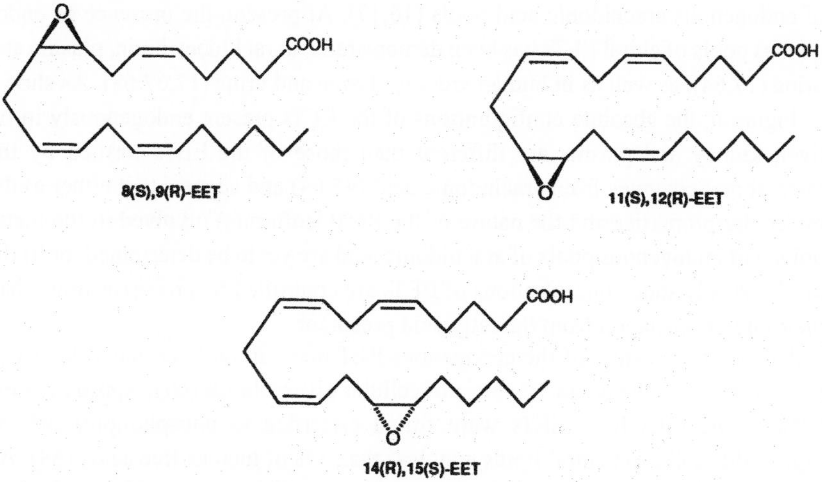

Figure 6. Absolute configuration of the epoxyeicosatrienoic acids (EETs) present endogenously in samples extracted for either rat liver, kidney, or plasma.

enzymes (65). Importantly, no significant changes in the regio- and stereoselectivity of the rat kidney microsomal epoxygenase were observed after animal treatment with either phenobarbital, β-naphtoflavone or deoxycorticosterone acetate suggesting that, at difference with liver, the kidney microsomal membrane houses a limited number of P450 isoforms active in arachidonic acid epoxidation (65).

These results demonstrated the enzymatic origin of the EETs present *in vivo* in rat liver and documented a novel metabolic function for P450 in the epoxygenation

Table 2. Effect of Animal Treatment with Phenobarbital on the Concentrations of Endogenous EETs (in ng/g of wet tissue) Present in Rat Liver. Values shown are means calculated from at least three different experiments with S.E. $\leq$ 10% of the mean for all cases.

Enantiomer	Controls	PB
8(R),9(S)-EET	58	19
8(S),9(R)-EET	221	1877
11(R),12(S)-EET	54	29
11(S),12(R)-EET	90	139
14(R),15(S)-EET	273	1127
14(S),15(R)-EET	83	72
Total	779	3263

of endogenous arachidonic acid pools (16,17). At present, the presence of endogenous pools of chiral EETs has been demonstrated in rat kidney brain, plasma and urine (17,64), as well as in human kidney, plasma and urine (17,67,68). As shown in Figure 6, the absolute configurations of the EETs present endogenously in rat liver, kidney and plasma are different than those of the EETs formed by the microsomal enzymes from arachidonic acid (45,64) and suggest that either a) the molecular properties and the nature of the P450 isoform(s) involved in the metabolism of endogenous pools of arachidonic acid are yet to be determined, or b) the *in vivo* steady state concentrations of EETs are controlled by processes other than their rate of formation from the fatty acid precursor.

A distinctive feature of the endogenous EET pools in rat liver and kidney was their presence in the organs as esters of cellular glycerolipids (69). Approximately 92% of the total liver EETs were found esterified to phospholipids, 4% to diglycerides, 4% to neutral lipids and, less than 1% of total as free acids (69). An analysis of the phospholipid pools showed that the EETs were esterified at the *sn-2* position (69). Studies of the mechanism of EET-phospholipid formation unraveled a multistep process initiated by the P450 enantioselective epoxidation of arachidonic acid, ATP-dependent activation to the corresponding EET-CoA derivatives and EET enantiomer-selective lysolipid acylation (69). Chiral analysis of the fatty acids at *sn-2* revealed an enantioselective preference for 8(S),9(R)-, 11(S),12(R)-, and 14(R),15(S)-epoxyeicosatrienoates in all three classes of phospholipids, with 55% of the total liver EETs in phosphatidylcholine, 32% in phosphatidylethanolamine and 12% in phosphatidylinositols (69). The asymmetric nature of the esterified EET demonstrated, unequivocally, that the rat liver biosynthesized these lipid from endogenous precursors, enzymatically and under normal physiological conditions (69). This *in vivo* esterification process for endogenous EETs appears to be unique since, although the esterification by isolated cell preparations of exogenously added HETEs and EETs has been demonstrated (70–72), most endogenously formed eicosanoids are either secreted, excreted or undergo oxidative metabolism and excretion.

The presence of endogenous pools of phospholipids containing esterified EET moieties in rat liver, kidney, brain, and plasma, and in human kidney and plasma (17), suggest new and potentially important functional roles for P450. As a participant in the arachidonate cascade, microsomal P450 may thus play a still undefined role(s) in the biosynthesis of novel cellular glycerolipids and, consequently, in the control of membrane physicochemical properties and/or the generation of new classes of lipid-derived mediators. The data discussed also indicates that, in contrast to most eicosanoids, cells have the potential for the generation of the bioactive EETs throughout hydrolytic reactions, independently of oxidative metabolism of arachidonic acid.

Functional Significance of the Epoxygenase Metabolites

Since the demonstration in 1983 of the EETs potent biological activities as *in vitro* mediators for the release of several peptide hormones (11,15–17), research from several groups has documented a wide variety of biological effects which are attributed to the different arachidonic acid metabolites generated by P450 (15–17). We will briefly discuss two areas of current interest in our laboratory.

Membrane biology. Evidence accumulated from studies of lipid peroxidation has demonstrated that the presence of oxidized phospholipids has profound consequences for the physicochemical properties and, therefore, for the structural and functional properties of many biological membranes. These include changes in membrane ion permeability (73), alteration in the activities of membrane bound enzymes (74), variations in membrane turnover (75) and, changes in the fluidity and fusogenic properties (75–77). The formation and incorporation of EETs into cellular lipids may provide the molecular basis underlining of some of the biological properties attributed to the EETs, many of which, can be interpreted in terms of the ability of these compounds to alter cell membrane permeability or, alternatively, its fusogenic properties. Published reports have demonstrated the ability of exogenously added EET to alter the cellular concentrations of ions such as Ca^{++}, Na^+, K^+ and H^+, the permeability of membranes to water or to peptide hormones (15–17). The lipid bilayer provides the matrix in which structural and functional membrane proteins fold or unfold, oscillate between functionally significant conformational states, rotate or move laterally, interact with each other or with their substrates and, ultimately carry out their cellular functions. Localized changes in membrane lipid composition, as well as, the asymmetry in the lipid distribution of many biological membranes allow cells to manipulate the physicochemical properties of membrane domains (76,77) albeit, in many situations with severe time and spatial limitations. The capability to epoxidize enzymatically, localized areas of the phospholipid bilayer could provide cells with a powerful tool for the rapid and efficient control of the structural properties of individual membrane domains. Based on these studies and, the capacity of synthetic 8,9-epoxyeicosatrienoyl-phosphocholine to alter the Ca^{++} permeability of synthetic liposomes, we proposed a functional role for microsomal P450 in the control of cell membrane microenvironment structure and, hence its functional properties (76–78). We envision a process in which, enzyme controlled EET acylation will induce localized changes in the fluidity of selected membrane microenvironments. The process could then be reversed by a lipase catalyzed hydrolysis of the acylated EET, followed by enzymatic hydration to the corresponding DHET. Of interest, under conditions which are optimal for EET esterification, no DHETs acylation could be detected (69).

Renal physiology and hypertension. Since the demonstration in 1984 that exogenously added 5,6-EET inhibited Na^+ and K^+ fluxes in the isolated rabbit cortical collecting tubule (43), the list of renal effects attributed to the products of the P450 arachidonate monooxygenase has expanded continuously (16). Among these we include: (a) Modulation of the renal Na^+/K^+ ATPases, (b) Changes in renal cell Ca^+ concentrations, (c) Changes in proximal tubule Na^+ transport, inhibition of water transport and, renal vasodilation and EET enantioselective renal vasoconstriction (15–17). The early proposal by McGiff and collaborators of a role for the P450 monooxygenase in the pathophysiology of hypertension (16), focused interest in the potential significance of this enzyme system to renal and body physiology. Although the complex and probably multigenic nature of human hypertension complicates most experimental approaches to its study, the potential physiological and clinical implications of this proposal are of pivotal importance and have and continue to stimulate a significant amount of research. Thus, the developmental phase of hypertension in the spontaneous hypertensive rat model (SHR/WKY model) was linked, by these authors, to alterations in the kidney microsomal arachidonic acid monooxygenase activities (16). More recently the relevance of this proposal has been strengthen by the demonstration of: (a) the EETs as endogenous constituents of rat and human kidney and urine (64,67,68), (b) increased urinary excretion of DHETs during pregnancy induced hypertension (68) and, (c) preferential expression of the P450 4A2 gene in the hypertensive (SHR) rat (79).

As a biochemical tool for the study of the functional relevance of the renal epoxygenase we have studied the regulation of this pathway by functionally significant protocols of animal manipulation. Increased salt intake results in increased renal salt excretion. This adaptive process prevents progressive salt retention, volume expansion and, one of its detrimental sequelae, hypertension. In rats, excess dietary salt results in marked increases in the urinary excretion of epoxygenase metabolites (64). Metabolic, immunological and nucleic acid hybridization studies have suggested that a high salt diet induces a P450 isoform which is either absent or present at very low concentrations in the kidneys of untreated animals (64). This association between increased salt intake and a marked, *in vivo*, induction of the epoxygenase activity, in conjunction with the known functional effects of the EETs as inhibitors of proximal and distal nephron Na^+ absorption (16), suggest that the salt inducible P450 arachidonic acid epoxygenase may be one of the functionally significant components of the kidney's adaptive response to an increased salt intake. Importantly, preliminary studies indicate that the intraperitoneal administration of the P450 inhibitor clotrimazole (twice in a 48 h period, 40 mg/kg body weight) to salt loaded rats results in marked increases in mean arterial blood pressure of the animals (80).

In conclusion, result from several groups, including ours, have demonstrated the central role that the cytochrome P450 system plays in the metabolism and the bioactivation of arachidonic acid. These studies have contributed not only to expand the list of biologically significant eicosanoids but also, documented new and functionally significant endogenous roles for microsomal P450. These observations may prove vital to our understanding of lipid metabolism, membrane biology and the physiological significance of P450 in the formation of lipid-derived mediators.

REFERENCES

1. Smith, W.L., Marnett, L.J., & DeWitt, D.L. (1991). Prostaglandin and thromboxane biosynthesis. Pharmacol. Ther. 49, 153–179.
2. Needleman, P., Turk J., Jakschik, B.A., Morrison, A.R., & Lefkowith, J.B. (1986). Arachidonic acid metabolism. Annu. Rev. Biochem. 55, 62–102.
3. Samuelsson, B., Haeggstrom, J.Z., & Wetterholm, A. (1991). Ann. New York Acad. Sci. 629, 89–99.
4. Namba T., Sugimoto, Y., Negishi, M., Irie, A., Ushikubi, F., Kakizuka, A., Ito, S., Ichikawa, A., & Narumiya, S. (1993). Alternative splicing of the C-terminal tail of prostaglandin E receptor determines G-protein specificity. Nature 365, 166–170.
5. Wakabayashi, K., & Shimazono, N. (1963). Studies of ω-oxidation of fatty acids *in vitro* I. Overall reaction and intermediate. Biochem. Biophys. Acta 70, 132–142.
6. Lu, A.Y.H., Junk, K., & Coon, M.J. (1969). Resolution of the cytochrome P-450-containing ω-hydroxylation system of liver microsomes into three components. J. Biol. Chem. 244, 3714–3721.
7. Capdevila, J.H., Parkhill, L., Chacos, N., Okita, R., Masters, B.S., & Estabrook, R.W. (1981). The oxidative metabolism of arachidonic acid by purified cytochromes P-450. Biochem. Biophys. Res. Commun. 101, 1357–1363.
8. Capdevila, J.H., Chacos, N., Werringloer, J., Prough, R.A., & Estabrook, R.W. (1981). Liver microsomal cytochrome P-450 and the oxidative metabolism of arachidonic acid. Proc. Natl. Acad. Sci. USA 78, 5362–5366.
9. Morrison, A.R., & Pascoe (1981). Metabolism of arachidonic acid through NADPH-dependent oxygenase of renal cortex. Proc. Natl. Acad. Sci. USA 78, 7375–7378.
10. Oliw, E.H., & Oates, J.A. (1981). Oxygenation of arachidonic acid by hepatic microsomes of the rabbit. Mechanism of biosynthesis of two vicinal diols. Biochim. Biophys. Acta 666, 327–340.
11. Capdevila, J.H., Chacos, N., Falck, J.R., Manna, S., Negro-Vilar, A., & Ojeda, S.R. (1983). Novel hypothalamic arachidonate products stimulate somatostatin release from the median eminence. Endocrinol. 113, 421–423.
12. Snyder, G.D., Capdevila, J.H., Chacos, N., Manna, S., & Falck, J.R. (1983). Action of luteinizing hormone-releasing hormone: Involvement of novel arachidonic acid metabolites. Proc. Natl. Acad. Sci. USA 80, 3504–3507.
13. Capdevila, J.H., Pramanik B., Napoli, J.L., Manna, S., & Falck, J.R. (1984). Arachidonic acid epoxidation: Epoxyeicosatrienoic acids are endogenous constituents of rat liver. Arch. Biochem. Biophys. 231, 511–517.
14. Karara, A., Dishman, E., Blair, I., Falck, J.R., & Capdevila, J.H. (1989). Endogenous epoxyeicosatrienoic acids. Cytochrome P-450 controlled stereoselectivity of the hepatic arachidonic acid epoxygenase. J. Biol. Chem. 264, 19822–19827.
15. Fitzpatrick F.A., & Murphy, R.C. (1989). Cytochrome P-450 metabolism of arachidonic acid: Formation and biological actions of "epoxygenase"-derived eicosanoids. Pharmacol. Rev. 40, 229–241.
16. McGiff, J.C. (1991). Cytochrome P-450 metabolism of arachidonic acid. Annul Rev. Pharmacol. Toxicol. 31, 339–369.
17. Capdevila, J.H., Falck, J.R., & Estabrook, R.W. (1992). Cytochrome P-450 and the arachidonate cascade FASEB J. 6, 731–736.
18. White, R.E., & Coon, M.G. (1980). Oxygena activation by cytochrome P-450. Ann. Rev. Biochem. 49, 315–356.

19. Rahimtula, A.D., & O'Brien, P.J. (1974). Hydroperoxide catalyzed liver microsomal aromatic hydroxylation reactions involving cytochrome P-450. Biochem. Biophys. Res. Commun. 60, 440–447

20. Capdevila, J.H. Saeki, Y., & Falck, J.R. (1984). The mechanistic plurality of cytochrome P-450 and its biological ramifications. Xenobiotica 14, 105–118.

21. Weiss, R.H., Arnold, J.L., & Estabrook, R.W. (1987). Transformation of an arachidonic acid hydroperoxide into epoxyhydroxy and trihydroxy fatty acids by liver microsomal cytochrome P-450. Arch. Biochem. Biophys. 252, 334–338.

22. Ullrich, V., Castle, L., & Haurand, M. (1982). Cytochrome P-450 as an oxene transferase In: Oxygenases and Oxygen Metabolism (Nozaki, M., Yamamoto, S., Ishimura, Y., Coon, M.J., Ernster, L. & Estabrook, R.W. eds.), p. 497. Academic Press, New York.

23. Ohashi, K., Ruan, K.H., Kulmacz, R.J., Wu, K.K., & Wang, L.H. (1992). Primary structure of human thromboxane synthase determined from the cDNA sequence. J. Biol. Chem. 267, 789–793.

24. Yokoyama, C., Miyata, A., Ihara, H., Ullrich, V., & Tanabe, T. (1991). Molecular cloning of human platelet thromboxane a synthase. Biochem. Biophys. Res. Commun,. 178, 1479–1484.

25. Hamilton, G.A. (1964). Oxidation by molecular oxygen. II. The oxygen atom transfer mechanism for mixed-function oxidases and the model mixed-function oxidases. Am. Chem. Soc. 86, 3391–3396.

26. Song, W.C., & Brash, A.R. (1991). Purification of an allene oxide synthase and identification of the enzyme as a cytochrome P-450. Science 253, 781–784.

27. Song, W.C., Funk, C.D., & Brash, A.R. (1993). Molecular cloning of an allene oxide synthase: A cytochrome P450 specialized for the metabolism of fatty acid hydroperoxides. Proc. Natl. Acad Sci. USA 90, 8519–8523.

28. Schwartzman. M.L., Falck, J.R., Yadagiri, P., & Escalante, B. (1989). Metabolism of 20-hydroxyeicosatetraenoic acid by cyclooxygenase. Formation and identification of novel endothelium-dependent vasoconstrictor metabolites. J. Biol. Chem. 264, 11658–11662.

29. Hamberg, M., & Samuelsson, B. (1966). Prostaglandins in human seminal plasma. Prostaglandins and related factors. J. Biol. Chem. 241, 257–263.

30. Kupfer, D. (1982). Endogenous substrates of monooxygenases: Fatty acids and prostaglandins. In Hepatic Cytochrome P-450 Monooxygenase System (Schenkman, J.B., & Kupfer, D., eds.), pp. 157–182. Pergamon Press, New York.

31. Gonzalez, F.J. (1989). The molecular biology of cytochromes P450s. Pharmacol. Revs. 40, 243–288 and cited references.

32. Yokotani, N., Bernhardt, R., Sogawa, K., Kusunose, E., Gotoh, O., Kusunose, M., & Fujii-Kuriyama, Y. (1989). Two forms of ω-hydroxylase toward prostaglandin A and laurate. cDNA cloning and their expression. J. Biol. Chem. 264, 21665–21669.

33. Imoaka, S., Nagashima, K., & Funae, Y. (1990). Characterization of three cytochrome P450s purified from renal microsomes of untreated male rats and comparison with human renal cytochrome P450. Arch. Biochem. Biophys. 276, 473–480.

34. Johnson, E.F., Walker, D.L., Griffin, K.J., Clark, I.E., Okita, R.T., Muerhoff, S., & Masters, B.S. (1990). Cloning and expression of three rabbit kidney cDNAs encoding lauric acid ω-hydroxylases. Biochemistry 29, 873–879.

35. Yokotani, N., Kusunose, E., Sogawa, K., Kawashima, H., Kinosaki, M., Kusunose, M., & Fujii-Kuriyama, Y. (1991). cDNA cloning and expression of the mRNA for cytochrome P-450$_{kd}$ which shows a fatty acid ω-hydroxylating activity. Eur. J. Biochem. 196, 531–536.

36. Muerhoff, A.S., Griffin, K.J., & Johnson, E.F. (1992). Characterization of a rabbit gene encoding a clofibrate-inducible fatty acid ω-hydroxylase: CYP4A6. Arch. Biochem. Biophys. 296, 66–72.

37. Kawashima, H., Kusunose, E., Kubota, I., Maekawa, M., & Kusunose, M. (1992). Purification and NH_2-terminal amino acid sequences of human and rat kidney fatty acid ω-hydroxylases. Biochem. Biophys. Acta 1123, 156–162.

38. Powell, W.S. (1978). ω-Oxidation of prostaglandins by lung and liver microsomes. Changes in enzyme activity induced by pregnancy, pseudopregnancy and progesterone treatment. J. Biol. Chem. 253, 6711–6716.

39. Lindgren, J.A., Hansson, G., Claesson, H.E., & Samuelson, B. (1982). Formation of novel biologically active leukotrienes by ω-oxidation in human leukocyte preparations. Adv. Prostaglandin Thromboxane Leukotriene. Res. 9, 53–60.

40. DiAgustine, R.P., & Fouts, J.R. (1969). The effects of unsaturated fatty acids on hepatic microsomal drug metabolism and cytochrome P-450. Biochem. J. 115, 547–554.

41. Pessayre, D., Mazel, P, Decatoire, V., Rogier, E., Feldmann, G., & Benhamou, J.P. (1979). Inhibition of hepatic drug-metabolizing enzymes by arachidonic acid. Xenobiotica 9, 301–310.

42. Cinti, D.L., & Feinstein, M.B. (1976). Platelet cytochrome P-450: A possible role in arachidonate-induced aggregation. Biochem. Biophys. Res. Commun. 73, 171–179.

43. Jacobson, H.R., Corona, S., Capdevila, J.H., Chacos, N., Manna, S., Womack, A., & Falk, J.R. (1984). In: Prostaglandins and Membrane Ion Transport (Braquet, P., Frolich, J.C., Nicosia, S., and Garay, R., eds.), p. 311. Raven Press, New York.

44. Falck, J.R., Lumin, S., Blair, I., Dishman, E., Martin, M.V., Waxman, D.J. Guengerich, F.P., & Capdevila, J.H. (1990). Cytochrome P-450-dependent oxidation of arachidonic acid to 16-, 17-, and 18-hydroxyeicosatetraenoic acids. J. Biol. Chem. 265, 10244–10249.

45. Capdevila, J.H., Karara, A., Waxman, D.J., Martin, M.V., Falck, J.R., & Guengerich, F.P. (1990). Cytochrome P-450 enzyme-specific control of the regio- and enantiofacial selectivity of the microsomal arachidonic acid epoxygenase. J. Biol. Chem. 265, 10865–10871.

46. Orellana, M., Valdes, E., Capdevila, J.H., & Gil, L. (1989). Nutritionally triggered alterations in the regiospecificity of arachidonic acid oxygenation by rat liver microsomal cytochrome P450. Arch Biochem. Biophys. 274, 251–258.

47. Capdevila, J.H., Yadagiri, P., Manna, S., & Falck, J.R. (1986). Absolute configuration of the hydroxyeicosatetraenoic acids (HETEs) formed during catalytic oxygenation of arachidonic acid by microsomal cytochrome P-450. Biochem. Biophys. Res. Commun. 141, 1007–1011.

48. Oliw, E.H. (1993). *bis*-Allylic hydroxylation of linoleic acid and arachidonic acid by human hepatic monooxygenases. Biochim. Biophys. Acta 1166, 258–263.

49. Yoshimoto, T., Miyamoto, Y., Ochi, K., & Yamamoto, S. (1982). Arachidonate 12-lipoxygenase of procine leukocyte with activity for 5-hydroxyeicosatetraenoic acid. Biochim. Biophys. Acta 713, 638.

50. Woollard, P.M. (1986). Biochem. Biophys. Res. Commun. 141, 1007–1011.

51. Schwartzman, M.L., Balazy, M., Masferrer, J., & Abraham, N.G. (1987). 12(R)-hydroxyeicosatetraenoic acid: A cytochrome P450-dependent arachidonate metabolite that inhibits Na^+, K^+-ATPase in the cornea. Proc. Natl. Acad. Sci. USA 84, 8125–8129.

52. Oliw, E.H. (1993). Biosynthesis of 12(S)-hydroxyeicosatetraenoic acid by bovine corneal epithelium. Acta Physiol. Scand. 147, 117–121.

53. Capdevila, J.H., Kim, Y.R., Martin-Wixtrom, C., Falck, J.R., Manna, S., & Estabrook, R.W. (1985). Influence of a fibric acid type of hypolipidemic agent on the oxidative metabolism of arachidonic acid by liver microsomal cytochrome P-450. Arch. Biochem. Biophys. 243, 8–19.

54. Masters, B.S., Okita, R.T., Muerhoff, A.S., Leithauser, M.T., Gee, A., Winquist, S., Roerig, D.L., Clark, J.E., Murphy, R.C., & Ortiz de Montellano (1989). Pulmonary P-450-mediated eicosanoid metabolism and regulation in the pregnant rabbit. Adv. Prostaglandin Thromboxane Leukotriene Res. 19, 335–338.

55. Kupfer, D., Jansson, I., Favreau, L.V., Theoharides, A.D., & Schenkman, J.B. (1988). Arch. Biochem. Biophys. 262, 186–195.

56. Sharma, R.K., Lake, B.G., Makowski, R., Bradshaw, T., Earnshaw, D., Dale, J.W., & Gibson, G.G. (1989). Differential induction of peroxisomal and microsomal fatty-acid-oxidising enzymes by peroxisome proliferators in rat liver and kidney. Eur. J. Biochem. 184, 69–78.

57. Oliw, E.H. (1989). Biosynthesis of 18(R)-hydroxyeicosatetraenoic acid from arachidonic acid by microsomes of monkey seminal vesicles. J. Biol. Chem. 264, 17845–17853.

58. Laethem, R.M., Balazy, M., Falck, J.R., Laethem, C.L., & Koop, D.R. (1993). Formation of 19(S)-, 19(R)-, and 18(R)-hydroxyeicosatetraenoic acids by alcohol-inducible cytochrome P450 2E1. J. Biol. Chem. 268, 12912–12918.

59. Schwartzman, M.L., Omata, K., Lin, F., Bhatt, R.K., Falck, J.R., & Abraham, N.G. (1991). Detection of 20-hydroxyeicosatetraenoic acid in rat urine. Biochem. Biophys. Res. Commun. 180, 445–449.

60. Prakash, C., Zhang, J.Y., Falck, J.R., Chauhan, K., & Blair, I.A. (1992). 20-hydroxyeicosatetraenoic acid is excreted as a glucoronide conjugate in human urine. Biochem. Biophys. Res. Commun 185, 728–733.

61. Borgeat, P., & Samuelsson, B. (1979). Arachidonic acid metabolism in polymorphonuclear leuko-cytes: Unstable intermediate in the formation of dihydroxy acids. Proc. Natl. Acad Sci. USA 76, 3213–3217.

62. Pace-Asciak, C.R., Granstrom, E., & Samuelsson, B. (1982). Arachidonic acid epoxides. Isolation and characterization of two hydroxy epoxide intermediates in the formation of 8,11,12- and 10,11,12-trihydroxyeicosatrienoic acids. J. Biol. Chem. 258, 6835–6840.

63. Pace-Asciak, C.R. (1984). Arachidonic acid epoxides. Demonstration through [^{18}O] oxygen studies of an intramolecular transfer of the terminal hydroxyl group of 12(S)-hydroxyeicosa-5,8,10,14-tetraenoic acid to form hydroxyepoxides. J. Biol. Chem. 259, 8332–8337.

64. Capdevila, J.H., Wei, S., Yan, Y., Karara, A., Jacobson, H.R., Falck, J.R., Guengerich, F.P., & DuBois, R.N. (1992). Cytochrome P-450 arachidonic acid epoxygenase. Regulatory control of the renal epoxygenase by dietary salt loading. J. Biol. Chem. 267, 21720–21726.

65. Karara, A., Makita, K., Jacobson, J.R., Falck, J.R., Guengerich, F.P., DuBois, R.N., & Capdevila, J.H. (1993). Molecular cloning, expression, and enzymatic characterization of the rat kidney cytochrome P-450 arachidonic acid epoxygenase. J. Biol. Chem. 268, 13565–13570.

66. Capdevila, J.H., Dishman, E., Karara, A., & Falck, J.R. (1991). Cytochrome P450 arachidonic acid epoxygenase: Stereochemical characterization of epoxyeicosatrienoic acids. Methods in Enzymol. 206, 441–453.

67. Karara, A., Dishman, Jacobson, H., Falck, J.R., & Capdevila, J.H. (1990). Arachidonic acid epoxygenase. Stereochemical analysis of the endogenous epoxyeicosatrienoic acids of human kidney cortex. FEBS Lett. 268, 227–230.

68. Catella, F., Lawson, J.A., Fitzgerald, D.J., & Fitzgerald, G.A. (1990). Endogenous biosynthesis of arachidonic acid epoxides in humans: increased formation in pregnancy induced hypertension. Proc. Natl. Acad Sci. USA 87, 5893–5897.

69. Karara, A., Dishman, E., Falck, J.R., & Capdevila, J.H. (1991). Endogenous epoxyeicosatrienoyl-phospholipids. A novel class of cellular glycerolipids containing epoxidized arachidonate moieties. J. Biol. Chem. 266, 7561–7569.

70. Brezinski, M., & Serhan, C.N. (1990). Selective incorporation of 15(S)-hydroxyeicosatetraenoic acid in phosphatidylinositol of human neutrophils: Agonist-induced deacylation and transformation of stored hydroxyeicosanoids. Proc. Natl. Acad. Sci. USA 87, 6248–6252.

71. Legrand, A.B., Lawson, J.A., Meyrick, B.O., Blair, I.A., & Oates, J.A. (1991). Substitution of 15-hydroxyeicosatetraenoic acid in the phosphoinositide signaling pathway. J. Biol. Chem. 266, 7570–7577.

72. Bernstrom, K., Kayganich, K., Murphy, R.C., & Fitzpatrick, F.A. (1992). Incorporation and distribution of epoxyeicosatrienoic acids into cellular phospholipids. J. Biol. Chem. 267, 3686–3690.

73. Frei, B., Winterhalter, K.H., & Richter, C. (1985). Quantitative and mechanistic aspects of the hydroperoxide-induced release of Ca^{+2} from rat liver mitochondria. Eur. J. Biochem. 149, 633–639.

74. Sevanian, A., & Hochstein, P. (1985). Mechanisms and consequences of lipid peroxidation in biological systems. Annul Rev. Nutr. 5, 365–390.

75. Gast, K., Zirwer, D., Ladhoff, M., Schreiber, J., Koelsch, R., & Kretschmer, K. (1982). Auto-oxidation-induced function of lipid vesicle. Biochim. Biophys. Acta 686, 99–109.
76. Hauser, H., & Poupart, G. (1992). Lipid structure. In: The Structure of Biological Membranes (Yeagle, P., ed.), pp. 3–71. CRC Press, Ann Arbor, MI.
77. Yeagle, P. (1992). The dynamics of membrane lipids. In: The Structure of Biological Membranes (Yeagle, P., ed.), pp. 157–174. CRC Press, Ann Arbor, MI.
78. Capdevila, J.H., Jin, Y., Karara, A., & Falk, J.R. (1993). Cytochrome P450 epoxygenase dependent formation of novel endogenous epoxyeicosatrienoyl phospholipids. In: Eicosanoids and Other Bioactive Lipids in Cancer, Inflammation and Radiation Injury (Nigam, S., Marnett, L.J., Honn, K.V., & Walden Jr., T.L., eds.), pp. 11–15. Klumer Academic Publishers, Boston, MA.
79. Iwai, N., & Inagami, T. (1990). Isolation of preferentially expressed genes in the kidneys of hypertensive rats. Hypertension 17, 161–169.
80. Capdevila, J.H., Takahashi, K., & Jacobson, H.R. (1993). Inhibition of the kidney P450 arachidonate epoxygenase causes hypertension in salt loaded rats. J. Am. Soc. Nephr. 4, 509. Abs. 119P.

STEROID HORMONES and OTHER PHYSIOLOGIC REGULATORS OF LIVER CYTOCHROMES P450:

METABOLIC REACTIONS AND REGULATORY PATHWAYS

David J. Waxman

Advances in Molecular and Cell Biology
Volume 14, pages 341–374.
Copyright © 1996 by JAI Press Inc.
All rights of reproduction in any form reserved.
ISBN: 0-7623-0113-9

I. LIVER P450s: DUAL ROLE IN XENOBIOTIC AND ENDOGENOUS SUBSTRATE METABOLISM

Liver cytochrome P450 (CYP[1] enzymes are integral membrane proteins of the endoplasmic reticulum that play an important role in NADPH-dependent oxygenation of a broad range of lipophilic compounds. Substrates for these hemeprotein monooxygenase catalysts include many drugs, carcinogens and other foreign chemicals found in the environment (Figure 1). Many endogenous steroids and other naturally-occurring lipophilic substances also serve as substrates for P450 enzymes found in liver and other tissues, in particular the primary steroidogenic tissues (Table 1). Protein purification and cDNA cloning and expression studies

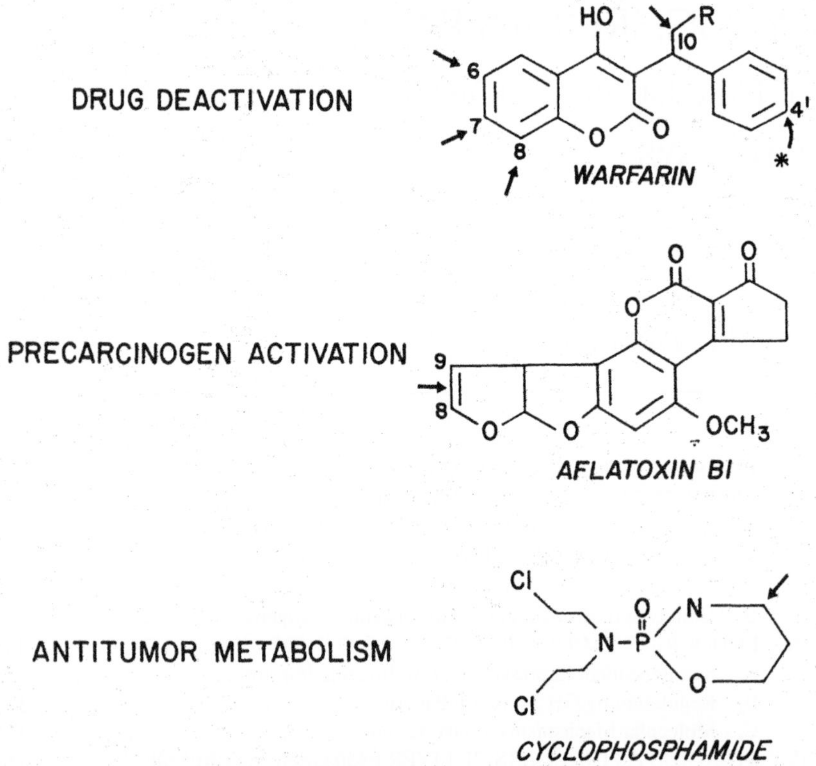

DRUG DEACTIVATION

WARFARIN

PRECARCINOGEN ACTIVATION

AFLATOXIN BI

ANTITUMOR METABOLISM

CYCLOPHOSPHAMIDE

Figure 1. P450-catalyzed xenobiotic metabolism. Hydroxylation of the anticoagulant drug warfarin at the sites marked by *arrows* leads to drug deactivation, whereas 4'-hydroxylation (*)yields a product that retains pharmacologic activity. Aflatoxin B1 is converted to a chemically reactive, DNA-alkylating metabolite by P450-catalyzed epoxidation of the Δ8–9 double bond, while the cancer chemotherapeutic drug cyclophosphamide is activated by liver P450-catalyzed 4-hydroxylation.

Table 1. Functional Properties of Cytochrome P450 Enzymes (CYPs)

Foreign Compound Metabolism (*Liver & Extrahepatics*)
- Toxification and detoxification of environmental chemicals
- Drug metabolism
- Carcinogen activation

Endogenous Substrate Metabolism (*Liver & Extrahepatics*)
- Hydroxylation of steroids, fatty acids, prostaglandins
- Conversion of cholesterol to bile acids (*Liver*)
- Vitamin D3 activation (*Liver, Kidney*)

Steroid Hormone Biosynthesis (*Adrenal, Gonads, Placenta*)
- Synthesis of all major steroid classes:
  Progesterone, Cortisol, Aldosterone (C21)
  Testosterone (C19), Estradiol (C18)

have led to the identification of more than two dozen individual P450s enzymes (P450 forms) that can be expressed in rat liver, a widely studied animal model (Nelson et al., 1993; Ryan and Levin, 1990; Waxman, 1988). A similar number of P450 forms is also found in human liver (Gonzalez et al., 1991; Guengerich and Shimada, 1991). These liver P450s exhibit broad and overlapping specificity profiles with foreign chemical substrates, which are typically converted to more polar (eg., hydroxylated) metabolites that can be conjugated and then eliminated. Detailed studies of these enzymes carried out over the past decade have helped to elucidate the general structural, biochemical and regulatory properties of mammal-

Table 2. General Properties of Mammalian Cytochromes P450

I. P450 Structure: Multiple Gene Families
- 14 different mammalian *P450* gene families
- Simple gene families (1–2 members; CYP 1A, 7, 11, 17, 19, 21) versus complex gene families and subfamilies (CYP 2A, 2B, 2C, 3A, 4A)
- Structural homology within a family or subfamily does not imply conservation of function or regulation

II. P450 Biochemistry: Multiple Enzyme Forms
- Broad and overlapping substrate specificities (liver P450s)
- Highly specific biosynthetic P450s (steroidogenic P450s)
- Microsomal (ER) P450s require a flavoprotein reductase for electron transfer from NADPH to the P450, versus Mitochondrial P450s receive electrons indirectly from a flavoprotein via an Fe-S protein

III. P450 Regulation: Multiple Mechanisms
- Drug induction (3-methylcholanthrene, phenobarbital, dexamethasone, clofibrate)
- Hormone regulation (growth hormone, gonadal steroids, thyroid hormones)
- Tissue-specific expression
- Genetic polymorphisms (CYP 2D6, 2C19, 21A others): mutant P450 genes, allelic variants

ian cytochrome P450s summarized in Table 2. Although P450-catalyzed xenobiotic metabolism often leads to detoxification, in some cases the lipophilic foreign chemical substrates are activated to deleterious, reactive electrophiles (Figure 2). Numerous studies document the roles that these electrophiles play in modification of cellular macromolecules, leading to mutation and the initiation of chemical carcinogenesis (Conney, 1982).

Steroid hormones, fatty acids, prostaglandins, cholesterol and other naturally-occurring lipophilic substances are metabolized by some of the same liver P450 enzymes that are active catalysts of foreign compound metabolism. In contrast to the foreign compound substrates, steroid hormones are metabolized by hepatic P450 enzymes with a high degree of regio- and stereoselectivity (for a review, see Waxman, 1988), suggesting that these endogenous lipophiles may serve as physiological P450 substrates. Indeed, nine of the twelve mammalian *P450* gene families described as of 1993 (Nelson et al., 1993) encode enzymes that catalyze steroid hydroxylations (Table 3). Four of these gene families encode the key P450 enzymes

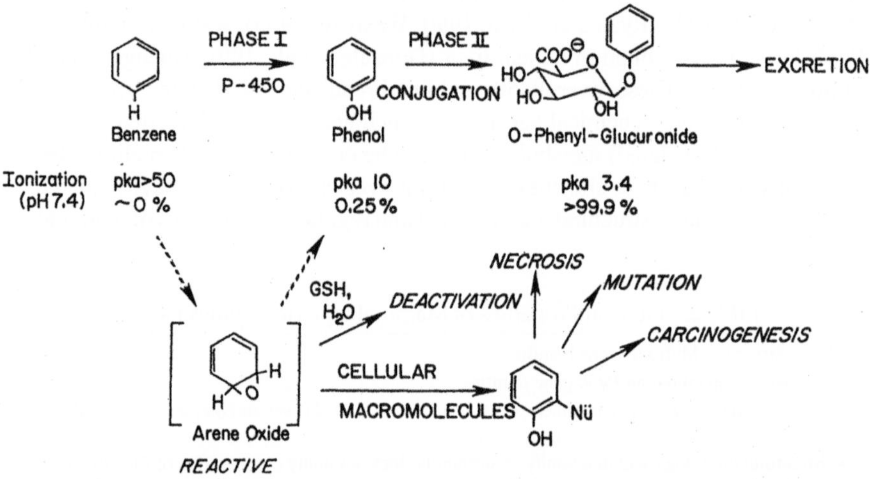

Figure 2. Role of P450 in activation of xenobiotics: the example of benzene. P450-catalyzed hydroxylation of benzene to yield phenol (*Phase I metabolism*) increases the foreign chemical's polarity and solubility, and also introduces a suitable functional group to permit subsequent conjugation catalyzed by a UDP-glucuronyl transferase (*Phase II metabolism*), yielding a water-soluble O-phenyl-glucuronide. Although the majority of benzene is successfully metabolized via this two-step detoxification scheme, the chemically reactive arene oxide intermediate shown in brackets will occasionally elude chemical or enzymatic conversion to deactivated metabolites, resulting in the formation of covalent adducts to cellular macromolecules. This may lead to necrosis, mutation and/or the initiation of chemical carcinogenesis. GSH—glutathione; Nu—nucleophile.

Table 3. Steroid Hydroxylase Cytochrome P450 (CYP) Gene Families

	Subcellular Localization	*Functional Importance*
Liver P450s[a]		
CYP2	ER[b]	Steroid hydroxylation, drug metabolism
CYP3	ER	Steroid and bile acid hydroxylation, drug metabolism
CYP7	ER	Bile acid biosynthesis (cholesterol 7α-hydroxylation)
CYP24	mitochondria	Steroid 24-hydroxylase
CYP 27[c]	mitochondria	Bile acid biosynthesis (steroid side chain oxidation)
Extrahepatic Steroidogenic P450s		
CYP11	mitochondria	Cholesterol side chain cleavage (CYP 11A), glucocorticoid and mineralocorticoid biosynthesis (CYP 11B)[d]
CYP17	ER	Glucocorticoid and androgen biosynthesis (17α-hydroxylase/lyase activities)
CYP19	ER	Estrogen biosynthesis (aromatase)
CYP21	ER	Glucocorticoid and mineralocorticoid biosynthesis

Notes: [a]Listing includes liver-specific steroid hydroxylase *P450* gene families that have been characterized at the cDNA or gene level. Other liver-specific steroid hydroxylase P450 enzymes, such as 7α-hydroxy-4-cholesten-3-one 12α-hydroxylase (Bjorkhem, 1985), have not been characterized in sufficient detail to establish whether they belong to one of the four liver *P450* gene families listed, or whether they belong to other gene families.

[b]ER—endoplasmic reticulum; ER localization of the encoded P450 protein is operationally defined by its association with the microsomal fraction, and by its acceptance of electrons from NADPH via the flavoprotein NADPH cytochrome P450 reductase, rather than via the flavoprotein-iron sulfur protein couple used by mitochondrial P450s.

[c]CYP 27 mRNA is also present in many extrahepatic tissues (Andersson et al., 1989), suggesting that it may be important for processes other than bile acid biosynthesis, e.g., formation of regulatory oxysterols.

required for steroid hormone biosynthesis from cholesterol (CYP 11, CYP 17, CYP 19, CYP 21) (see other chapters in this volume), one encodes a vitamin D 24-hydroxylase (CYP24), and two encode P450s that participate in the conversion of cholesterol to bile acids in the liver, where they contribute in a major way to cholesterol homeostasis (CYP 7, CYP 27) (for a review, see [Waxman, 1992]; also see chapter by Chiang, this volume). Two other P450 gene families, CYP 2 and CYP 3, together encode a large number of individual P450s (30 in the rat), many of which can be expressed at high levels in liver. Several of these latter P450s catalyze hydroxylation reactions that utilize a broad range of biologically important lipophilic substrates, including gonadal and adrenal steroid hormones, and other lipophilic compounds such as retinoic acid, bile acids, and fatty acids (for review, see [Zimniak and Waxman, 1993]), in addition to many drugs and other foreign chemicals. Liver expression of these steroid hydroxylases is under endocrine control (Waxman, 1988; Waxman, 1992; Westin et al., 1992; Zaphiropoulos et al.,

1989), with the gonadal steroids, as well as pituitary growth hormone (GH) and the thyroid hormones (T3, T4) playing important roles in regulating enzyme expression.

This chapter reviews current knowledge about the importance of liver cytochromes P450 for metabolism of endogenous steroid hormones. Also discussed is the regulation of liver P450 gene expression by gonadal steroids and other hormones, including pituitary growth hormone, which is a key determinant of the sex-dependent expression of multiple liver steroid hydroxylase P450s. Finally, this chapter reviews recent studies on the steroid- and receptor-dependent induction of liver P450 4A fatty acid hydroxylase enzymes, with particular emphasis on the role of the adrenal androgen dehydroepiandrosterone as a naturally-occurring steroidal inducer of these enzymes and the associated peroxisome proliferative response.

II.   HYDROXYLATION OF STEROID HORMONES BY MAMMALIAN LIVER P450s

A.   Hydroxylation of Androgens and Progesterone

The steroid hormones testosterone, androstenedione, and progesterone, are actively hydroxylated at multiple sites, and with a high degree of positional and stereochemical specificity by many individual liver P450s. These steroid hydroxylase reactions have been studied in detail with respect to their site specificities (Cheng and Schenkman, 1983; Swinney et al., 1987; Waxman et al., 1983; Wood et al., 1983) (Table 4). Unique patterns of hydroxysteroid metabolites are formed by a large number of liver P450 enzymes, making steroids very useful diagnostic, catalytic probes for characterizing individual liver P450s and for assessing the purity of isolated enzyme preparations (Waxman et al., 1985; Waxman et al., 1988b). These site-specific hydroxylations also provide catalytic monitors that differentiate between closely related liver P450s and can be used to assay for changes in their relative microsomal levels in response to drug exposure (Waxman, 1988; Waxman, 1991) or changes in hormonal status (see below). Thus, testosterone 15α-, 2α-, and 6β-hydroxylation serve as useful catalytic monitors of the adult male-specific P450 forms 2A2, 2C11 and 3A2 in uninduced rat liver microsomes (Sonderfan et al., 1987; Waxman et al., 1985; Waxman et al., 1988b; Yamazoe et al., 1988), while testosterone 7α-hydroxylase activity can be used to assay the levels of the female-predominant enzyme CYP 2A1 (Arlotto and Parkinson, 1989; Levin et al., 1987; Waxman et al., 1985) (Figure 3) (Table 4). Corresponding hydroxylation reactions in human liver microsomes (Waxman et al., 1988a; Waxman et al., 1991a; Wrighton et al., 1990) can also provide useful, non-invasive *in vivo* monitors for hepatic monooxygenase capacity attributable to specific P450s. In particular, cortisol 6β-hydroxylation catalyzed by one or more human CYP 3A enzymes appears to provide a useful catalytic monitor for these enzymes and their induction

Table 4. Gender-Dependent Rat Liver P450 Enzymes: Testosterone Hydroxylase Activities and Hormonal Regulation

| | | | Hormonal Regulation[d] | |
| | | Testosterone Hydroxylase Activities[c] | Androgenic Imprinting[e] | Thyroid Hormone[f] |
CYP Gene Designation [a]	Trivial Names[b]			
I. Male-specific				
2A2	RLM2,M2	15α	++	+/−
2C11	$2c^h$,UT-A M1,RLM5	$\underline{2\alpha}\ 16\alpha$	++	+/−
2C13[g]	^{g}RLM3,M3	$\underline{6\beta}^g,15\alpha$	++	ND
3A2	PCN2,6β-1	$\underline{6\beta}\ 2\beta$	++	├ − −
4A2	IVA2,K-5	$−^h$	ND	−
II. Female-Specific				
2C12	$2^{d,i}$, UT-I, F1	$15\beta^i$	− −	+/−
III. Female-predominant^j				
2A1	$3,^a$,UT-F	7α	ND	−
2C7	^{f}RLM5b	16α	ND	++
5α-reductase	5αR	−	− −	++

Notes: [a] *P450* gene designations are based on the systematic nomenclature of (Nelson et al., 1993). Table is taken from (Waxman, 1992).

[b] Designations given by various investigators to purified P450 protein preparations. See (Nelson et al., 1993; Ryan and Levin, 1990; Waxman, 1988) for more complete listings and references.

[c] The major sites of testosterone hydroxylation catalyzed by the individual P450 proteins are shown. Testosterone metabolites specific to the P450's activity in rat liver microsomal incubations are underlined. Based on (Ryan and Levin, 1990; Waxman, 1988; Waxman et al., 1985) and Refs. therein.

[d] See Figure 4 for a summary of the effects of GH secretory patterns on P450 enzyme expression. "++" indicates a positive effect on adult enzyme expression, while "−−" indicates a suppressive effect. "−" indicates a lesser degree of suppression, while "+/−" indicates no major effect. ND — not determined in a definitive manner.

[e] For further details see (Dannan et al., 1986a; McClellan et al., 1989; Waxman et al., 1988b).

[f] Based on (Ram and Waxman, 1990; Ram and Waxman, 1991; Sundseth and Waxman, 1992; Waxman et al., 1990; Yamazoe et al., 1990).

[g] Purified CYP 2C13 exhibits high testosterone hydroxylase activity in a purified enzyme system, but this enzyme makes only marginal contributions to liver microsomal testosterone hydroxylation (McClellan et al., 1987).

[h] P450 4A2 catalyzes fatty acid ω-hydroxylation, but does not catalyze testosterone hydroxylation.

[i] 15β-hydroxylation of steroid sulfates. CYP 2C12 also catalyzes weak testosterone 15αc- and 1α-hydroxylase activities.

[j] Liver expression of these enzymes is readily detectable in both male and female rats, but at a 3–10-fold higher level in females as compared to males.

status in human liver (Ged et al., 1989; Hunt et al., 1992; Saenger et al., 1981). Caution must be used when utilizing this approach to study other steroid substrates or other enzyme systems (e.g., drug-induced liver, extrahepatic tissues, or other species) since, for example, in drug-induced adult male rat liver, steroid 6β-hydroxylation can be catalyzed by both rat CYP 3A1 and CYP 3A2, whereas in uninduced liver CYP 3A1 is expressed at a very low level and the activity of CYP

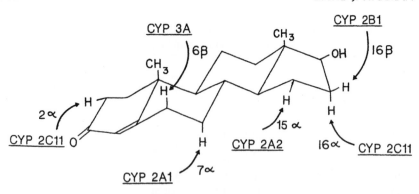

Figure 3. Site-specificity of testosterone hydroxylation catalyzed by rat liver P450 enzymes. See Table 4 and (Waxman, 1988) for additional details.

3A2 dominates. Moreover, whereas testosterone 7α-hydroxylation provides a specific catalytic monitor for CYP 2A1 activity in rat liver microsomes (Levin et al., 1987; Waxman et al., 1985), in the case of the closely related steroid androstenedione, both CYP 2A1 and CYP 2A2 catalyze the 7α-hydroxylation reaction. The relative contribution of each P450 to overall microsomal androstenedione 7α-hydroxylase activity is greatly affected by the rat's sex and by prior induction with phenobarbital, and thus this steroid hydroxylation reaction cannot be used as a specific catalytic monitor for CYP 2A1 (Waxman et al., 1990a).

The physiologic importance of these site-specific steroid hydroxylation reactions is not known with certainty, but may include the following possibilities: (a) inactivation of circulating steroid hormones; (b) targeting of the steroid for conjugation and elimination; and (c) synthesis of novel hydroxysteroids with unique biological properties or endocrine activities. One example of the latter is CYP 2A1-catalyzed formation of 7α-hydroxytestosterone, which may regulate testosterone production and metabolism in the testis (Inano et al., 1973; Mittler, 1985). (Also see the following section on estradiol hydroxylation.) Alternatively, some steroid hydroxylations may be incidental activities of liver P450s whose primary biological function is the hydroxylation of other endogenous substrates, or perhaps even xenobiotic metabolism. The broad substrate specificities that are inherent to steroid hydroxylase liver P450s (Guengerich, 1987; Ryan and Levin, 1990), are consistent with this latter suggestion. Finally, independent of their precise biological significance, these steroid hydroxylation reactions have served as useful analytical tools for monitoring the expression of individual P450 enzymes, as discussed above, and they have also provided useful catalytic probes for studies of P450 structure/function relationships, as revealed by the dramatic alterations in the activity and/or site-specificity of P450-catalyzed steroid hydroxylation that can occur with changes of as few as 1–3 amino acid residues (Aoyama et al., 1989; Kronbach et al., 1989; Negishi et al., 1992).

B. Conversion of 17β-Estradiol to Catechol Estrogens

Liver P450 enzymes hydroxylate 17β-estradiol primarily at the 2- and 4-positions, to yield the catechol estrogens 2-hydroxy-estradiol (2-OH-E2) and 4-hydroxy-estradiol (4-OH-E2). The catechol estrogens are chemically unstable, but can be chromatographed if suitably stabilized, e.g., by ascorbate. Assay methods for catechol estrogen formation include tritium release assays, assays based on enzymatic conversion of the primary products to the more stable methoxy derivatives by catechol-O-methyl transferase, and direct chromatography in the presence of ascorbate (Barnea et al., 1984; Kupfer et al., 1981; Numazawa et al., 1985). Catechol estrogens are rapidly and extensively metabolized *in vivo* by catechol-O-methyl transferase-catalyzed methylation, as well as by conjugation reactions that yield glucuronides and glutathione adducts. Catechol estrogens can also form thioether adducts with protein sulfhydryls, while other hydroxylated estrogens, in particular 16α-hydroxyestrone, can form stable ketoamine adducts via lysine amino groups (Bucala et al., 1982; Jellinck and Fishman, 1988). These latter reactions could contribute to some of the pathological conditions associated with increased estradiol metabolism, including systemic lupus erythematosus (Lahita et al., 1981) and estrogen-dependent carcinogenesis (Li and Li, 1987).

The biological functions of catechol estrogens are multiple (Barnea et al., 1984; Fishman, 1983), and include: (a) *estrogenic activity*—4-OH-E2, but not 2-OH-E2, stimulates uterine growth and exhibits other estrogenic activities that can be nearly as potent as those of 17,β-estradiol itself. By contrast, 2-OH-E2 can exhibit anti-estrogenic activity; (b) *interactions with estrogen receptor*—2-OH-E2 and 4-OH-E2 both bind to the estrogen receptor at 25% and 45% the relative binding affinity of estradiol, respectively; (Barnea et al., 1984), but only 4-OH-E2 can induce sustained nuclear receptor occupancy. This probably accounts for the high estrogenic activity of the latter compound; (c) *effects on catecholamine metabolism*—2-OH-E2 inhibits the activity of tyrosine hydroxylase, a rate-limiting enzyme of catecholamine biosynthesis, while both catechol estrogens can interfere with catecholamine metabolism by competition for catechol-O-methyl transferase. This can lead to significant effects on blood pressure and the regulation of local blood flow (Barnea et al., 1984).

Detailed studies on the roles of individual liver P450 enzymes in estradiol metabolism have been carried out in several species including rat, rabbit, and humans. Studies using purified and reconstituted enzymes have shown that while several rat liver P450s can carry out estradiol 2-hydroxylation, P450 forms 1A2 and 2C11 appear to be the most active (Dannan et al., 1986b; Ryan et al., 1982; Watanabe et al., 1991). In the case of rabbits, a CYP 2C enzyme, 2C5, also dominates estradiol 2-hydroxylation (Schwab and Johnson, 1985). Although purified CYP 1A2 is active in this reaction, the estradiol 2-hydroxylase activity of CYP 1A2 in rat liver microsomes is low, and appears to be masked by the inefficient interaction of this P450 with microsomal NADPH-P450 reductase (Graham et al., 1988). Accordingly, CYP 2C11 is the major catalyst of estradiol 2-hydroxylation in uninduced rats (Dannan et al., 1986b), particularly in adult males, where this male-specific P450 form is expressed at a high level (see below). This conclusion is in accord with the sex-differences in rat liver microsomal estradiol 2-hydroxylation reported in earlier studies (e.g., Brueggemeier, 1981; Bulger and Kupper, 1983; Theron et al., 1983) and can account for the dependence of microsomal 2-hydroxylation activity on postnatal developmental and hormonal factors (Dannan et al., 1986b; Quail and Jellinck, 1987).

CYP 2C11 can also catalyze estradiol 4-hydroxylation in rat liver microsomes, but in addition, a CYP 3A enzyme (CYP 3A1 and/or CYP 3A2) plays a major role in this microsomal activity. This is indicated by the high inducibility of microsomal estradiol 4-hydroxylation by the CYP 3A inducer pregnenolone 16α-carbonitrile, particularly in adult female rats (where there is no basal CYP 3A2 expression) and by the inhibitory effects of anti-CYP 3A antibodies (Dannan et al., 1986b) and the CYP 3A inhibitor troleandomycin (Fisher et al., 1990).

Analysis of the role of individual human P450s in estradiol 2- and 4-hydroxylation using a panel of cDNA-expressed human P450s has revealed that several human P450s, in particular CYP 1A2, CYP 3A4, and 3A5, as well as CYP 2C9 are catalytically competent with respect to estradiol 2-hydroxylation (Aoyama et al., 1990). Other studies have shown that CYP 1A1, which is not expressed at high levels in human liver, but can be induced in extrahepatic tissues by polycyclic aromatic hydrocarbons, can also catalyze estradiol 2-hydroxylation, in addition to hydroxylation at the 16α- and 15α-positions (Spink et al., 1992). CYP 1A2 and the CYP 3A P450s (but not CYP 2C9) also catalyze estradiol 4-hydroxylation, and as occurs in the rat, the ratio of 4-hydroxylation to 2-hydroxylation is highest with the CYP 3A enzymes. Moreover, antibody inhibition experiments indicate that CYP 3A enzymes (which are particularly abundant in human liver) make a major contribution to human liver microsome-catalyzed catechol estrogen formation (60–70% of the total), while CYP 1A2 and one or more CYP 2C enzymes play a role that is more minor, albeit significant (Aoyama et al., 1990; Ball et al., 1990). The extent to which CYP 1A2 contributes to estradiol 2-hydroxylation in humans can vary from liver to liver (Ball et al., 1990), and this may be a reflection of the inducibility of CYP 1A2 by polycyclic aromatic hydrocarbons. This possibility is consistent with the increased systemic estradiol 2-hydroxylation and the anti-estrogenic activity associated with cigarette smoking in women (Michnovicz et al., 1986).

In addition to catechol estrogen formation, estradiol also undergoes P450-catalyzed 16α-hydroxylation, an important pathway that competes with catechol estrogen formation and yields the highly estrogenic metabolite estriol. In humans, estradiol 2-hydroxylation predominates over 16α-hydroxylation by a factor of 2 to 4, with the extent of 2-hydroxylation greater in women than in men (Fishman et al., 1980). Increased estradiol 16α-hydroxylation is associated with several diseased states, including systemic lupus erythematosus and breast cancer, and is hypothesized to be a contributing factor in the development of these diseases (Lahita et al., 1981; Schneider et al., 1982). Estradiol 16α-hydroxylation is actively catalyzed by rat liver microsomes (Numazawa et al., 1985), but the specific rat or human P450 catalysts of this reaction have not been reported.

III. ENDOCRINE REGULATION OF STEROID HYDROXYLASE P450s

A. Sex-dependent Expression and Androgenic Imprinting

Several of the steroid hydroxylase P450s found in rat liver are expressed in a sex-specific manner, and are subject to a striking developmental regulation under endocrine control (Table 4) (Waxman, 1988; Zaphiropoulos et al., 1990). Corresponding regulation has also been observed for several mouse liver P450s (Harada and Negishi, 1988; Noshiro and Negishi, 1986; Squires and Negishi, 1988). In the case of the rat, CYP 2C11, the major male-specific androgen 16α- and 2α-hydroxylase in adult liver, is induced at puberty in males but not females (Morgan et al., 1985; Waxman, 1984). A similar developmental profile is found for two other male-specific rat steroid hydroxylase P450s, CYP 2A2 (Thummel et al., 1988; Waxman et al., 1988b) and CYP 2C13 (Bandiera et al., 1986; McClellan et al., 1989). The female-specific steroid sulfate 15β-hydroxylase CYP 2C12 is induced at puberty in female rat liver (MacGeoch et al., 1985; Waxman et al., 1985), as are two female-predominant liver enzymes: CYP 2C7 (Bandiera et al., 1986; Gonzalez et al., 1986a), which catalyzes retinoic acid 4-hydroxylation (Leo et al., 1984), and a cytochrome P450-independent enzyme, steroid 5α-reductase (Colby, 1980; Waxman et al., 1985). Other sex-dependent rat liver P450s exhibit more complex postnatal developmental profiles. The adult male-specific steroid 6β-hydroxylase CYP 3A2 is expressed at similar levels in both sexes shortly after birth, but is suppressed at puberty only in females (Gonzalez et al., 1986b; Sonderfan et al., 1987; Waxman et al., 1985; Yamazoe et al., 1988). CYP 2A1 is a female-predominant steroid 7α-hydroxylase (female:male CYP 2A1 ratio = 3–4 to 1 at adulthood) that is expressed in both sexes shortly after birth and is suppressed at puberty to a greater extent in males than in females (Arlotto and Parkinson, 1989; Waxman et al., 1985; Yamazoe et al., 1990).

Studies on the endocrine regulation of these liver P450s have provided insight into the hormone regulatory pathways and the underlying mechanisms that determine the sex-dependent expression of these enzymes. These findings are important for our understanding of the basic endocrine regulation of liver steroid metabolism. They also shed light on the underlying basis for the influence of hormonal status on the broad range of drug metabolism and carcinogen activation reactions that are also catalyzed by these liver P450s (Colby, 1980; Kato, 1974; Skett, 1987). As we have summarized elsewhere (Waxman, 1992), the following general conclusions can be drawn:

1.   Pretranslational control—The sex-dependent steroid hydroxylase P450s are regulated at a pretranslational level, with parallel changes in liver microsomal steroid hydroxylase activity, P450 protein, and P450 mRNA generally occurring in response to changes in hormonal status (e.g., Janeczko et al., 1990; Mode et al., 1989).

2.   Androgenic imprinting—The male-specific P450s are expressed in post-pubertal animals in response to postnatal androgen exposure, which *imprints* or *programs* the rat (Arnold, 1984; Gustafsson et al., 1983) for later developmental changes (Shimada et al., 1987; Waxman et al., 1985; Waxman et al., 1988b) (Table 4). In contrast, adult androgen exposure contributes in a *reversible* manner to the maintenance of full P450 enzyme expression at adulthood (Dannan et al., 1986a). The mechanism by which neonatal androgen exposure imprints liver gene expression is still obscure, but probably involves the hypothalamus and its regulation of pituitary growth hormone (GH) secretory patterns (Jansson et al., 1985), which play a key role in regulating expression of the sex-specific P450 enzymes (see below). Direct (GH-independent) effects of androgen on liver enzyme expression (i.e., effects of androgen in hypophysectomized rats) can be observed in some instances, but these are minor compared to the effects of GH.

3.   GH secretory profiles—Continuous plasma GH, a characteristic of adult female rats (Jansson et al., 1985), stimulates expression of the female-specific CYP 2C12, while intermittent GH pulsation, associated with adult male rats, induces the expression of CYP 2C11 (Kato et al., 1986; MacGeoch et al., 1985; Morgan et al., 1985; Waxman et al., 1991b) (Figure 4). Continuous GH can also stimulate the expression of several female-predominant enzymes, including steroid 5α-reductase, CYP 2A1 and CYP 2C7 (Mode et al., 1982; Ram and Waxman, 1990; Sasamura et al., 1990; Waxman et al., 1989a). The effect of intermittent GH exposure on other male-specific liver P450s (2A2, 2C13, 3A2, 4A2) is less clear. Expression of these CYP enzymes is not obligatorily dependent on GH pulses, when judged by their high level of expression in hypophysectomized rats of both sexes (McClellan et al., 1989; Sundseth et al., 1992; Waxman et al., 1988b; Yamazoe et al., 1988). On the other hand, the expression of these P450s in liver can be stimulated by intermittent GH pulses given to adult male rats depleted of circulating GH by neonatal monosodium glutamate treatment (Waxman et al., 1995a).

4.   GH suppression—GH can also have negative effects on liver steroid hydroxylase enzyme expression (Figure 4). Continuous infusion of GH markedly suppresses expression of the male-specific P450s. This suppression cannot be attributed solely to the destruction of pulsatile plasma GH profiles that occurs when intact male rats are given GH by continuous infusion, since suppression of the male-specific P450s can also be observed in hypophysectomized rats given con-

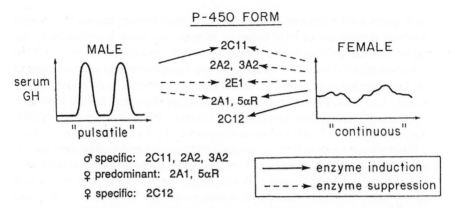

Figure 4. Role of GH secretory profiles in the expression of rat hepatic P450 enzymes and steroid 5α-reductase. Influence of pulsatile vs. continuous plasma GH on the expression of hepatic enzymes whose expression in adult rats is male specific, female predominant, or female specific. Stimulation of enzyme expression is indicated by a solid line, and suppression of enzyme expression by a dashed line. Other pituitary-determined hormones (e.g., thyroid hormone) may be required for the full effects of GH on some of these hepatic enzymes (see text). 5αR—Steroid 5α-reductase. Figure is modified from (Waxman et al., 1989a).

tinuous GH infusion. GH suppression also occurs in the case of some of the female-predominant enzymes, in response to the male pattern of intermittent plasma GH (Waxman et al., 1989a; Yamazoe et al., 1990). GH can also suppress the basal and/or the inducible levels of some of the xenobiotic inducible liver P450s, and probably is a key determinant of the lower responsiveness of female rats to phenobarbital induction of CYP2B1 (Shapiro et al., 1994; Yamazoe et al., 1987), and perhaps also the lower responsiveness of the females to clofibrate induction of CYP 4A (Sundseth and Waxman, 1992).

5. Thyroid hormone regulation of P450 and P450 reductase—Although GH is the major regulator of these liver P450s, thyroid hormone also plays a critical role: it positively regulates some (Ram and Waxman, 1990) but not all (Yamazoe et al., 1990) of the female-predominant enzymes, and it negatively regulates some of the male-specific enzymes (Ram and Waxman, 1991; Waxman et al., 1990) (Table 4). These effects of thyroid hormone are independent of the indirect effects that thyroid hormone has on liver P450 levels as a consequence of its effects on liver GH receptors (Hochberg et al., 1990) and its stimulation of GH gene transcription and GH secretion by the pituitary (Samuels et al., 1988).

Thyroid hormone is also required for full expression of NADPH-P450 reductase, a flavoenzyme that catalyzes electron transfer to all microsomal P450s. P450 reductase is an obligatory, and often rate-limiting electron-transfer protein that participates in all microsomal P450-catalyzed steroid hydroxylations. This thyroid hormone dependence of P450 reductase enzyme expression is evidenced by the major decrease ($\geq 80\%$) in liver microsomal P450 reductase activity and P450 reductase mRNA levels that occurs following hypophysectomy or in response to methimazole-induced hypothyroidism; it is further supported by the reversal of this activity loss when T4, but not GH or other pituitary-dependent factors, is given at a physiologic replacement dose (Ram and Waxman, 1992; Waxman et al., 1989b). Restoration of liver P450 reductase activity *in vivo* by T4 replacement also effects a substantial increase in liver microsomal P450 steroid hydroxylase activities. A similar effect can be achieved when liver microsomes isolated from hypophysec-tomized rats are supplemented with exogenous, purified P450 reductase, which preferentially stimulates steroid hydroxylation catalyzed by microsomes prepared from thyroid-deficient animals (Waxman et al., 1989b).

B. Regulation by GH Secretory Patterns

The precise mechanisms whereby GH regulates expression of the sex-dependent liver P450s are only partially understood. GH can act directly on the hepatocyte to regulate liver P450 expression, as demonstrated by the responsiveness of primary rat hepatocyte cultures to continuous GH-stimulated expression of CYP 2C12 (Guzelian et al., 1988). GH binds to its plasma membrane receptor (Leung et al., 1987; Roupas and Herington, 1989), which is a member of the cytokine receptor superfamily, and which transduces to the hepatocyte the effects of GH binding at the cell surface. This binding is associated with a sequential recruitment into the hormone-receptor complex of two GH receptor molecules, a process that is pro-posed to be essential for GH-induced, receptor-dependent signalling events (Cun-ningham et al., 1991). GH binding also leads to tyrosine phosphorylation of GH receptor and other associated proteins (Silva et al., 1993; Wang et al., 1993; Ram et al., 1996), as occurs with several other members of the cytokine receptor superfamily. This phosphorylation is not catalyzed by GH receptor itself, which lacks a tyrosine kinase domain, but rather is catalyzed by a GH receptor associated tyrosine kinase designated JAK2 (Argetsinger et al., 1993). This tyrosine kinase interacts in a similar manner with GH receptor and with erythropoietin receptor, suggesting that the intracellular signalling mechanisms used by GH may be shared with those of other members of the cytokine/hematopoietin receptor superfamily.

In vivo studies carried out in intact male rats have demonstrated that the GH receptor internalizes to an intracellular compartment coincident with its stimulation by plasma GH pulses, and reappears at the cell surface at the time of the next hormone pulse (Bick et al., 1989a; Bick et al., 1989b). Whether internalization of GH receptor is ligand-driven, and the importance of this internalization for transduction of the effects of GH on liver P450 expression is still undetermined. Studies of other GH responses indicate that activation of the GH receptor can lead to activation of protein kinase C (Doglio et al., 1989; Gurland et al., 1990; Johnson et al., 1990). Conceivably, activation of this pathway might also be important for the effects of GH on liver P450 expression (Tollet et al., 1991). IGF-I, which is produced in the liver in response to GH stimulation, does not mimic the effects of GH on liver P450s (Guzelian et al., 1988; Noshiro and Negishi, 1986; Tollet et al., 1990), suggesting that an autocrine mechanism does not apply.

Discrimination by the hepatocyte between male and female plasma GH profiles is required to achieve the two dramatically different patterns of liver *P450* gene expression that GH can elicit. This discrimination may occur at the cell surface, where a higher level of GH receptors is found in female as compared to male rats (Baxter and Zaltsman, 1984); see, however, (Mathews et al., 1989), or it may involve differences in the intracellular signaling pathways elicited by a chronic (female) versus an intermittent (male) pattern of GH stimulation. This latter possibility is strongly supported by the demonstration that, in a whole rat model, physiologic GH pulsation, but not continuous GH, leads to intermittent activation of the latent cytoplasmic signal transducer protein Stat 5 by a mechanism which involves GH-induced tyrosine phosphorylation (Waxman et al., 1995b). This JAK2 kinase-catalyzed phosphorylation is associated with translocation of Stat 5 from the cytosol to the nucleus where it exhibits specific DNA-binding activity. Stat 5 thus appears likely to be a key intracellular mediator of the stimulatory effects of GH pulses on male-specific liver CYP gene expression (Waxman et al., 1995b). GH can also activate two other liver Stat proteins, Stat 1 and Stat 3, but with a dependence on the concentration of GH and its temporal plasma profile that is distinct from Stat 5, and with a striking desensitization following a single hormone pulse that is not observed in the case of Stat 5 (Ram et al., 1996). GH activation of these two groups of Stats respectively leads to their selective binding to DNA response elements upstream of two distinct sets of genes: the *c-fos* gene (SIE element), in the case of Stat 1 and Stat 3 binding, and the *CYP2C11, CYP3A10,* and β-*casein* genes, in the case of liver Stat 5 binding. In addition to tyrosine phosphorylation, GH stimulates phosphorylation of all three Stats on serine or threonine in a manner that either enhances (Stat 1, Stat 3) or substantially alters (Stat 5) the binding of each Stat to its cognate DNA response element (Ram et al., 1996). Thus, hepatocytes display multiple Stat-dependent GH signaling pathways that can target distinct genes and thereby contribute to the diverse effects that GH and its sexually dimorphic plasma profile have on liver gene expression.

Studies have been carried out to determine which of the three descriptive features of a GH pulse—namely, hormone pulse duration, pulse height, and pulse frequency—is required for proper recognition of a GH pulse as masculine. Direct measurement of the actual plasma GH profiles achieved when GH is administered to hypophysectomized rats by twice daily s.c. GH injection [i.e., the intermittent GH replacement protocol most commonly used to stimulate CYP 2C11 expression] has revealed broad peaks of circulating GH, which last as long as 5–6 hr (Waxman et al., 1991b). Since these GH "pulses" do stimulate expression of the male-specific CYP 2C11, it is apparent that GH pulse duration need not be tightly regulated to elicit this response from the hepatocyte. Similarly, studies on the requirements for GH pulse height carried out using GH-deficient rats [either dwarf rats or rats depleted of circulating GH by monosodium glutamate treatment] have established that GH pulse height is also not a critical factor for stimulation of CYP 2C11 expression (Legraverend et al., 1992a; Shapiro et al., 1989). This finding can be understood in terms of the K_d Of the GH-GH receptor complex, which at 10^{-10} M (2ng/ml) (Leung et al., 1987), is only 1% of the peak plasma hormone level in adult male rats. Rather, GH pulse frequency appears to be the most critical determinant for GH stimulation of a male pattern of liver P450 expression. This finding has been established in studies of hypophysectomized rats given GH replacement pulsations at frequencies of 2, 4, 6, or 7 times per day (Waxman et al., 1991b). Analysis of liver CYP 2C11 levels in these rats after a 7 day period of GH pulsation revealed that a GH pulsation frequency of 6 pulses/day (which approximates the normal male plasma GH pulse frequency), as well as frequencies of only 2 or 4 times per day effectively stimulated a male pattern of liver *CYP 2C11* gene expression. By contrast, hypophysectomized rats given 7 daily GH pulses did not respond. Therefore, the hepatocyte no longer recognizes the pulse as "masculine" if GH pulsation becomes too frequent. Hepatocytes thus require a minimum GH off-time (~2.5 hr in the hypophysectomized rat model used in these studies), which implies a need for an obligatory recovery period, a condition not met in the case of hepatocytes exposed to GH continuously (female profile). This recovery period may be required to reset the Stat 5-dependent intracellular signaling apparatus (see above) or perhaps may provide time needed for replenishment of cell surface GH receptors.

C. Molecular Mechanisms of GH Action

As noted above, GH regulates steroid hydroxylase P450 expression at the level of steady-state mRNA (pretranslational regulation). In the case of CYP 2C12 mRNA, induction by continuous GH exposure requires ongoing protein synthesis, as indicated by the inhibitory effect of cycloheximide on CYP 2C12 mRNA accumulation in primary hepatocyte cultures (Tollet et al., 1990). Although it was initially concluded on the basis of quantitative discrepancies between CYP 2C steady-state mRNA levels and nuclear run-on transcription rates that post-transcriptional mechanisms play a role in the regulation of liver *P450* gene expression by GH secretory patterns [compare (Wright and Morgan, 1990) versus (Mode et al., 1989) and (Zaphiropoulos et al., 1990)], there is no direct evidence for post-transcriptional mechanisms, and it is now generally accepted that transcriptional mechanisms dominate. Thus, unprocessed, nuclear CYP 2C11 and 2C12 RNAs (hnRNA) respond to circulating GH profiles in a manner that is indistinguishable from the corresponding mature, cytoplasmic mRNAs (Sundseth et al., 1992). Consequently, transport of 2C11 and 2C12 mRNA to the cytoplasm, and cytoplasmic mRNA stability are unlikely to be important GH-regulated control points for sex-specific P450 RNA expression. These findings are supported by nuclear run-on transcription analyses, which, additionally, provided firm evidence that GH regulates the sex-specific expression of the *CYP 2C11* and *CYP 2C12* genes at the level of transcript initiation (Legraverend et al., 1992b; Sundseth et al., 1992). Transcription is also the major step for regulation of CYP 2A2 RNA, whose male-specific expression appears to be primarily a consequence of the suppressive effects of continuous GH exposure in adult female rats (Waxman et al., 1988b). Transcription initiation is thus the step at which three distinct effects of GH are operative: stimulation of 2C11 expression by pulsatile GH, suppression of 2A2 (and 2C11) expression by continuous GH, and stimulation of 2C12 expression by continuous GH (Sundseth et al., 1992). Recent studies have been directed to an analysis of cloned 5'-flanking DNA segments of the *CYP 2C11* gene (Morishima et al., 1987) and the *CYP 2C12* gene (Zaphiropoulos et al., 1990) in order to identify specific DNA segments on each gene that interact in a distinct manner with nuclear proteins (putative transcription factors) present in male versus female rat liver (Sundseth et al., 1992). Several GH status-dependent binding interactions have been observed, and these are hypothesized to contribute to the sex-specific transcription of the *CYP 2C11* and *CYP 2C12* genes (Zhao and Waxman, 1994). Presumably, one or more functional Stat 5 binding sites will be found to contribute to the GH pulse-induced transcriptional activation of *CYP 2C11*, as has been reported for the GH-regulated, male-specific hamster liver *CYP 3A10* (Subramanian et al., 1995). More detailed molecular studies will be required to identify the full range of positive and negative components of the 2C11 and 2C12 transcription machinery and their responsiveness to plasma GH patterns in order to fully elucidate the mechanisms by which GH regulates the sex-dependent expression of these liver *P450* genes.

IV. STEROIDS AS INDUCERS OF LIVER P450 GENE EXPRESSION

As noted earlier in this chapter, numerous drugs and other foreign chemicals have long been known to induce in liver the expression of P450 and other enzymes of drug and steroid metabolism (Conney, 1967). In many cases, individual P450 forms can be induced 50-fold or more by mechanisms that involve transcriptional activation of the target *P450* genes. Four general classes of P450 inducing agents are presently recognized, and for two of these classes specific intracellular receptor proteins that mediate the induction have been identified (Table 5). One is the Ah (aromatic hydrocarbon) receptor, a novel helix-loop helix protein that binds hydrocarbon inducers and heterodimerizes with a nuclear translocation factor to yield a complex which transcriptionally activates *CYP 1A1* genes (Burbach et al., 1992). The other is PPAR, peroxisome proliferator-activated receptor (Issemann and Green, 1990), a member of the steroid/thyroid receptor superfamily that activates *CYP 4A* genes, in addition to genes encoding several peroxisomal enzymes (see below). Both the Ah-receptor and PPAR are ligand-activatable transcription factors that reside in the nucleus in the activated state. In the case of the steroid/macrolide antibiotic inducers of *CYP 3A* genes (Table 5), it is still unclear whether the classical glucocorticoid receptor plays a role in the resultant CYP 3A induction process (Burger et al., 1992; Schuetz et al., 1984), while in the case of the phenobarbital inducer class it has yet to be established whether the mechanism of gene induction involves a specific receptor protein (Waxman and Azaroff, 1992) [also see (Shaw et al., 1993)]. Significant progress has recently been made toward an understanding of the mechanism of bacterial *P450* gene induction by phenobarbital, which does not involve steroid receptors (Shaw and Fulco, 1992; Shaw and Fulco, 1993), but the extent to which these findings are applicable to the induction of mammalian *P450* genes by phenobarbital is not yet clear. Although it is possible that these four P450 induction pathways (Table 5) respond to foreign chemicals exclusively, it seems more likely that the foreign compound inducers activate receptors and P450 induction events that are normally responsive to endogenous ligands that play a role in normal cell physiology (Waxman and Azaroff, 1992). Evidence for this hypothesis comes from studies identifying naturally-occurring inducers for two of the four P450 inducer classes. Thus, glucocorticoids can serve as endogenous regulators of *CYP 3A* gene expression (Schuetz et al., 1984), while certain endogenous steroid sulfates and fatty acids serve as inducers of *CYP 4A* genes, as discussed in the following section.

A. Receptor-dependent Induction of CYP 4A and Peroxisomal Enzymes

CYP 4A enzymes actively catalyze the oxygenation of a number of biologically important fatty acids and prostaglandins, including arachidonic acid and other eicosanoids (see chapter by Capdevilla, this volume). At least three *CYP 4A* genes, designated 4A1, 4A2, and 4A3 are expressed in rat liver and kidney, where their mRNA and protein levels are highly inducible by structurally diverse lipophilic drugs and foreign chemicals. Foreign compound inducers of *CYP 4A* gene expression include hypolipidemic fibrate drugs, phthalate ester plasticizers used in the

Table 5. Classification of Liver P450 Inducing Agents

Inducer Class	Rat CYP Genes Induced	Typical Foreign Chemical Inducers	Possible Naturally-occurring Inducers
Polycyclic Aromatic Hydrocarbons	1A1	3-Methylcholanthrene, TCDD[a]	?
Phenobarbital	2B1, 2B2, 2A1, 2C6	Phenobarbital, other drugs	?
Steroids/Macrolide Antibiotics	3A1,3A2	Dexamethasone, Troleandomycin	Glucocorticoids
Peroxisome Proliferators	4A1, 4A2, 4A3	Clofibric Acid, Wy-14,643	DHEA-S[b], Fatty Acids

Notes: [a] TCDD—2,3,7,8-tetrachlorodibenzo-p-dioxin
             [b] DHEA-S—dehydroepiandrosterone 3β-sulfate

medical and the chemical industries, and various environmental chemicals (Figure 5) (Rao and Reddy, 1987). These compounds are classified as peroxisome proliferators, since in addition to this induction of CYP 4A enzymes, they dramatically induce several peroxisomal enzymes that are active in the fatty acid β-oxidation pathway.

The overall induction process leads to peroxisome proliferation, i.e., a dramatic increase in both the size and the number of peroxisomes found in liver cells. This induction event, which involves transcriptional activation of both *CYP 4A* and peroxisomal enzymes genes (Hardwick et al., 1987; Reddy et al., 1986) is now known to be mediated by PPAR, peroxisome proliferator-activated receptor, a recently discovered orphan receptor protein that belongs to the steroid receptor gene superfamily (Dreyer et al., 1992; Green, 1992; Issemann and Green, 1990). Although foreign chemical activators of PPAR have been identified, none have been shown to bind directly to the receptor protein. Moreover, endogenous, natural ligands of PPAR have yet to be identified (see below). PPARs have been cloned from mouse and rat liver (Gottlicher et al., 1992; Issemann and Green, 1990) and from human liver and osteosarcoma cells (Schmidt et al., 1992; Sher et al., 1993). In addition, three PPARs have been cloned from *Xenopus* (Dreyer et al., 1992), thereby establishing the existence of multiple PPAR-related receptors. Structurally diverse peroxisome proliferators (Figure 5) activate these PPARs in transient trans-activation assays a manner that parallels their effectiveness at liver peroxisome proliferation (Dreyer et al., 1992; Issemann and Green, 1990). Synthetic fatty

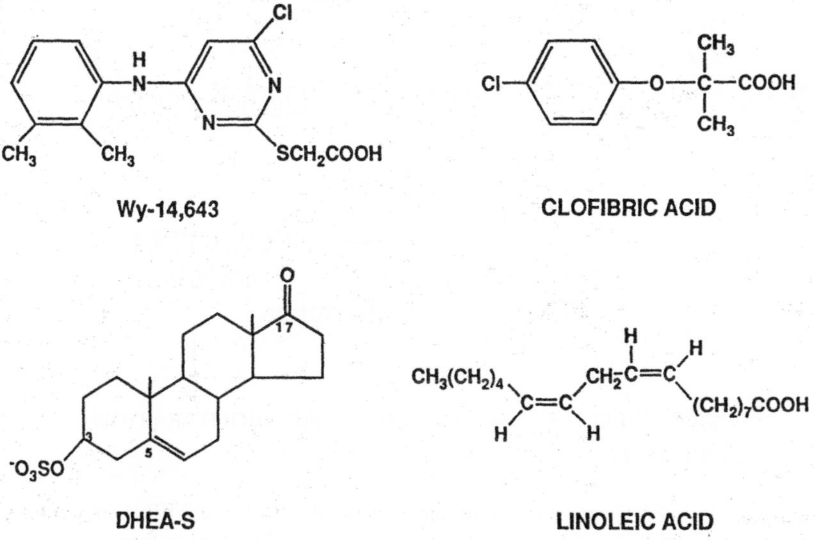

Figure 5. Chemical structures of peroxisome proliferators.

acids (e.g., 5,8,11,14-eicosatetraynoic acid) can also induce peroxisome proliferation (Keller et al., 1993), and interestingly, several naturally occurring fatty acids can also activate PPAR (Gottlicher et al., 1992; Schmidt et al., 1992). PPAR-dependent activation of peroxisomal and cytochrome *P450 4A* genes is mediated by peroxisome proliferator response elements (PPREs) that have been mapped to the 5′ flanks of these target genes (Dreyer et al., 1992; Muerhoff et al., 1992; Tugwood et al., 1992; Zhang et al., 1992) (Figure 6). While the physiological function(s) of PPAR are uncertain, the ontogenic profile of PPAR expression (Beck et al., 1992) suggests that these receptors may regulate peroxisome biogenesis during development (Stefanini et al., 1985) or in response to other physiological or environmental stimuli (Lazarow and Fujiki, 1985). This regulation could be achieved at the level of PPAR expression (Gebel et al., 1992) and/or by changes in the levels of endogenous PPAR activators.

The adrenal androgen DHEA (dehydroepiandrosterone) is a naturally occurring peroxisome proliferator (Wu et al., 1989; Yamada et al., 1991) and inducer of liver *CYP 4A* gene expression (Prough et al., 1994). It is distinguished from foreign

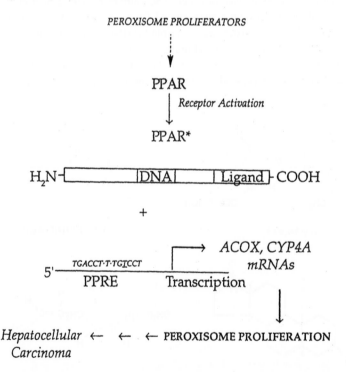

Figure 6. Role of peroxisome proliferator-activated receptor. (PPAR) in induction of CYP 4A and acyl CoA oxidase (*ACOX*) gene transcription leading to liver peroxisome proliferation and ultimately, development of hepatocellular carcinoma.

chemical peroxisome proliferators by its anti-carcinogenic and other therapeutic properties (Gordon et al., 1987; Schwartz et al., 1988) and by its apparent inability to activate both chimeric (Issemann and Green, 1990; Gottlicher et al., 1992) and wild type PPARα (Peters et al., 1996) in transient transfection experiments. The finding that DHEA is an active peroxisome proliferator when administered to intact animals but cannot activate PPAR in cultured cells suggests that this adrenal steroid may undergo metabolism *in vivo* to a more active derivative, which in turn mediates the activation of PPAR and the ensuing peroxisome proliferative response (Waxman, 1996). This possibility is supported by recent experiments showing that the 3β-sulfate of DHEA (DHEA-S) (Figure 5), which corresponds to the major circulating form of DHEA *in vivo*, readily induces the expression of CYP 4A and other genes associated with a peroxisome proliferative response in primary rat hepatocyte cultures (Ram and Waxman, 1994). It is significant that this effect occurs at physiologically relevant concentrations of DHEA-S (10μM) and under conditions where DHEA itself is inactive. DHEA-S could thus be a biologically important endogenous activator of PPAR. Moreover, given the PPAR multiplicity noted above, it is possible that only a subset of PPAR receptors is subject to activation by DHEA-S and related endogenous steroids. Indeed, despite its unreponsiveness to DHEA-S in cell culture, PPAR form α, the major liver-expressed form, *does* play an obligatory role in DHEA-S-induced CYP4A and peroxisomal enzyme induction, as demonstrated by the loss of these peroxisome proliferative responses in a PPARα gene knockout mouse model (Peters et al., 1996). Two other PPAR forms (PPARγ and PPAR Nucl/δ), whose genes are not disrupted in the PPARα-deficient mice (Lee et al., 1995), are apparently not required for a liver DHEA-S response. The observed inactivity of DHEA-S in PPARα *trans*-activation experiments, noted above, may reflect a requirement for liver cell-specific cellular uptake or perhaps a requirement for further metabolism that is met in intact animals *in vivo* but not in the heterologous cell culture transfection systems. Mammalian PPAR multiplicity could also provide a mechanism for the tissue-specific effects and the differential regulation of gene expression by the different structural classes of peroxisome proliferators (e.g., steroids vs. fatty acids vs. fibrate drugs and other classes of foreign chemicals).

Potent foreign chemical PPAR activators not only stimulate peroxisome proliferation in liver cells, but can induce hepatocellular carcinoma development following chronic exposure. The mechanistic basis for the proliferative actions of these carcinogenic peroxisome proliferators is poorly understood, but is hypothesized to involve their transcriptional activation of lipid-metabolizing enzymes, leading to the formation of DNA-damaging reduced oxygen species (Fahl et al., 1984; Reddy and Rao, 1989). Alternatively, the tumor-promoting activity of chemicals classified as peroxisome proliferators may not be directly related to their effects on peroxisomal metabolism, but perhaps may result from a more general effect on DNA replication or gene expression (Conway et al., 1989a; Conway et al., 1989b; Huber et al., 1991). Independent of which of these two models ultimately proves to be correct, it is clear that PPAR plays a central role in mediating the effects of structurally diverse peroxisome proliferators, including several potent hepatocarcinogens, on liver gene expression. Likely primary targets of PPAR are the acyl CoA oxidase gene (Osumi et al., 1991) and the cytochrome *P450 4A* genes (Kimura et al., 1989a). Both mRNAs encoded by the single acyl CoA oxidase gene (*ACOX-I* and *ACOX-II*) (Osumi et al., 1987) and three closely related rat *P450 4A* genes, designated 4A1, 4A2, and 4A3, are activated in liver and kidney by both fibrate (Kimura et al., 1989b; Sundseth and Waxman, 1992) and steroidal peroxisome proliferators (Ram and Waxman, 1994) via mechanisms that involve enhanced gene transcription (Hardwick et al., 1987; Prough et al., 1994; Reddy et al., 1986). The protein products of these genes catalyze fatty acid β-oxidation (acyl CoA oxidase) and the ω-hydroxylation of medium- and long-chain length fatty acid substrates (*P450 4A*) via pathways that can lead to formation of the dicarboxylic acids that may activate PPAR and play a crucial role in the early stages of peroxisome proliferation induced by fatty acyl compounds (Bell and Elcombe, 1991; Milton et al., 1990; Sharma et al., 1988). The expression of PPAR-related sequences in liver and kidney (Beck et al., 1992; Issemann and Green, 1990) parallels the fibrate inducibility of the acyl CoA oxidase and the *P450 4A* genes and peroxisome proliferation. This observation, together with the demonstrated inhibitory effect of cytochrome P450 4A inactivators on the fibrate induction of acyl CoA oxidase, but not *P450 4A* gene expression (Kaikaus et al., 1993) lends support to the concept of a mechanistic interrelationship between P450 4A induction, fatty acid metabolism, and peroxisome proliferation. On the other hand, the demonstrated ability of chemical inducers of peroxisome proliferation to stimulate reporter gene activity in heterologous systems via the binding of PPAR to an isolated PPRE derived from the 5'-flank of the acyl CoA oxidase gene (Dreyer et al., 1992; Kliewer et al., 1992; Tugwood et al., 1992) suggests that P450 4A metabolic activity might not be obligatory for PPAR activation and the peroxisome proliferative response. Further studies are needed to clarify some of these issues through an examination of the relationship between fibrate- and fatty acid-inducible peroxisome proliferations, and their relationship, in turn, to the steroid sulfate-inducible peroxisome proliferative pathway.

Important unanswered questions relating to the physiological effects of DHEA-S on CYP 4A and peroxisomal enzymes include:

1. Does DHEA-S, or one of its metabolites, correspond to an endogenous, high-affinity PPAR ligand? Does this putative PPAR activator play a regulatory role in peroxisome biogenesis?

2. Do auxiliary or accessory proteins modulate the responsiveness of PPAR to DHEA-S or other endogenous peroxisome proliferators compare to retinoid X-receptor (RXR) enhancement of PPAR binding to its cognate DNA response elements; (Kliewer et al., 1992)?

3. What is the relationship between endogenous steroidal and fatty acid PPAR activators? What role does PPAR multiplicity play in the biological response to these naturally-occurring CYP 4A and peroxisomal enzymes inducers?

4. Does DHEA-S bind directly to PPAR to stimulate *CYP 4A* gene expression, or does DHEA-S exert indirect effects on the cell, which perhaps lead to activation via covalent modification of PPAR, for instance by receptor phosphorylation (compare to phosphorylation leading to ligand-independent activation of the progesterone receptor [Beck et al., 1992; Denner et al., 1990])?

5. Is there a mechanistic link between PPAR activation, the resultant induction of cytochrome P450 4A enzyme activity and peroxisome proliferation? What roles do endogenous steroids play in this process? What is the underlying basis for the development of hepatocellular carcinoma in response to some peroxisome proliferators vs. the anticarcinogenic properties associated with endogenous steroidal peroxisome proliferators?

In conclusion, liver P450s catalyze the hydroxylation of steroid hormones and other important physiological regulators, in addition to their effects on cholesterol metabolism, bile acid synthesis and fatty acid metabolism. Each of the liver P450s enzymes and pathways is subject to unique regulatory controls. In the case of the sex-dependent steroid hormone hydroxylase P450s of rodent liver, GH secretory patterns and thyroid hormone levels are the most important endocrine regulators, but an underlying role for gonadal imprinting of adult hypothalamo-pituitary function is also evident. GH regulates the sex-specific expression of liver P450s through transcriptional mechanisms, while thyroid hormone exerts a multiplicity of effects: on P450 reductase, which is a rate-limiting component of the overall hydroxylation pathway, at the level of the pituitary, through its positive effects on GH secretion, and through its direct effects on expression of individual cytochrome *P450* genes. Finally, steroids such as DHEA-S can serve as endogenous activators of receptor proteins that mediate the induction of at least some classes of cytochrome P450 in liver tissue following exposure to foreign chemicals.

ABBREVIATIONS

CYP—cytochrome P450; 2-OH-E2—2-hydroxy-estradiol; 4-OH-E2—4-hydroxyestradiol; GH—growth hormone; PPAR—peroxisome proliferator-activated receptor; DHEA-S—dehydroepiandrosterone-3β-sulfate.

ACKNOWLEDGMENTS

Studies carried out in the author's laboratory were supported in part by N.I.H. grants DK33765 and ES07381.

REFERENCES

Andersson, S., Davis, D.L., Dahlback, H., Jornvall, H., & Russell, D.W. (1989). Cloning, structure, and expression of the mitochondrial cytochrome P450 sterol 26-hydroxylase, a bile acid biosynthetic enzyme. J. Biol. Chem. 264, 8222–8229.

Aoyama, T., Korzekwa, K., Nagata, K., Adesnik, M., Reiss, A., Lapenson, D.P., Gillette, J., Gelboin, H.V., Waxman, D.J., & Gonzalez, F.J. (1989). Sequence requirements for cytochrome P-450IIB1 catalytic activity. Alteration of the stereospecificity and regioselectivity of steroid hydroxylation by a simultaneous change of two hydrophobic amino acid residues to phenylalanine. J. Bio. Chem. 264, 21327–21333.

Aoyama, T., Korzekwa, K., Nagata, K., Gillette, J., Gelboin, H.V., & Gonzalez, F.J. (1990). Estradiol metabolism by complementary deoxyribonucleic acid-expressed human cytochrome P450s. Endocrinology 126, 3101–3106.

Argetsinger, L.S., Campbell, G.S., Yang, X., Witthuhn, B.A., Silvennoinen, O., Ihle, J.N., & Carter-Su, C. (1993). Identification of JAK2 as a growth hormone receptor-associated tyrosine kinase. Cell 74, 237–244.

Arlotto, M.P. & Parkinson, A. (1989). Identification of cytochrome P450a (P450IIA1) as the principal testosterone 7 alpha-hydroxylase in rat liver microsomes and its regulation by thyroid hormones. Arch. Biochem Biophys. 270, 458–471.

Arnold, A.P. (1984). Gonadal steroid induction of structural sex differences in the central nervous system. A. Rev. Neurosci. 7, 413–442.

Ball, S.E., Forrester, L.M., Wolf, C.R., & Back, D.J. (1990). Differences in the cytochrome P-450 isoenzymes involved in the 2-hydroxylation of oestradiol and 17 alpha-ethinyloestradiol. Relative activities of rat and human liver enzymes. Biochem. J. 267, 221–226.

Bandiera, S., Ryan, D.E., Levin, W., & Thomas, P.E. (1986). Age- and sex-related expression of cytochromes p450f and P450g in rat liver. Arch. Biochem. Biophys. 248, 658–676.

Barnea, E.R., MacLusky, N.J., & Naftolin, F. (1984). Catechol estrogens. In: Biochemical Actions of Hormones, Vol. 11, 267–307. 11. Academic Press, Inc., New Haven, CT.

Baxter, R.C., & Zaltsman, Z. (1984). Induction of hepatic receptors for growth hormone (GH) and prolactin by GH infusion is sex independent. Endocrinology 115, 2009–2014.

Beck, F., Plummer, S., Senior, P.V., Byrne, S., Green, S., & Brammar, W.J. (1992). The ontogeny of peroxisome-proliferator-activated receptor gene expression in the mouse and rat. Proc. R. Soc. Lond. [Biol.] 247, 83–87.

Bell, D.R., & Elcombe, C.R. (1991). Induction of acyl-CoA oxidase and cytochrome P450IVA1 RNA in rat primary hepatocyte culture by peroxisome proliferators. Biochem. J. 280, 249–253.

Bick, T., Youdim, M.B.H., & Hochberg, Z. (1989a). Adaptation of liver membrane somatogenic and lactogenic growth hormone (GH) binding to the spontaneous pulsation of GH secretion in the male rat. Endocrinology 125, 1711–1717.

Bick, T., Youdim, M.B.H., & Hochberg, Z. (1989b). The dynamics of somatogenic and lactogenic growth hormone binding: Internalization to Golgi fractions in the male rat. Endocrinology 125, 1718–1722.

Bjorkhem, I. (1985). Mechanism of bile acid biosynthesis in mammalian liver. In: Sterols and Bile Acids, (Danielsson, S.J., ed.), pp. 231–278. Elsevier, New York.

Brueggemeier, R.W. (1981). Kinetics of rat liver microsomal estrogen 2-hydroxylase. Evidence for sex differences at initial velocity conditions. J. Biol. Chem. 256, 10239–10242.

Bucala, R., Fishman, J., & Cerami, A. (1982). Formation of covalent adducts between cortisol and 16alpha-hydroxyestrone and protein: Possible role in the pathogenesis of cortisol toxicity and systemic lupus erythematosus. Proc. Natl. Acad. Sci. USA 79, 3320–3324.

Bulger, W. H., & Kupper, D. (1983). Effect of xenobiotic estrogens and structurally related compounds on 2-hydroxylation of estradiol and on other monooxygenase activities in rat liver. Biochem. Pharmacol. 32, 1005–1010.

Burbach, K.M., Poland, A., & Bradfield, C.A. (1992). Cloning of the Ah-receptor cDNA reveals a distinctive ligand-activated transcription factor. Proc. Natl. Acad. Sci. USA 89, 8185–8189.

Burger, H.J., Schuetz, J.D., Schuetz, E.G., & Guzelian, P.S. (1992). Paradoxical transcriptional activation of rat liver cytochrome P-450 3A1 by dexamethasone and the antiglucocorticoid pregnenolone 16 alpha-carbonitrile: Analysis by transient transfection into primary monolayer cultures of adult rat hepatocytes. Proc. Natl. Acad. Sci. USA 89, 2145–2149.

Cheng, K.C. & Schenkman, J.B. (1983). Testosterone metabolism by cytochrome P-450 isozymes RLM3 and RLM5 and by microsomes. Metabolite identification. J. Biol. Chem. 258, 11738–11744.

Colby, H.D. (1980). Regulation of hepatic drug and steroid metabolism by androgens and estrogens. Adv. Sex Horm. Res. 4, 27–71.

Conney, A.H. (1967). Pharmacological implications of microsomal enzyme induction. Pharmacol. Rev. 19, 317–366.

Conney, A.H. (1982). Induction of microsomal enzymes by foreign chemicals & carcinogenesis by polycyclic aromatic hydrocarbons: G.H.A. Clowes memorial lecture. Cancer Res. 42, 4875–4917.

Conway, J.G., Cattley, R.C., Popp, J.A., & Butterworth, B.E. (1989a). Possible mechanisms in hepatocarcinogenesis by the peroxisome proliferator di(2-ethylhexyl)phthalate. Drug Metab. Rev. 21, 65–102.

Conway, J.G., Tomaszewski, K.E., Olson, M.J., Cattley, R.C., Marsman, D.S., & Popp, J.A. (1989b). Relationship of oxidative damage to the hepatocarcinogenicity of the peroxisome proliferators di(2-ethylhexyl)phthalate and Wy-14,643. Carcinogenesis 10, 513–519.

Cunningham, B.C., Ultsch, M., de Vos, A.M., Mulkerrin, M.G., Clauser, K.R., & Wells, J.A. (1991). Dimerization of the extracellular domain of the human growth hormone receptor by a single hormone molecule. Science 254, 821–825.

Dannan, G.A., Guengerich, F.P., & Waxman, D.J. (1986a). Hormonal regulation of rat liver microsomal enzymes. Role of gonadal steroids in programming, maintenance, and suppression of delta 4-steroid 5 alpha-reductase, flavin-containing monooxygenase, and sex-specific cytochromes P-450. J. Biol. Chem. 261, 10728–10735.

Dannan, G.A., Porubek, D.J., Nelson, S.D., Waxman, D.J., & Guengerich, F.P. (1986b). 17 beta-estradiol 2- and 4-hydroxylation catalyzed by rat hepatic cytochrome P-450: Roles of individual forms, inductive effects, developmental patterns, and alterations by gonadectomy and hormone replacement. Endocrinology 118, 1952–1960.

Denner, L.A., Weigel, N.L., Maxwell, B.L., Schrader, W.T., & O'Malley, B.W. (1990). Regulation of progesterone receptor-mediated transcription by phosphorylation. Science 250, 1740–1743.

Doglio, A., Dani, C., Grimaldi, P., & Ailhaud, G. (1989). Growth hormone stimulates c-fos gene expression by means of protein kinase C without increasing inositol lipid turnover. Proc. Natl. Acad. Sci. USA 86, 1148–1152.

Dreyer, C., Krey, G., Keller, H., Givel, F., Helftenbein, G., & Wahli, W. (1992). Control of the peroxisomal beta-oxidation pathway by a novel family of nuclear hormone receptors. Cell 68, 879–887.

Fahl, W.E., Lalwani, N.D., Watanabe, T., Goel, S.K., & Reddy, J.K. (1984). DNA damage related to increased hydrogen peroxide generation by hypolipidemic drug-induced liver peroxisomes. Proc. Natl. Acad. Sci. USA 81, 7827–7830.

Fisher, D., Labbe, G., Berson, A., Tinel, M., Loeper, J., Larrey, D., & Pessayre, D. (1990). Inhibition of rat liver estrogen 2/4-hydroxylase activity by troleandomycin: Comparison with erythromycin and roxithromycin. J. Pharmacol. Exp. Ther. 254, 1120–1127.

Fishman, J. (1983). Aromatic hydroxylation of estrogens. Ann. Rev. Physiol. 45, 61–72.

Fishman, J., Bradlow, H.L., Schneider, J., Anderson, K.E., & Kappas, A. (1980). Radiometric analysis of biological oxidations in man: Sex differences in estradiol metabolism. Proc. Natl. Acad. Sci. USA 77, 4957–4960.

Gebel, T., Arand, M., & Oesch, F. (1992). Induction of the peroxisome proliferator activated receptor by fenofibrate in rat liver. FEBS Lett. 309, 37–40.

Ged, C., Rouillon, J.M., Pichard, L., Combalbert, J., Bressot, N., Bories, P., Michel, H., Beaune, P., & Maurel, P. (1989). The increase in urinary excretion of 6 beta-hydroxycortisol as a marker of human hepatic cytochrome P450IIIA induction. Br. J. Clin. Pharmacol. 28, 373–387.

Gonzalez, F.J., Crespi, C.L., & Gelboin, H.V. (1991). cDNA-expressed human cytochrome P450s: a new age of molecular toxicology and human risk assessment. Mutation Research 247, 113–127.

Gonzalez, F.J., Kimura, S., Song, B.J., Pastewka, J., Gelboin, H.V., & Hardwick, J.P. (1986a). Sequence of two related P-450 mRNAs transcriptionally increased during rat development. An R.dre.1 sequence occupies the complete 3′ untranslated region of a liver mRNA. J. Biol. Chem. 261, 10667–10672.

Gonzalez, F.J., Song, B.J., & Hardwick, J.P. (1986b). Pregnenolone 16 alpha-carbonitrile-inducible P450 gene family: Gene conversion and differential regulation. Mol. Cell Biol. 6, 2969–2976.

Gordon, G.B., Shantz, L.M., & Talalay, P. (1987). Modulation of growth, differentiation and carcinogenesis by dehydroepiandrosterone. Adv. Enz. Reg. 26, 355–382.

Gottlicher, M., Widmark, E., Li, Q., & Gustafsson, J.A. (1992). Fatty acids activate a chimera of the clofibric acid-activated receptor and the glucocorticoid receptor. Proc. Natl. Acad. Sci. USA 89, 4653–4657.

Graham, M.J., Lucier, G.W., Linko, P., Maronpot, R.R., & Goldstein, J.A. (1988). Increases in cytochrome P-450 mediated 17 beta-estradiol 2-hydroxylase activity in rat liver microsomes after both acute administration and subchronic administration of 2,3,7,8-tetrachlorodibenzo-p-dioxin in a two-stage hepatocarcinogenesis model. Carcinogenesis 9, 1935–1941.

Green, S. (1992). Receptor-mediated mechanisms of peroxisome proliferators. Biochem. Pharmacol. 43, 393–401.

Guengerich, F.P. (1987). Enzymology of rat liver cytochromes P450. In: Mammalian Cytochromes P450 (Guengerich, F.P., ed.), pp. 1–54. CRC Press, Boca Raton, FL.

Guengerich, F.P. & Shimada, T. (1991). Oxidation of toxic and carcinogenic chemicals by human cytochrome P-450 enzymes. Chem. Res. Toxicol. 4, 391–407.

Gurland, G., Ashcom, G., Cochran, B.H., & Schartz, J. (1990). Rapid events in growth hormone action. Induction of c-fos and c-jun transcription in 3T3-F442A preadipocytes. Endocrinology 127, 3187–3195.

Gustafsson, J.A., Mode, A., Norstedt, G., & Skett, P. (1983). Sex steroid induced changes in hepatic enzymes. A Rev. Physiol. 45, 51–60.

Guzelian, P.S., Li, D., Schuetz, E.G., Thomas, P., Levin, W., Mode, A., & Gustafsson, J.A. (1988). Sex change in cytochrome P-450 phenotype by growth hormone treatment of adult rat hepatocytes maintained in a culture system on matrigel. Proc. Natl. Acad. Sci. USA 85, 9783–9787.

Harada, N., & Negishi, M. (1988). Substrate specificities of cytochrome P-450, C-P-450(16)alpha and P-450(15)alpha, and contribution to steroid hydroxylase activities in mouse liver microsomes. Biochem. Pharmacol. 37, 4778–4780.

Hardwick, J.P., Song, B.J., Huberman, E., & Gonzalez, F.J. (1987). Isolation, complementary DNA sequence, and regulation of rat hepatic lauric acid omega-hydroxylase (cytochrome P450 LA omega): Identification of a new cytochrome P450 gene family. J. Biol. Chem. 262, 801–810.

Hochberg, Z., Bick, T., & Harel, Z. (1990). Alterations of human growth hormone binding by rat liver membranes during hypo- and hyperthyroidism. Endocrinology 126, 325–329.

Huber, W., Kraupp, G.B., Esterbauer, H., & Schulte, H.R. (1991). Role of oxidative stress in age dependent hepatocarcinogenesis by the peroxisome proliferator nafenopin in the rat. Cancer Res. 51, 1789–1792.

Hunt, C.M., Watkins, P.B., Saenger, P., Stave, G.M., Barlascini, N., Watlington, C.O., Wright, J.J., & Guzelian, P.S. (1992). Heterogeneity of CYP3A isoforms metabolizing erythromycin and cortisol. Clin. Pharmacol. Ther. 51, 18–23.

Inano, H., Suzuki, K., Wakabayashi, K., & Tamaoki, B.I. (1973). Biological activities of 7 alpha-hydroxylated C19-steroids and changes in rat testicular 7 alpha-hydroxylase activity with gonadal status. Endocrinology 92, 22–30.

Issemann, I., & Green, S. (1990). Activation of a member of the steroid hormone receptor superfamily by peroxisome proliferators. Nature 347, 645–650.

Janeczko, R., Waxman, D.J., Le, B.G., Morville, A., & Adesnik, M. (1990). Hormonal regulation of levels of the messenger RNA encoding hepatic P450 2c (IIC11) a constitutive male-specific form of cytochrome P450. Mol. Endocrinol. 4, 295–303.

Jansson, J.O., Ekberg, S., & Isaksson, O. (1985). Sexual dimorphism in the control of growth hormone secretion. Endocrine Rev. 6, 128–150.

Jellinck, P.H., & Fishman, J. (1988). Activation and irreversible binding of regiospecifically labeled catechol estrogen by rat liver microsomes: Evidence for differential cytochrome P-450 catalyzed oxidations. Biochemistry 27, 6111–6116.

Johnson, R.M., Napier, M.A., Cronin, M.J., & King, K.L. (1990). Growth hormone stimulates the formation of sn-1,2-diacylglycerol in rat hepatocytes. Endocrinology 127, 2099–2103.

Kaikaus, R.M., Chan, W.K., Lysenko, N., Ray, R., Ortiz de Montellano, P.R., & Bass, N.M. (1993). Induction of peroxisomal fatty acid beta-oxidation and liver fatty acid binding protein by peroxisome proliferators: Mediation via the cytochrome P450IVA1 omega-hydroxylase pathway. J. Biol. Chem. 268, 9593–9603.

Kato, R. (1974). Sex-related differences in drug metabolism. Drug Metab. Rev. 3, 1–32.

Kato, R., Yamazoe, Y., Shimada, M., Murayama, N., & Kamataki, T. (1986). Effect of growth hormone and ectopic transplantation of pituitary gland on sex-specific forms of cytochrome P450 and testosterone and drug oxidations in rat liver. J Biochem. 100, 895–902.

Keller, H., Dreyer, C., Medin, J., Mahfoudi, A., Ozato, K., & Wahli, W. (1993). Fatty acids and retinoids control lipid metabolism through activation of peroxisome proliferator-activated receptor-retinoid X receptor heterodimers. Proc. Natl. Acad. Sci. USA 90, 2160–2164.

Kimura, S., Hanioka, N., Matsunaga, E., & Gonzalez, F.J. (1989a). The rat clofibrate-inducible CYP4A gene subfamily I. Complete intron and exon sequence of the CYP4A1 and CYP4A2 genes, unique exon organization, and identification of a conserved 19-bp upstream element. DNA 8, 503–516.

Kimura, S., Hardwick, J.P., Kozak, C.A., & Gonzalez, F.J. (1989b). The rat clofibrate-inducible CYP4A subfamily II. cDNA sequence of IVA3, mapping of the Cyp4a locus to mouse chromosome 4, and coordinate and tissue-specific regulation of the CYP4A genes. DNA 8, 517–525.

Kliewer, S.A., Umesono, K., Noonan, D.J., Heyman, R.A., & Evans, R.M. (1992). Convergence of 9-cis retinoic acid and peroxisome proliferator signalling pathways through heterodimer formation of their receptors. Nature 358, 771–774.

Kronbach, T., Larabee, T.M., & Johnson, E.F. (1989). Hybrid cytochromes P-450 identify a substrate binding domain in P-450IIC5 and P-450IIC4. Proc. Natl. Acad. Sci. USA 86, 8262–8265.

Kupfer, D., Miranda, G.K., & Bulger, W.H. (1981). A facile assay for 2-hydroxylation of estradiol by liver microsomes. Anal. Biochem. 116, 27–34.

Lahita, R.G., Bradlow, H.L., Kunkel, H.G., & Fishman, J. (1981). Increased 16alpha-hydroxylation of estradiol in systemic lupus erythematosus. J Clin. Endocrinol. Metab. 53, 174–178.

Lazarow, P.B., & Fujiki, Y. (1985). Biogenesis of peroxisomes. Ann. Rev. Cell Biol. 1, 489–530.

Lee, S.S., Pineau, T., Drago, J., Lee, E.J., Owens, J.W., Kroetz, D.L., Fernandez-Salguero, P.M., Westphal, H., & Gonzalez, F.J. (1995). Targeted disruption of the alpha isoform of the peroxisome proliferator-activated receptor gene in mice results in abolishment of the pleiotropic effects of peroxisome proliferators. Mol. Cell. Biol. 15, 3012–3022.

Legraverend, C., Mode, A., Wells, T., Robinson, I., & Gustafsson, J.A. (1992a). Hepatic steroid hydroxylating enzymes are controlled by the sexually dimorphic pattern of growth hormone secretion in normal and dwarf rats. Faseb. J. 6, 711–718.

Legraverend, C., Mode, A., Westin, S., Strom, A., Eguchi, H., Zaphiropoulos, P.G., & Gustafsson, J.A. (1992b). Transcriptional regulation of rat P-450 2C gene subfamily members by the sexually dimorphic pattern of growth hormone secretion. Mol. Endocrinol. 6, 259–266.

Leo, M.A., Iida, S., & Lieber, C.S. (1984). Retinoic acid metabolism by a system reconstituted with cytochrome P-450. Arch. Biochem. Biophys. 234, 305–312.

Leung, D.W., Spencer, S.A., Cachianes, G., Hammonds, R.G., Collins, C., Henzel, W.J., Barnard, R., Waters, M.J., & Wood, W.I. (1987). Growth hormone receptor and serum binding protein: Purification, cloning and expression. Nature 330, 537–543.

Levin, W., Thomas, P.E., Ryan, D.E., & Wood, A.W. (1987). Isozyme specificity of testosterone 7 alpha-hydroxylation in rat hepatic microsomes: Is cytochrome P-450a the sole catalyst? Arch. Biochem. Biophys. 258, 630–635.

Li, J.J., & Li, S.A. (1987). Estrogen carcinogenesis in Syrian hamster tissues: Role of metabolism. Fed. Proc. 46, 1858–1863.

MacGeoch, C., Morgan, E.T., & Gustafsson, J.A. (1985). Hypothalamo-pituitary regulation of cytochrome P450(15) beta apoprotein levels in rat liver. Endocrinology 117, 2085–2092.

Malee, M.P., & Mellon, S.H. (1991). Zone-specific regulation of two messenger RNAs for P450c11 in the adrenals of pregnant and nonpregnant rats. Proc. Natl. Acad. Sci. USA 88, 4731–4735.

Mathews, L.S., Enberg, B., & Norstedt, G. (1989). Regulation of rat growth hormone receptor gene expression. J. Biol. Chem. 264, 9905–9910.

McClellan, G.P., Linko, P., Yeowell, H.N., & Goldstein, J.A. (1989). Hormonal regulation of male-specific rat hepatic cytochrome P-450g (P-450IIC13) by androgens and the pituitary. J. Biol. Chem. 264, 18960–18965.

McClellan, G.P., Waxman, D.J., Caveness, M., & Goldstein, J.A. (1987). Phenotypic differences in expression of cytochrome P-450g but not its mRNA in outbred male Sprague-Dawley rats. Arch. Biochem. Biophys. 253, 13–25.

Michnovicz, J.J., Hershcopf, R.J., Naganuma, H., Bradlow, H.L., & Fishman, J. (1986). Increased 2-hydroxylation of estradiol as a possible mechanism for the anti-estrogenic effect of cigarette smoking. New Eng. J. Med. 315, 1305–1309.

Milton, M.N., Elcombe, C.R., & Gibson, G.G. (1990). On the mechanism of induction of microsomal cytochrome P450IVA1 and peroxisome proliferation in rat liver by clofibrate. Biochem. Pharmacol. 40, 2727–2732.

Mittler, J.C. (1985). Studies on the kinetics of the interaction of 7 alpha-hydroxy-testosterone with the steroid 5 alpha-reductase. Steroids 45, 153–239.

Mode, A., Gustafsson, J.A., Jansson, J.O., Eden, S., & Isaksson, O. (1982). Association between plasma level of growth hormone and sex differentiation of hepatic steroid metabolism in the rat. Endocrinology 111, 1692–1697.

Mode, A., Wiersma-Larsson, E., & Gustafsson, J.A. (1989). Transcriptional and posttranscriptional regulation of sexually differentiated rat liver cytochrome P450 by growth hormone. Molec. Endocr. 3, 1142–1147.

Morgan, E.T., MacGeoch, C., & Gustafsson, J.A. (1985). Hormonal and developmental regulation of expression of the hepatic microsomal steroid 16alpha-hydroxylase cytochrome P-450 apoprotein in the rat. J. Biol. Chem. 260, 11895–11898.

Morishima, N., Yoshioka, H., Higashi, Y., Sogawa, K., & Fujii-Kuriyama, Y. (1987). Gene structure of cytochrome P450(M-1) specifically expressed in male rat liver. Biochemistry 26,

Muerhoff, A.S., Griffin, K.J., & Johnson, E.F. (1992). The peroxisome proliferator-activated receptor mediates the induction of CYP4A6, a cytochrome P450 fatty acid w-hydroxylase, by clofibric acid. J. Biol. Chem. 267, 19051–19053.

Mukai, K., Imai, M., Shimada, H., & Ishimura, Y. (1993). Isolation and characterization of rat CYP11B genes involved in late steps of mineralo- and glucocorticoid syntheses. J. Biol. Chem. 268, 9130–9137.

Negishi, M., Iwasaki, M., Juvonen, R.O., & Aida, K. (1992). Alteration of the substrate specificity of mouse 2A P450s by the identity of residue-209: Steroid-binding site and orientation. J. Steroid Biochem. Molec. Biol. 43, 1031–1036.

Nelson, D.R., Kamataki, T., Waxman, D.J., Guengerich, F.P., Estabrook, R.W., Feyereisen, R., Gonzalez, F.J., Coon, M.J., Gunsalus, I.C., Gotoh, O., et al. (1993). The P450 superfamily: Update on new sequences, gene mapping, accession numbers, early trivial names of enzymes, and nomenclature. DNA Cell Biol. 12, 1–51.

Noshiro, M., & Negishi, M. (1986). Pretranslational regulation of sex-dependent testosterone hydroxylases by growth hormone in mouse liver. J. Biol. Chem. 261, 15,923–927.

Numazawa, M., Satoh, S., Ogura, Y., & Nagaoka, M. (1985). Determination of estradiol 2- and 16alpha-hydroxylase activities in rat liver microsomes using high-performance liquid chromatography. Anal. Biochem. 149, 409–414.

Ogishima, T., Shibata, H., Shimada, H., Mitani, F., Suzuki, H., Saruta, T., & Ishimura, Y. (1991). Aldosterone synthase cytochrome P450 expressed in the adrenals of patients with primary aldosteronism. J. Biol. Chem. 266, 10731–10734.

Osumi, T., Ishii, N., Miyazawa, S., & Hashimoto, T. (1987). Isolation and structural characterization of the rat acyl-CoA oxidase gene. J. Biol. Chem. 262, 8138–8143.

Osumi, T., Wen, J.K., & Hashimoto, T. (1991). Two cis-acting regulatory sequences in the peroxisome proliferator-responsive enhancer region of rat acyl-CoA oxidase gene. Biochem. Biophys. Res. Commun. 175, 866–871.

Peters, J.M., Zhou, Y.-C., Ram, P.A., Lee, S.S.T., Gonzalez, F.J., & Waxman, D.J. (1996). Peroxisome proliferator-activated receptor alpha mediates gene induction by dehydroepiandrosterone 3beta-sulfate. Molec. Pharm. (in press).

Prough, R.A., Webb, S.J., Wu, H.Q., Lapenson, D.P., & Waxman, D.J. (1994). Induction of microsomal and peroxisomal enzymes by dehydroepiandrosterone and its reduced metabolite in rats. Cancer Res. 54, 2878–2886.

Quail, J.A., & Jellinck, P.H. (1987). Modulation of catechol estrogen synthesis by rat liver microsomes: Effects of treatment with growth hormone or testosterone. Endocrinology 121, 987–992.

Ram, P.A., & Waxman, D.J. (1990). Pretranslational control by thyroid hormone of rat liver steroid 5 alpha-reductase and comparison to the thyroid dependence of two growth hormone-regulated CYP2C mRNAs. J. Biol. Chem. 265, 19223–19229.

Ram, P.A., & Waxman, D.J. (1991). Hepatic P450 expression in hypothyroid rats: Differential responsiveness of male-specific P450 forms 2a (IIIA2), 2c (IIC11), and RLM2 (IIA2) to thyroid hormone. Mol. Endocrinol. 5, 13–20.

Ram, P.A., & Waxman, D.J. (1992). Thyroid hormone stimulation of NADPH P450 reductase expression in liver and extrahepatic tissues. Regulation by multiple mechanisms. J. Biol. Chem. 267, 3294–3301.

Ram, P.A., & Waxman, D.J. (1993). Induction of cytochrome P450 4A (CYP) and acyl CoA oxidase mRNAs by dehydroepiandrosterone sulfate (DHEA-S) in primary rat hepatocyte culture. Endocrin. Soc. (Abstr.), in press.

Ram, P.A., & Waxman, D.J. (1994). Dehydroepiandrosterone 3 beta-sulphate is an endogenous activator of the peroxisome-proliferation pathway: induction of cytochrome P-450 4A and acyl-CoA oxidase mRNAs in primary rat hepatocyte culture and inhibitory effects of Ca(2+)-channel blockers. Biochem. J. 301, 753–758.

Ram, P.A., Park, S.-H., Cho, H.K., & Waxman, D.J. (1996). Growth hormone activation of Stat 1, Stat 3, and Stat 5 in rat liver. Differential kinetics of hormone desensitization and growth hormone stimulation of both tyrosine phosphorylation and serine/threonine phosphorylation. J. Biol. Chem. 271, 5929–5940.

Rao, M.S., & Reddy, J.K. (1987). Peroxisome proliferation and hepatocarcinogenesis. Carcinogenesis 8, 631–636.

Reddy, J.K., Goel, S.K., Nemali, M.R., Carrino, J.J., Laffler, T.G., Reddy, M.K., Sperbeck, S.J., Osumi, T., Hashimoto, T., Lalwani, N.D., & Rao, M.S. (1986). Transcriptional regulation of peroxisomal fatty acyl-CoA oxidase and enoyl-CoA hydratase/3-hydroxyacyl-CoA dehydrogenase in rat liver by peroxisome proliferators. Proc. Natl. Acad. Sci. USA 83, 1747–1751.

Reddy, J.K., & Rao, M.S. (1989). Oxidative DNA damage caused by persistent peroxisome proliferation: Its role in hepatocarcinogenesis. Mutat. Res. 214, 63–68.

Roupas, P., & Herington, A.C. (1989). Cellular mechanisms in the processing of growth hormone and its receptor. Molec. Cell. Endocr. 61, 1–12.

Ryan, D.E., & Levin, W. (1990). Purification and characterization of hepatic microsomal cytochrome P-450. Pharmacol. Ther. 45, 153–239.

Ryan, D.E., Thomas, P.E., Reik, L.M., & Levin, W. (1982). Purification, characterization and regulation of five rat hepatic microsomal cytochrome P450 isozymes. Xenobiotica 12, 727–744.

Saenger, P., Forster, E., & Kream, J. (1981). 6-Beta-hydroxycortisol: A noninvasive indicator of enzyme induction. J. Clin. Endocr. Metab. 52, 381–384.

Samuels, H.H., Forman, B.M., Horowitz, Z.D., & Ye, Z.S. (1988). Regulation of gene expression by thyroid hormone. J. Clin. Invest. 81, 957–967.

Sasamura, H., Nagata, K., Yamazoe, Y., Shimada, M., Saruta, T., & Kato, R. (1990). Effect of growth hormone on rat hepatic cytochrome P-450f mRNA: a new mode of regulation. Mol. Cell Endocrinol. 68, 53–60.

Schmidt, A., Endo, N., Rutledge, S.J., Vogel, R., Shinar, D., & Rodan, G.A. (1992). Identification of a new member of the steroid hormone receptor superfamily that is activated by a peroxisome proliferator and fatty acids. Mol. Endocrinol. 6, 1634–1641.

Schneider, J., Kinne, D., Fracchia, A., Pierce, V., Anderson, K.E., Bradlow, H.L., & Fishman, J. (1982). Abnormal oxidative metabolism of estradiol in women with breast cancer. Proc. Natl. Acad. Sci. USA 79, 3047–3051.

Schuetz, E.G., Wrighton, S.A., Barwick, J.L., & Guzelian, P.S. (1984). Induction of cytochrome P450 by glucocorticoids in rat liver. I. Evidence that glucocorticoids and pregnenolone 16 alpha-car-bonitrile regulate de novo synthesis of a common form of cytochrome P450 in cultures of adult rat hepatocytes and in the liver in vivo. J. Biol. Chem. 259, 1999–2006.

Schwab, G.E., & Johnson, E.F. (1985). Variation in hepatic microsomal cytochrome P-450 1 concentration among untreated rabbits alters the efficiency of estradiol hydroxylation. Arch. Biochem. Biophys. 237, 17–26.

Schwartz, A.G., Whitcomb, J.M., Nyce, J.W., Lewbar, M.L., & Pashko, L.L. (1988). Dehydroepian-drosterone and structural analogs: A new class of cancer chemopreventive agents. Adv. Cancer Res. 51, 391–424.

Shapiro, B.H., MacLeod, J.N., Pampori, N.A., Morrissey, J.J., Lapenson, D.P., & Waxman, D.J. (1989). Signaling elements in the ultradian rhythm of circulating growth hormone regulating expression of sex-dependent forms of hepatic cytochrome P450. Endocrinology 125, 2935–2944.

Shapiro, B.H., Pampori, N.A., Lapenson, D.P., & Waxman, D.J. (1994). Growth hormone-dependent and -independent sexually dimorphic regulation of phenobarbital-induced hepatic cytochromes P450 2B1 and 2B2. Arch. Biochem. Biophys. 312, 234–239.

Sharma, R., Lake, B.G., Foster, J., & Gibson, G.G. (1988). Microsomal cytochrome P452 induction and peroxisome proliferation by hypolipidaemic agents in rat liver. A mechanistic inter-relationship. Biochem. Pharmacol. 37, 1193–1201.

Shaw, G.C., & Fulco, A.J. (1992). Barbiturate-mediated regulation of expression of the cytochrome P450BM-3 gene of Bacillus megaterium by Bm3R1 protein. J. Biol. Chem. 267, 5515–5526.

Shaw, G.C., & Fulco, A.J. (1993). Inhibition by barbiturates of the binding of Bm3R1 repressor to its operator site on the barbiturate-inducible cytochrome P450BM-3 gene of Bacillus megaterium. J. Biol. Chem. 268, 2997–3004.

Shaw, P.M., Adesnik, M., Weiss, M.C., & Corcos, L. (1993). The phenobarbital-induced transcriptional activation of cytochrome P450 genes is blocked by the glucocorticoid-progesterone antagonist RU486. Molec. Pharmacol. 44, 775–783.

Sher, T., Yi, H.F., McBride, O.W., & Gonzalez, F.J. (1993). cDNA cloning, chromosomal mapping, and functional characterization of the human peroxisome proliferator activated receptor. Biochemistry 32, 5598–5604.

Shimada, M., Murayama, N., Yamazoe, Y., Kamataki, T., & Kato, R. (1987). Further studies on the persistence of neonatal androgen imprinting on sex-specific cytochrome P450, testosterone and drug oxidations. Jpn. J. Pharmacol. 45, 467–478.

Silva, C.M., Weber, M.J., & Thorner, M.O. (1993). Stimulation of tyrosine phosphorylation in human cells by activation of the growth hormone receptor. Endocrinology 132, 101–108.

Skett, P. (1987). Hormonal regulation and sex differences of xenobiotic metabolism. Prog. Drug Metab. 10, 85–139.

Sonderfan, A.J., Arlotto, M.P., Dutton, D.R., McMillen, S.K., & Parkinson, A. (1987). Regulation of testosterone hydroxylation by rat liver microsomal cytochrome P-450. Arch. Biochem. Biophys. 255, 27–41.

Spink, D.C., Eugster, H.P., Lincoln II, D.W., Schuetz, J.D., Schuetz, E.G., Johnson, J.A., Kaminsky, L.S., & Gierthy, J.F. (1992). 17beta-estradiol hydroxylation catalyzed by human cytochrome P450 1A1: A comparison of the activities induced by 2,3,7,8-tetrachlorodibenzo-p-dioxin in MCF-7 cells with those from heterologous expression of the cDNA. Arch. Biochem. Biophy. 293, 342–348.

Squires, E. J., & Negishi, M. (1988). Reciprocal regulation of sex-dependent expression of testosterone 15 alpha-hydroxylase (P450(15 alpha)) in liver and kidney of male mice by androgen. Evidence for a single gene. J. Biol. Chem. 263, 4166–4171.

Stefanini, S., Ferrace, M.G., & Argento, M.P.C. (1985). Differentiation of liver peroxisomes in the foetal and newborn rat. Cytochemistry of catalase and D-aminoacid oxidase. J. Embryol. Exp. Morph. 88, 151–163.

Subramanian, A., Teixeira, J., J., W., & Gil, G. (1995). A STAT factor mediates the sexually dimorphic regulation of hepatic cytochrome P450 3A10/lithocholic acid 6beta-hydroxylase gene expression by growth hormone. Molec. Cell Biol. 15, 4672–4682.

Sundseth, S.S., Alberta, J.A., & Waxman, D.J. (1992). Sex-specific, growth hormone-regulated transcription of the cytochrome P450 2C11 and 2C12 genes. J. Biol. Chem. 267, 3907–3914.

Sundseth, S.S., & Waxman, D.J. (1992). Sex-dependent expression and clofibrate inducibility of cytochrome P450 4A fatty acid omega-hydroxylases. Male specificity of liver and kidney CYP4A2 mRNA and tissue-specific regulation by growth hormone and testosterone. J. Biol. Chem. 267, 3915–3921.

Swinney, D.C., Ryan, D.E., Thomas, P.E., & Levin, W. (1987). Regioselective progesterone hydroxylation catalyzed by eleven rat hepatic cytochrome P-450 isozymes. Biochemistry 26, 7073–7083.

Theron, C.N., Neethling, A.C., & Taljard, J.J.F. (1983). Ontogeny and sexual dimorphism of estradiol-2-hydroxylase activity in rat liver microsomes. J. Steroid Biochem. 19, 1185–1190.

Thummel, K.E., Favreau, L.V., Mole, J.E., & Schenkman, J.B. (1988). Further characterization of RLM2 and comparison with a related form of cytochrome P450, RLM2b. Archs. Biochem. Biophys. 266, 319–333.

Tollet, P., Enberg, B., & Mode, A. (1990). Growth hormone (GH) regulation of cytochrome P-450IIC12, insulin-like growth factor-I (IGF-I), and GH receptor messenger RNA expression in primary rat hepatocytes: A hormonal interplay with insulin, IGF-I, and thyroid hormone. Mol. Endocrinol 4, 1934–1942.

Tollet, P., Legraverend, C., Gustafsson, J.A., & Mode, A. (1991). A role for protein kinases in the growth hormone regulation of cytochrome P4502C12 and insulin-like growth factor-I messenger RNA expression in primary adult rat hepatocytes. Mol. Endocrinol 5, 1351–1358.

Tugwood, J.D., Issemann, I., Anderson, R.G., Bundell, K.R., McPheat, W.L., & Green, S. (1992). The mouse peroxisome proliferator activated receptor recognizes a response element in the 5′ flanking sequence of the rat acyl CoA oxidase gene. EMBO J. 11, 433–439.

Wang, X., Moller, C., Norstedt, G., & Carter-Su, C. (1993). Growth hormone-promoted tyrosyl phosphorylation of a 121-kDa growth hormone receptor-associated protein. J. Biol. Chem. 268, 3573–3579.

Watanabe, K., Takanashi, K., Imaoka, S., Funae, Y., Kawano, S., Inoue, K., Kamataki, T., Takagi, H., & Yoshizawa, I. (1991). Comparison of cytochrome P-450 species which catalyze the hydroxylations of the aromatic ring of estradiol and estradiol 17-sulfate. J. Steroid Biochem. Mol. Biol. 38, 737–743.

Waxman, D., Ko, A., & Walsh, C. (1983). Regioselectivity and stereospecificity of androgen hydroxylations catalyzed by cytochrome P450 isozymes purified from phenobarbital-induced rat liver. J. Biol. Chem. 258, 11937–11947.

Waxman, D.J. (1984). Rat hepatic cytochrome P-450 isoenzyme 2c. Identification as a male-specific, developmentally induced steroid 16 alpha-hydroxylase and comparison to a female-specific cytochrome P450 isoenzyme. J. Biol. Chem. 259, 15481–15490.

Waxman, D.J. (1988). Interactions of hepatic cytochromes P-450 with steroid hormones. Regioselectivity and stereospecificity of steroid metabolism and hormonal regulation of rat P-450 enzyme expression. Biochem. Pharmacol. 37, 71–84.

Waxman, D.J. (1991). Rat hepatic P450IIA and P450IIC subfamily expression using catalytic, immunochemical, and molecular probes. Methods Enzymol. 206, 249–267.

Waxman, D.J. (1992). Regulation of liver-specific steroid metabolizing cytochromes P450: Cholesterol 7alpha-hydroxylase, bile acid 6beta-hydroxylase, and growth hormone-responsive steroid hormone hydroxylases. J. Ster. Biochem. Molec. Biol. 43, 1055–1072.

Waxman, D.J. (1996). Role of metabolism in the activation of dehyroepiandrosterone as a peroxisome proliferator. J. Endocrinol. (in press).

Waxman, D.J., Attisano, C., Guengerich, F.P., & Lapenson, D.P. (1988a). Human liver microsomal steroid metabolism: Identification of the major microsomal steroid hormone 6 beta-hydroxylase cytochrome P-450 enzyme. Arch. Biochem. Biophys. 263, 424–436.

Waxman, D.J., & Azaroff, L. (1992). Phenobarbital induction of cytochrome P-450 gene expression. Biochem. J. 281, 577–592.

Waxman, D.J., Dannan, G.A., & Guengerich, F.P. (1985). Regulation of rat hepatic cytochrome P-450: Age-dependent expression, hormonal imprinting, and xenobiotic inducibility of sex-specific isoenzymes. Biochemistry 24, 4409–4417.

Waxman, D.J., Lapenson, D.P., Nagata, K., & Conlon, H.D. (1990). Participation of two structurally related enzymes in rat hepatic microsomal androstenedione 7 alpha-hydroxylation. Biochem. J. 265, 187–194.

Waxman, D.J., Lapenson, D.P., Aoyama, T., Gelboin, H.V., Gonzalez, F.J., & Korzekwa, K. (1991a). Steroid hormone hydroxylase specificities of eleven cDNA-expressed human cytochrome P450s. Arch. Biochem. Biophys. 290, 160–166.

Waxman, D.J., LeBlanc, G.A., Morrissey, J.J., Staunton, J., & Lapenson, D.P. (1988b). Adult male-specific and neonatally programmed rat hepatic P-450 forms RLM2 and 2a are not dependent on pulsatile plasma growth hormone for expression. J. Biol. Chem. 263, 11396–11406.

Waxman, D.J., Morrissey, J.J., & LeBlanc, G.A. (1989a). Female-predominant rat hepatic P-450 forms j (IIE1) and 3 (IIA1) are under hormonal regulatory controls distinct from those of the sex-specific P-450 forms. Endocrinology 124, 2954–2966.

Waxman, D.J., Morrissey, J.J., & LeBlanc, G.A. (1989b). Hypophysectomy differentially alters P-450 protein levels and enzyme activities in rat liver: Pituitary control of hepatic NADPH cytochrome P-450 reductase. Mol. Pharmacol. 35, 519–525.

Waxman, D.J., Pampori, N.A., Ram, P.A., Agrawal, A.K., & Shapiro, B.H. (1991b). Interpulse interval in circulating growth hormone patterns regulates sexually dimorphic expression of hepatic cytochrome P450. Proc. Natl. Acad. Sci. USA 88, 6868–6872.

Waxman, D.J., Ram, P.A., Notani, G., LeBlanc, G.A., Alberta, J.A., Morrissey, J.J., & Sundseth, S.S. (1990). Pituitary regulation of the male-specific steroid 6 beta-hydroxylase P-450 2a (gene product IIIA2) in adult rat liver. Suppressive influence of growth hormone and thyroxine acting at a pretranslational level. Mol. Endocrinol. 4, 447–454.

Waxman, D.J., Ram, P.A., Pampori, N.A., & Shapiro, B.H. (1995a). Growth hormone regulation of male-specific rat liver P450s 2A2 and 3A2: induction by intermittent growth hormone pulses in male but not female rats rendered growth hormone deficient by neonatal monosodium glutamate. Mol. Pharmacol. 48, 790–797.

Waxman, D.J., Ram, P.A., Park, S.H., & Choi, H.K. (1995b). Intermittent plasma growth hormone triggers tyrosine phosphorylation and nuclear translocation of a liver-expressed, Stat 5-related DNA binding protein. Proposed role as an intracellular regulator of male-specific liver gene transcription. J. Biol. Chem. 270, 13262–13270.

Westin, S., Tollet, P., Strom, A., Mode, A., & Gustafsson, J.A. (1992). The role and mechanism of growth hormone in the regulation of sexually dimorphic P450 enzymes in rat liver. J. Steroid. Biochem. Molec. Biol. 43, 1045–1053.

Wood, A.W., Ryan, D.E., Thomas, P.E., & Levin, W. (1983). Regio- and stereoselective metabolism of two C19 steroids by five highly purified and reconstituted rat hepatic cytochrome P-450 isozymes. J. Biol. Chem. 258, 8839–8847.

Wright, K., & Morgan, E.T. (1990). Transcriptional and post-transcriptional suppression of P450IIC11 and P450IIC12 by inflammation. FEBS Lett. 271, 59–61.

Wrighton, S.A., Brian, W.R., Sari, M.A., Iwasaki, M., Guengerich, F.P., Raucy, J.L., Molowa, D.T., & Vandenbranden, M. (1990). Studies on the expression and metabolic capabilities of human liver cytochrome P450IIIA5 (HLp3). Mol. Pharmacol. 38, 207–213.

Wu, H.Q., Masset, B.J., Tweedie, D.J., Milewich, L., Frenkel, R.A., Martin, W.C., Estabrook, R.W., & Prough, R.A. (1989). Induction of microsomal NADPH-cytochrome P-450 reductase and cytochrome P-450IVA1 (P-450LA omega) by dehydroepiandrosterone in rats: A possible peroxisomal proliferator. Cancer Res. 49, 2337–2343.

Yamada, J., Sakuma, M., Ikeda, T., Fukuda, K., & Suga, T. (1991). Characteristics of dehydroepiandrosterone as a peroxisome proliferator. Biochim. Biophys. Acta. 1092, 233–243.

Yamazoe, Y., Ling, X., Murayama, N., Gong, D., Nagata, K., & Kato, R. (1990). Modulation of hepatic level of microsomal testosterone 7 alpha-hydroxylase, P-450a (P450IIA), by thyroid hormone and growth hormone in rat liver. J. Biochem. (Tokyo) 108, 599–603.

Yamazoe, Y., Murayama, N., Shimada, M., Yamauchi, K., Nagata, K., Imaoka, S., Funae, Y., & Kato, R. (1988). A sex-specific form of cytochrome P-450 catalyzing propoxycoumarin O-depropylation and its identity with testosterone 6 beta-hydroxylase in untreated rat livers: reconstitution of the activity with microsomal lipids. J. Biochem. (Tokyo) 104, 785–790.

Yamazoe, Y., Shimada, M., Murayama, N., & Kato, R. (1987). Suppression of levels of phenobarbitalinducible rat liver cytochrome P-450 by pituitary hormone. J. Biol. Chem. 262, 7423–7428.

Zaphiropoulos, P.G., Mode, A., Norstedt, G., & Gustafsson, J.A. (1989). Regulation of sexual differentiation in drug and steroid metabolism. Trends Pharmacol. Sci. 10, 149–153.

Zaphiropoulos, P.G., Strom, A., Robertson, J.A., & Gustafsson, J.A. (1990). Structural and regulatory analysis of the male-specific rat liver cytochrome P-450 g: Repression by continuous growth hormone administration. Mol. Endocrinol. 4, 53–58.

Zhang, B., Marcus, S.L., Sajjadi, F.G., Alvares, K., Reddy, J.K., Subramani, S., Rachubinski, R.A., & Capone, J.P. (1992). Identification of a peroxisome proliferator-responsive element upstream of the gene encoding rat peroxisomal enoyl-CoA hydratase/3-hydroxyacyl-CoA dehydrogenase. Proc. Natl. Acad. Sci. USA 89, 7541–7545.

Zhao, S., & Waxman, D.J. (1994). Interaction of sex- and growth hormone (GH)-dependent liver nuclear factors with CYP2C12 promoter. FASEB J. 8, A1250.

Zimniak, P., & Waxman, D.J. (1993). Liver cytochrome P450 metabolism of endogenous steroid hormones, bile acids, and fatty acids. In: Handbook of Experimental Pharmacology, Vol. 105: Cytochrome P450 (Schenkman, J.B., & Greim, H., eds.), pp. 123–44. 105. Springer-Verlag, Berlin.

INDEX

375

J A I P R E S S

Advances in Molecular and Cell Biology

Edited by **E. Edward Bittar**, *Department of Physiology, University of Wisconsin, Madison*

Cell biology is a rapidly developing discipline which brings together many of the branches of the biological sciences that were separate in the past. The interrelations between cell structure and function at the molecular and subcellular level are the central connecting theme of this series. "As the twentieth century nears its close", writes Kenneth Miller, "the development of an enormous range of tools and techniques, some physical, some chemical, some biological, has changed the situation forever. Cell biology today crosses the boundary, links the molecule with the organelle, associates the cellular response with the larger organism." Most of the contributors are cell and molecular biologists who are currently engaged in research and have made important contributions to our present understanding of cell function.

Volume 13, Cell Cycle
1995, 211 pp. $97.50
ISBN 1-55938-949-4

Edited by **Michael Whitaker**, *University of Newcastle, Newcastle on Tyne*

CONTENTS: Preface, *Michael N. Whitaker*. Centrosomes and the Cell Cycle, *Greenfield Sluder*. Regulation of the Centrosome Function During Mitosis, *Brigitte Buendia and Eric Karsenti*. The Essential Role of Calcium During Mitosis, *Robert M. Tombes and Gary G. Borisy*. Calcium and Calmodulin Regulation of the Nuclear Division Cycle of Aspergillus Nidulans, *Kung Ping Lu and Anthony R. Means*. Cell Cycle Control by Protein Phosphatase Genes, *Mitsuhiro Yanagida*. The cdc25 Phosphatase: Biochemistry and Regulation in the Eukaryotic Cell Cycle, *Ingrid Hoffmann, Paul R. Clarke, and Giulio Draetta*. Control of Nuclear Lamina Assembly/Disassembly by Phosphorylation, *Matthias Peter and Erich A. Nigg*. Dissection of the Cell Cycle Using Cell-Free Extracts from Xenopus laevis, *C.C. Ford and H. Lindsay*. Subject Index.

Also Available:
Volumes 1-12 (1987-1995) $97.50

Printed and bound by CPI Group (UK) Ltd, Croydon, CR0 4YY

03/10/2024

01040431-0015